Handbook of Human Factors in Air Transportation Systems

T0175717

Human Factors and Ergonomics

Series Editors

Waldemar Karwowski

University of Central Florida,
Orlando, USA

Pamela McCauley

University of Central Florida,
Orlando, USA

PUBLISHED TITLES

Around the Patient Bed: Human Factors and Safety in Health Care
Y. Donchin and D. Gopher

Cognitive Neuroscience of Human Systems: Work and Everyday Life
C. Forsythe, H. Liao, M. Trumbo, and R. E. Cardona-Rivera

Conceptual Foundations of Human Factors Measurement
D. Meister

Content Preparation Guidelines for the Web and Information Appliances:
Cross-Cultural Comparisons
H. Liao, Y. Guo, A. Savoy, and G. Salvendy

Cross-Cultural Design for IT Products and Services
P. Rau, T. Plocher and Y. Choong

Data Mining: Theories, Algorithms, and Examples
Nong Ye

Designing for Accessibility: A Business Guide to Countering Design Exclusion
S. Keates

Ergonomics in Design: Methods and Techniques
Marcelo M. Soares and Francisco Rebelo

Ergonomic Workplace Design for Health, Wellness, and Productivity
Alan Hedge

Fall Prevention and Protection: Principles, Guidelines, and Practices
Hongwei Hsiao

Handbook of Cognitive Task Design
E. Hollnagel

The Handbook of Data Mining
N. Ye

Handbook of Digital Human Modeling: Research for Applied Ergonomics
and Human Factors Engineering
V. G. Duffy

Handbook of Human Factors and Ergonomics in Health Care and Patient Safety
Second Edition
P. Carayon

Handbook of Human Factors in Air Transportation Systems
Steven James Landry

Handbook of Human Factors in Web Design, Second Edition
K. Vu and R. Proctor

Handbook of Occupational Safety and Health
D. Koradecka

Handbook of Standards and Guidelines in Ergonomics and Human Factors
W. Karwowski

Handbook of Virtual Environments: Design, Implementation,
and Applications, Second Edition,
K. S. Hale and K M. Stanney

PUBLISHED TITLES (CONTINUED)

Handbook of Warnings
M. Wogalter

Human–Computer Interaction: Designing for Diverse Users and Domains
A. Sears and J. A. Jacko

Human–Computer Interaction: Design Issues, Solutions, and Applications
A. Sears and J. A. Jacko

Human–Computer Interaction: Development Process
A. Sears and J. A. Jacko

Human–Computer Interaction: Fundamentals
A. Sears and J. A. Jacko

The Human–Computer Interaction Handbook: Fundamentals
Evolving Technologies, and Emerging Applications, Third Edition
J. A. Jacko

Human Factors in System Design, Development, and Testing
D. Meister and T. Enderwick

Introduction to Human Factors and Ergonomics for Engineers, Second Edition
M. R. Lehto

Macroergonomics: Theory, Methods and Applications
H. Hendrick and B. Kleiner

Modeling Sociocultural Influences on Decision Making:
Understanding Conflict, Enabling Stability
Joseph V. Cohn, Sae Schatz, Hannah Freeman, and David J. Y. Combs

Practical Speech User Interface Design
James R. Lewis

The Science of Footwear
R. S. Goonetilleke

Skill Training in Multimodal Virtual Environments
M. Bergamsco, B. Bardy, and D. Gopher

Smart Clothing: Technology and Applications
Gilsoo Cho

Theories and Practice in Interaction Design
S. Bagnara and G. Crampton-Smith

The Universal Access Handbook
C. Stephanidis

Usability and Internationalization of Information Technology
N. Aykin

User Interfaces for All: Concepts, Methods, and Tools
C. Stephanidis

Variability in Human Performance
T. Smith, R. Henning, and M. Wade

Handbook of Human Factors in Air Transportation Systems

Edited by

Steven J. Landry

CRC Press
Taylor & Francis Group
Boca Raton London New York

CRC Press is an imprint of the
Taylor & Francis Group, an **informa** business

CRC Press
Taylor & Francis Group
6000 Broken Sound Parkway NW, Suite 300
Boca Raton, FL 33487-2742

First issued in paperback 2019

ISBN-13: 978-1-4665-7264-5 (hbk)
ISBN-13: 978-0-367-86787-4 (pbk)

Library of Congress Cataloging-in-Publication Data

Names: Landry, Steven J., editor.
Title: Handbook of human factors in air transportation systems / edited by
Steven James Landry.
Description: Boca Raton : CRC Press, Taylor & Francis, [2018] | Includes
bibliographical references and index.
Identifiers: LCCN 2017019535| ISBN 9781466572645 (hardback : alk. paper) |
ISBN 9781138747395 (pbk. : alk. paper) | ISBN 9781315116549 (ebook)
Subjects: LCSH: Aeronautics--Human factors--Handbooks, manuals, etc.
Classification: LCC TL553.6 .H355 2018 | DDC 629.13--dc23
LC record available at https://lccn.loc.gov/2017019535

Visit the Taylor & Francis Web site at
http://www.taylorandfrancis.com

and the CRC Press Web site at
http://www.crcpress.com

Contents

Preface...ix
Acknowledgments...xi
About the Editor... xiii
Contributors xv

SECTION I   Operations and Design

Chapter 1   Operational Details of the Air Transportation System 3

   Steven J. Landry

Chapter 2   The Design of the Flight Deck.. 13

   Lisa C. Thomas

Chapter 3   The Design of the Air Traffic Control System.. 31

   Michael Nolan

Chapter 4   Maintenance 51

   William L. Rankin and Bill Johnson

Chapter 5   Distributed Work in the National Airspace System: Traffic Flow
   Management and Airline Operations Control... 107

   Philip J. Smith and Ellen J. Bass

Chapter 6   Accident Investigation 115

   William J. Bramble, Jr.

SECTION II   Regulations and Certification

Chapter 7   An Overview of Flight Deck Human Factors in the U.S. Aviation
   Regulatory System 137

   Kathy Abbott

Chapter 8   Certification of Aircraft and Aircraft Systems 147

   Robert E. Joslin

Chapter 9 Safety Systems and Practices 195

Alan Stolzer

Chapter 10 Security 209

Douglas Harris

SECTION III   Research and Development

Chapter 11 Data Sources and Research Tools for Human Factors 239

Mark Wiggins, Shayne Loft, and Johanna Westbrook

Chapter 12 Flight Deck Human Factors 259

Michelle Yeh, Thomas R. Chidester, and Thomas E. Nesthus

Chapter 13 Situation Awareness in Air Traffic Control...301

Kim-Phuong L. Vu and Dan L. Chiappe

Chapter 14 Error in Air Transportation 321

Cees Jan Meeuwis and Sidney W. A. Dekker

Chapter 15 The Systems Perspective in Air Transportation...341

Lance Sherry

SECTION IV   The Future

Chapter 16 A Personal Journey in the Design and Management of the Growing Airspace.......363

Guy André Boy

Chapter 17 Remotely Piloted Aircraft 379

Alan Hobbs

Author Index..397

Subject Index 403

Preface

This book is intended to be a reference for human factors researchers, practitioners, and students on the air transportation system, rather than be a reference on air transportation human factors. Many people who do work or conduct research on the air transportation system start out with little knowledge on how the system works, and finding that information is extremely difficult. There are few college courses that comprehensively discuss the air transportation system, and books on the subject usually assume operational knowledge and focus on research being conducted on the performance of the system or its agents. It is hoped that this book can fill this gap.

A broad range of international experts were recruited to author the chapters of this book—from industry, academia, and government agencies. This was done to ensure that a broad perspective was obtained on how things happen within the air transportation system. The authors all have extensive experience in air transportation, and come from organizations that deal directly with the topic of their respective chapters.

Section I describes operational aspects of the air transportation system. In Chapter 1, the reader is introduced to the system, including its main components and how an aircraft moves from one place to another within the system. In Chapters 2 and 3, readers can learn about how flight decks are designed, and how the air traffic control system works. Chapter 4 discusses the maintenance of aircraft within the system, which is becoming a leading cause of accidents as the system itself becomes ever more safe. Chapter 5 discusses pilot–air traffic controller and traffic manager–airline dispatcher interactions as a distributed work system. Chapter 6 discusses the accident investigation process, including the history of accident investigations in the air transportation system.

Section II describes the management environment of the air transportation system. Chapter 7 describes regulatory practices regarding risk management and human factors in the air transportation system. Chapter 8 discusses how systems become certified for use in the air transportation system. Chapter 9 provides the reader with information on how safety is monitored and recorded within the air transportation system. The topic of Chapter 10 is security, with a focus on how human factors issues factor into security efforts and research.

Section III describes research and development on the air transportation system. Chapter 11 provides a reference on data sources and research tools that are available to human factors researchers and practitioners. Chapters 12 and 13 deal with research and development that have gone on in the flight deck and air traffic control domains. Chapter 13 discusses how the concept of error in air transportation has been considered and studied, and Chapter 15 discusses how researchers approach problems in the air transportation system as *systems problems*.

Section IV briefly discusses aspects related to the present and future of air transportation. In Chapter 16, the reader can learn about the European efforts to modernize the air transportation system. Finally, in Chapter 17, the concept of unmanned aerial vehicles in the air transportation system is discussed.

I hope the reader finds this book informative and interesting. As an educator in air transportation human factors, I further hope that this handbook becomes a reference that makes future researchers and practitioners more knowledgeable and productive.

Acknowledgments

This handbook could not have been completed without the assistance, direct or indirect, of many people other than the authors and I would remiss in not indicating my thanks to them. First and foremost, the editorial team at CRC Press, most notably Cindy Carelli, was extraordinarily helpful and patient. In addition, these people have been invaluable to me over the years, both personally and professionally, thereby contributing to this book toward its completion, and they therefore deserve my sincere gratitude: My colleagues at Purdue University (including my students, from whom I have learned more than I learned in all my years in school), my colleagues at NASA Ames Research Center, those with whom I worked and learned from at the Federal Aviation Administration, and, perhaps most important, Michael Mashtare.

About the Editor

Steven J. Landry, Ph.D., is an associate professor and the associate head in the School of Industrial Engineering at Purdue University, West Lafayette, Indiana, with a courtesy appointment in the Purdue University School of Aeronautics and Astronautics. At Purdue, Dr. Landry conducts research in air transportation systems engineering and human factors, and teaches undergraduate and graduate courses in human factors, statistics, and industrial engineering. Dr. Landry's research has been primarily focused on systems analysis and human factors for the Next Generation Air Transportation System. Dr. Landry has published more than 80 peer-reviewed journal articles, conference papers, and book chapters, including book chapters on human–computer interaction in aerospace, flight deck automation, and aviation human factors. Prior to joining the faculty at Purdue, Dr. Landry was an aeronautics engineer for NASA at the Ames Research Center, working on air traffic control automation. Previously, Dr. Landry was also a C-141B aircraft commander, instructor, and flight examiner with the U.S. Air Force. He has more than 2,500 heavy jet flying hours, including extensive international and aerial refueling experience.

- BS, Department of Electrical Engineering, Worcester Polytechnic Institute, Worcester, Massachusetts, 1987.
- MS, Department of Aeronautics and Astronautics, Massachusetts Institute of Technology, Cambridge, Massachusetts 1999.
- Ph.D., H. Milton Stewart School of Industrial and Systems Engineering, Georgia Institute of Technology, Atlanta, Georgia, 2004.

Contributors

Kathy Abbott
Office of Aviation Safety
U.S. Federal Aviation Administration
Washington, DC

Ellen J. Bass
Department of Information Science
College of Computing and Informatics and
    Health Systems and Services Research
College of Nursing and Health Professions
Drexel University
Philadelphia, Pennsylvania

Guy André Boy
School of Human-Centered Design,
    Innovation and Art
Florida Institute of Technology
Melbourne, Florida

William J. Bramble, Jr.
Senior Human Performance Investigator
U.S. National Transportation Safety Board
Washington, DC

Dan L. Chiappe
Department of Psychology
California State University, Long Beach
Long Beach, California

Thomas R. Chidester
Civil Aerospace Medical Institute
U.S. Federal Aviation Administration
Oklahoma City, Oklahoma

Sidney W. A. Dekker
Safety Science Innovation Lab
Griffith University
Nathan, QLD, Australia

Douglas Harris (Retired)
Anacapa Sciences, Inc.
Port Townsend, Washington

Alan Hobbs
San Jose State University
Human Systems Integration Division
NASA Ames Research Center
Mountain View, California

Bill Johnson
Office of Aviation Safety
U.S. Federal Aviation Administration
Washington, DC

Robert E. Joslin
Office of Aviation Safety
U.S. Federal Aviation Administration
Washington, DC

Steven J. Landry
School of Industrial Engineering
Purdue University
West Lafayette, Indiana

Shayne Loft
University of Western Australia
Perth, Western Australia, Australia

Cees Jan Meeuwis
Safety and Compliance
Voqx Innovative Safety
Rotterdam, the Netherlands

Thomas E. Nesthus
Civil Aerospace Medical Institute
U.S. Federal Aviation Administration
Oklahoma City, Oklahoma

Michael Nolan
School of Aviation and Transportation
    Technology
Purdue University
West Lafayette, Indiana

William L. Rankin
The Boeing Company (retired)
Kirkland, Washington

Lance Sherry
System Engineering and Operations Research
George Mason University
Fairfax, Virginia

Philip J. Smith
Department of Integrated Systems Engineering
The Ohio State University
Columbus, Ohio

Alan Stolzer
College of Aviation
Embry-Riddle Aeronautical University
Daytona Beach, Florida

Lisa C. Thomas
Boeing Commercial Airplanes
The Boeing Company
Kirkland, Washington

Kim-Phuong L. Vu
Department of Psychology
California State University, Long Beach
Long Beach, California

Johanna Westbrook
Department of Psychology
Macquarie University
North Ryde, NSW, Australia

Mark Wiggins
Macquarie University
North Ryde, NSW, Australia

Michelle Yeh
U.S. Federal Aviation Administration
Washington, DC

Section I

Operations and Design

1 Operational Details of the Air Transportation System

Steven J. Landry

CONTENTS

Introduction 3
Ground Operations 4
    The "Gate" (A, B) 5
    Pushback (C) 5
    Taxi Out (D) 6
    Taking the Runway (E, F) 6
    Takeoff (G) 6
    Arrival Ground Operations (S–V) 7
Airborne Operations 7
    Climbout (H–K) 8
    Cruise (L) 8
    Descent (M, N, O) 9
    Arrival (P, Q) 9
    Landing and Go-Around (R) 10
Conclusion 10
References 10

INTRODUCTION

Air transportation systems worldwide consist of several common features—aircraft are parked at airports, taxi from the parking area to a runway, use the runway to take off, then fly to some destination airport where the aircraft lands at a runway, taxis back to a parking area, and parks. There are substantial variations in the operational details of how this works under different conditions, including the experience of a passenger on a commercial flight, which is how most people experience the system, passengers on modern private aircraft, and passengers or pilots in older general aviation aircraft. Although much of the current chapter is focused on the common commercial flight experience, notations are provided as to what differences one can expect across different types of operations.

Much more of the technical details of the air traffic control system is contained in Chapter 3. This chapter gives an overview of what is involved operationally in moving an aircraft from one place to another within an air transportation system, to bring readers *up to speed* on the various elements of the system, and how they go together in an operational sense. This chapter, of course, is not a replacement for the actual air traffic control regulations, which are more complete, up-to-date, and authoritative. See especially Federal Aviation Administration (2016).

One can generally separate an aircraft's ground operations, shown schematically in Figure 1.1, from its airborne operations, shown schematically in Figure 1.2, which appears after the next section. These figures are labeled; those labels correspond to the references in the following subsections.

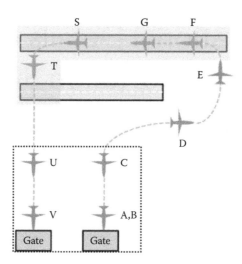

FIGURE 1.1   Ground operations overview.

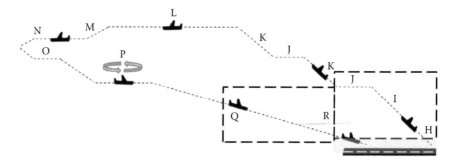

FIGURE 1.2   Airborne operations overview.

GROUND OPERATIONS

Controlled airports, where air transportation authorities manage the operations of the airport, typically have one or more *ground controllers* who perform the functions that will be discussed in the next few sections. At very busy airports, there will be numerous ground controllers, each with different responsibilities; at less busy airports there may be one or even none. These ground controllers typically sit in an air traffic control tower, where they can watch the ground operations visually. Sometimes, there is more than one *tower*, with separate towers located where ground controllers can see parking areas that would otherwise be obscured from the main tower.

At very busy airports, where one commercial airline has a majority presence, some parking areas may be controlled by a specialized *ramp controller* who may be employed by the airline. That controller's responsibility is to manage the gate and initial taxi process in that particular ramp area.

Moreover at controlled airports there are *tower controllers*, who manage operations on the runway, including approving landings and takeoffs. These controllers are usually separate from the ground controllers and focus their entire attention on the runway environment. These controllers are in the tall control tower seen at all busy airports, and operate almost exclusively visually, eschewing most computer-based decision support systems that would distract them from seeing operations at the runway.

At airports that are not controlled, there may be no controllers at all. In such cases, pilots are responsible for decisions regarding taxi and takeoff.

The "Gate" (A, B)

Most people have experienced operations at the gate. In commercial operations, the gate is where a commercial aircraft is parked to be serviced and load or offload passengers. Servicing includes being fueled; inspected; having perishable items, such as food and drinks, restocked; having the latrine serviced; and having cargo and/or passenger bags offloaded and onloaded. At the gate, prior to departure, pilots also perform preflight checks and prepare the aircraft for departure.

Private aircraft are typically not parked at a gate, but are parked in a spot on the ramp, and passengers and crewmembers onload or offload through a door and stairs, and need to walk across the ramp to get to the terminal building. Often, at airports with both commercial and private service, this area and terminal building is a separate area from the commercial area.

Depending on the capabilities of the aircraft, at the gate pilots may need to load or check the flight plan into the aircraft's flight management computer, if the aircraft is so equipped. (Most commercial aircraft do have such a system installed, whereas many older small aircraft do not.) The flight management computer, described in more detail in Chapter 2, interfaces with the autopilot system to help navigate the aircraft to its destination.

The efficiency of the servicing and onloading/offloading of passengers is important to commercial air carriers, as it determines the minimum amount of time required between an arrival of a flight and a departure of the same aircraft. In addition, passengers on commercial flights prefer not to sit awaiting departure any longer than they need to, so customer satisfaction with the commercial flight experience is somewhat dependent on the efficiency of this process. For that reason, researchers and practitioners have studied this process and made recommendations for how it should be done (Bazargan, 2007; Hapsari, Clemes, & Dean, 2017; Kuo, 2014; Steffen, 2008; Steffen & Hotchkiss, 2012). Consensus on how to accomplish the passenger-loading process has not been achieved, as most passengers can attest, and is complicated by, among other things, airlines' marketing decisions regarding providing preferential treatment to frequent fliers and other commercially important passengers. Passengers and crew on private or general aviation flights, in which the number of passengers is very low or even zero, experience a much less-regulated process, loading whenever the aircraft is ready by walking out to the aircraft.

Pushback (C)

Pilots of aircraft flying under the control of air traffic controllers, that is, under *instrument flight rules*, must obtain a clearance from air traffic controllers before they are allowed to depart. Pilots flying under *visual flight rules* do not need a clearance. (See Chapter 3 for more information on the different types of flight rules and what they mean.) This clearance identifies for the pilots the route they must use to depart the general environs of the airport, that is, the *departure* to use along with the cleared route all the way to destination. An air traffic controller, referred to as *clearance delivery*, typically delivers this clearance verbally over a radio frequency to the pilots, and the pilots read back the clearance to the controller to ensure they have recorded it correctly. Usually this clearance contains some instructions for immediately after departure and then just identifies any changes to the preplanned and prefiled flight plan the airline (or pilots) have sent in to the relevant air traffic control authorities.

Once pilots have obtained a clearance and all preflight operations, including passenger loading if applicable, have been completed, pilots of aircraft parked at a gate can request a *pushback*. As aircraft at a gate need to go backward, and can only go in reverse with great difficulty, a truck is used to push the aircraft back away from the gate.

This request, and the pushback, initiates the *taxi* portion of the flight, where the aircraft transits from the parking area to the runway. Aircraft not parked at gate, or those who have already received a pushback away from the gate, then request permission from a ground or ramp controller to taxi.

Taxi Out (D)

At controlled airports, permission is required to taxi, whereas at uncontrolled airports pilots do so on their own authority. Permission to taxi typically includes specific instructions about what taxiways to use and what to do when the aircraft gets to the runway.

During taxi, pilots are responsible for ensuring that they remain on the correct taxiways and stay clear of obstacles, vehicles, and other aircraft. At busy airports, markings visible from the pilot's seat are provided to indicate taxiway identification, which is usually by letters such as taxiway K, closed taxiways, taxiway centerline indications, and markings indicating the intersection of taxiways with runways. Sometimes, these markings are accompanied by a lighting system providing these same indications but making them more visible at night or in poor visibility conditions. Uncontrolled airports typically have few or even no such markings, relying on pilots to navigate the taxiway system without assistance.

Taking the Runway (E, F)

As indicated in the previous section, taxi instructions, when provided, usually include specifics on what the pilot should do when reaching the runway. Typically, this is an instruction for the pilot to hold short of the runway and contact the tower controller.

After contacting the tower controller, the pilot is given instructions to continue holding, to taxi on to the runway and hold their position, or is cleared to takeoff. This decision is dependent on whether an aircraft is still on, or about to land on, the runway, as only one aircraft is allowed to use the runway for takeoff or landing at any one time.

This decision may also reflect a desire to keep the departing aircraft from encountering the *wake vortex* of the aircraft that just departed. Wake vortices are not fully understood, as they are relatively complicated fluid dynamics phenomena, and have been modeled in a number of ways (Rossow, 2002; Shortle & Jeddi, 2007). However, as interactions with wake vortices are extremely dangerous and have caused fatal accidents, several minutes between departures (and arrivals) on the same runway are required when the aircraft involved are very heavy, such as large commercial aircraft, which can also affect runway capacity (Burnham, Hallock, & Greene, 2002).

Takeoff (G)

Once cleared for takeoff, pilots advance power and accelerate to achieve a speed at which the aircraft can become airborne and fly safely. This process is actually one of the more dangerous parts of the flight, because, should a malfunction occur during the takeoff roll, pilots must make determinations as to whether it is safer to try to stop the aircraft on the runway or take the aircraft into the air with the malfunction. This is not a trivial decision, as either choice could lead to a serious, even fatal, accident.

This decision is referred to as the *go–no go* decision. At least for commercial aircraft, a speed is identified during preflight at which the aircraft must *go* regardless of the malfunction, short of the aircraft becoming nonairworthy. (All commercial aircraft are engineered such that they can fly with one engine not operating.) For large commercial aircraft, this speed is typically referred to as "V1," which is short for "velocity #1." This decision has been the subject of several academic papers, mostly on whether any automation assistance is possible in this case (Balachandran, Ozay, & Atkins, 2016; Bove & Andersen, 2002).

Another speed is identified at which the aircraft's nose should be lifted so that the aircraft will lift off and fly off of the runway. This speed is referred to as "V2."

ARRIVAL GROUND OPERATIONS (S–V)

At controlled airports, the departure ground operations (gate, taxi, and takeoff) are reversed on arrival. Approaching aircraft contact the tower for approval to land; such approval includes specifically for which runway the approval is given. Pilots are approved only for landing on that runway unless other instructions are given. After landing and before departing the runway, the aircraft is instructed to contact the ground controller on a separate radio frequency. At uncontrolled airports, pilots land and taxi on their own authority, visually clearing the area of any other aircraft or vehicles that would pose an obstruction or collision risk.

From there, again at controlled airports, the aircraft is given specific instructions to taxi to the parking area. At busy commercial airports, this includes a gate at which the aircraft is cleared to park. The gate assignment is typically given by the airline that controls that gate, although airlines can *lend* gates, although this is done rarely, if ever.

AIRBORNE OPERATIONS

After departure, aircraft talk to a series of air traffic controllers who manage volumes of airspace and control all aircraft within those volumes that are following instrument flight rules. See Chapters 3 and 15 for more information on airspace segmentation.

Initially, pilots talk to *departure* controllers, who are part of the *terminal radio detection and ranging (RADAR) approach control* facility, or TRACON. Close to controlled airports, within about 50 miles and up to about 10,000 feet, pilots talk to these departure controllers. The dashed box in Figure 1.3 is intended to depict the TRACON area of control, although the figure is not to scale. Pilots again talk to *arrival* controllers, also part of the TRACON, when approaching their destination.

The largest airport regions, about 27 of them, have TRACONS that are entirely separate from the control tower, although some, or even many, of these facilities are located in the basement

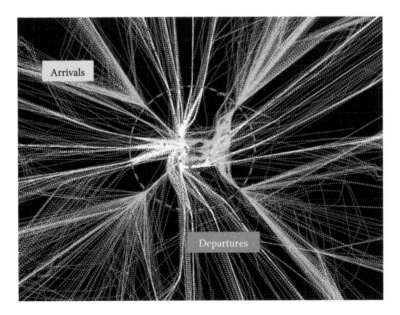

FIGURE 1.3   Atlanta arrival and departure routes (From Air Traffic Atlanta, http://www.airtrafficatlanta.com/channels.php? encid=8).

of the air traffic control tower building. The largest few regions, such as near Washington, D.C. and Los Angeles, have consolidated multiple TRACONs that control the arrival and departure traffic going into more than one airport. Such *consolidated TRACONs* are located in a separate building away from any of the airports they control.

For smaller airports, the tower controllers perform the arrival and departure control responsibilities. It is possible, in regions of the country where TRACON airspaces are contiguous, for aircraft to fly between airports only talking to tower controllers. Such operations are referred to as *tower enroute* operations.

No TRACON services are provided for uncontrolled airports. In such cases, pilots typically announce their position and intentions on a *common frequency* so that any other aircraft in the vicinity will know where other aircraft are by means other than visually acquiring them.

CLIMBOUT (H–K)

Climbout is the portion of flight that transitions the aircraft from the immediate vicinity of the airport to high-altitude cruise flight. For small general aviation aircraft, *high-altitude* may be a few thousand feet; for commercial aircraft, cruise flight typically occurs between 29,000 and 41,000 feet. (Some small business jets can fly at slightly higher altitude, and the now-defunct supersonic Concorde would cruise at up to 60,000 feet.)

Pilots are expected to follow the instructions given to them as part of their clearance, unless departure controllers give them different instructions, which they typically do. These instructions are designed to expedite the aircraft's climb while keeping aircraft in the departure/arrival area safely separated from one another.

At busy airports, this departure process is very challenging for both pilots and controllers, as it involves separating climbing/accelerating and descending/decelerating aircraft, who are trying to go in roughly opposite directions. The problem is somewhat simplified procedurally by having arrival and departure aircraft fly along different *corridors*, which limits where these aircraft routes intersect. An example, for the Atlanta TRACON, is shown in Figure 1.3. As can be seen in Figure 1.3, however, these arrival and departure routes are very complex; aircraft are often given numerous climb, level-off, and heading instructions.

Further complicating matters, controllers who are providing approval for aircraft to climb can only estimate their climb performance. Although experience provides them with some guidance, unknown weights and idiosyncrasies among pilots make this estimate error-prone. Controllers must be conservative in allowing departures to climb, as it is difficult to be sure that a climbing aircraft will reach a certain altitude prior to the intersection of that departing aircraft's flight path with that of an arriving aircraft. Departing aircraft are, therefore, often given intermediate level-off altitudes, as shown in Figure 1.2 with letter J, which makes the climbout less efficient, and which results in less cruise time, where the aircraft is the most fuel efficient.

Climbout is considered one of the higher workload periods for pilots of large commercial aircraft. Pilots experience a large number of communications during this period, must make many changes to the flight path of the aircraft either manually or to the automation that is controlling the aircraft, and must remain vigilant due to the high density of aircraft in the area.

CRUISE (L)

After leaving the TRACON, aircraft flying under instrument flight rules are *handed off* to the first of a sequence of *enroute controllers*. These controllers sit in a facility that is separate physically from TRACONs and/or towers. These Air Route Traffic Control Centers control large volumes of airspace, in which the Center volume is broken up into irregular volumes of airspace called *sectors*. Groups of sectors on which individual controllers are trained are called *areas*.

Sectors are shaped to conform to the typical flows of aircraft that transit the sectors. Some sectors primarily handle arrival and/or departure traffic to/from TRACONS and are oriented accordingly – typically long and slender in the predominant direction of flight for the arriving/departing aircraft. Sectors that handle cruise aircraft in which traffic density is low are typically large and shaped more regularly, such as a large box shape.

Above 20,000 feet where most large commercial aircraft fly, aircraft at cruise are separated into *flight levels* (FL) corresponding to the altitude in hundreds of feet, such that aircraft flying generally eastbound (0°–179°) are required to fly at *odd* FL(FL210 or 21,000 feet, FL230, FL250, FL270, FL290, FL330, FL370, etc.), and aircraft flying westbound are required to fly at *even* FL (FL220, FL240, FL260, FL280, FL310, FL350, etc.). These rules vary internationally, including FL being defined by meters instead of feet in some Asian and Eastern European countries.

The 2,000 versus 1,000-feet change in the difference between eastbound and westbound altitudes the reader might note above FL290 is reduced to a 1,000-feet difference in areas in which *reduced vertical separation minima* airspace has been introduced. Only certain properly equipped aircraft are allowed to use these altitudes in that airspace.

Cruise flight can comprise the majority of the time a flight is airborne. It is generally considered a low-risk flight regime, due to little climbing and descending and long periods in low-density airspace. It is also considered low-workload, as for properly equipped aircraft the autopilot is typically engaged, often coupled to the flight management system, leaving pilots the task of communicating infrequently with air traffic controllers and monitoring the autopilot and other systems.

Descent (M, N, O)

When getting within a few hundred miles of their destination, depending on cruise altitude, aircraft may begin to descend. This descent may be initiated by controllers or be initiated by the pilots. A typical descent profile consists of a set of descents to lower altitudes, a level-off for a period of time, followed by another descent. This is referred to as a *staircase* type descent because of its resemblance to a staircase.

The level-off portions of such descents are inefficient, as pilots must increase power to keep the aircraft level. Air traffic authorities and researchers have been investigating, and in some cases even implementing, *continuous descent* procedures for commercial operations that eliminates these inefficiencies (Clarke et al., 2004; Coppenbarger, Mead, & Sweet, 2007; Dinges, 2007; Green & Vivona, 2001; Landry, 2009; Robinson & Kamgarpour, 2010; Staigle & Nagle, 2006; Tong, Boyle, & Warren, 2006; White & Clarke, 2006). Such procedures have the aircraft cruise to a point in which they can continuously descent, at near idle power, to very close to runway touchdown, thereby eliminating the inefficient level-off periods. Of course, such aircraft still have to be separated from other arriving and departing aircraft, making the operation challenging to execute. As of this writing, such operations are limited to low-demand periods or to airports in which the operations can be carefully controlled.

Descent again begins a higher workload period, in which there is typically a higher communication workload and more maneuvering. This period is also more risky than cruise flight, due to more complex interactions with other climbing, descending, and cruising aircraft.

Arrival (P, Q)

Workload and risk increases further as the aircraft gets closer to the airport. Center controllers descend the aircraft and deliver it to a point of entrance (a *fix*) to the TRACON at a particular altitude based on agreements made between the relevant Center and TRACON. Center controllers *hand off* the aircraft to the arriving TRACON, in which aircraft talk to an arrival controller. Different TRACONS handle incoming aircraft differently, either by having a set of arrival *sectors* much similar to those found in Center airspace, or by having one or more arrival controllers that handle aircraft from TRACON entry to Tower handoff.

Arrival controllers can refuse to accept handoffs from Center controllers if the TRACON is too busy. In such cases, aircraft must then enter a holding pattern, which is a racetrack pattern consisting of straightaway *legs* of 10 or 20 nautical miles.

If the TRACON controller continues to refuse aircraft entry, subsequent aircraft are stacked in holding patterns at altitudes above the first. When the Center controller runs out of vertical space, aircraft are held further out from the TRACON, and so on. In particularly egregious situations, such holding aircraft can even block departure routes from the (arriving) runway, resulting in *gridlock* in which departing aircraft cannot leave their gates due to blocked airspace and arriving aircraft cannot arrive due to blocked gates. This is fortunately rare.

Aircraft under instrument flight rules are most commonly given headings (*vectors*) and altitudes to fly to get aligned for landing. However, aircraft could also be cleared for a particular arrival procedure, which requires them to fly a particular set of route legs, speeds, and altitudes. Aircraft are expected to eventually take over navigating to the runway, either by visually acquiring the runway and positioning themselves to land, or by following ground-based navigation systems such as an instrument landing system that provides both lateral and vertical (glidepath) guidance.

Smaller, general aviation aircraft must find and align themselves to land based on whatever equipment the aircraft possesses, which may be very little. Most commonly, such aircraft use visual references to find and land at the airport.

LANDING AND GO-AROUND (R)

Using such guidance, aircraft can position themselves to land, obtaining permission to do so from the tower, if the airport is controlled. However, if the aircraft for some reason cannot land safely, such as not having sufficient visual references to identify the runway environment, they must execute a *go-around* in which the aircraft climbs away from the runway and back into the approach environment. Other reasons for go-arounds include vehicles/aircraft on the runway and aircraft malfunctions. There is commonly a designated procedure for go-arounds that aircraft are expected to follow, not only to ensure predictability but also to ensure safe separation from ground obstructions.

Go-arounds are very risky and high workload periods. Go-arounds are not executed very often, so pilots are initiating a procedure with which they are not entirely familiar. There is also high demand on them, with configuration changes (landing gear, flaps) occurring, communication requirements, and navigation requirements. Moreover, most go-arounds are flown manually, exacerbating the workload issue.

CONCLUSION

Aircraft operations are fairly complex, with a great deal of information that can be difficult to find if one does not have actual operational experience. Hopefully, the current chapter has provided an overview of how aircraft move through the air transportation system. With reference to the more detailed and comprehensive information in the subsequent chapters, readers should be able to become sufficiently knowledgeable about the system to be productive and effective researchers and practitioners.

REFERENCES

Balachandran, S., Ozay, N., & Atkins, E. M. (2016). Verification guided refinement of flight safety assessment and management system for takeoff. *Journal of Aerospace Computing, Information and Communication*, *13*, 357–369.

Bazargan, M. (2007). A linear programming approach for aircraft boarding strategy. *European Journal of Operational Research*, *183*, 394–411.

Bove, T., & Andersen, H. B. (2002). The effect of an advisory system on pilots' go/no-go decision during take-off. *Reliability Engineering and System Safety, 75*, 179–191.

Burnham, D. C., Hallock, J. N., & Greene, G. C. (2002). Wake turbulence limits on paired approaches to parallel runways. *Journal of Aircraft, 39*, 630–637.

Clarke, J.-P., Ho, N. T., Ren, L., Brown, J. A., Elmer, K. R., Zou, K., … Warren, A. W. (2004). Continuous descent approach: Design and flight test for Louisville International Airport. *Journal of Aircraft, 41*, 1054–1066.

Coppenbarger, R., Mead, R., & Sweet, D. (2007, September). *Field evaluation of the tailored arrivals concept for datalink-enabled continuous descent approach.* Paper presented at the 7th AIAA aviation technology, integration, and operations conference, Belfast, Northern Ireland.

Dinges, E. (2007, July). *Determining the environmental benefits of implementing continuous descent arrival operations at Atlanta and Miami (paper no. 594).* Paper presented at the 7th USA/Europe seminar on air traffic management research and development, Barcelona, Spain.

Federal Aviation Administration. (2016). *Aeronautical information manual: Official guide to basic flight information and ATC procedures*, Washington, DC: U.S. Department of Transportation, Federal Aviation Administration.

Green, S. M., & Vivona, R. A. (2001, August). *En route descent advisor concept for arrival metering.* Paper presented at the AIAA guidance, navigation, and control conference, Montreal, Canada.

Hapsari, R., Clemes, M. D., & Dean, D. (2017). The impact of service quality and customer engagement and marketing constructs on airline passenger loyalty. *International Journal of Quality and Service Sciences, 9*, 2–29.

Kuo, C.-C. (2014). A review of heuristic and optimal aircraft boarding strategies in the U.S. airline industry. In M. A. Warkentin (Ed.), *Trends and research in the decision sciences*, pp. 313–330. Upper Saddle River, NJ: Decision Sciences Institute.

Landry, S. J. (2009). *Mid-term CDA operations to ATL: Delay requirements under nominal and off-nominal conditions.* Paper presented at the NASA airspace technical interchange Meeting, San Antonio, TX.

Robinson III, J. E., & Kamgarpour, M. (2010, September). *Benefits of continuous descent operations in high-density terminal airspace under scheduling constraints.* Paper presented at the 10th AIAA aviation technology, integration, and operations conference, Fort Worth, TX.

Rossow, V. J. (2002). Reduction of uncertainties in prediction of wake vortex locations. *Journal of Aircraft, 39*, 587–596.

Shortle, J., & Jeddi, B. (2007). Using multilateration data in probabilistic analysis of wake vortex hazards for landing aircraft. *Transportation Research Record: Journal of the Transportation Research Board, 2007*, 90–96.

Staigle, T., & Nagle, G. (2006, September). *Details and status of CDA procedures for early morning arrivals at Hartsfield-Jackson Atlanta International Airport (ATL).* Paper presented at the CDA Workshop No. 3, Atlanta, GA.

Steffen, J. H. (2008). Optimal boarding method for airline passengers. *Journal of Air Transport Management, 14*, 146–150.

Steffen, J. H., & Hotchkiss, J. (2012). Experimental test of airplane boarding methods. *Journal of Air Transport Management, 18*, 64–67.

Tong, K.-O., Boyle, D. A., & Warren, A. W. (2006, September). *Development of continuous descent arrival (CDA) procedures for dual-runway operations at Houston Intercontinental.* Paper presented at the 6th AIAA aviation technology, integration, and operations conference, Wichita, KS.

White, W., & Clarke, J.-P. B. (2006, September). *Details and status of CDA procedures at Los Angeles international airport (LAX).* Paper presented at the CDA workshop No. 3, Atlanta, GA.

2 The Design of the Flight Deck

Lisa C. Thomas

CONTENTS

Introduction..13
If You Were to Design a Flight Deck from Scratch, Where Would You Look for Inspiration?..............14
    Flight Crews Then and Now 15
    The Evolving Flight Deck..18
    Automation in the Flight Deck...19
    Aviation Is a Unique Context for Product Design...20
    The Design Process..20
    Strategies for Design Evaluation..22
        Paper-Based Evaluations 23
        Low and Medium Fidelity Mockups, Part Task Trials 23
        High Fidelity Mockup in Integrated (Flight Simulator) Setting Using Detailed Test Plan..........24
    Does the Design Pass or Fail?..25
    How Does the Design Get on the Airplane? 27
    Commonality, Legacy, and the Future...28
    Summary of Recommendations 28
Conclusion 29
References...30

INTRODUCTION

The goal of the current chapter is to describe the designer's role in development, implementation, and evolution of the flight deck instruments and how the designs relate to the pilot's existing, new, and evolving roles and responsibilities. The first half provides a brief overview of the history of flight deck design and relevant factors that influence design. The second half contains recommendations for design process and evaluation strategies, as well as a description of when and how new designs can be integrated into a flight deck.

The flight deck supports the three primary pilot task categories of Aviate, Navigate, and Communicate (ANC; Federal Aviation Administration (FAA), 2015). At this fundamental level, the flight deck instruments, including displays, controls, and automation, are designed to provide the capability for the pilot to safely and efficiently fly from point A to point B. All the necessary and relevant information needs to be displayed in a way that is immediately accessible, unambiguous, and easily interpreted, without causing clutter that could create a distraction. The controls need to be designed so that each one has a clear function, easy operability, and, when possible, is located within reach of at least one of the pilots. Any system automation should be implemented to work in concert with the pilots to reduce and preferably avoid instances of confusion over how the aircraft is operating. Finally, the displays, controls, and automation must be evaluated collectively to ensure that pilot performance is optimally supported, and that no unintended consequences emerge as a function of combining disparate designs.

Although the examples provided in this chapter come primarily from an aircraft manufacturer, flight deck designs are also created by avionics manufacturers, government research agencies, university labs, small consulting and design firms, aftermarket modification centers, and private citizens. Each of these sources has contributed in some way to the components in a flight deck; the opportunity to innovate is not limited to the major plane builders.

IF YOU WERE TO DESIGN A FLIGHT DECK FROM SCRATCH, WHERE WOULD YOU LOOK FOR INSPIRATION?

Ideally, a designer first considers what the pilot needs to *know* and how it should be presented as well as what the pilot needs to *do* to operate the aircraft. In the simplest of aircraft, an unpowered glider or sailplane, the pilot just needs to know altitude, airspeed, and heading information. Without an engine, there is only a need for this minimum set of displays: altimeter, airspeed indicator, and compass (FAA, 2013). Additional instruments make flying a glider a little easier and more efficient by providing the pilot with more information about the aircraft state and the operating environment, but the primary goal of continued safe flight in a glider is supported by that minimum set of three instruments. Once aloft, the pilot has to be able to control the movable surfaces (rudder, ailerons, elevators, spoilers) to direct the aircraft along a desired flight path (FAA, 2013). The controls for these surfaces, rudder pedals, control stick, and spoiler handle, form the minimum set of controls to affect aircraft state.

Consider how much more the pilot needs to know if the aircraft has one or more engines. In addition to the minimum set of instruments, the pilot needs information about each engine's performance, fuel level, consumption rate, and other related factors. In addition, the pilot now needs to have some level of control over the operation of the engine(s), the fuel, and the resultant thrust.

Consider further an aircraft that will be operated without reference to outside visual references. The aircraft now needs exterior and interior lighting, as well as instrumentation that allows for identifying and understanding the aircraft's position and attitude in space when it cannot be externally verified. Further, the displays must be salient and comprehensible enough to compensate for, and be trusted, when the pilot's own vestibular and kinesthetic senses fail to provide an accurate idea of the aircraft's state.

At this point, the designer is creating a flight deck for an aircraft that can be operated under its own power across a wide range of environmental conditions. There is a large array of information types and required controls that need to be integrated into a flight deck layout that a crew can be trained to operate.

Given that more than a century of design efforts have been made, no aircraft is created truly from scratch today. There are federal regulations that dictate a number of constraints about the design of the flight deck that must be followed to result in a certifiable product. For the larger transport category aircraft, these are primarily found in the following:

- *Code of Federal Regulations*: Title 14, Aeronautics and Space
- *Chapter I*: Federal Aviation Administration, Department of Transportation
- *Subchapter C*: Aircraft
- *Part 25*: Airworthiness Standards: Transport Category Airplanes.

Other types of aircraft (rotorcraft, commuter, etc.) are regulated under different Part numbers within Subchapter C, but the examples in this book will draw primarily on Part 25 aircraft.

The regulations vary in their specificity depending on the area of the flight deck being addressed; control-related regulations tend to be far more prescriptive whereas regulations on layout and information display are less well defined. To illustrate, code of federal regulations (CFR) 25.779 defines specific motions and effects for a number of controls, and CFR 25.781 dictates the shapes (complete with technical drawings) that the various controls must conform to. There is little room to vary the design of an individual control, so controls for the various flight surfaces and operating mechanisms look virtually the same across any flight deck. This standardization enables immediate recognition of the critical controls in terms of both what they control and how they operate.

Given that a minimum set of instruments for aviating and navigating are prescribed by the regulations, however, the layout and design of those information displays can involve more variety. For example, as long as the core pieces of information are presented in a way that pilots can be trained to identify and interpret accurately, the specific design may change from, say, a stand-alone airspeed gage to a digital *speed-tape* integrated within the primary flight display.

For human factor issues associated with the design of individual flight deck elements, more specific design guidance and recommendations that have been derived from existing regulatory, standards, and other relevant guidance material can be found in the FAA report "Human Factors Considerations in the Design and Evaluation of Flight Deck Displays and Controls" (Yeh, Jo, Donovan, & Gabree, 2013).

FLIGHT CREWS THEN AND NOW

Currently the default assumption about the minimum flight crew on a commercial airplane is that it consists of two pilots. The two-pilot crew was not always the standard. After World War I, as civilian aviation transportation began to develop, the size of the crew was often dictated by the size and complexity of the aircraft and its mission requirements (Fadden, Morton, Taylor, & Lindberg, 2014). Aircraft were operated by anywhere from one to five pilots, and the traditional breakdown of operating responsibilities (aviate, navigate, communicate, and manage systems) were divided among crew members dedicated to each of those areas. Pilots handled the aviating duties, whereas the flight engineer managed systems, the navigator aided in navigation, and a radio operator handled communications. Given the crew sizes and respective duties, flight decks were designed with the instruments for each area (usually analog gages and mechanical controls) set up in distinct crew stations within the cockpit space (Figure 2.1).

Over the decades, advances in radio, navigation, and engine technologies allowed for simplification and consolidation of controls such that the pilots eventually began assuming more responsibilities for these roles. Evolutionary improvement in designs and layouts and the emergence of new technologies that made previously unthought-of, but significantly more efficient instrument configurations possible combined to reduce the workload across a variety of tasks.

For example, the engine instruments and related alerting indicators were originally located at the flight engineer's station. With the advent of a consolidated warning and caution panel, and later the engine indicator and crew alerting system (EICAS), the relevant alerts could now be placed in

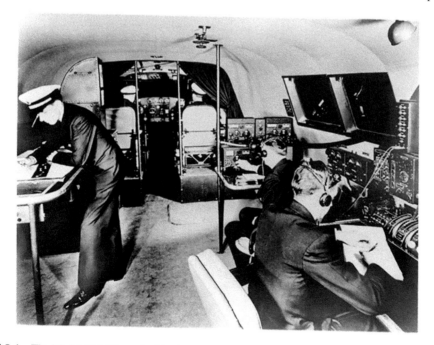

FIGURE 2.1 The Model 314 Clipper had its first flight in 1938 and came outfitted with crew stations for (from left to right) navigator, pilots, radio operator, and flight engineer. (From Boeing images, www.boeingimages.com.)

front of the pilots, to be acted on by the pilots as needed. Although this gave the pilots the additional role of systems management (previously the domain of the flight engineer), concurrent improvements in technology and automation reduced the number of tasks associated with managing systems while in flight. In fact, the reduction in systems management tasks contributed to the omission of *Manage Systems* from today's Aviate-Navigate-Communicate task prioritization (Fadden et al., 2014). Similarly, radio controls were consolidated into panels accessible by the pilots; the non-flying pilot could operate the radio as needed.

Fadden et al. (2014) noted that "[t]he transition to digital computation radically altered what and how information was displayed in commercial flight decks. The original [Boeing] 737 design exploited transistor technology to implement automated and redundant controllers for many airplane systems and to coordinate caution and warning information" (Figure 2.2).

"The 757 and 767 airplanes went much further introducing cathode ray tube displays as primary flight instruments and creating the digital computer-based flight management system to integrate flight planning, navigation, performance management, and guidance. The handling of caution, warning, and advisory information was significantly enhanced [with the introduction of EICAS]. For the first time engine operational and health information was managed and selectively displayed to facilitate timely and accurate crew response" (Fadden et al., 2014).

However, the question of total workload remained: Could one or both pilots perform their own tasks as well as all of the remaining, nonautomated monitoring and tasks that used to be done by additional dedicated crew members?

Through the 1970s, some groups insisted that even the most current (at the time) flight decks still required at least three full-time crew members to manage the total workload, whereas others felt that these same flight decks functioned at least as well with two pilots. Airlines and pilot groups were split, with some airlines flying one aircraft type with a two-pilot crew and other airlines flying that same aircraft type with a three-pilot crew. There was a startling amount of controversy over the

FIGURE 2.2 In 1967, the 737 had its first flight with consolidated displays and controls for the pilots.

reduction in regulated minimum crew size from three crew members to two that lasted for decades and is chronicled in the book *The Third Man* (Komons, 1987). Three important design lessons derive from that era:

1. **Crew complement should be considered a design specification; two crew members are sufficient for a flight deck designed for two, whereas three crew members are sufficient for a flight deck designed for three (Komons, 1987).**

Throughout and especially toward the end of the crew complement controversy period (1947–1981), airlines were sometimes required to staff crews with a third crew member on aircraft with flight decks designed for two pilots only. Without a dedicated crew stations for the third crew member, he was relegated to an observer seat behind the pilots in the flight deck. It became clear that the third crew member not only did not perform any critical or unique tasks but was also unable to perform many secondary or support tasks from that position. To put it another way, if a crew member is expected to serve a vital role in flight operations, he must have a flight deck design that allows him the capability to do so.

2. **The best crew size is the absolute minimum number of people it takes to complete the work (Fadden et al., 2014).**

Even before the Presidential Task Force on crew complement declared in 1981 that a two-pilot crew was sufficient, the design of emerging flight decks included a layout of the displays and controls centered around two pilots. The position that the presence of the third crew member improved safety by providing a third brain to discuss, to observe, and to contribute to decision making was not supported by the actual practice; in reality, the third crew member often could not see much of what the pilots could, particularly out the windows in a see-and-avoid capacity, and so any contributions from the third crew member would be at best minimal and at worst misleading due to a lack of shared situation awareness (Komons, 1987; Fadden et al., 2014).

3. **External forces can put sufficient pressure on manufacturers to slow the progress of technological advancement (Komons, 1987).**

This is potentially the most important lesson of the three. As mentioned earlier, there were several key advances in technology that allowed a consolidation of displays and controls and their related tasks, which in turn enabled the pilots to take on additional responsibilities that previously had required one or more dedicated crew members. However, given the significant pressure on the airlines to maintain a three-person flight crew, manufacturers continued to build flight decks with dedicated flight engineer stations to meet their customers' operational requirements.

Customer airlines must also consider the effect that a new, untested, or significantly different flight deck design will have on their pilot populations in terms of training time and cost as well as fleet management. New aircraft are expensive and are expected to operate for decades, which makes upgrading any part of the fleet a huge investment on the part of the customer. Manufacturers use the selling point of flight deck commonality from one model to the next to encourage customers to invest in their aircraft, because it is easier, faster, and cheaper to transition pilots from the previous model to the newer, almost-the-same model than to undergo an extensive training program to learn how to operate a significantly different aircraft. But, to keep the commonality, flight deck design changes become primarily evolutionary, not revolutionary. A radical change must buy its way into the flight deck by virtue of providing such an improvement or benefit that the customers are willing to support it. The glass cockpit was one such change (Figure 2.3).

FIGURE 2.3 In 1981, the Boeing 767's first flight demonstrated a flight deck updated with cathode ray tube displays and the engine indicators and crew alerting system, representing the new *glass cockpit* concept. The Boeing 757 flew the following year with a nearly identical flight deck.

THE EVOLVING FLIGHT DECK

Since the emergence of the glass cockpit, information can be presented in new and different ways. As Fadden et al. (2014) put it, "[i]n one generation of aircraft, the flight deck designers' job had shifted from determining how to get the most pilot performance from the least amount of displayed information to one of selecting the most appropriate information for display in support of each flight task."

Each generation of displays moves a little further away from directly replicating the original analog, electromechanical gages and toward a mode of communicating information that is more integrated and lends itself to better understanding and diagnosis of the aircraft state (Figure 2.4). The move to fly-by-wire systems also led to a set of controls that is decoupled from the systems they act

FIGURE 2.4 Comparison of the flight decks of the 747-100 (right half, debuted in 1969) and 747-400 (left half, debuted in 1988).

on. Controls no longer directly operate the cables connected to the hydraulic actuators that manipulate the flight surfaces or other equipment but are now the initiators of electronic impulses, though they remain physical objects due to legacy and commonality and, in some cases, regulations.

At the same time that displays and controls were finding new forms, pilot roles and responsibilities changed. The specific tasks performed by pilots have changed over the years as the flight deck and the aircraft evolved. Pilots since the late 1980s are trained in Crew Resource Management (CRM) to help enable the kind of teamwork they need to accomplish all of the tasks and identify and troubleshoot any problems that might emerge. The cooperative nature of CRM means that the less senior or lower ranking pilot is encouraged to challenge the more senior or higher ranking pilot if the situation warrants, rather than deferring purely on the basis of rank. This is primarily accomplished by prompting both pilots to be aware of and involved in any changes to aircraft state, which is supported by providing both pilots with essentially identical information via the displays and controls.

The reduction in controls, indicators, and displays in the flight deck from enough to populate four or five crew stations to the two-pilot dashboard setup familiar to today's crews was due to three types of technological improvements: improved system reliability and performance, the paradigm shift of information display from individual gages to integrated large-screen displays, and the increased amount of automation on the flight deck.

Automation on the flight deck simplified the crew's interaction with systems to the point where the flight engineer's position was no longer critical for most flight decks. The controls and indicators could be reduced and consolidated, and the systems they were associated with had much fewer problems than previous generations, and of those problems, only a relatively small number still had to be managed directly by the pilots rather than by automation.

AUTOMATION IN THE FLIGHT DECK

Beyond systems management, automation began to be incorporated to control aspects of aviating and navigating, resulting in an array of capabilities including autopilot, auto-throttles, autoland, and various flight modes. During the course of a typical revenue flight today, the pilots manually operate the airplane until just after takeoff and, other than making periodic adjustments to the mode control panel in response to occasional air traffic control requests, do not usually resume manual control until just before landing. The autopilot maintains course or heading, speed, and altitude as dictated by the flight plan and the selected modes.

For all the advances that have been made in automation design, however, the same fundamental issues come up again and again. The following provides just a brief snapshot of concerns about poor automation design over the past few decades; the topic is a perennial favorite in research because of the persistence of these issues. In 1983, Bainbridge wrote an article on "The ironies of automation" that described a variety of issues associated with increased automation, in particular that it reduces the human's role to monitoring for unusual events, which is not one of our strong suits (Bainbridge, 1983). In 1990, Norman published a paper on the challenges of integrating automation into systems in ways that allow humans to use it well (Norman, 1990). The FAA Human Factors Team compiled a comprehensive report with a review of automation-related issues and recommendations aimed at improving the regulations and certification process to help clarify how automation should be designed and certified (Federal Aviation Administration, 1996). A study by Mumaw, Sarter, & Wickens (2001) demonstrated that even experienced pilots had difficulties using some familiar automation features accurately during unexpected events. Parasuraman and Wickens (2008) wrote a metareview of automation studies and provided a set of recommendations for improved automation design. Finally, the most recent edition of *Engineering Psychology and Human Performance* contains a chapter dedicated to automation design (Wickens, Hollands, Banbury, & Parasuraman, 2016), identifying the potential performance issues associated with poor design and providing recommendations for improved approaches to create effective automated systems.

Clearly there are opportunities for design improvements whenever automation is integrated into an environment in which humans must retain some role. One that is particularly relevant to increased use of automation in the flight deck is the preservation of fundamental pilot skills and airmanship, particularly in the younger pilot population. Senior pilots tend to have much more hand-flying experience in their backgrounds and so may be more immune to skill loss due to the lack of opportunity to practice. Junior pilots, on the other hand, may have trained more on simulators and small planes outfitted with automated features, with a much lower emphasis on hand-flying prior to operating highly automated, large transport category aircraft that also do not provide much hand-flying time, and so do not have the depth of experience to draw on in the event of an automation failure. As the senior pilots retire out of the workforce, it is the current population of junior pilots, with their level of hand-flying experience and expectations of automation that will drive flight deck design.

AVIATION IS A UNIQUE CONTEXT FOR PRODUCT DESIGN

Designing for an aircraft flight deck is unlike design of most consumer products. The market for transport category aircraft is defined by a limited customer base and extremely high costs per unit, highly trained operators, and very long product life spans. The designer must take the long view and consider the long-term implications of any new design or design change.

As each aircraft manufacturer has a limited set of customers that can afford its products, there is incredible competition for their business. The manufacturer must try to keep costs low and consumer value high while still developing a quality product that will last for the expected decades of service and building enough customer loyalty and investment to ensure future business. The way that this affects design is to place a strong emphasis on commonality and legacy while also promoting, to the extent possible, features and functions that differentiate one manufacturer's product from the competitors'. Successful designs are those that push the envelope just enough to improve the product while conforming to long-established expectations of familiarity and avoiding the introduction of additional costs to the customer. Increased pilot training resulting from changes to the flight deck is a significant cost concern for customers.

Consider the user population; pilots are required to have a certain amount of training before they even obtain an airline transport pilot certificate, including at least 1,500 flight hours. Pilots hired by the major U.S. airlines typically have about 4,000 flight hours (Bureau of Labor Statistics, 2015). This is a select and highly trained user population who tend to specialize in a small number of aircraft types and so are very familiar with the aircraft and very sensitive to any changes that might be made. Flight operations tend to be safe and routine; millions of flight hours pass without incidents, much less accidents. However, when unusual or emergency situations arise, the pilots must be able to react quickly and accurately and the flight deck must support all the information and actions needed to do so.

Once an airline takes delivery of a new airplane, it expects to keep that aircraft in service for decades with only slight modifications made to any part of the plane (including the flight deck) in all that time. Each individual aircraft type needs to be designed with longevity in mind. Although upgrades and retrofits provide opportunities to make incremental improvements on a flight deck, they are expensive, time-consuming propositions for airlines. This can make it a challenge to sell the airlines on new designs for an established aircraft unless the new designs come with a number of benefits to offset those costs (Figure 2.5).

THE DESIGN PROCESS

The following process description represents a suggested best practice for designing flight deck features that can be used effectively and certified to fly on an aircraft (Figure 2.6). Each component of the process offers a chance to evaluate whether the design is needed, and if it is, whether the proposed design solution is successful. Ultimately, each manufacturer conducts its design process

FIGURE 2.5 The 777's first flight was in 1994. It was the first aircraft to be designed entirely on computers.

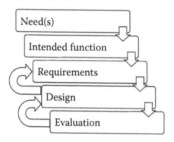

FIGURE 2.6 Flowchart of the recommended steps in the design process.

according to internally developed procedures and criteria, so this process should be considered a high-level set of guidelines to keep in mind as the design work proceeds.

The first step in developing a design for a flight deck is to *identify the need it will satisfy*. What is the problem to be solved? Is it a gap in the way information is presented? Is it a shortcoming of an existing design? Is it a customer request? Is it a required safety enhancement? The first key to a design that will make it into a flight deck is discriminating between a request, or want, and a need. Customers may want certain improvements to the flight deck, but if they do not offer enough value to offset the cost of research, development, and implementation, the design is unlikely to go forward. If there is truly a need for the design, it has a higher chance of becoming a part of the flight deck.

Identifying the fundamental need or needs helps one to *elicit the intended function* of the new (or modified) design—what does the new design do that is not available on the current flight deck? What is the value or the benefit that it provides? The intended function is the succinct explanation for why the design exists; it is not a design description. For certification of a design, the intended function is held as the basis for evaluating the design. If the end result of the design (the product) does not meet its intended function, it will not be certified. This may seem obvious, but there are many cases in which a design is proposed to take advantage of some new, emerging technology or capability, but ultimately is shelved because the product did not do anything beneficial from an operational standpoint.

A good rule to remember is as follows: Just because something can be done, does not mean it should be. Technological improvements have enabled many advances in the flight deck, but not all technologies are good matches for the flight deck for a variety of reasons. In some cases, the technology is not mature enough to meet the rigors of the development and certification process. In other cases, the technology cannot be incorporated in a way that produces a measurable benefit to flight operations. Finally, a technology may be appropriate for some transportation contexts but not for the unique environment of an aircraft flight deck.

Once the intended function is identified and composed in a way that fully addresses the need, the designer's next step is to *develop requirements for the design* that will fulfill the intended function. The requirements must reflect the design philosophy for the flight deck to ensure a design that is consistent with the other features and functionality of the flight deck. For example, if the design philosophy describes a *quiet and dark* approach, whereby most indications are not present to signify that systems are operating within expected parameters, then a new design that proposes signaling normal functioning with attention getting lights or aurals would violate that philosophy. The requirements should be written in such a way to prevent a design philosophy violation from being considered when design solutions are being formulated. In the example given, a good requirement to have would be that the designed feature shall not use lights or aurals to indicate normal functioning.

High-level requirements are those that state generally what the design will accomplish. Lower level requirements provide more details about how it will accomplish each piece and under what constraints. Requirements do not need to be developed solely by the designer; a team that includes a variety of related disciplines (engineers for the system associated with the flight deck feature, the avionics team, pilots, maintenance representatives, etc.) should be convened to evaluate and refine the requirements. Customer input and regulations should also be taken into consideration when creating requirements as well.

Note that so far in this process, a specific design has not been created. A common approach is to start with a proposed design solution and then reverse-engineer a purported need for it. This is not an automatically unsuccessful route, but if it turns out that the solution does not really fit a real operational need, then it will likely be shelved and all the work that went into creating that design solution would have been better spent on considering *first* what needs exist and *then* how a new design could meet those needs. Although it is tempting to try to capitalize on the latest technologies, decisions about what designs get into a flight deck are pragmatic and not often swayed by flashy solutions, no matter how clever. One recent nonaviation example of a technology without a clearly defined need was Google Glass, a wearable augmented reality system that was taken off the market when it failed to find an audience. Altman (2015) wrote that "Google Glass didn't fail because of the technology, rather because it wasn't clear to the customer what problem it solved or why they needed it."

Once the requirements are written, the designer can *begin designing a solution* that satisfies the criteria established in the requirements. As the design is iterated, it should be vetted against a set of questions: Can it be certified? Does it do what it is expected to do? Does it interfere with any other task? Does it introduce complexity into a system? Does it affect training or commonality? The answers to these questions can help identify whether the requirements are complete, realistic, appropriate, and written well enough to avoid ambiguity.

Another area to define in detail is the success criteria, which derive from the earlier step of defining the intended function. How are the success criteria quantified to determine that a design solution is effective? Does it reduce head-down time and increase situation awareness? Does it support better CRM or improved task performance? Identifying these criteria early on will help guide the evaluation of the final design further along in the process.

STRATEGIES FOR DESIGN EVALUATION

At some point, the design will be far enough along that it can be prototyped for more in-depth evaluation outside of the requirements development team. The evaluation process spans a continuum from quick paper-based surveys to complete test plans for conducting well-controlled experiments in a high fidelity simulator (Figure 2.7).

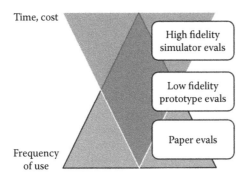

FIGURE 2.7 Depiction of the three phases of evaluation and the associated tradeoffs of their respective costs and frequency of use.

Independent of what evaluation strategies are used, it is strongly recommended that the designer recruit pilots to participate in any evaluation as they represent the target user population. Experienced pilots can help guide the requirements and initial design features and capabilities and, in fact, should be part of the requirements team. Pilots who represent the intended user population (e.g., general aviation, commercial revenue, or cargo operator) should be recruited to provide feedback on the design prototype.

There are three general categories of evaluation strategies that can also be considered consecutive phases of the evaluation process:

Paper-Based Evaluations

This approach is typically quick and cost effective but has limited usefulness. Little-to-no interaction is represented, whereas pilots do not get as thorough a sense of the design's role within the larger flight deck. As there has been a relatively low investment in the design by this point, and typically, the design is not very mature, this is a good opportunity to develop multiple alternatives and have the pilots evaluate the options. By comparing alternatives side by side, pilots can provide more targeted feedback about the relative desirability of each component of the design. If constraints on the design prevent the ideal solution from materializing, pilots can help determine the best compromises that can be made.

This approach is best to use to identify the optimal set of features and functions and can also be used as an initial step before the prototype is tested in a higher fidelity environment, for example, when the prototype is still in the drawing or drafting stage rather than a working piece of software or physical component.

Low and Medium Fidelity Mockups, Part Task Trials

Typically at this point, there is a single, mostly complete design that meets the requirements, and it can now be evaluated to determine how well it performs the intended function. A low or medium fidelity mockup is a working prototype that can be evaluated through at least some level of interaction by an operator. The more features in the mockup that are interactive, the more thorough the evaluation can be.

Evaluating a prototype within a larger context involves ensuring that the prototype is sufficiently advanced in design to emulate its features accurately. More time and effort is needed to create a working prototype than a paper one, though again technology is improving how quickly something can be prototyped. New display designs now typically consist of software changes, which can be mocked up fairly quickly by an experienced programmer. New control or other physical designs can be modeled in a computer aided design (CAD) format and quickly fashioned using additive manufacturing (otherwise known as 3-D printing).

At this stage, it is useful to evaluate the design within an environment that represents some aspects of the flight deck. For example, if the design is a new display feature or layout, it should be presented on a screen similar to one that a pilot would use in a flight deck. Similarly, if the design is a physical control, it should be evaluated in approximately the configuration the pilot would expect to find it within the flight deck.

The data obtained in this type of evaluation are typically more helpful in refining the design than that from the first type of approach because it is based on interacting with the design, not just reviewing a description of how it works or what it looks like. However, as the prototype is still being evaluated more or less in isolation, this approach is best used for stand-alone features and specific tasks requiring interaction, or as preparation for more thorough evaluations conducted in a flight simulator or real flight environment.

High Fidelity Mockup in Integrated (Flight Simulator) Setting Using Detailed Test Plan

This is the most expensive and time consuming option for evaluating a design, but it provides the best opportunity to fully investigate whether the design functions as intended and also does not introduce unintended consequences as a result of its integration into the flight deck (Figure 2.8). At this stage, the prototype is typically very refined and as close to fully functional as possible. It is evaluated in a flight simulator so that the more complex features or tasks, those that may interfere with existing operations, habits, and expectations, and those that are expected to require the highest levels of training, can be tested. For that reason, the design under evaluation should be investigated within the context of as complete a simulated operation as possible, with pilots performing other typical tasks before, during, and after the task involving the new design as they would if the design were implemented and in use on an actual flight deck.

Due to the complexity of this type of approach, it is recommended to develop a complete test plan based on a well-developed experimental design. Performance using the new design is typically compared against performance using the current flight deck to determine whether, and how, the new design supports any benefits to flight operations. Benefits may include improved efficiency, increased situation awareness or system understanding, reduced error rates, faster response times, or whatever success criteria were established based on the original need and intended function.

At all stages in the concept development, and especially in stage 3, the evaluation participants should be interviewed to get their take on the design and its features and functionality. It has been well established that subjective opinions and objective performance do not always align. For designs intended to improve performance, objective performance-based data should be given more weight. However, the value of subjective feedback should not be underestimated. Many studies include analyses in which the pilots' responses collected during post-trial interviews have (a) clarified the

FIGURE 2.8   The flight simulator for the Boeing 787, which had its first flight in 2009.

objective data when it otherwise would have appeared confounded, (b) indicated that the pilots do not agree with the point of the design but still use it flawlessly, (c) provided evidence of a range of pilot opinion that might influence performance depending on circumstances, and (d) identified flaws in the evaluation that prevented some aspects of the design from being fully utilized or understood. In each case, the pilot feedback gave a better idea of the varying contexts within which the design must be considered and either led to redesign of the prototype, the evaluation methodology, or the training that would accompany the design.

At the end of any evaluation, the data (both objective and subjective) are reviewed and analyzed, conclusions are drawn, and any recommended changes are applied to the design and in some cases the requirements driving the design. Once the design is finalized, it is handed off to the appropriate product developers.

DOES THE DESIGN PASS OR FAIL?

At any point in the aforementioned design process, the decision may be made to iterate, start over, or completely discontinue work on a design because it fails one or more of the success criteria. There are a variety of reasons why a new design or design change fails to make it onto a flight deck.

Sometimes a design was implemented before the application of good human factors and is kept because its service history demonstrates that the theoretical human factors violation has no negative impact on pilot understanding or performance. Although it might seem like a target for redesign to bring it in line with current human factors practice, a change in the existing design might introduce unintended consequences. For example, when the 737 landing gear is in transit (retracting after take-off or extending prior to landing), the in-transit indicators illuminate with red lights. That design was established prior to the development and application of the current flight deck color philosophy, which dictates that the color red is reserved for time critical warning situations. A transiting gear is a completely normal operation and occurs twice during a standard flight. If it were designed today, the indicator color would be something other than red. However, because the existing gear light design has been in use for decades without any safety issues, it might introduce unintended consequences if the red indicators were replaced with another color, because the vast experience of thousands of trained 737 pilots drive their expectation that that specific indicator is red.

Sometimes a design may seem like a good target for change, but the resulting design could lead to the loss of secondary information. Noise reduction in the flight deck is one such example. Although the flight deck is a noisy environment and excessive noise should be reduced as appropriate, the further reduction from what is experienced as the *normal* noise level, whether through improved sound-proofing materials or the use of noise-canceling headphones, can mask certain aural cues (such as variations in engine noise) that tip the pilots off to a non-normal aircraft condition.

Sometimes there are no strong population stereotypes to guide the design of a feature, which lead to conflicting or divergent designs. Over the decades of airplane development, attitude indicators have taken different forms. These variations can require pilots to adapt to new cues. For example, inside-out versus outside-in attitude display indicators have not been found to be associated with any strong performance trends to support one design format over another, but transferring between the designs can lead to significant negative transfer effects.

Currently, Western-built airplanes show changes in attitude by moving the horizon against a static airplane symbol. These are known as moving-horizon indicators or *inside-out* displays. Earlier Russian-built airplanes maintained a stationary horizon and moved the airplane symbol as attitude changed. These are known as moving-airplane or *outside-in* displays. An early study (Gardner & Lacey, 1954) comparing these two styles of presenting attitude information found that for novices, the moving-airplane displays led to fewer control input errors as they better link the pilot actions on the wheel with the effect on the airplane. Experienced pilots, who had trained on moving horizon style indicators, performed less well when using the unfamiliar moving airplane style than when using the familiar moving horizon format. However, in postinterview questionnaires, both the novices and the experienced pilots indicated a preference for the moving airplane format over the moving horizon format.

The issue with these conflicting designs is that they display nearly identical information and at first glance appear interchangeable, but they indicate movement in opposing ways. Misinterpreting the movement in one type of display as meaning the same as the other type of display can lead to significant errors.

Sometimes the same task is interpreted in two very different ways, resulting in significantly different pilot behaviors but no overall impact to performance. One well-known difference between Boeing and Airbus commercial flight decks is the manner of controlling the flight surfaces. Boeing flight decks have used a column and wheel setup since the very first airplanes were built. The wheel controls the differential wing surfaces (ailerons and spoilers), whereas the column controls the elevators on the tail. Both pilots' columns and wheels are linked so that inputs made to one are replicated in the other; moving one wheel or column moves the other in synchrony. Further, flight control changes made while on autopilot are manifested in the analogous movements of the columns and wheels (they are *back-driven*), such that if autopilot commands a turn, the wheels rotate correspondingly, providing a visual and potentially tactile cue to the pilots that the aircraft is changing attitude. The columns and wheels take up quite a bit of real estate between the pilots and the forward displays and require seats to be designed to accommodate a column in the full-aft position (involving a substantial cutout in the seat front). Designs for the Boeing forward panels must account for the full range of motion of the yoke-and-column.

By contrast, Airbus flight decks use a side stick, which operates like a control stick and is positioned on the outboard side of each pilot's position. The current generation of Airbus side sticks are not linked or back-driven and do not provide tactile or physical motion cues as to the inputs acting on the flight controls; neither pilot can see or feel what the other, or autopilot, is doing to control the flight surfaces. The location of the side stick, however, frees up the space directly in front of the pilots and enables a different configuration of the flight deck and the pilot seats, with fewer constraints on the physical space compared with the Boeing flight deck.

New versions of side sticks incorporate active inceptors to provide the tactile feedback and visual cues missing from the current passive side sticks (Warwick, 2015), but the location of the stick on the pilot's outboard side diminishes the effectiveness of those visual back-driven motion cues as they are outside of the pilot's typical scan pattern, which focuses on the areas in front (dedicated displays) and slightly inboard (shared displays) of the pilot's position. This type of stick is not currently implemented in Airbus flight decks.

Despite the very different manifestations between the two flight surface control designs, pilots can easily adjust to one or the other type as the inputs are intuitive for each. Preference for one style over the other appears driven more by personal opinion than by objective performance. These significantly different control designs do not lend themselves to negative transfer of training because there is little to no expectation that one is controlled using the same physical manipulations as the other; they simply do not afford the same actions.

Sometimes there are reasons why a design evolved the way it did and the implementation became so widely used that even though better (according to some objective metrics) designs exist, widespread adoption of those alternatives will never be embraced. Even if the basis for an original design is no longer valid, the designer must consider whether that design is so entrenched that it has become standard. Further, it is possible that the nature of the design offers informal or secondary cues or information that, if altered, might be lost altogether, which might decrease the overall usefulness of the altered design.

The QWERTY keyboard illustrates how a design was influenced by several technological factors that are no longer relevant. The original key layout was possibly structured in part to prevent mechanical typebars from jamming by separating commonly used letters (Stamp, 2013), but successive layout iterations were primarily based on input from telegraph operators transcribing Morse code into typed messages (Yasuoka & Yasuoka, 2011). Soon after its layout was finalized, the QWERTY keyboard became the standard typewriter format across multiple brands and its ubiquity continued long after mechanical typebars and Morse code were no longer in common use.

Competing keyboard layouts such as Dvorak may produce faster typing by some metrics but are unlikely to ever replace QWERTY as the standard.

Sometimes a design works well in theory and fails in practice. In this case, the design appears to meet all the requirements and perform the intended function, at least in early evaluations, and yet does not establish itself as useful, desirable, or necessary to the end user, or is far too complicated to operate with too little return, or can only operate under a very limited set of conditions. There are many examples of designs that were pushed into the marketplace only to disappear quickly due to lack of consumer interest or widespread adoption. Before its release in 2001, the Segway scooter was expected to change how people traveled around an urban environment, and its technology was very impressive. Due to its operating limitations, hefty price tag, and logistical challenges, only a relatively small number (far fewer than predicted) have been sold and that was to niche market customers rather than the general public (Golson, 2015).

The point of failure may not occur until the simulator-based evaluations, if it is determined that the design introduces some unexpected complication when it is integrated into the flight deck. Although this type of failure occurs very late in the design process, it emphasizes the importance of testing in as close to the real environment as possible, as it is better to find out on the ground before the design is cleared to fly. In addition, it may be that a further refinement of the design, based on the failure feedback, is enough to result in a successful design.

How Does the Design Get on the Airplane?

There are several opportunities to implement updated, improved, and new features into a flight deck. These primarily occur at the introduction of a new aircraft model, during the development of a new derivative of an existing aircraft model, or during the release of a new block point for the model (Figure 2.9).

First, block point updates occur regularly throughout the lifespan of an aircraft and provide an opportunity to include updated software that may fix existing issues or provide incremental improvements to an existing system. These are typically extensive procedures that take the aircraft out of service for some amount of time, especially if hardware changes are included. In some cases, new designs that are ready in time for a block point will be deferred to a later delivery date if the logistics do not work out. Given the amount of time to service all the aircraft in an airline's fleet, block point updates result in a staggered delivery that means some aircraft receive the changes much sooner than others. This is referred to as a mixed fleet and can mean difficulties in pilot training, scheduling, and expectations of the individual aircraft when they do not have consistent flight deck features.

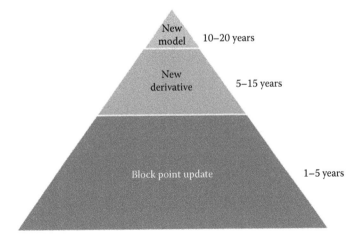

FIGURE 2.9 Schematic of the relative opportunities to get a new or modified design onto a flight deck.

When a new derivative of an existing model goes into production (e.g., 737 MAX, 777–9), typically it provides a chance to incorporate at least a minimal set of new and modified features (both hardware and software) into the flight deck. The reason that the number of changes is kept relatively small is that the new model is intended to have the majority of features in common with previous models to simplify both the certification effort as well as the resultant differences training pilots receive when they become type-rated on the new model. In some cases, changes that are introduced on a new model are sometimes retrofitted onto older models if the benefits of including those features on the older models outweigh the costs of taking otherwise fully functioning aircraft off the revenue line for several days or longer.

Significant design changes to a flight deck do not tend to get implemented until a major new aircraft model is announced. For example, master caution and warning system debuted in the 737, whereas the glass cockpit-enabling cathode ray tube displays as well as the EICAS were first flown in the 757 and 767. The electronic checklist was first introduced on the Boeing 777 (Myers, 2016), which was also the first Boeing model to incorporate *fly by wire* features. Those features have since become standard on most of the Boeing platforms. Among other features, large screen displays were first flown in the 787 and will be implemented on the future derivatives of the 767, 777, and 737.

There are, of course, exceptions to the usual processes. Features and capabilities that are mandated by the regulators via new regulations and amendments to existing regulations and fixes needed to issues that are only discovered once the plane is in service or mandated by Airworthiness Directives may need to be implemented outside of an entry-into-service schedule.

COMMONALITY, LEGACY, AND THE FUTURE

Although flight deck commonality across a manufacturer's fleet tends to be a design goal, designs that work well for some platforms are not developed for other platforms for a variety of reasons.

To illustrate, EICAS has been adopted by all of the wide body Boeing planes from the 757/767 on but has not been implemented into the 737 fleet. The main reason is that the 737's entry into service preceded the development of EICAS by several decades, so it has a different format for providing similar engine and alerting information to the pilots. It would require extensive retrofitting of existing 737s or redesign of future models to accommodate the EICAS structure. Another reason is that the 737 is the single most common platform operating today, with a pilot population that is acutely familiar with the operation of the flight deck as it currently exists. Producing a 737 with a significantly modified flight deck would likely have significant repercussions on customer investment and pilot training.

The designer is advised to consider the history of the decisions that drove a current flight deck's design to better understand where, and whether, design changes to that flight deck, or a new derivative's flight deck, or even a new model's flight deck, are warranted. The new design must fit appropriately within the flight deck and align with the philosophy that underlies its configuration.

SUMMARY OF RECOMMENDATIONS

The design of any individual display or control, whether it is a modification or improvement on an existing design or the introduction of a completely new design, must be incorporated into the flight deck as a whole. Manufacturers such as Boeing and Airbus have flight deck philosophies that provide guidance into how the pilots interact with the aircraft. Further, there are styleguides that provide recommendations on details such as color use, font type and size, and layout and organization of different categories of content. Finally, there are requirements, regulations, and standards that dictate how various designs may be instantiated. The designer must deal with this set of constraints to design a new flight deck feature that satisfies a true need, fits well within the flight deck, and can be certified. Special care must be taken so that a new or modified feature does not interfere with existing displays, controls, procedures, or other operational factors.

The design of the flight deck must take into account the pilot population who will be using it. This means understanding what the pilots' background in aviation consists of the following: what training and experience they might be expected or required to have, what part of the world they operate in, what airline culture they participate in, and any other factor that could influence how effectively the pilots interpret the information presented in the design of a display or a control. Gathering information on what features are used during typical revenue operations and how they are used or even thought of may help guide design decisions.

In the last century, the flight deck design efforts focused on reducing and simplifying the displays and controls provided to the flight crew. More recently, flight decks have become increasingly filled with new information and advanced automation, and understanding the evolving role of the pilot is crucial for good design. The pilot role has slowly changed over the decades from active control operator in a risky environment to mostly passive systems manager in a very safe environment. The average pilot has changed from ex-military or long-time general aviation pilot with thousands of hours of aviating experience across multiple platforms to someone who may have only accrued enough hours (and most of those in a simulator or as a relief cruise pilot with limited duties) to become type rated before being assigned to a regular carrier route. Although we want to preserve piloting skills in the face of increasing reliance on automation, we must also acknowledge that the average new pilot's airmanship skills are not the same as those of last century's pilots. Despite significant decreases over the years, human error is still a source of accidents and incidents and will continue to be dependent, at least in part, on the quality of flight deck design (Yeh et al., 2013).

The designer is encouraged to take a very long view of flight deck design and consider its evolutionary path. There are opportunities to design to improve the current models, to design for the next new aircraft, and to design for the one after that. In some cases, the designer may invent a way to capitalize on newly introduced technology (e.g., large screen displays, touch screen capability, improved optics) that is not possible or practical in the near future but may be an ideal solution in a decade or two.

One last recommendation for any designer: Do not get too attached to a solution. It may be the optimal design, or it may benefit from iteration. It may not be the solution that solves the problem after all and may have to be scrapped. A professional detachment to the product of your work is recommended to keep your objectivity about it. Too often designs are pursued because they are clever or someone expresses a strong preference for a particular design. Ultimately, if the design does not meet a need or accomplish the intended function, it will only remain a clever idea.

CONCLUSION

The constraints around a modern flight deck are significant: Legacy. Commonality. Preservation of critical information. Regulations.

Displays and controls must conform to pilot expectations, built over years of experience, regarding their content and their actions. Most, if not all, aircraft flown today are designed to be operated by two pilots and so real estate on the flight deck is constrained by the physical environment they occupy. Further, most, if not all, aircraft can be flown by one pilot when absolutely necessary, as in the case of pilot incapacitation. This requires not only a significant amount of redundancy in the displays and controls but also that the critical controls be located such that they can be operated from either pilot's seat.

Despite these constraints, there are and will continue to be opportunities for improvements and new designs. "What has gone before is really only prologue; the art and fascination of flight deck design is alive and well, continuously improved in all classes of airplanes, and form a rich tapestry of career opportunities for future practitioners" (Fadden et al., 2014).

REFERENCES

Abbott, K. (2015). Chapter 15: Human factors engineering and flight deck design. In C. Spitzer, U. Ferrell, & T. Ferrell (Eds.), *The digital avionics handbook*, 3rd ed. Boca Raton, FL: CRC Press, Taylor & Francis Group.

Altman, I. (2015). Why Google Glass failed and why apple watch could too. *Forbes.* Retrieved March 7, 2016, from http://www.forbes.com/sites/ianaltman/2015/04/28/why-google-glass-failed-and-why-apple-watch-could-too/#4e1ec6af58ec.

Bainbridge, L. (1983). The ironies of automation. *Automatica*, *19*(6), 775–779.

Bureau of Labor Statistics. (2015). *Occupational outlook handbook: Airline and commercial pilots.* Retrieved March 15, 2016, from http://www.bls.gov/ooh/transportation-and-material-moving/mobile/airline-and-commercial-pilots.htm.

FAA Flight Standards Service. (2013). *Glider flying handbook, FAA-H-8083-13A.* US Department of Transportation, Federal Aviation Administration. Washington, DC: Federal Aviation Administration.

FAA Safety Briefing. (2015) *Fly the aircraft first.* FAA Aviation Safety. Retrieved March 2, 2016, from http://www.faa.gov/news/safety_briefing/2015/media/SE_Topic_15_01.pdf.

Fadden, D. M., Morton, P. M., Taylor, R. W., & Lindberg, T. (2014). *First-Hand: Evolution of the 2-person crew jet transport flight deck.* Retrieved July 30, 2014, from http://ethw.org/First-Hand:Evolution_of_the_2-Person_Crew_Jet_Transport_Flight_Deck.

Federal Aviation Administration. (1996). *The human factors team report on: The interfaces between flightcrews and modern flight deck systems.* US Department of Transportation, Federal Aviation Administration. Washington, DC: Federal Aviation Administration.

Gardner, J. F., & Lacey, R. J. (1954). An experimental comparison of five different attitude indicators. *USAF Wright Air Development Center Technical* (Report 54–32). Dayton, OH: McGregor & Werner.

Golson, J. (2015). *Well, that didn't work: The Segway is a technological marvel. Too bad it doesn't make any sense. Wired.com.* Retrieved March 28, 2016, from http://www.wired.com/2015/01/well-didnt-work-segway-technological-marvel-bad-doesnt-make-sense/.

Komons, N. A. (1987). *The third man: A history of crew complement controversy 1947–1981.* Washington, DC: US Department of Transportation, Federal Aviation Administration. US Government Printing Office: 1987–181-783/74703.

Mumaw, R. J., Sarter, N. B., & Wickens, C. D. (2001). Analysis of pilots' monitoring and performance on an automated flight deck. In *Proceedings of the 11th international symposium on aviation psychology.* Columbus, OH: The Ohio State University.

Myers, P. L. (2016). Commercial aircraft electronic checklists: Benefits and challenges (Literature Review). *International Journal of Aviation, Aeronautics, and Aerospace, 3*(1). http://dx.doi.org/10.15394/ijaaa.2016.1112.

Norman, D. (1990). The problem of automation: Inappropriate feedback and interaction, not over-automation. In *Philosophical Transactions of the Royal Society of London.* London, UK: The Royal Society.

Parasuraman, R., & Wickens, C. D. (2008). Humans: Still vital after all these years of automation. *Human Factors, 50*(3), 511–520.

Stamp, J. (2013). Fact of fiction? The legend of the QWERTY keyboard. *Smithsonian.com.* Retrieved March 29, 2016, from http://www.smithsonianmag.com/arts-culture/fact-of-fiction-the-legend-of-the-qwerty-keyboard-49863249/?no-ist.

Warwick, G. (2015). Active sidestick controls make commercial debut. *Aviation Week and Space Technology.* Retrieved January 5, 2015, from http://aviationweek.com/technology/active-sidestick-controls-make-commercial-debut.

Wickens, C. D., Hollands, J., Banbury, S., & Parasuraman, R. (2016). *Engineering psychology and human performance,* 4th ed. New York, NY: Routledge (Taylor & Francis Group).

Yasuoka, K., & Yasuoka, M. (2011). On the prehistory of QWERTY. *ZINBUN, 42*: 161–174. Retrieved March 29, 2016, from http://hdl.handle.net/2433/139379.

Yeh, M., Jo, Y. J., Donovan, C., & Gabree, S. (2013). Human factors considerations in the design and evaluation of flight deck displays and controls, version 1.0. *FAA Final Report DOT/FAA/TC-13/44.* US Department of Transportation, Federal Aviation Administration. Washington, DC: Federal Aviation Administration.

3 The Design of the Air Traffic Control System

Michael Nolan

CONTENTS

Introduction..32
Aircraft Operations 32
Airspace Classes 32
Air Traffic Control Providers 33
Air Traffic Control Assignments...34
Air Traffic Control Services..34
Air Traffic Control Services Offered within Each Type of Airspace 35
Aeronautical Navigation Aids...35
Global Navigation Satellite System 37
Radar Surveillance in Air Traffic Control 37
Aircraft Separation in an Air Traffic Control System 37
Nonradar Separation 39
Radar Separation..40
Radar System Limitations..40
Additional Radar Services 41
Radar Identification of Aircraft...41
Radar Separation Criteria..41
Current Trends in Automation 42
Airborne Systems...43
Conflict Alert/Visual Flight Rules Intruder..43
Traffic Management Systems...43
    Airport Capacity Restrictions..44
Air Traffic Control System Overloads 44
Pilot/Controller Communications-Radio Systems..45
Voice Communication Procedures...46
Electronic Data Communications 47
Controller Coordination 47
Flight Progress Strips...48
Flight Information Automation 49
Controller Responsibilities in the Air Traffic Control System.....................................49
Future Enhancements to Air Traffic Control Systems 49
Suggested Reading...50
Other FAA Publications...50

INTRODUCTION

The primary function of an air traffic control (ATC) system is to keep aircraft participating in the system separated from one another. Secondary reasons for the operation of an ATC system are to make more efficient use of airspace and to provide additional service to pilots such as traffic information, weather avoidance, and navigational assistance.

Not every aircraft may be required to participate in an ATC system, however. Each nation's regulations only obligate certain aircraft to participate in the ATC system. ATC participation in each country may range from mandatory participation of all aircraft to no-ATC services offered at all.

The level of ATC services provided is usually based on each nation's priorities, technical abilities, weather conditions, and traffic complexity. To more specifically define and describe the services that can be offered by an ATC system, the International Civil Aviation Organization (ICAO) has defined different aircraft operations and classes of airspace within which aircraft may operate. Different rules and regulations apply to each type of aircraft operation, and these rules vary depending on the type of airspace within which the flight is conducted. Although ICAO publishes very specific guidelines for the classification of airspace, it is the responsibility of each country's aviation regulatory agency to categorize its national airspace.

AIRCRAFT OPERATIONS

Visual meteorological conditions (VMC) are defined as weather conditions in which pilots are able to see and avoid other aircraft. In general, pilots flying in VMC conditions comply with visual flight rules (VFR). VFR generally require that three to five miles of flight visibility be maintained at all times, that the aircraft remain clear of clouds, and that pilots have the responsibility to see and avoid other aircraft. Pilots provide their own air traffic separation. The ATC system may assist the pilots and may offer additional services, but the pilot has the ultimate responsibility to avoid other air traffic.

Instrument meteorological conditions (IMC) age generally defined as weather conditions in which the visibility is below that required for VMC or whenever the pilot cannot remain clear of clouds. Pilots operating in IMC must comply with instrument flight rules (IFR), which require the filing of a flight plan, and ATC normally provides air traffic separation. Pilots may operate under IFR when flying in VMC conditions. Under these circumstances, ATC will separate only those aircraft complying with IRF. VFR aircraft provide their own separation, and IFR aircraft have the responsibility to see and avoid VFR aircraft.

AIRSPACE CLASSES

National governments define the extent to which they wish to offer ATC services to pilots. In general, ICAO recommendations suggest three general classes of airspace within which different services are provided to VFR and IFR pilots. These three general classes are uncontrolled, controlled, and positive controlled airspace.

Uncontrolled airspace is that within which absolutely no aircraft separation is provided by ATC, regardless of weather conditions. Uncontrolled airspace is normally that airspace with little commercial aviation activity.

Controlled airspace is that within which ATC separation services may be provided to certain select categories of aircraft (usually those complying with IFR). In controlled airspace, pilots flying VFR must remain in VMC and are not normally provided ATC separation and therefore must see and avoid all other aircraft. Aircraft who wish to utilize ATC services in controlled airspace must file a flight plan and comply with IFR. IFR aircraft are permitted to operate in VMC and IMC. When operating within controlled airspace, IFR aircraft are separated by ATC from other aircraft operating under IFR. When operating in VMC in controlled airspace, IFR pilots must see and avoid aircraft operating under VFR.

TABLE 3.1

Requirements for Operation and ATC Services Provided to Flight Operations within General Airspace Categories

	Uncontrolled Airspace (Class G Airspace)	Controlled Airspace (Class C, D and E Airspace)	Positive Controlled Airspace (Class A and B Airspace)
VFR flight operations	Must remain in VMC (VMC minima are fairly low, typically clear of clouds and one mile visibility) If VMC conditions exist, VFR operations are permitted and no ATC clearance is required If IMC conditions exist, VFR operations are not authorized No ATC separation services are provided Pilots responsibility to see and avoid both IFR and VFR aircraft	Must remain in VMC (VMC minima are higher, typically a specified distance from clouds and three to five miles visibility) If VMC conditions exist, VFR operations are permitted but an ATC clearance may be required to operate in certain areas If IMC conditions exist, VFR operations are not authorized No ATC separation services are provided to VFR aircraft Pilot's responsibility to see and avoid both IFR and other VFR aircraft VFR aircraft operating in controlled airspace may be required to meet additional, class-specific operating rules and procedures	VFR flight operations might not be permitted If permitted, VFR aircraft must remain in VMC (VMC minima are higher, typically a specified distance from clouds and three to five miles visibility) If VMC conditions exist, VFR operations may be permitted but an ATC clearance would be required If IMC conditions exist, VFR operations are not authorized Separation services are provided to all aircraft All aircraft will be separated by ATC VFR aircraft operating in positive controlled airspace may be required to meet additional, class-specific operating rules and procedures
IFR flight operations	IFR operations permitted with no ATC clearance required, nor will it be issued ATC separation services not provided Pilots responsibility to see and avoid both IFR and VFR aircraft	ATC clearance required ATC separation will be provided between IFR aircraft; IFR pilots must see and avoid VFR aircraft while in VMC IFR aircraft operating in controlled airspace may be required to meet additional, class-specific operating rules and procedures	ATC clearance required ATC separation will be provided between all aircraft All aircraft operating in positive controlled airspace may be required to meet additional, class-specific operating rules and procedures

In positive controlled airspace, all aircraft, whether IFR or VFR, are separated by ATC. All aircraft operations require an ATC clearance. VFR pilots must remain in VMC conditions but are separated by ATC from both VFR and IFR aircraft. IFR aircraft are also separated from both IFR and VFR aircraft. Table 3.1 describes the general rules that both IFR and VFR pilots must comply with when operating in these three classes of airspace.

AIR TRAFFIC CONTROL PROVIDERS

In most countries, a branch of the national government normally provides ATC services. The ATC provider may be a civilian, military, or a combination of both. Some national ATC services are now being operated by private corporations funded primarily by user fees. Other governments are experimenting with ATC system privatization. Some of these initiatives purpose to transfer all ATC responsibility to private agencies, whereas other propose to transfer only certain functions, such as weather dissemination and the operation of low-activity control towers, to private or semipublic entities.

Privatized ATC is a fairly recent historical development with roots tracing back to the 1930s. When an ATC system was first started in the United States, control towers were operated by the municipalities that owned the airports. Enrooted ATC was provided through a consortium of airlines. Only in the 1940s was ATC taken over and operated by the national government.

The concept behind privatized ATC is that if freed from cumbersome government procurement requirements, employment regulations, and legislative pressures, private corporations might provide service at less cost, be more efficient, and be more responsive to users' needs because they would be funded and controlled by the users. Possible disadvantages of such a system include lack of governmental oversight and responsibility, possible conflict of interest between system users and operators, little incentive to assist military aviation activities, and restricted access to the capital funding needed to upgrade and operate such a complex system.

AIR TRAFFIC CONTROL ASSIGNMENTS

Every nation is responsible for providing ATC services within its national borders. To provide for a common method of ATC, ICAO promulgates standardized procedures that most countries generally adhere to. These standards include universally accepted navigation systems, a common ATC language (English), and general ATC separation standards. ICAO is a voluntary organization of which most countries are members. Every ICAO signatory nation agrees to provide ATC services to all aircraft operating within its boundaries and agrees to accept that their pilots abide by other national ATC systems when operating within foreign countries.

Every nation's ATC procedures can and do occasionally deviate from ICAO recommended practices. Each operational procedure that deviates from ICAO standards is published by the national ATC service provider in the Aeronautical Information Publication.

ICAO has been granted the responsibility for providing ATC services in international airspace, which mostly comprises oceanic and polar airspace. ICAO has assigned separation responsibility in those areas to individual states both willing and able to accept that responsibility. Some countries that have accepted this responsibility include the United States, the United Kingdom, Canada, Australia, Japan, Portugal, and the Philippines.

AIR TRAFFIC CONTROL SERVICES

Airspace with little or no potential traffic conflicts requires little in the way of sophisticated ATC systems. If air traffic density increases, if aircraft operations increase in complexity, or if special, more hazardous operations are routinely conducted; additional control of aircraft is usually required to maintain an acceptable level of safety. The easiest method of defining these increasing ATC system requirements and their associated operating rules is to define different classes of airspace within which different ATC services and requirements exist.

Standard ICAO airspace classifications include classes labeled A, B, C, D, E, F, and G. In general, Class A airspace is positive controlled, in which ATC services are mandatory for all aircraft. Class G is uncontrolled airspace in which no ATC services are provided to either IFR or VFR aircraft. Classes B, C, D, E, and F provide declining levels of ATC services and requirements.

It is each nation's responsibility to describe, define, explain, and chart the various areas of airspace within their respective boundaries. In general, areas with either high-density traffic or a mix of different aircraft operations are classified as class A, B, or C airspace. Areas of low traffic density are usually designated as class D, E, F, or G.

AIR TRAFFIC CONTROL SERVICES OFFERED WITHIN EACH TYPE OF AIRSPACE

The requirements to enter each airspace classification and the level of ATC services offered within each area are listed here:

Class A airspace: All operations must be conducted under IFR and are subject to ATC clearances and instructions. ATC separation is provided to all aircraft. Radar surveillance of aircraft is usually provided.

Class B airspace: Operations may be conducted under IFR or VFR. However, all aircraft are subject to ATC clearances and instructions. ATC separation is provided to all aircraft. Radar surveillance of aircraft is usually provided.

Class C airspace: Operations may be conducted under IFR or VFR; however, all aircraft are subject to ATC clearances and instructions. ATC separation is provided to all aircraft operating under IFR and, as necessary, to any aircraft operating under VFR when any aircraft operating under IFR is involved. All VFR operations will be provided with safety alerts and, on request, conflict resolution instructions. Radar surveillance of aircraft is usually provided.

Class D airspace: Operations may be conducted under IFR or VFR; however, all aircraft are subject to ATC clearances and instructions. ATC separation is provided to aircraft operating under IFR. All aircraft receive safety alerts and, on pilot request, conflict resolution instructions. Radar surveillance of aircraft is not normally provided.

Class E airspace: Operations may be conducted under IFR or VFR. ATC separation is provided only to aircraft operating under IFR within a surface area. As far as practical, ATC may provide safety alerts to aircraft operating under VFR. Radar surveillance of aircraft may be provided if available.

Class F airspace: Operations may be conducted under IFR or VFR. ATC separation will be provided, so far as practical, to aircraft operating under IFR. Radar surveillance of aircraft is not normally provided.

Class G airspace: Operations may be conducted under IFR or VFR. Radar surveillance of aircraft is not normally provided.

AERONAUTICAL NAVIGATION AIDS

Air traffic separation can only be accomplished if the location of an aircraft can be accurately determined. Therefore, an ATC system is only as accurate as its ability to determine an aircraft's position. The navigation systems currently in use were developed in the 1950s but are undergoing a rapid change in both technology and cost. As integrated circuitry and computer technology continue to become more robust and inexpensive, the newly certified global navigation satellite system (GNSS), global positioning system, promises unprecedented navigational performance at a relatively low cost. ICAO has just recently affirmed its preference for GNSS as the future primary international navigation standard. Various experts predict that existing navigation systems will be either decommissioned or relegated to a GNSS backup role by the turn of the century.

In general, the accuracy of existing navigation aids is a function of system cost and/or aircraft distance from the transmitter. Relatively inexpensive navigation systems are generally fairly inaccurate. The most accurate systems tend to be the most expensive. Table 3.2 describes the type, general cost, advantages, and disadvantages of many common aeronautical navigation systems.

TABLE 3.2
Navigation System Capabilities, Advantages, and Disadvantages

System	General Cost	Effective Range	Accuracy	Ease of Use	Advantages	Disadvantages
Nondirectional beacon	Inexpensive transmitter. Inexpensive receiver	50–1,000 nautical miles	Fairly inaccurate	Somewhat difficult to use	Inexpensive and easy transmitter installation	Susceptible to atmospheric interference. No data transmission capability
VORTAC	Moderately expensive transmitter. Fairly inexpensive receiver	25–200 nautical miles	Fairly accurate	Fairly easy to use	Current enroute international standard. Can provide distance information if aircraft suitably equipped	Large number of transmitters required to provide adequate coverage. Primarily a point-to-point navigation system
Inertial Navigation Systems (INS)	No transmitters required. Very expensive receivers	Unlimited	Fairly accurate	Fairly easy to use	Very independent operation as no ground transmitters required	Expensive. Needs to be programmed. Accuracy deteriorates over time without external input
Global Navigation Satellite system	Extremely expensive, space-based transmitters. Inexpensive receivers	Worldwide	Very accurate. Can be made more accurate with augmentation	Fairly easy to use	Inexpensive, world-wide coverage with point-to-point navigation capability	Only one or two independent systems currently available, other countries considering systems but very expensive to create, operate and maintain
Instrument Landing System (ILS)	Expensive transmitter. Fairly inexpensive receiver	10–30 nautical miles	Fairly accurate	Fairly easy to use	Current world-wide standard for precision approaches	Limited frequencies available. Only one approach path provided by each transmitter. Extensive and expensive site preparation might be required

GLOBAL NAVIGATION SATELLITE SYSTEM

GNSSs have just recently been adopted as the future navigation standard by ICAO. Currently, GNSS systems are as accurate as most current enroute navigation systems. Inherent inaccuracies (and some intentional signal degradation) require that GNSS be augmented if it is to replace instrument landing system (ILS) and/or microwave landing system (MLS) as a precision navigation system. Satellite accuracy augmentation (both space and ground based) has been proposed as one method to provide general improvements to accuracy that may permit GNSS to replace ILS as the precision approach standard. Ground-based augmentation may be required before GNSS will be sufficiently accurate for all-weather automatic landings. Which system or combination of systems will be eventually used is still undermined.

RADAR SURVEILLANCE IN AIR TRAFFIC CONTROL

Radar is used by air traffic controllers to monitor aircraft position, detect navigational blunders, reduce separation if possible, and make more efficient use of airspace. Controllers can utilize radar to provide aircraft navigational assistance during both the enroute and approach phases of flight. If radar is able to provide more accurate aircraft positional information than existing navigation systems can provide, it may be possible to reduce the required separation between aircraft.

Three different types of radar are used in ATC systems. Primary surveillance radar was first developed during World War II and can detect aircraft without requiring onboard aircraft equipment. Secondary surveillance radar (SSR) requires an interrogator on the ground and an airborne transponder in each aircraft. SSR provides more accurate aircraft identification and position, and can transmit aircraft altitude to the controller. Mode-S secondary radar is a recent improvement to secondary radar systems that will provide unique aircraft identification and the ability to transmit flight information to the controller, and ATC instruction and other information directly to the aircraft. Table 3.3 lists the functional advantages and disadvantages of each radar surveillance system.

Automatic dependent surveillance (ADS) is a scheme whereby GPS-derived position data are transmitted to the ground (and eventually nearby aircraft) using a discreet frequency. ADS data can be used in lieu of or in combination with radar-derived position data to accurately locate aircraft.

AIRCRAFT SEPARATION IN AN AIR TRAFFIC CONTROL SYSTEM

The airspace within which ATC services are provided is normally divided into three-dimensional blocks of airspace known as sectors. Sectors have well-defined lateral and vertical limits and normally are shaped according to traffic flow and airspace structure. Only one controller has ultimate responsibility for the separation of aircraft within a particular sector. The controller may be assisted by other controllers, but is the one person who makes the decisions (in accordance with approved procedures), concerning the separation of aircraft within that particular sector.

If pilots of participating aircraft within the sector can see other nearby aircraft, the pilots can simply *see and avoid* nearby aircraft. Or if a controller can see one or both aircraft, the controller may issue heading and/or altitude instructions that will keep the aircraft separated. This informal but effective method of aircraft separation is known as visual separation. Although a simple concept, it is very effective and efficient when properly used. As long as aircraft can be spotted and remain identified, the use of visual separation permits aircraft to operate in much closer proximity than if the aircraft cannot be seen. Most airports utilize visual separation and visual approaches during busy traffic periods. If weather conditions permit visual separation to be applied, the capacity of most major airports can be significantly increased.

Visual separation can only be employed if one pilot sees the other aircraft, or if the controller can see both aircraft. The primary disadvantage of visual separation is that it can only be employed when aircraft are flying fairly slowly. It would be next to impossible to utilize visual separation

TABLE 3.3
Air Traffic Control Radar Systems

Radar System	Operational Theory	Information Provided	Advantages	Disadvantages
Primary surveillance radar	Very powerful electrical transmission is reflected by aircraft back to radar receiver which is then displayed to ATC personnel	Range and azimuth	Detects all aircraft within range regardless of aircraft equipment	Also detects unwanted objects. Weather and terrain can reflect and block signal. System prone to numerous false targets
Secondary surveillance radar, (also known as the Air Traffic Control Radar Beacon System or ATCRBS)	Low powered electrical signal transmitted from ground station triggers response from airborne equipment	Range and azimuth, assigned aircraft code and altitude	Detects only aircraft. If ground system is properly equipped, aircraft identity and altitude can be displayed to ATC	System requires aircraft to be equipped with operable transponder. Operation restricted to common frequency which can be overwhelmed if too many aircraft respond
Mode S	Selective low-powered signal transmitted from ground triggers response from individual aircraft	Range, azimuth, aircraft identity and altitude. Capability exists to transmit additional data both to and from aircraft	Detects only those aircraft specifically interrogated by the ground equipment	Requires all aircraft to be equipped with Mode-S capable transponder
Automatic Dependent Surveillance (ADS)	Signal transmitted from ground triggers response from individual aircraft. ADS does permit aircraft-to-aircraft communication	ADS signals include aircraft identity and altitude. Future iterations could include aircraft flight data and ATC clearances	Identification signal is unique to each aircraft. Much greater bandwidth for data transmission than other data schemes	Requires all aircraft to be equipped with ADS equipment. Mode-S is compatible with ADS systems

during high-altitude, high-speed cruising conditions common to modern aircraft. Visual separation can therefore only be effectively employed within the immediate vicinity of airports. The use of visual separation near airports requires that aircraft remain continuously in sight of one another. This is a difficult proposition at best during the approach to landing or departure phase of flight because these are two of the busiest times for pilots.

NONRADAR SEPARATION

When visual separation cannot be employed, controllers must use either radar or nonradar separation techniques. Due to range and curvature of the earth limitations inherent to radar, there are many situations in which radar cannot be used to identify and separate aircraft. Radar coverage exists near most medium- and high-density airports, and at altitudes of 5,000 feet or above in the continental United States and Europe. Outside of these areas, and over the ocean, radar surveillance may not exist, and the controller must employ some form of nonradar separation to provide ATC.

Nonradar separation depends on accurate position determination and the transmittal of that information to the controller. Due to navigation and communication system limitations, ATC is unable to precisely plot the position of each aircraft in real time. As navigation systems have inherent inaccuracies, it is impossible to know exactly where each aircraft is at any given time. Nonradar separation therefore assumes that every aircraft is located within a three-dimensional block of airspace. The dimensions of the airspace are predicated on the speed of the aircraft and the accuracy of the navigation system being used. In general, if VHF (very high frequency) omnidirectional range is being utilized for aircraft navigation, the airspace assigned to each aircraft may have lateral width of about eight nautical miles, vertical height of 1,000 feet (2,000 feet when operating above 29,000 feet), and a longitudinal length that varies depending upon the speed of the aircraft. In general, the longitudinal extent of the airspace box extends about 10 nautical miles of flight time in front of the aircraft. Depending of the speed of the aircraft, this longitudinal dimension could extend from 10 to 100 miles in front of the aircraft.

As neither the controller nor the pilot knows exactly where within the assigned airspace box each aircraft is actually located, the controller must assume that aircraft might be located anywhere within the box. The only way to insure that aircraft do not collide is to insure that airspace boxes assigned to different aircraft never overlap. Airspace boxes are permitted to get close to one another, but as long as they never overlap, aircraft separation is assured.

Nonradar separation is accomplished by assigning aircraft either different altitudes or nonoverlapping routes. If aircraft need to operate on the same route at the same altitude, they must be spaced accordingly to prevent longitudinal overlap. Controllers may separate potentially conflicting aircraft either through the use of nonoverlapping holding patterns, or by delaying departing aircraft on the ground. If there is a sufficient speed differential between two conflicting aircraft, the controller can normally permit the faster aircraft to lead the slower aircraft using the same route and the same altitude. Depending on the speed difference between the aircraft, the longitudinal separation criteria can normally be reduced.

The controller uses flight progress strips to visualize the aircraft's position and therefore effect nonradar separation. Pertinent data are written on a flight strip as the aircraft progresses through each controller's sector. The controller may request that the pilot make various position and altitude reports, and these reports are written on the flight strip.

The primary disadvantage of nonradar separation is that its application depends on the pilot's ability to accurately determine and promptly report the aircraft's position, and the controller's ability to accurately visualize each aircraft's position. To reduce the probability of an in-flight collision occurring to an acceptably low level, the separation criteria must take into account these inherent inaccuracies and built-in communication delays. This requires that fairly large areas of airspace be assigned to each aircraft. An aircraft traveling at 500 knots might be assigned a block of airspace 1,000 feet in height, covering close to 400 square miles. This is hardly an efficient use of airspace.

RADAR SEPARATION

Radar can be utilized in ATC to augment nonradar separation, possibly reducing the expanse of airspace assigned to each aircraft. Radar's design history causes it to operate in ways that are not always advantageous to ATC, however. Primary radar was developed during World War II as a defensive, antiaerial invasion system. It was also used to locate enemy aircraft and direct friendly aircraft on an intercept course. It was essentially designed to bring aircraft together, not to keep them apart.

Primary radar is a system that transmits high-intensity electromagnetic pulses focused along a narrow path. If the pulse is reflected off an aircraft, the position of the aircraft is displayed as a bright blip, or target, on a display screen known as a plan position indicator. This system is known as primary surveillance radar.

The radar antenna rotates slowly to scan in all directions around the radar site. Most radars require 5 to 15 seconds to make one revolution. This means that once an aircraft's position has been plotted by radar, it will not be updated until the radar completes another revolution. If an aircraft is moving at 600 knots, it might move two to three miles before it is replotted on the radar display.

Primary radar is limited in a range based on the curvature of the earth, the antenna rotational speed, and the power level of the radar pulse. Radars used by approach control facilities have an effective range of about 75 nautical miles. Radars utilized to separate enroute aircraft have a range of about 300 nautical miles.

SSR is a direct descendent of a system also developed in World War II known as identification friend or foe. Secondary radar enhances the radar target and can be integrated with a ground-based computer to display the aircraft's identity, altitude, and ground speed. This alleviates the need for the controller to constantly refer to flight progress strips to correlate this information. However, flight progress strips are still used by radar controllers to maintain other information, and as a backup system utilized in the case of radar system failure.

Although one might think that radar dramatically reduces aircraft separation, in fact it only normally significantly reduces the longitudinal size of the airspace box assigned to each aircraft. The vertical dimension of the airspace box remains 1,000 feet, the lateral dimension may be reduced from eight to five nautical miles (sometimes three miles), but longitudinal separation is reduced from 10 flying minutes, to three to five nautical miles.

RADAR SYSTEM LIMITATIONS

There are various physical phenomena that hamper primary radar effectiveness. Weather and terrain can block radar waves, and natural weather conditions such as temperature inversions can cause fake or false targets to be displayed by the system. Radar also tracks all moving targets near the airport, which may include highway, train, and in some cases, ship traffic. While controlling air traffic, the controller can be distracted and even momentarily confused when nonaircraft targets such as these are displayed on the radar. It is difficult for the controller to quickly determine whether a displayed target is a *false target* or an actual aircraft.

Another major limitation of radar is its positional accuracy. As the radar beam is angular in nature (usually about ½-degree wide), the beam widens as it travels away from the transmitter. At extreme ranges, the radar beam can be miles wide. This makes it difficult to accurately position aircraft located far from the antenna and makes it impossible to differentiate between two aircraft operating close to one another. As radar system accuracy decreases as the aircraft distance from the radar antenna increases, aircraft close to the radar antenna (less than about 40 miles) can be laterally or longitudinally separated by three miles. Once the aircraft is greater than 40 miles from the radar antenna, five nautical miles of separation must be used. The size of the airspace box using radar is still not reduced vertically but can now be as little as nine square miles (compared with 600 when using nonradar separations).

ADDITIONAL RADAR SERVICES

Radar can also be used by the controller to navigate aircraft to provide a more efficient flow of traffic. During the terminal phase of flight, as the aircraft align themselves with the runway for landing, radar can be used by the controller to provide navigational commands (vectors) that position each aircraft at the optimal distance from one another, something impossible to do if radar surveillance is not available. This capability of radar is at least as important as the ability to reduce the airspace box assigned to each aircraft.

Air-traffic controllers can also utilize radar to assist the pilot to avoid severe weather, although the radar used in ATC does not optimally display weather. The controller can also advise the pilot of nearby aircraft or terrain. In an emergency, the controller can guide an aircraft to the nearest airport and can guide the pilot through an instrument approach. All these services are secondary to the primary purpose of radar, which is to safely separate aircraft participating in the ATC system.

RADAR IDENTIFICATION OF AIRCRAFT

Before controllers can utilize radar for ATC separation, they must positively identify the target on the radar. Due to possible false target generation, unknown aircraft in the vicinity, and weather-induced false targets, it is possible for a controller to be unsure of the identity of any particular radar target. Therefore, the controller must use one or more techniques to positively verify the identity of any target before radar-separation criteria can be utilized. If positive identity cannot be ascertained, nonradar separation techniques must be utilized.

Controllers can verify the identity of a particular target using either primary or secondary radar. Primary methods require that the controller correlate the pilot's reported position with a target on the radar, or by asking the pilot to make a series of turns and watching for a target to make similar turns. Secondary radar identification can be established by asking the pilot to transmit an IDENT signal (which causes a distinct blossoming of the radar target), or, if the radar equipment is so equipped, asking the pilot to set the transponder to a particular code, and verifying that the radar displays that code (or the aircraft identification) next to the target symbol on the radar.

None of these methods are foolproof, and all have the potential for aircraft misidentification. During the identification process, the wrong pilot may respond to a controller's request, equipment may malfunction, or multiple aircraft may follow the controller's instruction. If an aircraft is flying too low or is outside the limits of the radar display, the target may not even show up on the radar scope. Once identified, the controller may rely completely on radar positioning information when applying separation, so multiple methods of radar identification are usually utilized to insure that a potentially disastrous misidentification does not occur and that the aircraft remains identified. If positive radar identification or detection is lost at any time, the controller must immediately revert to nonradar separation rules and procedures until aircraft identify can be reestablished.

RADAR SEPARATION CRITERIA

Radar accuracy is inversely proportional to the aircraft's distance from the radar antenna. The further away an aircraft is, the less accurate is the radar positioning of that aircraft. Radar separation criteria have been developed with this limitation in mind. One set of criteria has been developed for aircraft that are less than 40 nautical miles from the radar site. An additional set of criteria has been developed for aircraft 40 or more nautical miles from the antenna. As the display system used in air route traffic control centers uses multiple radar sites, controllers using this equipment must always assume that aircraft might be 40 miles or farther from the radar site when applying separation criteria. Table 3.4 describes the separation criteria utilized by air-traffic controllers when using radar. The controller must utilize at least one form of separation.

TABLE 3.4
Radar Separation Criteria

Aircraft Distance from Radar Antenna	Vertical Separation	Lateral Separation	Longitudinal Separation
Less than 40 nautical miles	1,000 feet	Three nautical miles	Three nautical miles. Additional separation may be required for wake turbulence avoidance
40 nautical miles or greater	1,000 feet	Five nautical miles	Five nautical miles. Additional separation may be required for wake turbulence avoidance

As stated previously, radar serves only to reduce the nonradar-separation criteria previously described. It does nothing to reduce the vertical separation between aircraft. Radar primarily serves to reduce lateral and longitudinal separation. Nonradar lateral separation is normally eight nautical miles, but the use of radar permits lateral separation to be reduced to three to five nautical miles. Radar is especially effective when reducing longitudinal separation, however. Nonradar longitudinal separation requires 5–100 nautical miles, whereas radar longitudinal separation is three to five nautical miles. It is this separation reduction that is most effective in maximizing the efficiency of the ATC system. Instead of lining up aircraft on airways 10–50 miles in trail, controllers using radar can reduce the separation to three to five miles, therefore increasing the airway capacity 200%–500%. While under radar surveillance, pilots are relieved of the responsibility of making routine position and altitude reports. This dramatically reduces frequency congestion and pilot–controller miscommunications.

Another advantage of radar is that controllers are no longer restricted to assigning fixed, inflexible routes to aircraft. As aircraft position can be accurately determined in near real time, controllers can assign new routes to aircraft that may shorten the pilot's flight, using the surrounding airspace more efficiently.

Radar vectors such as these are most effective in a terminal environment in which aircraft are converging on one or more major airports and are in a flight transitional mode in which they are constantly changing altitude and airspeed. A controller using radar is in a position to monitor the aircraft in the terminal airspace and can make overall adjustments to traffic flow by vectoring aircraft for better spacing, or by issuing speed instructions to pilots to close or widen gaps between aircraft. It is because of these advantages that most national ATC organizations first install radar in the vicinity of busy terminals. Only later (if at all) are enroute navigation route provided by radar monitoring.

CURRENT TRENDS IN AUTOMATION

Early forms of radar provided a display of all moving targets within the radar's area of coverage. This included not only aircraft, but weather, birds, vehicular traffic, and other atmospheric anomalies. By using technology developed in World War II, air traffic controllers have been able to track and identify aircraft using the ATC radar beacon system (ATCRBS). ATCRBS, sometimes known as SSR, or simply secondary radar, requires a ground-based interrogator and an airborne transponder installed in each aircraft. When interrogated by the ground station, the transponder replies with a unique code that can be used to identify the aircraft, and if so equipped, it can also transmit the aircraft's altitude to the controller.

This system is tremendously beneficial to the controller because all aircraft can easily be identified. Nonpertinent aircraft and other phenomena observed by the radar can be ignored by the controller. If the ground-based radar is properly equipped, aircraft identity and altitude can also be constantly displayed on the radar screen, relieving the controller of mentally trying to keep each radar target properly identified.

The ground-based component of the secondary radar system has since been modified to perform additional tasks that benefit the air traffic controller. If the ground radar is properly equipped, and the computer knows the transponder code that a particular aircraft is using, the aircraft can be tracked and flight information can be computer processed and disseminated. As the radar system tracks each aircraft, basic flight information can be transmitted to subsequent controllers automatically as the aircraft nears each controller's airspace boundary. Future aircraft position can also be projected on the basis of past performance, and possible conflicts with other aircraft and with the ground can be predicted and prevented. These last two systems (known as conflict alert for aircraft–aircraft conflicts and minimum safe altitude warning for aircraft–terrain conflicts) only provide the controller with a warning when aircraft are projected to be in danger. The system does not provide the controller with any possible remediation of the impending problem. Future enhancements to the computer system should provide the controller with options that can be selected to resolve the problem. This future system is to be known as conflict resolution advisories.

AIRBORNE SYSTEMS

Engineers and researchers have experimented with aircraft-based traffic avoidance systems since the 1960s. These prototype systems were not designed to replace but rather to augment and back up the current ground-based ATC system. This device is known as threat collision and avoidance system (TCAS). TCAS was developed with three different levels of service and capabilities.

TCAS is an aircraft-based system that monitors and tracks nearby transponder-equipped aircraft. This position and relative altitude of nearby aircraft are constantly displayed on a TCAS display located in the cockpit of each aircraft. TCAS I provides proximity warning only to assist the pilot in the visual acquisition of intruder aircraft. No recommended avoidance maneuvers are provided nor authorized as a direct result of a TCAS I warning. It is intended for use by smaller commuter aircraft holding 10 to 30 passenger seats, and general aviation aircraft. TCAS II provides traffic advisories and resolution advisories. Resolution advisories provide recommended maneuvers in a vertical direction (climb or descent only) to avoid conflicting traffic. Airline aircraft, and larger commuter and business aircraft holding 31 passenger seats or more, use TCAS II equipment. TCAS III provides all the capabilities of TCAS II but adds the capability to provide horizontal maneuver commands. All three versions of TCAS monitor the location of nearby transponder-equipped aircraft. Current technology does not permit TCAS to monitor aircraft not transponder equipped.

CONFLICT ALERT/VISUAL FLIGHT RULES INTRUDER

The ATCRBS has recently been enhanced with a new conflict alert program known as conflict alert/VFR intruder. The old conflict alert program only advised the controller of impending collisions between participating IFR aircraft. It did not track nonparticipating aircraft such as those operating under VFR. Conflict alert/VFR intruder tracks all IFR and VFR aircraft equipped with transponders and alerts the controller if a separation error between the VFR and a participating IFR aircraft is predicted. The controller can then advise the pilot of the IFR aircraft and suggest alternatives to reduce the risk of collision.

TRAFFIC MANAGEMENT SYSTEMS

It has become recently apparent that the current ATC system may not be able to handle peak traffic created in a hub-and-spoke airline system. Much of this is due to inherent limitations of the ATC system. ATC system expansion is planned in many countries, but until it is completed, other methods of ensuring aircraft safety have been developed. To preserve an acceptable level of safety, special traffic management programs have been developed to assist the controllers in their primary function, the safe separation of aircraft.

AIRPORT CAPACITY RESTRICTIONS

During hub-and-spoke airport operations, traffic can become intense for fairly short periods of time. During these intense traffic periods, if optimal weather and/or airport conditions do not exist, more aircraft may be scheduled to arrive than the airport and airspace can safely handle. In the past, this traffic overload would be handled through the use of airborne holding of aircraft. Controllers would try to land as many aircraft as possible, with all excess aircraft assigned to nearby holding patterns until space became available.

This method of smoothing out the traffic flow has many disadvantages. The primary disadvantage is that while holding, aircraft consume airspace and fuel. In today's highly competitive marketplace, airlines can ill afford to have aircraft circle an airport for an extended period of time.

In an attempt to reduce the amount of airborne holding, the Federal Aviation Administration (FAA) has instituted a number of new traffic management programs. One program seeks to predict near-terms airport acceptance rates (AARs) and match arriving aircraft to that number. One program in use is the controlled departure program. This program predicts an airport's acceptance rate over the next 6–12 hours and matches the inbound flow of aircraft to that rate. Aircraft flow is adjusted through the delaying of departures at remote airports.

Overall delay factors are calculated, and every affected aircraft is issued a delayed departure time that will coordinate its arrival to the airport's acceptance rate.

The primary disadvantage of such a system is twofold. First, it is very difficult to predict 6–12 hours in advance conditions that will affect a particular airport's acceptance rate. These conditions include runway closures, adverse weather, and so on. As unforeseen events occur that require short-term traffic adjustments, many inbound aircraft are already airborne and therefore cannot be delayed on the ground. This means that the only aircraft that can be delayed are those that have not yet departed and are still on the ground at nearby airports. This system inadvertently penalizes airports located close to hub airports because they absorb the brunt of these unpredictable delays. In other situations, traffic managers may delay aircraft due to forecasted circumstances that do not develop. In these situations, aircraft end up being delayed unnecessarily. On the contrary, once an aircraft has been delayed, that time can never be made up.

Once aircraft are airborne, newer traffic flow management programs attempt to match real-time airport arrivals to the AAR. These programs are known as aircraft metering. Metering is a dynamic attempt to make short-term adjustments to the inbound traffic flow to match the AAR. In general terms, a metering program determines the number of aircraft that can land at an airport during a 5- to 10-minute period, and it also determines and then applies a delay factor to each inbound aircraft so that they land in sequence with proper spacing. The metering program dynamically calculates the appropriate delay factor and reports this to the controller as a specific time that each aircraft should cross a specific airway intersection. The controller monitors the progress of each flight and issues speed restrictions to ensure that every aircraft crosses the appropriate metering fix at the computer-specified time. This should, in theory, ensure that aircraft arrive at the arrival airport in proper order and sequence.

AIR TRAFFIC CONTROL SYSTEM OVERLOADS

Due to the procedural limitations placed upon aircraft participating in the ATC system, many ATC sectors far away from major airports can become temporarily overloaded with aircraft. In these situations, controllers would be required to separate more aircraft than they could mentally handle. This is a major limitation to the expansion of many ATC systems.

Various programs are being researched to counteract this problem. A prototype system has been developed in the United States known as enroute sector loading. The enroute sector loading computer program calculates every sector's current and predicted traffic load and alerts ATC personnel whenever it predicts that a particular sector may become overloaded. When this occurs, management personnel determine whether traffic should be rerouted around the affected sector. This particular program is successful at predicting both systemic overloads and transient overloads due to adverse weather and traffic conditions.

PILOT/CONTROLLER COMMUNICATIONS-RADIO SYSTEMS

Most ATC instructions, pilot acknowledgments, and requests are transmitted via voice-radio communications. By international agreement, voice communication in ATC is usually conducted in the English language using standardized phraseology. This phraseology is specified in ICAO documents and is designed to formalize phrases used by all pilots and controllers, regardless of the native language. This agreement permits pilots from the international community to be able to fly to and from virtually any airport in the world with few communication problems.

Voice communications between pilots and controllers are accomplished using two different formats and multiple frequency bands. The most common form of voice communication in ATC is simplex communications, in which the controller talks to the pilot and vice versa utilizing a single radio frequency. This method makes more efficient use of the narrow radio frequency bands assigned to aviation but has many inherent disadvantages. As one frequency is used for both sides of the conversation, when one person is transmitting, the frequency is unavailable to others for use. To prevent radio system overload, simplex radios are designed to turn off their receiver whenever transmitting.

These conditions make it difficult for a controller to issue instructions in a timely manner when using simplex communications. If the frequency is in use, the controller must wait until a break in communications occurs. More problematic is the occasion when two or more people transmit at the same time or if someone's transmitter is inadvertently stuck on. Due to the way radios operate, if two people try to transmit at the same time, no one will be able to understand the transmission, and neither of the individuals transmitting would be aware of the problem, because their receivers are turned off when transmitting.

Duplex transmission utilizes two frequencies—one for controller-to-pilot communications and the other for pilot-to-controller communications. This communication method is similar to that utilized during telephone conversations. Both individuals can communicate simultaneously and independently are able to interrupt one another and can listen while talking. Duplex transmission schemes have one major disadvantage, however. To prevent signal overlap, two discrete frequencies must be assigned to every controller–pilot communication. This essentially requires that double the number of communications frequencies be made available for ATC. Due to the limited frequencies available for aeronautical communications, duplex transmissions can seldom be used in ATC.

Most short-range communications in ATC utilize the VHF radio band located just above those used by commercial FM radio stations. Just as FM radio stations, aeronautical VHF is not affected by lightning and other electrical distortion, but it is known as a line-of-sight frequency band, which means that the radio signal travels in a straight line and does not follow the curvature of the Earth.

Airborne VHF radios must be above the horizon line if they are to receive any ground-based transmissions. If an aircraft is below the horizon, it will be unable to receive transmissions from the controller and vice versa.

This problem is solved in the ATC system through the use of remote communications outlets (RCOs). RCOs are transmitter/receivers located some distance from the ATC facility. Whenever a controller transmits, the transmission is first sent to the RCO using land-based telephone lines, and then it is transmitted to the aircraft. Aircraft transmissions are relayed from the RCO to the controller in the same manner. Each RCO is assigned a separate frequency to prevent signal inter-ference. This system permits a single controller to communicate with aircraft over a wide area but requires the controller to monitor and operate multiple radio frequencies. The use of RCOs not only extends the controller's communications range but also makes the ATC communications system vulnerable to ground-based telephone systems that may malfunction or be damaged, thereby caus-ing serious ATC communication problems.

Most civil aircraft utilize VHF communications equipment. Military aircraft utilize ultra-high-frequency (UHF) band transmitters. UHF is located above the VHF band. UHF communications systems are preferred by most military organizations because UHF antennas and radios can be made smaller and more compact than those utilized for VHF. UHF is also a line-of-sight communications system. Most ATC facilities are equipped with both VHF and UHF radio communications systems.

Extended-range communication is not possible with VHF/UHF transmitters. RCOs can help one extend the range of the controller but need solid ground on which to be installed. VHF/UHF radios are unusable over the ocean, the poles, or in sparsely populated areas. For long-range, over-ocean radio communications, high-frequency (HF) radios are used. HF uses radio frequencies just above the medium-wave or AM radio band. HF radios can communicate with line-of-sight limita-tions, as far as 3,000 miles in some instances, but can be greatly affected by sunspots, atmospheric conditions, and thunderstorm activities. This interference is hard to predict and depends on the time of day, season, sunspot activity, local and distant weather, and the specific frequency in use. HF radio communication requires the use of multiple frequencies, with the hope that at least one interference-free frequency can be found for communications at any particular time. If controllers cannot directly communicate with aircraft, they may be required to use alternate means of com-munications, such as using the airline operations offices to act as communication intermediaries. This limitation requires that controllers who rely on HF communications not place the aircraft in a position where immediate communications may be required.

Experiments have been conducted using satellite transmitters and receivers to try to overcome the limitations of HF/VHF/UHF transmission systems. Satellite transmitters utilize frequencies located well above UHF and are also line-of-sight. But if sufficient satellites can be placed in orbit, communications anywhere in the world will be virtually assured. Satellite communications have already been successfully tested on overseas flights and should become commonplace within a few years.

VOICE COMMUNICATION PROCEDURES

As previously stated, virtually every ATC communication is currently conducted by voice. Initial clearances, taxi and runway instructions, pilot requests, and controller instructions are all primarily conducted utilizing voice. This type of communication is fairly unreliable due to both the previ-ously mentioned technical complications and communications problems inherent in the use of one common language in ATC. Although every air traffic controller utilizes English, they may not be conversationally fluent in the language. In addition, different cultures pronounce words and letters in different ways. Many languages do not even use the English alphabet. Moreover, every controller has idioms and accents peculiar to their own language and culture. All these factors inhibit com-munications and add uncertainty to ATC communications.

When using voice-radio communications, it can be very difficult for a controller to insure that correct and accurate communication with the pilot has occurred. Pilots normally read back all instructions, but this does not solve miscommunication problem. Informal and formal surveys lead experts to believe that there are literally millions of miscommunications worldwide in ATC every year. Obviously, most of these are immediately identified and corrected, but some are not, leading to potential problems in the ATC system.

ELECTRONIC DATA COMMUNICATIONS

In an attempt to minimize many of these communication problems, various schemes of nonvoice data transmission have been tried in ATC. The most rudimentary method still in use is the ATCRBS transponder. If the aircraft is properly equipped, its identity and altitude will be transmitted to the ground station. Existing ATCRBS equipment is currently incapable of transmitting information from the controller to the aircraft. The new Mode-S transponder system will be able to transmit more information in both directions. This information might include aircraft heading, rate of climb/descent, airspeed, and rate of turn, for example. Mode-S should also be able to transmit pilot requests and controller instructions. Mode-S is slowly being installed on the ground, and airborne equipment is gradually being upgraded. Until a sufficient number of aircraft have Mode-S capability, the ATCRBS system will still be utilized. Full Mode-S implementation will probably be completed sometime after the year 2000.

An intra-airline data-communications system known as the aircraft communications addressing and reporting system (ACARS) has been utilized by the airlines for years to send information to and from properly equipped aircraft. ACARS essentially consists of a keyboard and printer located on the aircraft and corresponding equipment in the airline's flight operations center. ACARS is currently used by the airlines to transmit flight planning and load information. A few ATC facilities are now equipped to transmit initial ATC clearances to aircraft using ACARS. This limited service will probably be expanded until Mode-S becomes widespread.

CONTROLLER COORDINATION

As controllers are responsible for the separation of aircraft within their own sector, they must coordinate the transfer of aircraft as they pass from one sector to another. In most situations, this coordination is accomplished using voice communications between controllers. In most cases, unless the controllers are sitting next to each other within the same facility, coordination is accomplished using the telephone.

Hand-offs are one form of coordination and consist of the transfer of identification, communications, and control from one controller to the next. During a hand-off, the controller with responsibility for the aircraft contacts the next controller, identifies the aircraft, and negotiates permission for the aircraft to cross the sector boundary at a specific location and altitude. This is known as the transfer of identification. Once this has been accomplished, and all traffic conflicts are resolved, the first controller advises the pilot to contact the receiving controller on a specific radio frequency. This is known as the transfer of communication. Separation responsibility still remains with the first controller until the aircraft crosses the sector boundary. Once the aircraft crosses the boundary, separation becomes the responsibility of the receiving controller. This is known as the transfer of control.

To simplify hand-offs, standardized procedures and predefined altitudes and routes are published in a document known as a *letter of agreement* (LOA). LOAs simplify the coordination process because both controllers already know what altitude and route the aircraft will be utilizing. If the controllers wish to deviate from these procedures, they must agree to an *approval request* (*appreq*).

The transferring controller usually initiates an appreq verbally, requesting a different route and/or altitude for the aircraft to cross the boundary. If the receiving controller approves the appreq, the transferring controller may deviate from the procedures outlined in the LOA. If the receiving controller does not approve the deviation, the transferring controller must amend the aircraft's route/altitude to conform to those specified in the LOA.

There are many problems inherent in this system of verbal communication/coordination. When both controllers are busy, it is very difficult to find a time when both are not communicating with aircraft. Controllers are also creatures of habit and may sometimes *hear* things that were not said. There are many situations in ATC in which aircraft are delayed or rerouted, not due to conflicting traffic, but because the required coordination could not be accomplished in a timely manner.

Automated hand-offs have been developed in an attempt to reduce these communication/coordination problems. An automated hand-off can be accomplished if the two sectors are connected by computer, and the routes, altitudes, and procedures specified in the LOA can be complied with. During an automated hand-off, as the aircraft nears the sector boundary, the transferring controller initiates a computer program that causes the aircraft information to be transferred and start to flash on the receiving controller's radar display. This is a request for a hand-off and implies that all LOA procedures will be complied with. If the receiving controller determines that the hand-off can be accepted, computer commands are entered that cause the radar target to flash on the transferring controller's display.

This implies that the hand-off has been accepted, and the first controller then advises the pilot to contact the next controller on the appropriate frequency. Although this procedure may seem quite complex, in reality it is very simple, efficient, and reduces voice coordination between controllers significantly. Its primary disadvantage is that the route and altitudes permissible are reduced, and the ATC system becomes less flexible overall.

FLIGHT PROGRESS STRIPS

Virtually, all verbal communications are written down for reference on paper flight progress strips. Flight strips contain most of the pertinent information concerning each aircraft. When a controller verbally issues or amends a clearance or appreqs a procedural change with another controller, this information is hand written on the appropriate flight progress strip. Flight progress strips are utilized so that controllers do not need to rely on their own memory for critical information. Flight strips also make it easier for other controllers to ascertain aircraft information if the working controller needs assistance or when a new controller comes on duty. Due to differences in each controller's handwriting, very specific symbology is used to delineate this information. Figure 3.1 contains examples of some common flight strip symbologies.

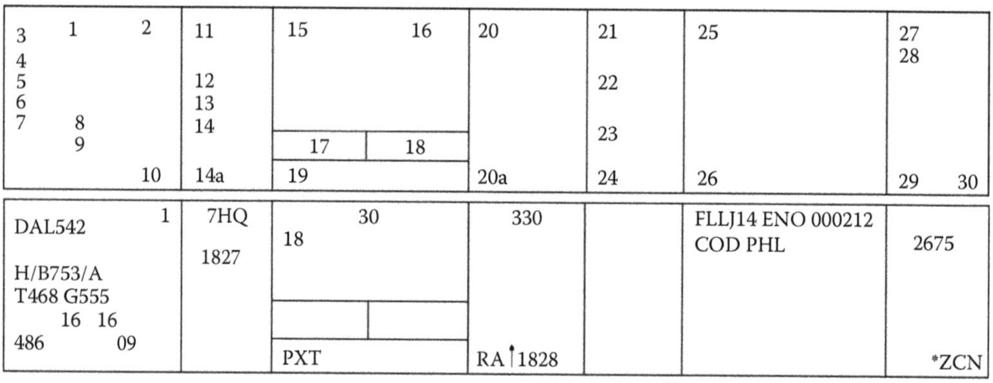

3 1 2	11	15 16	20	21	25	27 28
4						
5	12			22		
6	13					
7 8	14			23		
9		17 18				
10	14a	19	20a	24	26	29 30

DAL542 1	7HQ	30	330		FLLJ14 ENO 000212	
		18			COD PHL	2675
H/B753/A	1827					
T468 G555						
16 16						
486 09						
		PXT	RA↑1828			*ZCN

FIGURE 3.1   Flight strip symbology.

FLIGHT INFORMATION AUTOMATION

The constant updating of flight progress strips and the manual transferring of information consume much of a controller's time and may necessitate the addition of another controller to the sector to keep up with this essential paperwork. This process is forecast to become somewhat more automated in the future. Future ATC systems have been designed with flight strips displayed on video screens. It is theoretically possible that as controllers issue verbal commands, these commands will be automatically interpreted and the electronic flight strips will be updated. Future enhancements may make it possible for the controller to update an electronic flight strip, and that information might be automatically and electronically transmitted to the pilot or even to the aircraft's flight control system.

CONTROLLER RESPONSIBILITIES IN THE AIR TRAFFIC CONTROL SYSTEM

Controllers are responsible for the separation of participating aircraft within their own sector. They also provide additional services to aircraft, such as navigational assistance and providing weather advisories. Additional responsibilities placed on the controller include maximizing the use of the airspace and complying with air traffic management procedures.

To accomplish these tasks, the controller must constantly monitor both actual and predicted aircraft positions. Due to rapidly changing conditions, a controller's plan of action must remain flexible and subject to constant change. The controller must continuously evaluate traffic flow, plan for the future, evaluate the problems that may occur, determine appropriate corrective action, and implement this plan of action. In the recent past, when traffic moved relatively slowly and the airspace was not quite as crowded, a controller might have minutes to evaluate situations and decide on a plan of action. As aircraft speeds have increased, and the airspace has become more congested, controllers must now make these decisions in seconds. As in many other career fields, experts feel that the current system may have reached its effective limit and increased ATC system expansion will not be possible until many of the previously mentioned tasks become automated.

FUTURE ENHANCEMENTS TO AIR TRAFFIC CONTROL SYSTEMS

ICAO has agreed that GNSS should become the primary aircraft-positioning system. It appears at this time that uncorrected GNSS systems should supplant VORTAC as both an enroute and non-precision instrument approach aid. Space-based augmentation should permit GNSS to be used as a CAT I precision approach replacement for ILS. Ground-based augmentation should correct GNSS to meet CAT II and possibly CAT III ILS standards.

The GNSS system can be modified to permit the retransmission of aircraft position back to ATC facilities. This system, known as ADS, should supplant radar as a primary aircraft surveillance tool. Not only is this system more accurate than radar surveillance, but also it does not have the range and altitude limitations of radar and is able to transmit additional data both to and from the controller. This might include pilot requests, weather information, traffic information, and more.

Many other changes are planned. ICAO has completed a future air navigation system (FANS) that defines changes to navigation, communication, and surveillance systems. FANS is a blueprint for the future of international aviation and ATC. Table 3.5 summarizes FANS.

Once these improvements have taken place, automated ATC systems can be introduced. Various research programs into automation have been initiated by many ATC organizations, with the eventual goal of developing an air traffic management system called *free flight*.

The concept of free flight has been discussed since the early 1980s. Free flight proposes to change ATC separation standards from a static, fixed set of standards to dynamic separation that takes into

TABLE 3.5
Future ATC and Navigation System Improvements

Function	Type	Current Standard	Future Standard
Navigation	Enroute	VORTAC	Skeleton VORTAC and ILS backup system
	Approach	ILS	Augmented GNSS
		GNSS	
Communication	Short-range	VHF and UHF	VHF and UHF
	Long-range	HF	Satellite
Surveillance		Radar and ADS	Radar and automatic dependent surveillance
Datalink		ATCRBS	Mode-S and ADSs

account aircraft speed, navigational capability, and nearby traffic. Based on these parameters, each aircraft will be assigned a *protected* zone that will extend ahead, to the sides and above and below the aircraft. This zone will be the only separation area protected for each aircraft. This differs from the current system that assigns fixed airway dimensions and routes for separation.

Assuming that each aircraft is equipped with an accurate flight management system, free flight proposes that each aircraft transmit to ground controllers its flight management system–derived position. On the ground, computer workstations will evaluate the positional data to determine whether any aircraft conflicts are predicted to exist, and if so, offer a resolution instruction to the air traffic controller. The controller may then evaluate this information and pass along appropriate separation instructions to the aircraft involved.

The free flight concept is still being developed, but if found feasible, will soon be implemented at high altitudes within the U.S. airspace structure. As confidence in the system is gained, it will likely be extended overseas and into the low-altitude flight structure.

SUGGESTED READING

Federal Aviation Administration. (1976). *Takeoff at Mid-century*. Washington, DC: Department of Transportation.
Federal Aviation Administration. (1978). *Bonfires to Beacons*. Washington, DC: Department of Transportation.
Federal Aviation Administration. (1979). *Turbulence Aloft*. Washington, DC: Department of Transportation.
Federal Aviation Administration. (1980). *Safe, Separated and Soaring*. Washington, DC: Department of Transportation.
Federal Aviation Administration. (1987). *Troubled Passage*. Washington, DC: Department of Transportation.
International Civil Aviation Organization. (1953–2013). *ICAO Bulletin*. Montreal, Canada.
International Civil Aviation Organization. (various). *Annexes to the Convention of International Civil Aviation*. Montreal, Canada: Author.
Jackson, W. E. (1970). *The Federal Airways System*. Institute of Electrical and Electronic Engineers.
Nolan, M. S. (2010). *Fundamentals of Air Traffic Control*. Cengage.

OTHER FAA PUBLICATIONS

Aeronautical Information Publication (2016).
Air Traffic Handbook (2016).
National Airspace System Plan (2015).

4 Maintenance

William L. Rankin and Bill Johnson

CONTENTS

Introduction 52
Overview and History 53
    HF Historical Perspective 53
    Define HF and Major HF Disciplines 53
        Three Types of HF Professionals 54
        Types of Educational Disciplines of HF Professionals 55
        Maintenance HF 55
    Safety Statistics and Maintenance Human Factors 55
    Major Human Factors Research and Development Initiatives and Regulations 57
        U.S. Legislation Drives HF Research 57
        Other Regulatory Influence on Maintenance Human Factors 60
Human Factors Models Used in the Industry 61
    Introduction 61
    SHELL Model 61
    Skills–Rules–Knowledge-Based Performance Model 62
        Skill-Based Performance 63
        Rule-Based Performance 63
        Knowledge-Based Performance 63
        Summary 64
    Dirty Dozen 64
    Swiss Cheese Model 64
    Human Factors Analysis and Classification System 65
    Threat and Error Management Model 67
        Threats 68
        Errors 68
        TEM Model 69
    PEAR Model 70
        People 70
        Environment—Physical and Organizational 71
        Actions 71
        Resources 72
        Safety Management System and PEAR 72
Maintenance Human Factors Programs/Processes in Use Today 72
    Maintenance Resource Management/Maintenance Human Factors Training 73
    The Human Factors Guide and the Operator's Manual for Human Factors in Maintenance 74
        Event Investigation 75
        Documentation 75
        Human Factors Training 76
        Shift/Task Turnover 76
        Fatigue Management 76

Sustainability and Cost Justification 76
Summary 76
Maintenance Error Decision Aid Event Investigation Process... 76
Maintenance Error Decision Aid Philosophy and Model... 78
Contributing Factors.. 79
Maintenance System Failures.. 80
Event... 80
Internal Hazard Reporting System 81
Maintenance Line Operations Safety Assessment 82
Maintenance Fatigue Risk-Management System... 86
Safety Culture/Just Culture 89
Research Findings Regarding Maintenance Human Factors Issues 90
Documentation and Failure-to-Follow Procedures 90
Documentation 90
Failure-to-Follow Procedures... 91
Research on Failure-to-Follow Procedures 91
Summary 94
A Systems View of Dealing with Failure-to-Follow Procedures 95
Summary 96
Visual Inspections 97
Types of Visual Inspections.. 97
Inspection Process... 98
Factors Affecting Visual Inspections 100
Task Factors... 101
Inspector Factors 101
Equipment Factors.. 103
Environment Factors 103
Social Factors 103
Summary 103
Projections on the Future of Maintenance Human Factors.. 103
The Impact of Safety Management Systems.. 103
The Impact of Human Factors on Unmanned Aerial Systems 104
References.. 104

INTRODUCTION

The purpose of the current chapter is twofold. First, it provides a brief historical overview of the development of the field of Maintenance Human Factors (MHF) in aviation. Second, this chapter provides an in-depth discussion of Human Factors (HF), in general, the human performance (error) models used by the industry in its MHF journey, and the types of MHF programs/processes that are now available for all segments of the aviation industry, including maintenance, repair, and overhaul (MRO) organizations.

The current chapter will cover the following areas:

1. Overview and history
   a. HF historical perspective
   b. Define HF and major HF disciplines
   c. Safety statistics and MHF
   d. Major HF research and development initiatives and regulations

2. Describe the various HF models that have been used in the industry, including the following:
   a. SHELL model
   b. Skill–rule–knowledge-based performance model
   c. Dirty dozen
   d. Swiss Cheese model
   e. Human factors analysis and classification system (HFACS)
   f. Threat and error management model (TEM)
   g. PEAR model
3. Describe HF programs/processes that are being used in maintenance, including the following:
   a. Maintenance resource management
   b. Operator's manual: HF in maintenance
   c. Maintenance error decision aid (MEDA) event investigation process
   d. Internal hazard reporting process
   e. Maintenance line operations safety assessment (M-LOSA) observation process
   f. Fatigue risk-management system (FRMS)
   g. Safety culture/just culture
4. Summarize MHF research findings regarding the following:
   a. Documentation
   b. Failure to follow procedures
   c. Visual inspections.
5. Finally, we offer a projection on the future of MHF.

OVERVIEW AND HISTORY

HF HISTORICAL PERSPECTIVE

A variety of sources would contend that HF dates back to the early drawings and inventions of Leonardo Da Vinci. However, as a formal science, HF started somewhere near the turn of the twentieth century. Two industrial engineers, Frank and Lillian Gilbreth, applied a formal task analytic approach to raising the efficiency of surgeons. The classic example of the surgeon requesting scalpel and the assistant repeating the request and providing the scalpel is an early example of HF at work. This procedure permitted the doctor to concentrate on surgery rather than on finding the correct instrument. The verbal challenge response, of course, is used in all cockpits today (Johnson, 1998). The application of HF techniques and principles was used by the major airframe Original Equipment Manufacturers (OEMs), on the military side, starting in the 1950s and on the commercial aircraft side starting in the 1960s.

DEFINE HF AND MAJOR HF DISCIPLINES

The U.S. Human Factors Society was established in 1957 (www.hfes.org). Its name was later changed to the Human Factors and Ergonomics Society. Human Factors and Ergonomics Society is a member of the International Ergonomics Association, which is a federation of the world's ergonomics societies. The following definition was adopted by the International Ergonomics Association in August 2000:

Ergonomics (or HF) is the scientific discipline concerned with the understanding of interactions among humans and other elements of a system, and the profession that applies theory, principles, data, and other methods to design in order to optimize human well-being and overall system performance.

To optimize human well-being and overall system performance, Ergonomists/HF professionals contribute to the design and evaluation of tasks, jobs, products, environments, and systems to make them compatible with the needs, abilities, and limitations of people.

The Boeing Company definition of HF closely follows the previous definition:

HF is a technical discipline aimed at optimizing human performance within a system by contributing to the planning, design, and evaluation of tasks, jobs, products, organizations, environments, and systems to make them compatible with the needs, abilities, and limitations of people.

Three Types of HF Professionals

There are, at least, three types of HF professionals—cognitive, physical, and organizational.

Cognitive HF: Concerned with mental processes, such as perception, memory, reasoning, and motor (muscle) response as they affect interactions among humans and other elements of a system. Relevant topics include:
- Mental workload
- Decision making
- Human reliability (skilled performance/errors)
- Training methods
- Human–computer interaction
- Situation awareness
- Work stress.

as these may relate to human-system design.

Physical HF (ergonomics): Concerned with human anatomical, anthropometric, physiological, and biomechanical characteristics as they relate to *physical activity*. Relevant topics include
- Working postures
- Materials handling
- Repetitive movements
- Vibrating equipment
- Workplace layout
- Safety and health
- Work-related musculoskeletal disorders
- Design for maintainability.
- Design for usability.

Organizational HF: Concerned with *optimization of sociotechnical systems*, including their organizational structures, policies, and processes. Relevant topics include
- Leadership
- Communication
- Crew resource management (CRM)
- Work design
- Design of working schedules
- Teamwork
- Participatory design
- Cooperative work
- New work paradigms
- Virtual organizations
- Telework
- Quality management.

There are several different terms used for these types of professionals, including HF, Human Factors Engineering, Human Engineering, and Ergonomics. We will consider all these terms to mean the same thing, although ergonomics is most often used to specifically refer to physical HF.

Types of Educational Disciplines of HF Professionals

Educational disciplines of HF professionals include psychologists (e.g., experimental psychology, industrial/organizational psychology, educational psychology, and clinical psychology), engineers (e.g., industrial engineering, safety engineering, and anthropometric engineering), cognitive science, and medical science, all of which have a common interest in designing systems and equipment to be safe and effective for the people who operate and maintain them.

It was not until the 1980s that one could get a degree in HF. Even then, the degree was typically awarded through the Industrial Engineering Department or the Psychology Department. Colleges and universities now offer degrees, including Ph.D., in HF.

Maintenance HF

So, what is Maintenance HF? Simply, it is the application of HF principles and techniques in preventing and solving human performance issues relevant to aircraft maintenance.

The development and implementation of MHF processes in the aviation industry has taken the following path. First, HF challenges/hazards are identified on the basis of data from the maintenance organizations. Those data can range from analytic means to a hunch that there are opportunities for improvements. In ideal situations, a committee is formed, which includes degreed HF personnel and experienced airline industry staff with subject matter expertise. This group develops a product, such as the training materials and instructions on how to implement a Maintenance Fatigue Risk Management Program at an aircraft maintenance organization. Then the implementation and evaluation is done almost by existing maintenance organization staff, sometimes assisted by HF professionals.

SAFETY STATISTICS AND MAINTENANCE HUMAN FACTORS

When an aircraft is not maintained properly, this can result in safety-of-flight related incidents and accidents. Moreover, incorrectly performed maintenance can lead to increased costs due, for example, to flight delays and cancellations. However, the safety-of-flight case is most compelling as the industry strives to become safer and safer.

Incorrectly performed aircraft maintenance, due to *errors* and/or not following policies, processes, and procedures (*violations*), can affect safety of flight in two different ways.

1. It can be the *Primary Cause* of an accident. Here the accident is due to the maintenance/ inspection failure. The accident is not in any way due to flight crew action.
2. It can be a *Contributing Factor* to an accident. The accident chain begins with an on-aircraft failure that is incorrectly handled by the flight crew, ultimately ending up as an accident (Primary Cause is pilot error).

One very good study on the relationship of incorrectly performed maintenance being a primary cause of accidents was from data obtained from European Aviation Safety Agency (EASA) Safety Analysis & Research. It is based on airliner accidents only, covering the period 1990 to January 2006. Only reports in which causal factors have been positively identified were included in the analysis. As shown in Table 4.1, incorrectly performed maintenance was the primary cause of 8% of the accidents from 1990 to January 2006.

TABLE 4.1

Incorrectly Performed Maintenance as a Primary Cause of Aircraft Accidents

Primary Cause of HF Related Accidents	No. of Accidents	No. of Total
Design	135	3%
Production	101	2
Operations (flight crew related)	3038	58
Maintenance	**416**	**8**
Air traffic management	66	1
Dispatch	18	0
Loading	*129*	2
Total HF-related accidents	3903	75
Non-HF-related accidents	*1320*	*25*
TOTAL	5223	100

Source: EASA Safety Analysis & Research group.

The International Air Transport Association (IATA) publishes an annual Safety Report. In the Safety Reports through 2008, IATA did not distinguish between maintenance as a primary cause or as a contributing factor to aircraft accidents—counting those accidents with *technical faults*. In 2003, 26% of the accidents had a technical fault, in 2006 (40%), in 2007 (20%), and in 2008 (15%). The average for these four years was 25%.

In 2009, IATA began to distinguish between maintenance as a primary cause and as a contributing factor. As shown in Table 4.2, Technical Faults were the primary cause of from 9% to 12% of the accidents from 2009 to 2011 or were involved in 29%–39% of the total accidents as a primary cause or contributing factor.

As discussed earlier, improperly performed maintenance can have an effect on safety of flight. However, improperly performed maintenance can also be costly, even if it does not lead to a safety-of-flight incident or accident. For example, improperly performed maintenance leads to

- 20%–30% of engine-in-flight shutdowns at a cost of U.S. $500,000 per in-flight shutdown.
- 50% of flight delays due to engine problems at an average cost of U.S. $9000 per hour.
- 50% of flight cancelations due to engine problems at an average cost of U.S. $66,000 per cancelation.

Thus, HF needs to be applied to aircraft maintenance both to improve safety of flight and to reduce costs.

TABLE 4.2

International Air Transport Association (IATA) Data from 2009 to 2011 on Aircraft Accidents That Were Because of Technical Faults

Accidents with Technical Faults	2009	2010	2011
Maintenance issues as primary cause	10	11	8
Percent of annual total	11%	12%	9%
Total number of accidents with technical faults	26	36	26
Percent of annual total	29%	39%	30%

MAJOR HUMAN FACTORS RESEARCH AND DEVELOPMENT INITIATIVES AND REGULATIONS

Research on and application of HF in commercial flight operations began in the 1970s and was strongly supported by the U.S. Federal Aviation Administration (FAA). Interest on HF research and application in commercial aviation maintenance began with the Aloha Airlines Flight 243 accident in Hawaii on April 28, 1988 (Figure 4.1). A major portion of the upper crown and fuselage skin of Section 43 of the 737-200 separated from the aircraft structure in-flight causing an explosive decompression of the cabin. The in-flight separation was due to that fact that there were 240 cracks in that portion of the upper crown that had gone undetected during a visual inspection of the aircraft just several weeks earlier (National Transportation Safety Board [NTSB], 1989). This event started the Aging Aircraft Research Program and also fueled passage and full funding of the Aviation Safety Research Act of 1988, passed into law in November 1988.

U.S. Legislation Drives HF Research

The Aviation Safety Research Act (1988) had a very significant impact on the aviation HF program in place today. It mandated HF research. Section 3 of the Act, "Research on Relationship Between Human Factors and Air Safety and on Dynamic Simulation Modeling," stated that

> The Administrator shall undertake or supervise research to develop a better understanding of the relationship between HF and aviation accidents and between HF and air safety, to enhance air traffic controller and mechanic and flight crew performance, to develop a human-factor analysis of the hazards associated with new technologies to be used by air traffic controllers, mechanics, and flight crews, and to identify innovative and effective corrective measures for human errors which adversely affect air safety (P.L 100-591).

The Aviation Safety Research Act also mandated the establishment of the Research, Engineering, and Development Advisory Committee (FAA, 2010). That committee began in 1990 and still exists today providing guidance regarding emerging safety trends and suggested research activity. After the Act was published FAA established a Government-Industry panel to write the National Plan for Aviation Human Factors Research and Development (Johnson, 1991). That plan was the basis for 20 years plus of HF Research and Development (R&D).

The 1988 Act had a profound impact on maintenance-related research and development. It permitted congress to specifically earmark funding for MHF activity. The FAA Office of Aviation

FIGURE 4.1 The fuselage of Aloha Airlines Flight 243 after suffering an explosive decompression while in flight on April 28, 1988. (This image is a work of a National Transportation Safety Board employee, taken or made as part of an employee's official duties. As a work of the U.S. federal government, all NTSB images are in the public domain.)

Medicine appointed Dr. William Shepherd to lead the MHF R&D effort. As an Aeronautical Engineer and experienced pilot, he had the vision to assemble experts from the aviation maintenance community to define immediate and long-term plans for applied research that would have value to users *in the field*. He was ably assisted by Ms. Jean Watson, who guided the program from 1995 to 2005, after Dr. Shepherd's retirement. From 2005 to 2016 the MHF research is guided by the Aircraft Maintenance Division of the Flight Standards Service, in cooperation with the Chief Scientific and Technical Advisors (CSTAs) program.

The MHF research started by hosting a Government-Industry conference that became known as the Annual FAA-Industry MHF Conference. The first meeting had 40 delegates of which 12 were speakers. From 1988 to 2011, FAA had a variety of U.S. industry and international regulatory partners. Most notably were the partnerships with Transport Canada (TC) and the United Kingdom Civil Aviation Authority (UK CAA). Both shared the responsibility of hosting annual meetings in their countries. The U.S. Air Transport Association (ATA) (now called Airlines for America [A4A]) partnered with FAA for many of the U.S.-based meetings. From 1988 to 2011, there were approximately 24 annual meetings. The 2011 meeting had 800 + delegates. All of the meeting proceedings are available on the current FAA Website (www.humanfactorsinfo.com).

From 1989 to 2016, FAA conducted some of the research itself, mostly at the Civil Aerospace Medical Institute in Oklahoma City or at the FAA Technical Center in Atlantic City, New Jersey. The program relied heavily on industry, involving most U.S. airlines, manufacturers, and maintenance service providers. From 1988 to 2016, numerous universities/colleges, research consulting companies, professional groups, organized labor groups, airlines, and MROs participated in the FAA R&D (Table 4.3).

TABLE 4.3

A Partial List of Universities/Colleges, Research Companies, Professional Groups, Organized Labor Groups, and Airlines/ MROs Involved in FAA MHF R&D from 1988 to 2016

Universities/Colleges
Clemson University
Embry-Riddle Aeronautical University
Florida Institute of Technology
Purdue University
St. Louis University, Parks College
State University of New York, Buffalo
San Jose State
University of Illinois
University of Southern California
University of Texas
Wichita State University

Research Companies
Applied Ergonomics Group, Inc.
Biotech Inc.
Booze Allen Corp
Galaxy Scientific Corporation
Pulsar Informatics
Sisyphus Incorporated

(Continued)

TABLE 4.3 (*Continued*)
A Partial List of Universities/Colleges, Research Companies, Professional Groups, Organized Labor Groups, and Airlines/ MROs Involved in FAA MHF R&D from 1988 to 2016

Professional Groups
The Boeing Company
The LOSA Collaborative
Situation Awareness Solutions
Xyant

Organized Labor
Aviation Mechanics Fraternal Association (AMFA)
International Association of Machinists and Aerospace Workers (IAM)
Teamsters
Transport Workers Union (TWU)

Airlines and MROs
AAR Corp
Air Evac
Air Tran
Air Wisconsin
Alaska Airlines
American Airlines
Air Canada
Atlas Air
BF Goodrich Aerospace (now Aviation Technical Services)
British Airways
Centurion/Sky Lease
Continental Airlines (now United Airlines)
Delta Air Lines
Duncan Aviation
Eastern Airlines (No longer)
Emirates
Everest Air
Express Jet
Garuda Indonesia
HAECO
Horizon Airlines
Lufthansa Airlines
Lufthansa Technic
Northwest (Now Delta)
Pan American Airlines (No Longer)
PEMCO
Piedmont Airlines
Polar Air Cargo
Southwest Airlines
Sky West
TIMCO (Now HAECO)
Trans World Airlines (Now American)
United Airlines
U.S. Airways/U.S. Air (Now American)

Other Regulatory Influence on Maintenance Human Factors

In the late 1990s International Civil Aviation Organization (ICAO) developed a standard for training and application of HF in maintenance. The standard read

> The maintenance organization shall ensure that all maintenance personnel receive initial and continuation training appropriate to their assigned tasks and responsibilities. The training programme established by the maintenance organization shall *include training in knowledge and skills related to human performance,* including coordination with other maintenance personnel and flight crew. (ICAO 8.7.6.4)

Publication of this standard was closely followed by regulations by TC, the European Joint Aviation Authority (JAA), and many of the other National Aviation Authorities (NAAs) around the world. For example, TC required MHF training and maintenance-caused event investigation.

The European JAA required by regulation a 10-part MHF program that included:

1. The development of a Safety Culture in the maintenance organization
2. Incident investigation and internal/external reporting of findings
3. Design/maintenance interface (what to do about poor maintenance data)
4. MHF training
5. Procedural noncompliance
6. Planning of tasks, equipment and spares
7. Fatigue
8. Shift handover and task handover
9. Error capturing (duplicate inspections, etc.)
10. Signing off tasks not seen nor checked.

The JAA also specified what should be covered in the MHF training program:

1. General/introduction to HF
2. Safety culture/organizational factors
3. Human error
4. Human performance and limitations
5. Effect of the environment on human performance
6. Procedures, information, tools, and practices
7. Communication
8. Teamwork
9. Professionalism and integrity
10. Organization's HF Program.

When the EASA took over the regulatory role from the JAA in the early 2000s, it maintained the MHF regulatory requirements specified previously. This had a major impact on the implementation of MHF training and processes around the world, as any NAA that followed the EASA regulations or any MRO that had an EASA 145 certification was now covered by this regulation. Many of the NAAs that developed a regulation based on the ICAO standard required at least MHF training, and sometimes event investigation. The UK CAA prepared an HF Guidance that heavily influenced EASA regulations. The document, "Aviation Maintenance Human Factors Guidance Material on the UK CAA Interpretation of Part-145 Human Factors and Error Management Requirements," remains relevant today (UK CAA 2003). The Civil Aviation Safety Authority (CASA) of Australia was instrumental in providing the world with the "Safety Behaviors: Human Factors for Engineers. A Multi-Media Training System" (CASA, 2013). These training materials, including videos, are an absolute must for MHF trainers.

FAA has been the leader in MHF research and development of training and advisory materials and job aids. For the most part, FAA has been very strong on guidance but not reliant on HF regulations. With minimal regulation, the FAA has convinced the U.S. commercial airline industry and MROs to voluntarily agree to MHF training and event investigation. Current Safety Management

Systems (SMS) are expected to identify additional organizational-specific HF hazards. Therefore, SMS shall influence new HF interventions including training.

HUMAN FACTORS MODELS USED IN THE INDUSTRY

INTRODUCTION

Early interest in MHF focused on existing models of human performance, including the ICAO supported Software-Hardware-Environment-Liveware-Liveware (SHELL) model and the Jens Rasmussen Skill-Rule-Knowledge-Based Behaviors model. But in the early 1990s, several new models emerged that were specific to aircraft maintenance: Gordon Dupont of TC developed the Dirty Dozen model, Dr. James Reason of Manchester University in England developed the failed barriers error model now more commonly known as the *Swiss Cheese* error model, and The Boeing Company, working with its industry partners, developed the Maintenance Error Decision Aid (MEDA) event investigation process based on the *contributing factors to error* model (MEDA will be discussed in the HF processes section). In the late 1990s/early 2000s, Dr. Scott Shappell and Dr. Doug Wiegmann developed the HFACS, based on the Reason Swiss Cheese model, to standardize the terminology used in accident/incident investigation. The Threat and Error Management (TEM) model was developed in the 1990s Dr. Robert Helmreich and colleagues at the University of Texas, Austin. Working with Captain Dan Maurino of ICAO, and Captain Frank Tullo of Continental Airlines, the TEM model was used to develop the Line Operations Safety Audit (LOSA) process. In the early 2010s an industry committee headed by the U.S. ATA (now called Airlines for American or A4A), with strong input from the FAA, developed and made available to industry the Maintenance Line Operations Safety Assessment (M-LOSA) observation process (this will be discussed in the HF processes section). Dr. Bill Johnson and Dr. Mike Maddox launched the People-Environment-Actions-Resources (PEAR) model in that same timeframe.

When airline organization staff become interested in the field of HF, they are often first encouraged to study HF performance/error models to understand what to be concerned about when doing a HF-related project. The major models that influenced the aircraft maintenance industry are briefly discussed in the following.

SHELL MODEL

The *Software/Hardware/Environment/Liveware* (SHEL) model was first developed by Edwards (1972) and modified by Hawkins (1993) in 1975 when the second L was added (Figure 4.2). ICAO

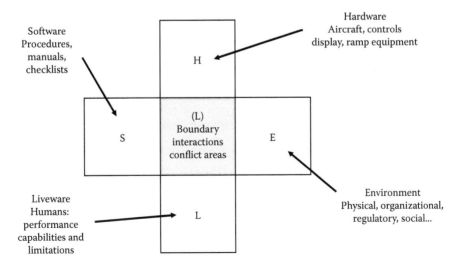

FIGURE 4.2 SHELL model.

adopted the model and published ICAO circular 216-AN/131, which was subsequently republished by the UK CAA as CAP 719 Fundamental Human Factors Concepts.

> *Software* (this model came out long before computers became an everyday tool) is the non-physical, intangible aspects of the aviation system that govern how the system operates and how information is organized. This includes rules, instructions, regulations, policies, norms, safety procedures, SOPs, customs, practices, and collections of documents.
>
> *Hardware* includes the physical elements of the aviation system, such as aircraft (structures, systems, seating, etc.), operator equipment, tools, materials, buildings, vehicles, computers, conveyer belts, and so on.
>
> *Environment* is the context in which aircraft and aviation system resources operate, made up of physical, organizational, economic, regulatory, political, and social variables that may impact on the worker/operator. Environment also includes work environments, such as cockpits, cabin areas, ramp areas, and maintenance facilities.
>
> *Liveware* is the human element or people in the aviation system, including cockpit, cabin, maintenance, ramp, and dispatch crews, management, and administrative staff.

The second *L* was added to model because of the importance of the human and its interfaces with other parts of the model, including (especially) other humans. The 4 components of the model interact with the Liveware component (2nd L placed in middle—see Figure 4.2).

1. Liveware-Software (L-S)
2. Liveware-Hardware (L-H)
3. Liveware-Environment (L-E)
4. Liveware-Liveware (L-L)

A mismatch at any of these interfaces can be a source of human error or system vulnerability.

Skills–Rules–Knowledge-Based Performance Model

The Skills–Rules–Knowledge-Based Performance model was developed by Rasmussen (1983). His idea was that we carry out tasks using three different types of behavior

1. *Skill-based*: Behavior involves highly practiced actions in very familiar situations. These are usually executed from memory without significant conscious thought or attention. It is learned through practice and is *easily performed once learned*. Also called a psychomotor or sensorimotor behavior.
2. *Rule-based*: Behavior is characterized by the *use of rules and procedures* to select a course of action in a familiar work situation.
3. *Knowledge-based*: Behavior is a response to a *totally unfamiliar situation* (no skill or rule recognizable to the individual). The person must rely on his or her understanding and knowledge of the system, the system's present state, and the scientific principles and fundamental theory related to the system.

In this model, some behavior

1. Can be performed without significant conscious thought or attention (*Automatic*)
2. Requires significant thought or attention (*Conscious*).

Figure 4.3 shows how the three types of behavior and the two types of performance interact. For example, skill-based behavior is automatic after enough practice. However, knowledge-based behavior will almost always require conscious performance.

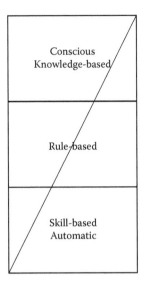

FIGURE 4.3   Jens Rasmussen's SRK model showing how knowledge-, rule-, and skill-based behaviors interact with conscious and automatic performance.

Skill-Based Performance

Here actions are controlled unconsciously by human instinct. Examples would include writing one's signature, using a hand tool, hanging a tag, lock wiring, and replacing a part. About 90% of a mechanic's daily activity is skill based, but it is estimated that only about 25% of the errors made by mechanics are skill based.

The error mode for skill-based performance is *inattention*. The errors are primarily execution errors involving slips and lapses in attention (concentration). The errors are triggered by simple human variability or by not recognizing changes in task requirements, system response, or system conditions. Also, too great a familiarity with a task can lead to underestimation of risk, which can lead to risk-taking behaviors (violations).

Rule-Based Performance

Here actions are controlled by an If/Then logic. In the simplest situation, one follows the procedure. However, the situation, although possibly familiar, can also be unanticipated. Then one must, for example, decide what to do if a problem is found, recognize that a situation has changed therefore requiring a different procedure, or decide whether to replace a bearing following an inspection.

As rule-based performance requires interpretation using an If/Then logic, the prevalent error mode is *misinterpretation*. Errors involve deviating from an approved procedure, applying the wrong response to a work situation, or applying the correct procedure to the wrong situation. A study in the nuclear industry found that ~60% of all errors are rule based.

Knowledge-Based Performance

This involves a response to a totally unfamiliar situation in which no procedural guidance exists and no skill applies. It requires diagnosis and problem solving. May need to *think through* different paths and pick the best option. May perform a *trial and error* approach to the problem.

Decision-making is erroneous if problem-solving is based on inaccurate information. Most decisions are made with limited information and assumptions. Consequently, the prevalent error mode is an *inaccurate mental model* of the system, process, or aircraft status.

Summary

This model has been somewhat useful in thinking through different types of errors. For example, skill-based behavior may be in error because of lack of practice or inattention, whereas rule-based behavior may be in error because of not reading/following a procedure.

DIRTY DOZEN

The Dirty Dozen was developed by Gordon Dupont in 1993 while he was working for TC and developing an MHF type of course. The Dirty Dozen refers to 12 common precursors to maintenance errors that lead to incidents and accidents. [They would be called *contributing factors* in MEDA and *hazards* in SMS.] That is, they influence maintenance staff to make errors.

The Dirty Dozen list includes

1. Lack of communication
2. Distraction
3. Lack of resources
4. Stress
5. Complacency
6. Lack of teamwork
7. Pressure
8. Lack of awareness
9. Lack of knowledge
10. Fatigue
11. Lack of assertiveness
12. Norms

The Dirty Dozen come in a poster format. Each poster illustrates in cartoon format one of the Dirty Dozen and then lists countermeasures to take in order not to succumb to that particular Dirty Dozen. The Dirty Dozen is certainly not a comprehensive list of accident precursors (see the MEDA discussion for a more comprehensive list). However, it has served as the basis of many MHF training courses around the world. In addition, Dirty Dozen posters can be seen in maintenance facilities worldwide.

SWISS CHEESE MODEL

The Swiss Cheese model of maintenance error was developed by Reason (1995, 1997), who was a professor at the University of Manchester, England. He believed that most accidents could be traced to one of four failure areas.

- *Organizational influences*: Resource management, organizational climate, and organizational policies.
- *Supervision*: Inadequate, planned inappropriate operations, failed to correct known problem, and so on.
- *Preconditions for unsafe acts*: Condition of operators, personnel factors, and environmental factors, and so on.
- *Unsafe acts*: Errors and violations.

In his theory, accidents just do not happen—they come from a chain of failures (conditions) that have accumulated with a final event occurs triggering the accident (Figure 4.4). These failures include

- *Active failures*: (errors and violations that have an immediate adverse effect)
- *Latent failures (conditions)*: Situations or conditions, which are placed in the system by decisions or actions of those (typically) at some distance from the operation. They lie dormant for some period of time, and then can become part of an accident chain (poor design, gaps in supervision, undetected defects, lack of training, and inadequate resources).

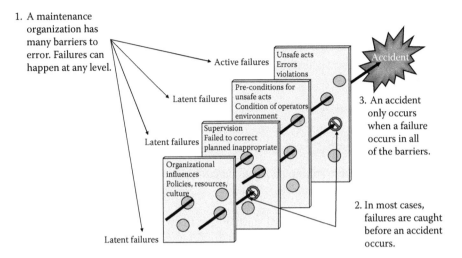

1. A maintenance organization has many barriers to error. Failures can happen at any level.

Active failures

Latent failures

Latent failures

Latent failures

Latent failures

Unsafe acts
Errors
violations

Pre-conditions for unsafe acts
Condition of operators environment

Supervision
Failed to correct
planned inappropriate

Organizational influences
Policies, resources, culture

Accident

3. An accident only occurs when a failure occurs in all of the barriers.

2. In most cases, failures are caught before an accident occurs.

FIGURE 4.4   How an accident occurs using the Reason Swiss Cheese model.

The Reason Swiss Cheese model has had a great impact on MHF. A discussion of it is included in much of the MHF training. Searching for "the holes in the Swiss Cheese" has become a common jargon and endeavor in aviation.

Reason also defined three types of errors:

1. *SLIP*: An error in *EXECUTING* the steps of a task. *Example*: The mechanic knows how to install a pump, but turns the wrench too hard and breaks a fitting. Also called an *error of commission*.
2. *LAPSE*: An error in *RETRIEVING* information about a task. *Example*: A mechanic is called to help on a different task after torqueing three of five bolts. When he comes back to his original job he forgets that he had two bolts left to torque and moves on to the next task. Also called an *error of omission*.
3. *MISTAKE*: An error in *PLANNING* a task. *Example*: "I do not need to do the fault isolation, because I have seen this problem before! I will replace this box."

Although the Reason definitions of slip and lapse have been commonly accepted in the airline industry, his specific meaning of a mistake has not. Most of the time, the word mistake is used synonymously as the word error.

HUMAN FACTORS ANALYSIS AND CLASSIFICATION SYSTEM

Scott Shappell and Doug Wiegmann developed HFACS to define the *holes in the Swiss Cheese* and to facilitate the application of this model to accident/incident/event investigation and analysis (Shappell and Weigmann, 2000; Wiegmann and Shappell, 2003). It was originally developed for classifying causal data on aircraft accidents. Now many airline Flight Operations departments in the United States and abroad use HFACS to classify their investigation/reported hazard data. Although HFACS is designed from a pilot's perspective, it is easy to take the pilot examples and apply them to, for example, maintenance, ramp operations, and cabin attendants.

HFACS is not used much in the aircraft maintenance business. However, it is included in this chapter on MHF for two reasons. First, as the Swiss Cheese model has had a large impact on MHF, in general, it is useful to provide a more in-depth breakdown of the Reason classification. Second, it has become a very useful tool for an airline SMS because it is a *classification system* that allows you to classify the findings from different types of event investigations (e.g., flight operations, maintenance, ground operations, and cabin crew) into one schema. Therefore, we will provide a brief description of HFACS in this chapter.

HFACS is based on the four components of the production system:

- Organizational influences
- Supervision
- Preconditions for unsafe acts
- Unsafe acts

Organizational influences: This area focuses on fallible decisions of upper level management, including the following:

- Resource management
  - Human resources (selection, staffing, and training)
  - Monetary/budget resources (lack of budget, excessive cost cutting)
  - Equipment/facility resources (poor aircraft/cockpit design, purchasing of unsuitable equipment, failure to correct known design flaws)
- Organizational climate
  - Structure (chain of command, communication, and delegation of authority)
  - Policies (promotion, hiring, firing, retention, drugs, and alcohol)
  - Culture (norms and rules, values, beliefs, attitudes, and organizational customs)
- Organizational processes
  - Operations (operational tempo, incentives, and time pressure)
  - Procedures (performance standards, clearly defined objectives, procedures/instructions about procedures)
  - Oversight (established safety program/risk management programs, management's monitoring and checking of resources, climate, and processes to ensure a safe work environment)

Supervision: The focus here is on the things that supervisors do not do (that they should do) or that they do incorrectly, including

- Inadequate supervision
  - Failed to provide proper training
  - Failed to provide professional guidance/oversight
  - Failed to provide current publications/adequate technical data and/or procedures
  - Lack of accountability
  - Perceived lack of authority
  - Failed to provide operational doctrine
  - Loss of supervisory situation awareness
- Planned inappropriate operations
  - Poor crew pairing
  - Failed to provide adequate brief time
  - Risk outweighs benefit
  - Failed to provide adequate opportunity for crew rest
  - Excessive tasking/workload
- Failed to correct a known problem
  - Failed to correct inappropriate behavior/identify risky behavior
  - Failed to correct a safety hazard
  - Failed to initiate corrective action
  - Failed to report unsafe tendencies

- Supervisory violations
  - Authorized unqualified crew for flight
  - Failed to enforce rules and regulations
  - Violated procedures
  - Authorized unnecessary hazard
  - Willful disregard for authority by supervisors
  - Inadequate documentation
  - Fraudulent documentation

Preconditions for unsafe acts: This area focuses on *why* unsafe acts are done, including

- Conditions of operators
  - Adverse mental states (loss of situation awareness, stress, fatigue)
  - Adverse physiological states (illness, hypoxia)
  - Physical/mental limitations (visual limitations, insufficient reaction time, information overload)
- Personnel factors
  - Crew resource management (failed to conduct adequate briefing, lack of assertiveness, lack of teamwork, failure of leadership)
  - Personal readiness (failure to adhere to crew rest requirements, inadequate training, self-medicating, overextension while off duty)
- Environmental factors
  - Physical environment (weather, altitude, terrain, lighting)
  - Technological environment (equipment/controls design, checklist layout, automation).

Unsafe acts: These are acts that are carried out by the staff that were causal in the accident—errors and violations of policies, processes, and procedures, including the following:

- Errors
  - Skill-based errors (breakdown of visual scan, inadvertent use of flight controls, failure to see and avoid)
  - Decision errors (inadequate maneuver/procedure, inadequate knowledge of systems/processes)
  - Perceptual errors (misjudged distance/clearance)
- Violations
  - Routine (inadequate flight briefing, failed to use ATC radar advisories, violations of orders, regulations, or SOPs, failed to comply with departmental manuals)
  - Exceptional (performed unauthorized acrobatic maneuver, improper takeoff technique, failed to complete performance computations for flight, accepted unnecessary hazard)

In summary, HFACS is an analysis and classification scheme that is based on the Reason Swiss Cheese model. It can be used to classify information gained during incident/event investigations. It can also be used to standardize investigation findings across an airline, which is very useful for SMS implementation.

THREAT AND ERROR MANAGEMENT MODEL

The idea for the TEM model came from Dr. Helmreich and others (1995) at the University of Texas in the 1990s. It was applied to the commercial aviation flight operations working with Captain Tullo (Continental Airlines), and Captain Dan Maurino (ICAO). They were, at first, focused on trying to decide how much training would be required for *perfect pilot performance*. When talking to pilots about errors that they had made during flight, they found that most of the errors were discovered and corrected before they could lead to a bad outcome. As pilot training was quite expensive anyway (if you wanted so much training that you got perfect pilot performance), they decided to focus on training pilots to watch out for things that could lead to errors (threats), coming up with strategies to deal with these threats, and learning to catch your errors if they did get through. This leads to the development of the LOSA process.

LOSA is an observational process whereby a trained observer pilot sits in the jump seat of an aircraft during a normal flight to observe the two pilots. They are observing whether the pilots detect and deal with threats in the environment and detect and deal with any errors that they made. A LOSA Consortium, started by Dr. Helmreich, served as a source of LOSA-trained pilot observers and a data storage organization. LOSA became such a respected process that the IATA requires that member airlines carry out the LOSA process every few years to maintain their IATA membership. The reason that TEM is important in this chapter is that the model was used in 2010 (www.humanfactorsinfo.com) by the U.S. ATA (now Airlines for America) team, which included several working members from the FAA, as well as other industry maintenance and MHF experts, to develop a Maintenance LOSA observation process. Therefore, we will now briefly discuss the TEM model, whereas a discussion of the Maintenance LOSA process will be provided later in the chapter.

Threats

A threat is any condition that increases the complexity of the (flight, maintenance, ramp) operation. Threats, if not managed properly, can decrease safety margins and can lead to errors. When you observe a threat, this should serve as a *red flag* to you to "Watch out! Something bad can happen!"

There are two types of threats:

- *External threats*: Those are threats that are outside of your control (e.g., weather, lack of equipment, hard to understand documentation, system errors, and inadequate lighting)
- *Internal (Human) threats*: Those are threats that are within our control (fatigue, loss of situation awareness, stress, and disregard for following procedures).

In this theory, operators should always be engaged in Threat Management. There are two aspects to Threat Management:

1. Recognizing that a threat exists.
2. Coming up with a strategy to deal with the threat, so that it does not reduce safety margins or contribute to an error.

However, observing threats is not always easy. For example, the LOSA Consortium informed the ATA Task Force that only 25% of pilot errors had an observable threat that lead to the error. The other 75% of errors had no observable threat prior to the error.

Errors

An error is the mistake that is made when threats are mismanaged. There are five types of errors:

1. Intentional noncompliance errors
2. Procedural errors

3. Communication errors
4. Proficiency errors
5. Operational decision errors

In this theory, operators should also always be engaged in Error Management. Error Management is the mitigation or reduction in seriousness of the outcome of an error. There are two aspects to Error Management:

1. *Manage the past*: The resist and resolve filters or defense mechanisms may be applied to an existing error to correct it before it becomes consequential to safety.
2. *Manage the future*: By applying the resist and resolve filters in the analysis of an error, you may improve strategies or counter-measures to identify and manage both internal and external threats, such as fatigue, condition of ground equipment, and so on.

TEM Model

The TEM model is presented in Figure 4.5. The model shows threats in the environment entering the system. If the threat is observed, then the operator can develop strategies to deal with the threat before it turns into an error. If the threat is not seen or the strategy is not effective, then an error occurs. Now the operator has to deal with the error. The error can be dealt with through the Resist *filters*—things that are already in the system to protect against errors—or the Resolve process, whereby the operator catches his/her own error and then corrects it. If the Resist and Resolve functions fail, then the error gets through and has consequences.

There are three types of error consequences:

1. *Inconsequential*: The error has no immediate effect on safety.
2. *Undesired state*: Risk or unsafe operational conditions are increased.
3. *Additional error*: The error causes another error(s).

As stated earlier, the TEM model serves as the basis for the M-LOSA observational process. This process is discussed in more detail later in this chapter.

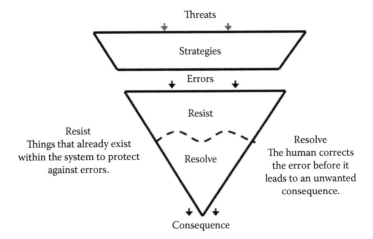

FIGURE 4.5   Threat and error management (TEM) model.

PEAR Model

The PEAR model was developed by William B. Johnson and Michael E. Maddox (Johnson and Maddox, 2007) in the 1990s. One of the major benefits of this generic model is that it was initially designed to be understood and used by the aviation maintenance and engineering (M&E) audience. PEAR remains as the main HF training paradigm for FAA inspector training (Johnson, 2015a) as well as for the widely used CASA Australia (Johnson, 2015b) training program. In addition, PEAR is the basis for MHF training at many airlines and MROs.

For over two decades, the term *PEAR* has been used as memory jogger, or mnemonic, to characterize HF in aviation maintenance. PEAR prompts recall of the four important considerations for HF programs:

- *P*eople who do the job
- *E*nvironment in which they work
- *A*ctions they perform
- *R*esources necessary to complete a job

People

Aviation MHF programs focus on the people who perform the work, and address physical, physiological, psychological, and psychosocial factors (Figure 4.6). It must focus on individuals, their physical capabilities, and the factors that affect them. It should also consider their mental state, cognitive capacity, and conditions that may affect their interaction with others.

In most cases, MHF programs are designed around the people in the organization's workforce. You cannot apply identical strength, size, endurance, experience, motivation, and certification standards equally to all employees. The company must match the physical characteristics of each person to the tasks they perform. The company must consider factors such as each person's size, strength, age, eyesight, and more, to ensure that they are physically capable of performing all the tasks that make up the job. A good MHF program will consider the limitations of humans and design the job accordingly.

Attention to the individual does not stop at physical abilities. A good MHF program must address physiological and psychological factors that affect performance. Companies should do their best to foster good physical and mental health. Offering educational programs on health and fitness is one way to encourage good health. Many companies have reduced sick leave and increased productivity by making healthy meals, snacks, and drinks available to their employees. Companies should also have programs to address issues associated with chemical dependence, including tobacco and alcohol.

Another People issue involves teamwork and communication. Safe and efficient companies find ways to foster communication and cooperation among the workers, managers, and owners. For example, workers should be rewarded for finding ways to improve the system, eliminate waste, and help ensure continuing safety.

Physical factors	Psychological factors
Physical size	Workload
Sex	Experience
Age	Knowledge
Strength	Training
Sensory limitations	Attitude
	Mental or emotional state

Physiological factors	Psychosocial factors
Nutritional factors	Interpersonal conflicts
Health	Personal loss
Lifestyle	Financial hardships
Fatigue	Recent divorce
Chemical dependency	

FIGURE 4.6　People factors in the PEAR.

Environment—Physical and Organizational

There are two environments in aviation maintenance. There is the physical environment on the ramp, in the hangar, or in the shop. There is also the organizational environment that exists within the company. An MHF program must pay attention to both environments (Figure 4.7).

The physical environment is obvious. It includes ranges of temperature, humidity, lighting, noise control, cleanliness, and workplace design. Companies must acknowledge these conditions and cooperate with the workforce to either accommodate or change the physical environment. It takes a corporate commitment to address the physical environment. This topic overlaps with the *Resources* component of PEAR when it comes to providing portable heaters, coolers, lighting, clothing, and workplace and task design.

The second, less tangible, environment is the organizational one. The important factors in an organizational environment are typically related to cooperation, communication, shared values, mutual respect, and the culture of the company. An excellent organizational environment is promoted with leadership, communication, and shared goals associated with safety, profitability, and other key factors. The best companies guide and support their people and foster a culture of safety.

An example program that has a notable positive effect on corporate organizational culture is the FAA's Aviation Safety Action Program (ASAP) (FAA, 2002) (www.faa.gov/safety/programs_initiatives/aircraft_aviation/asap/policy). The ASAP program is a cooperative arrangement in which the FAA joins with company management and its labor representation to report and correct errors as they occur. The result is a new level of teamwork that promotes nonpunitive event reporting and clear communication to manage error and cost while ensuring continuing safety.

Actions

Successful HF programs carefully analyze all the actions people must perform to complete a job efficiently and safely (Figure 4.8). Job task analysis (JTA) is the standard HF approach to identify the knowledge, skills, and attitudes that are necessary to perform each task in a given job. The JTA helps identify what instructions, tools, and other resources are necessary. Adherence to the JTA helps ensure that each worker is properly trained and that each workplace has the necessary equipment and other resources to perform the job. Many regulatory authorities require that the JTA serve as the basis for the company's general maintenance manual and training plan.

Physical	Organizational
Weather	Personnel
Location inside/outside	Supervision
Workspace	Labor-mangement relations
Shift	Pressures
Lighting	Crew structure
Sound level	Size of company
Safety	Profitability
	Morale
	Corporate culture

FIGURE 4.7   Environment factors in PEAR.

Steps to perform a task	Knowledge requirements
Sequence of activity	Skill requirements
Number of people involved	Attitude requirements
Communicatin requirements	Certification requirements
Information control requirements	Inspection requirements

FIGURE 4.8   Action factors in PEAR.

Procedures/work cards	Ground handling equipment
Technical manuals	Work stands and lifts
Other people	Fixtures
Test equipment	Materials
Tools	Task lighting
Computers/software	Training
Paperwork/signoffs	Quality systems

FIGURE 4.9 Resource factors in PEAR.

Many HF challenges associated with use of job cards and technical documentation fall under Actions. A crystal clear understanding and documentation of actions ensures that instructions and checklists are correct and useable.

Resources

The final PEAR letter is *R* for *R*esources. Again, it is sometimes difficult to separate *R*esources from the other elements of PEAR. In general, the characteristics of the *P*eople, *E*nvironment, and *A*ctions dictate the *R*esources. Many resources are tangible, such as lifts, tools, test equipment, computers, technical manuals, and so forth (Figure 4.9). Other resources are less tangible. Examples are the number and qualifications of staff to complete a job, the amount of time allocated, and the level of communication among the crew, supervisors, vendors, and others.

Resources should be viewed (and defined) from a broad perspective. A resource is anything a mechanic (or anyone else) needs to get the job done. For example, protective clothing is a resource. A mobile phone can be a resource. Rivets can be resources. What is important to the *R*esource element in PEAR is that you focus on identifying the need for additional resources.

Safety Management System and PEAR

All national regulatory authorities now require that airlines implement a SMS. Regardless of the regulatory requirement, it makes good sense for an organization to have a formal process that identifies potential hazards and their associated levels of risk.

As part of a company's SMS, polices must be established, hazards must be identified and analyzed for risk, unacceptable risk must be mitigated, and the system must be monitored for acceptable safety. An SMS must be formalized, documented, and become the key element of a company's safety culture. The HF program, exemplified by PEAR, provides methods for identifying and controlling many of the potential hazards within an organization and should be an integral part of your company's SMS program.

PEAR is an easy means to remember the four key elements of any HF program. The primary premise of the PEAR is that an MHF program does not have to be complex, expensive, or a burden to the organization. Simply apply PEAR to identify HF issues and facilitate your efforts to develop and support your SMS. Although SMS is discussed next, it is worthy noting to mention that PEAR provides an excellent high-level means to categorize hazard identification and subsequent organizational management for SMS.

MAINTENANCE HUMAN FACTORS PROGRAMS/PROCESSES IN USE TODAY

The FAA-sponsored R&D activity was meant to be applied. Although reports were always published, the success of the work was based on its direct applicability to those who maintained aircraft. All segments of the aviation industry were served, but the worked favored the airline and large maintenance organizations. The R&D products included, but were not limited to, computer-based and traditional training programs, handbooks, and other guidance materials; software tools; and the support to implement and measure the value of the MHF interventions.

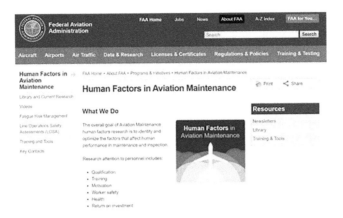

FIGURE 4.10 The 2016 homepage for FAA maintenance human factors.

The FAA made a significant effort to distribute the materials. For over a decade, starting in 1990, FAA created thousands of CDs and DVDs containing multimedia, reports, and tools. By 1994, FAA had launched the MHF website. That website, pictured in Figure 4.10, remains as a legacy for all the reports and many of the products created in the almost 30-year period from 1988 to the time of this edited book. Similar to this chapter, that website offers the history and evolution on the aviation MHF programs that we know today. Site address as publication time is www.humanfactorsinfo.com or go to the FAA website for HF. This section discusses selected FAA R&D activity and products, as well as MHF processes developed by the industry.

MAINTENANCE RESOURCE MANAGEMENT/MAINTENANCE HUMAN FACTORS TRAINING

In the 1980s, a type of training called Crew Resource Management (CRM) was developed and implemented for commercial airline pilots in response to several deadly accidents due to poor communication and situation awareness by flight crews. As the interest in MHF began in the late 1980s/early 1990s, attention was turned to developing a version of CRM for maintenance that was called Maintenance Resource Management (MRM). As MRM was born out of CRM, it took on some of the same types of training, regarding, for example, communication skills, situation awareness, and a high level of emphasis on the organization.

The types of MRM training implemented by some of the large U.S. airlines in the late 1990s/early 2000s differed greatly from airline to airline. However, an FAA Advisory Circular 120-72 (FAA, 2000) gave the general outline of what MRM programs typically looked like. MRM is described as a general process for improving communication, effectiveness, and safety in M&E organizations to reduce maintenance errors, improve individual and team performance, and develop a safety culture and positive safety attitudes. MRM programs do this by reducing maintenance errors through improved coordination, communication, and increased awareness. An important part of MRM is collecting data to determine whether the MRM program is effective at reducing errors, injuries, equipment damage, and so on. As with any type of safety program, MRM needs positive, explicit, and demonstrated support of senior management.

A typical MRM training would consist of the following:

- HF knowledge
- Understanding the maintenance operations as a system
- Identifying and understanding basic HF issues
- Recognizing contributing causes to human errors
- Communication skills

- Basic communication skills
- Inter- and intra-team communication
- Assertiveness
- Peer-to-peer work performance feedback techniques
- Teamwork skills
- Basic team concepts
- Coordination
- Decision-making
- Norms
- Performance management (leadership)/situation awareness
- Stress management
- Fatigue
- Complacency
- Team situation awareness and error chain recognition
- Leadership

There is large overlap in subject matter between the MRM topics listed previously and the EASA MHF training requirements discussed earlier in the Introduction. However, because of the explicit EASA regulation regarding MHF training, maintenance organizations (in Europe and worldwide) are required to follow the EASA MHF training requirements. As the United States has no specific regulation on MHF training (as of 2016), M&E organizations under a FAA 145 certificate will typically provide some type of MHF training, but there is no requirement that it follow the earlier MRM training guidelines. Moreover, several of the big U.S. airline M&E organizations hold both an FAA 145 certificate and an EASA 145 certificate. For these organizations, their overall MHF training would typically follow the EASA guidance. There has been a gradual abandonment of the term MRM training, which has been replaced by simply using the term MHF training.

The Human Factors Guide and the Operator's Manual for Human Factors in Maintenance

In 2004, FAA created the CSTA for Human Factors in Aircraft Maintenance Systems. By that time, FAA management, U.S. airline M&E organizations, and other segments of the aviation maintenance industry were asking for increased applied MHF information. They wanted to implement MHF programs and wanted to know how to do that. They wanted FAA research to transition from the laboratory to the field. Bill Johnson, as the first CSTA in MHF, commenced to collect and publish useful information.

The first step was to rewrite a 1995 version of the Human Factors Guide for Aviation Maintenance and Inspection (Maddox, 2007). That document provided not only the detailed scientific/engineering basis for MHF but also suggestions for implementation. The 300plus pages of that document covered topics shown in Table 4.4.

As the 2nd Edition of the *Human Factors Guide* was completed, it became clear that there was an additional requirement for a short *How to* manual for industry HF practitioners. To specify and fulfill that requirement, the CSTA called together a panel of highly experienced industry and government personnel. After considerable deliberation, the panel identified the top MHF issues that should be addressed. In 2006, experts from that team, and others, wrote the 1st Edition of the *Operator's Manual for Human Factors in Aviation Maintenance* (Johnson, 2005) (Op's Manual). The chapters were as follows:

1. Event investigation
2. Documentation
3. MHF training
4. Shift/task turnover
5. Fatigue management
6. Sustainability and cost justification

TABLE 4.4

**Chapters of the 2007 Human Factors Guide
for Aviation Maintenance and Inspection**

1. Human Factors
2. Establishing a Human Factors Program
3. Fatigue and Fitness for Duty
4. Facilities Design/Usability
5. Shiftwork and Scheduling
6. Procedures and Technical Documentation
7. Error and Error Reporting
8. Technical Training
9. Workplace Safety
10. Communications
11. Ethics in Maintenance

These six topics are critical because they are based on operational data and practical experience from the United States and other countries. TC, UK CAA, and EASA regulations contributed to selecting these areas and to writing this manual.

For each of the six topics that contribute to the success of any MHF program, the Op's Manual offered the following:

- Why the topic is important.
- How you implement it.
- How you know it is working.

Similar to any good operator's manual, the document told the reader what to do without excessive description of why it should be done. The first edition of the manual assumed that the reader already recognized the importance of HF. For detailed information, there are *Three Key References* at the end of each topic. These straightforward suggestions provide the key components for implementing a successful MHF program.

Event Investigation

The purpose of an event investigation process is to manage the risks from events caused by human actions that affect flight safety, personal injury, and equipment damage. An error is a human action that unintentionally deviates from the required, intended, and expected action. A violation is a human action that intentionally deviates from company or regulatory policies or procedures.

Event investigations help organizations identify and understand multiple contributing factors to errors and deviations. Examples include hard-to-understand procedures, time pressure, task interruption, poor communication, and a variety of additional workplace and life conditions. The identification of these contributing factors provides an organization with a specific focus to prevent future events.

Documentation

Documentation runs the gamut from manufacturer technical manuals, to internal job cards, to publications from the regulator. Numerous postevent investigations have clearly demonstrated that failure to use the documentation appropriately was a major contributing factor to these events. Issues associated with maintenance documents are related to the way the information is written, to where and how it is accessed in an organization, and to the motivation of the aviation maintenance technician to use required data at all times. There are often numerous and diverse reasons that documentation is not used properly. A good MHF program must address and correct documentation challenges.

Human Factors Training

Training is a critical and primary means to ensure that company personnel are prepared to conduct assigned work. Research has shown that MHF training can address many of the issues that contribute to adverse events. It has been demonstrated that such training can reduce costs associated with human performance issues.

Shift/Task Turnover

Shift and task turnover are critical moments in aircraft maintenance activities. Efficient and effective turnover requires policy, procedures, planning, teamwork, and effective communication when workers relay crucial information for ending a shift and starting another. The classic challenges associated with fatigue, distraction, false assumptions, and failure to properly document can negatively affect the quality of shift turnover as well as task turnovers within shifts. Accidents and incidents have shown us that inadequate information exchange during shift and task turnovers can have serious consequences.

Fatigue Management

Most aviation maintenance personnel that have worked extensive hours to recover an aircraft on ground remember the fatigue more than the complexity of the technical repair. Maintenance technicians, operations people, cabin crew, dispatchers, and managers are susceptible to error induced by fatigue. Modern considerations of fatigue issues are termed *Alertness* to indicate that there is a broad range of human readiness for work in which fatigue is merely one level on the continuum.

FAA research in 2001 showed that the maintenance workforce does not get enough sleep. Further, the workforce does not recognize the challenges associated with fatigue and does not compensate accordingly. Fatigue management programs are generally voluntary. Nevertheless, fatigue management is an important component of a MHF program.

Sustainability and Cost Justification

The first five topic areas of this document recommended specific actions. The topics of Program Sustainability and Cost Justification are general and applied to all aspects of an MHF program. These programs often get off to a good start but then struggle over time. Challenges to program sustainability included changes in policies and projects when management changes, a lack of cost justification, and limited program integration. The ideas presented help sustain multiple MHF initiatives and provide a straightforward consideration of cost justification.

Summary

The *Operator's Manual: Human Factors in Maintenance* was written by an FAA-led panel of experts to provide U.S. airline M&E organizations with a simple and manageable list of actions to implement an MHF program. The manual was easy to read and focused on how to implement the program elements. The first edition of the Op's Manual was originally written in English but was also translated into Spanish and Mandarin.

A second edition of the Op's Manual was published in 2014 (Johnson and Avers, 2014). All sections were rewritten and a seventh Chapter was added. The titles of the chapters are shown in the Cover of the book shown in Figure 4.11.

MAINTENANCE ERROR DECISION AID EVENT INVESTIGATION PROCESS

The MEDA process was developed in the early 1990s by the Boeing Company working with some of its customer airlines and industry mechanic labor union groups (Rankin and Allen, 1996;

September 2014

Operator's Manual
Human Factors in Aviation Maintenance

FIGURE 4.11   Cover of the second edition of the operator's manual.

Rankin et al., 2000; Rankin, 2007). It is an event investigation process for events that were caused by incorrectly performed maintenance. Boeing now estimates that more than 900 aircraft maintenance organizations around the world have implemented the MEDA process. The reasons for the success of MEDA were summarized for an FAA-sponsored industry meeting on MHF and are provided next. Another important reason for MEDA use is that it is a reactive hazard identification process that is needed for an M&E organization's SMS.

Airline Input: Boeing partnered with nine airlines and two unions in the development of the MEDA process. This helped one to make sure that the process was something that an airline maintenance organization would find useful and that the labor groups would not oppose (or even support) the use of the process. Three products were developed:
1. The MEDA Results Form, which is used to collect the contributing factor information during the investigation.
2. The MEDA User's Guide, which is a *how to manual* for carrying out a MEDA investigation.
3. Training presentations. The project resulted in the development of two training presentations. First, the MEDA Investigator Training presentation, which is used to train airline staff in how to use the Results Form and how to carry out the investigative interview. Second, the MEDA Management Training, which is used to educate maintenance management on the use of MEDA and to inform the management on how to implement the process.

Field test: Before releasing the process to industry, the MEDA process was tried out by nine airlines. This was supported by a contract with the FAA. Feedback from the airlines and analysis of the filled-out MEDA Results Forms were instrumental in making changes to the process/forms, so that MEDA would be as useful as possible to aircraft maintenance organizations.

Organizational support: Starting in 1995, Boeing has provided MEDA implementation support to its airline customers. This greatly helped one to get the word out to the airline community about MEDA and its benefits. Moreover, Boeing supported presentations regarding MEDA at international conferences, which greatly increased the number of aircraft maintenance organizations that requested MEDA implementation support.

Regulatory requirements: ICAO recognized the MEDA process in one of its late 1990s publications on HF and recommended its use. Later in the 1990s, TC and the JAA of Europe wrote regulatory requirements for maintenance event investigation, partly based on knowing that such a process already existed. These regulatory requirements certainly increased the number of aircraft maintenance organizations that requested the MEDA training and, overall, also increased the requests for MEDA implementation support outside of the areas affected by the regulation.

Continual refinement of the product: Based on airline input, the MEDA Results Form is updated about every two years, which also drives a change to the MEDA User's Guide. A major change to MEDA in the early 2000s added the concept of violations of company policies, processes, and procedures as causal in MEDA investigations.

Industry cultural change: Over the 18 years since MEDA has been offered, the airline industry has changed dramatically and has made the implementation of MEDA easier. The change has been to move from a Punishment Culture, in which mechanics were punished for their errors, to a Just Culture, in which decisions about punishment are based on a finding of reckless behavior, which is rare. It is much easier to implement MEDA in a Just Culture environment than in a Punishment Culture environment. Numerous maintenance organizations failed in their first attempts at implementing MEDA, because MEDA was perceived by the mechanics as a way of determining *how much punishment*. It took a transformation to a Just Culture in many of these organizations in order for MEDA to be successfully implemented.

Maintenance Error Decision Aid Philosophy and Model

The MEDA philosophy is as follows:

- A maintenance-related event can be caused by an error, by not following company policies, processes, and procedures (a violation), or by an error/violation combination.
- Maintenance errors are not made on purpose.
- Errors result from a series of contributing factors in the workplace.
- Violations, while intentional, are also caused by contributing factors.
- Most of the contributing factors to errors and violations are under management control.
- Therefore, improvements can be made to these contributing factors to reduce the probability that they will contribute to future events.
- The maintenance organization must be viewed as a system, in which the mechanic/inspector is one part of the system.
- Addressing lower level events helps prevent more serious events.

The MEDA theory is that contributing factors in the workplace contribute to errors and violations, which can lead to a maintenance system failure, which can lead to an event. The MEDA model is presented in Figure 4.12.

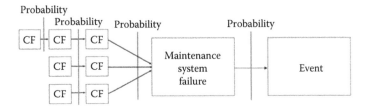

FIGURE 4.12 MEDA error model showing contributing factors (CFs) leading to maintenance system failures leading to an event.

This model illustrates several points. First, it is a probabilistic model. A given contributing factor (CF) does not *guarantee* a maintenance system failure; it only increases the probability of that maintenance system failure. Second, there are typically more than one CFs to a failure (Boeing estimates that the average number is four to five). Third, there are CFs to CFs. For example, a mechanic may be fatigued, which lead to an error. However, why was he fatigued? (As he only got four hours of sleep the night before.) Why did he only get four hours of sleep? (As he was asked to work a double shift, and it was an hour drive home and an hour drive back to work the next morning.) Why did he work a double shift? (As the organization was shorthanded.) So, the *original* CF is understaffing, which leads to mechanic fatigue.

Contributing Factors

Contributing factors are the same thing as *performance shaping factors* in the HF literature. Boeing called them *contributing factors* because they contribute to the error. The MEDA process takes a sociotechnical view of the maintenance operation and the various contributing factors that can be found within the model. Their model has the mechanic/inspector working in their immediate work environment under supervision and under the company's policies, hiring practices, training practices, processes, and procedures. There are 10 different categories of CFs on the Results Form, including the following:

1. Information (documentation): This includes the Instructions for Continued Airworthiness as provided by the airframe OEMs as well as company-developed Work Orders/Task Orders.
2. Ground support equipment/tools/safety equipment
3. Aircraft design/configuration/parts/equipment/consumables
4. Job/task
5. Knowledge/skills
6. Individual factors
7. Environment/facilities
8. Organizational factors
9. Leadership/supervision
10. Communication.

Figure 4.13 shows the Information section from the MEDA Results Form. Potential CFs for these sections are listed on the form so that the investigator does not need to memorize them. Moreover, the investigator is asked to specifically state how the selected CF contributed to the system failure and is for recommendations in dealing with the CF so that its potential for causing error in the future is reduced or eliminated.

Section VI—Contributing factors checklist		
N/A __	**A. Information (e.g., work cards, maintenance manuals, service bulletins, maintenance tips, non-routines, illustrated parts catalogs, etc.)**	

__ 1. Not understandable   __ 4. Too much/conflicting information   __ 7. Information not used
__ 2. Unavailable/inaccessible __ 5. Update process is too long/complicated   __ 8. Inadequate
__ 3. Incorrect   __ 6. Incorrectly modified manufacturer's MM/SB __ 9. Uncontrolled
                                                                   __ 10. Other (explain below)

Describe specifically how the selected information factor(s) contributed to the system failure.

Recommendations to correct the contributing factors listed above.

FIGURE 4.13   A screen shot of the information section from the MEDA results form.

Maintenance System Failures

In the MEDA theory, the CFs lead to errors and violations, which are called Maintenance System Failures on the MEDA Results Form. See Figure 4.14 for a screen shot of this section of the MEDA Results Form. The different kinds of failures that can be investigated using the MEDA process include the following:

1. Installation failure
2. Servicing failure
3. (On-aircraft) Repair failure
4. Fault isolation/test/inspection failure
5. Foreign object damage/debris
6. Airplane/equipment damage
7. Personal injury
8. Maintenance control failure.

Within these categories are more specific examples of the failure on the form. For instance, as can be seen in Figure 4.14, there are 14 separate ways that the installation was incorrect (wrong part, wrong location, wrong orientation, etc.). In addition, there is always an *other* in case the specific system failure was not listed. Also note that the investigator is to determine whether the system failure flew out on a flight or not.

Event

Finally, these Maintenance System Failures can lead to an event (which is what starts off the investigation). These events include the following:

1. Operations process event
   a. Flight delay
   b. Flight cancelation
   c. Gate return
   d. In-flight (engine) shutdown
   e. Air turn back
   f. Diversion
   g. Smoke/fumes/odor event

Section III—Maintenance System Failure

Please select the maintenance system failure(s) that caused the event:

1. Installation Failure
( ) a. Equipment/part not installed
( ) b. Wrong equipment/part installed
( ) c. Wrong orientation
( ) d. Improper location
( ) e. Incomplete installation
( ) f. Extra parts installed
( ) g. Access not closed
( ) h. System/equipment not reactivated/deactivated
( ) i. Damaged on remove/replace
( ) j. Cross connection
( ) k. Mis-rigging (controls, doors, etc.)
( ) l. Consumable not used
( ) m. Wrong consumable used
( ) n. Unserviceable part installed
( ) o. Other (explain below)

2. Servicing Failure
( ) a. Not enough fluid
( ) b. Too much fluid
( ) c. Wrong fluid type
( ) d. Required servicing not performed
( ) e. Access not closed
( ) f. System/equipment not deactivated/reactivated
( ) g. Other (explain below)

3. Repair Failure (e.g., component or structural repair)
( ) a. Incorrect

( ) b. Unapproved
( ) c. Incomplete
( ) d. Other (explain below)

4. Fault Isolation/Test/Inspection failure
( ) a. Did not detect fault
( ) b. Not found by fault isolation
( ) c. Not found by operational/functional test
( ) d. Not found by task inspection
( ) e. Access not closed
( ) f. System/equipment not deactivated/reactivated
( ) g. Not found by part inspection
( ) h. Not found by visual inspection
( ) i. Technical log oversight
( ) j. Other (explain below)

5. Foreign Object Damage/Debris
( ) a. Tooling/equipment left in aircraft/engine
( ) b. Debris on ramp
( ) c. Debris falling into open systems
( ) d. Other (explain below)

6. Airplane/Equipment Damage
( ) a. Tools/equipment used improperly
( ) b. Defective tools/equipment used
( ) c. Struck by/against
( ) d. Pulled/pushed/drove into
( ) e. Fire/smoke
( ) f. Other (explain below)

7. Personal Injury
( ) a. Slip/trip/fall
( ) b. Caught in/on/between
( ) c. Struck by/against
( ) d. Hazard contacted (e.g., electricity, hot or cold surfaces, and sharp surfaces)
( ) e. Hazardous substance exposure (e.g., toxic or noxious substances)
( ) f. Hazardous thermal environment exposure (heat, cold, or humidity)
( ) g. Other (explain below)

8. Maintenance Control Failure
( ) a. Scheduled task omitted/late/incorrect
( ) b. MEL interpretation/application/removal
( ) c. CDL interpretation/application/removal
( ) d. Incorrectly deferred/controlled defect
( ) e. Airworthiness data interpretation
( ) f. Technical log oversight
( ) g. Airworthiness Directive overrun
( ) h. Modification control
( ) i. Configuration control
( ) j. Records control
( ) k. Component robbery control
( ) l. Mx information system (entry or update)
( ) m. Time expired part on board aircraft
( ) n. Tooling control
( ) o. Mx task not correctly documented
( ) p. Not authorized/qualified/certified to do task
( ) q Other (explain below)

( ) **9. Other** (explain below)

Did the Maintenance System Failure "fly" on the aircraft? ( ) Yes ( ) No
Describe the specific maintenance failure (e.g., auto pressure controller installed in wrong location).

FIGURE 4.14 Screen shot of the maintenance system failure portion of the MEDA results form.

2. Aircraft damage event
3. Personal injury event
4. Rework (on-aircraft)
5. Airworthiness control
6. Found during maintenance
7. Found during flight

Although *Found During Maintenance* and *Found During Flight* are not really events, they were added to the MEDA Results Form for maintenance organizations that wanted to investigate why failures occurred that were found during maintenance by a mechanic or during flight by a pilot, even though the failures did not lead to an *event*.

Finally, there are two other sections on the MEDA Results Form. One is the Chronological Summary of the Event Section. The second is the Summary of Recommendations Section.

The MEDA Results Form changed dramatically from its inception in 1995 to the present day based on experience and requests from maintenance organizations and regulatory authorities. The Boeing Company makes the Results Form and User's Guide available to anyone who asks for them.

- Send an email request to MHF@Boeing.com
- MEDA User Guide on FAA website http://www.faa.gov/about/initiatives/maintenance_hf/library/documents/media/media/meda_users_guide_updated_09-25-13.pdf
- MEDA Results Form on FAA website http://www.faa.gov/about/initiatives/maintenance_hf/library/documents/media/media/meda_results_form_revl.pdf

INTERNAL HAZARD REPORTING SYSTEM

Internal Hazard Reporting System (ICAO) has recommended that an aviation company's SMS should include three types of hazard identification processes: reactive, proactive, and predictive. MEDA is the reactive process, and the M-LOSA process that we discuss later is the predictive process. A common proactive process used by M&E organizations is an Internal Hazard Reporting System of some sort. This system allows employees to inform management of hazards in the work place. These types of systems have been used in M&E for decades but were initially aimed at personal injury safety. Now these systems should also be used for reporting safety of flight hazards in M&E.

First, let us briefly discuss the difference between *confidential* reporting versus *anonymous* reporting. Confidential means that you do not release the information that was reported. Anonymous means you do not pair the reporter's name with a report. As we often want to make the reporting information available to many, so that we can all learn from it, internal hazard reporting systems promise anonymity, not confidentiality.

An effective Hazard Reporting System should include the following:

- Various reporting methods
  - Paper and pencil
  - Telephone
  - Intranet or
  - Texting
- A policy that encourages the reporting of hazards
- Training on how to use the system
- Training on what type of hazards to report—typically personal injury hazards and safety of flight hazards relevant to the job
- A shared staff/management responsibility to promote the system
- A process that provides absolute protection regarding anonymity
- A process to provide feedback to the reporter
- A process for secure deidentification of confidential reports
- A tracking process of action taken in response to reports
- A communication process to ensure that relevant personnel are informed of potential operating hazards through dissemination of deidentified report information

A common process for an Internal Hazard Reporting System is as follows:

1. A report is received by the organization in charge.
2. The reporter is given feedback that the report was received.
3. The reporter is contacted to get additional information, if needed.
4. The organization in charge contacts the relevant organization about the issue and asks for resolution of it. Some action is taken or a decision is made not to carry out any action.
5. The reporter is given feedback on the action taken.
6. The report is *closed*.
7. Then the reporter's name is removed from the report/dataset.
8. The data are summarized for relevant uses, such as Tool Box briefings, reports to upper management, and so on.

Internal Hazard Reporting Systems have been in use for many years in M&E organizations. They began with a paper form and a box to drop it in and have morphed to computerized and texting systems to keep up with the latest forms of communication. The main issue today with regard to SMS is getting mechanics to report safety-of-flight hazards as well as personal injury hazards.

Maintenance Line Operations Safety Assessment

M-LOSA is based upon the Threat and Error Management (TEM) model. It was developed by an ATA working group following the very successful use of the pilot LOSA process. It is an observation process in which mechanics observe other mechanics carrying out maintenance tasks. The intent is to discover procedural and systemic threats and errors (hazards in SMS terminology) that could lead to safety-of-flight risks and personal injury risks. In the overall M-LOSA process, observations are first carried out to find the threats and errors, then corrective actions are implemented to address the most significant threats and errors, and then further observations are made to determine whether the corrective actions were successful in reducing the threats and errors.

The following are the M-LOSA key characteristics:

1. *Fly on the wall* observations by *peers* during normal maintenance operations.
2. Joint management/maintenance staff sponsorship.
3. Voluntary crew participation.
4. Deidentified, confidential and safety-minded data collection.
5. Targeted observation instrument (M-LOSA Observation Form).
6. Trusted, trained, and calibrated observers who come from the maintenance staff.
7. Data verification roundtables to find data inaccuracies—for example, due to conflicting beliefs about existing policies/processes/procedures.
8. After a series of observations, the data are analyzed and presented to crews and management.
9. Data-derived targets for improvement.
10. Feedback of results to the maintenance crews.

The following steps are recommended to implement the M-LOSA process:

1. Form an initial development team/M-LOSA Steering Committee made up of management, safety staff, and maintenance staff. This committee will handle planning, scheduling, observer support, and data verification.
2. Gather information from other companies using M-LOSA and get their input on benefits, their process, and implementation issues.
3. Modify the observation forms, if desired.
4. Train all maintenance staff on Maintenance M-LOSA and its characteristics. Emphasize that M-LOSA is not for disciplinary purposes and that the observation data collection forms process guarantees the anonymity of those being observed.
5. Select and train credible and respected M-LOSA observers from among the maintenance staff.
6. Identify problem areas to observe.
7. Determine how many maintenance tasks to observe and where.
8. Schedule the observation dates.
9. Carry out the observations. Observers will record and code:
   1. Threats.
   2. Whether the threats were managed or mismanaged.
   3. Errors.
   4. The outcomes of the errors.
   (NOTE: Observers do not intervene unless there is an imminent safety issue or if an aircraft would be flown in a nonairworthy condition.)
10. Carry out data verification, as necessary.
11. Analyze the data.
12. Write a report and give to the Steering Committee, management, maintenance training organization, maintenance standards organization, and maintenance safety. The report is to list problems, not solutions.
13. Develop enhanced policies, procedures, training, and so on based on the findings.

The M-LOSA observation form consists of 10 sections:

Demographics
A. Planning
B.1 Prepare for removal
B.2 Removal
B.3 Prepare to install
B.4 Install
B.5 Installation test
B.6 Close-up/complete restore
C. Fault isolation/troubleshooting/deferral
D. Servicing

The first section of the form is for demographic information. Some of this could be collected through observation, but other information in this section would be collected through a brief interview with the mechanic following the observation. Collection of demographic information is not required but could be useful.

Section A is for Planning. Then Sections B.1 through B.6 cover the full extent of a *remove and replace* task. As the forms were designed so that an observer could begin the observation anywhere in the task, there is great overlap among these forms (e.g., "Personal protective equipment used" is on each of these six forms). An example of part of the *B.2 Removal* form is shown in Figure 4.15 to

B.2 Removal

Observation number: _____   ☐ Did not observe this section

		Safety risk N/A, Safe (S), At Risk (AR), Didn't Observe (DNO)	Threat code (See Threat Codes List)	Threat effectively managed Y/N	Error outcome 1.Inconsequential 2.Undesired state 3.Additional error & Remarks
	Safety				
1	Notes, cautions, and warnings reviewed				
2	Notes, cautions, and warnings followed				
3	Personal protective equipment (PPE) used				
4	Collective protective equipment (e.g., yellowblack streamers, flags) used				
5	Personnel use correct manual handling, ergonomics (e.g., proper lifting techniques)				
6	Personnel exhibit appropriate work behavior (e.g., no "horseplay")				
	Personnel				
7	Required personnel available				
	Procedures				
8	Current documentation (e.g., task cards, AMM, service bulletins) available and reviewed				
9	Task identified				
10	Task prioritized				
11	Task delegated				
12	Removal procedures followed				
	Threat management				
13	Strategies developed for identified threats				
14	Generated non-routines for work-not-specified in the tech publications				

FIGURE 4.15   A screen shot of part of the B.2 removal M-LOSA observation form.

give the reader an idea of what the forms look like. Finally, there are separate forms for observing fault isolation/troubleshooting/deferral and for servicing.

As shown in Figure 4.15, the observer is to record for each line of the form (e.g., "1. Notes/cautions/warnings reviewed") whether that observation is "N/A, (Not applicable)," "Safe, that is, done correctly," "AR (At Risk), that is, not done correctly," or "DNO (Didn't Observe)," which means it was an applicable thing to observe, but it was not observed. Then there are columns for the Threat Code, whether the threat was effectively managed (Yes or No), and, if it was not effectively managed, what the outcome of the error was (inconsequential, undesired state, or caused an additional error).

The Threat Codes are on an additional set of forms. The categories of threats include the following:

A. Information
B. Equipment/tools/safety equipment
C. Aircraft design/configuration/parts
D. Job/task
E. Knowledge/skills
F. Individual factors
G. Environment/facilities
H. Organizational factors
I. Leadership/supervision
J. Communication
K. Quality control
L. Other contributing factors

You will note that the threat codes, except for K and L, are exactly the same as the MEDA Results Form contributing factors categories. Figure 4.16 is a screen shot from the Threat Code form.

Mx/B. Equipment/Tools/Safety equipment

Mx/B1. Defective
Mx/B2. Unsafe
Mx/B3. Unreliable
Mx/B4   Layout of controls or displays
Mx/B5. Miscalibrated
Mx/B6. Unavailable
Mx/B7. Incomplete
Mx/B8. Inappropriate for the task
Mx/B9. Cannot use in intended environment
Mx/B10. No instructions
Mx/B11. Too complicated
Mx/B12. Incorrectly labeled
Mx/B13. Incorrectly used (including unsafely)
Mx/B14. Inadequate
Mx/B15. Not used (e.g., personal protection equipment)
Mx/B16. Other (explain below)

Mx/C. Aircraft design/Configuration/Parts

Mx/C1. Complex
Mx/C2. Inaccessible
Mx/C3. Aircraft configuration variability
Mx/C4. Parts unavailable
Mx/C5. Parts incorrectly labeled/certified
Mx/C6. Easy to install incorrectly
Mx/C7. Parts not used
Mx/C8. Other (explain below)

FIGURE 4.16   Screen shot of a portion of the M-LOSA threat code form.

M-LOSA is a relatively new hazard identification process. It was modeled on the hugely successful LOSA for pilots. The M-LOSA materials are available on the FAA website www.humanfactors-info.com.

MAINTENANCE FATIGUE RISK-MANAGEMENT SYSTEM

Fatigue is recognized as a contributor to error in the aviation industry. Pilots and flight attendants have to conform to rigorous duty time limitations to be alert while flying. However, on the maintenance side, regulators are not taking a duty time limitation approach with regard to maintenance technician error. Rather they are taking a Fatigue Risk-Management System (FRMS) approach. EASA and TC now require that M&E organizations implement an FRMS. The U.S. FAA recommends such a program (Avers and Johnson, 2011). The FAA has a significant amount of FRMS available on the website at www.humanfactorsinfo.com.

A successful FRMS would contain the following components:

- Policies and procedures
- Responsibilities
- Training and education
- Controls and action plans
- Provide sufficient sleep opportunity
- Assess actual sleep
- Assess symptoms of fatigue
- Fatigue proofing
- Report incidents and accidents.

Policies and procedures: For this component, you would outline the commitment of high-level organizational management to manage fatigue-related risk and write detailed procedures for managing fatigue at the operational level.

Responsibilities: For this component, you would
- List personnel responsible for FRMS design, implementation, and maintenance.
- Document organizational responsibilities
  - Comply with any regulations/legislation
  - Develop policies
  - Provide training and education
  - Develop error/incident reporting systems
  - Assess work schedules/tasks for fatigue-related risk.
- Document individual responsibilities
  - Use nonwork time away from work to get adequate sleep—*fit for duty*
  - Report potential risks to manager if feeling fatigued
  - Report fatigue-related errors and incidents.

Training and education: For this component
- Employees are provided training on the organization's fatigue management policies and procedures.
- Employees are trained on sleep and fatigue and how to identify and manage risks associated with fatigue at both a personal and an organizational level.
- Managers and employees are trained on their responsibilities in managing fatigue.
- Managers are trained on their responsibilities and how to implement appropriate fatigue-reduction strategies where necessary.

Controls—Provide sufficient sleep opportunity: For this component you would
- Assess work schedules for adequate sleep opportunity.
- Use fatigue modeling to help develop schedules.

- Consider ways to maximize sleep opportunity.
  - Limit night shifts.
  - Restrict shift length to 12 hours or less.
  - Limit early morning starts.
  - Limit extended duty/overtime.
  - Ensure appropriate breaks.
  - Create a napping policy and napping facilities.

Controls—Assess actual sleep: For this component you would
- Provide employees with methods for assessing whether they have gotten adequate sleep before coming to work.
- Assess whether policies to provide adequate sleep are working.

Controls—Assess symptoms of fatigue: For this component you would
- Include fatigue-related symptom checklists within the FRMS.
- Assess employees for fatigue-related symptoms.
- Counsel employees regarding major sleep disorders:....
  - Insomnia
  - Sleep apnea
  - Restless leg syndrome
  - Periodic limb movement disorder

Controls—Fatigue proofing: For this component, you want to target the areas of highest fatigue in the schedule with fatigue-proofing strategies such as
- "Double checking" to increase probability of finding errors
  - Close supervision
  - Task rotation
  - Checklists
  - Working in pairs
- Work environment
  - Self-selected breaks
  - Appropriate break facilities and healthy snacks
  - Good lighting
  - Temperature control
  - Carpooling
- Scheduling less complex or less safety-critical tasks
  - High risk/highly complex activities done during the day
  - Rotate tasks
  - Avoid boring tasks at night
- Training programs
  - Fatigue awareness
  - How to maximize sleep and alertness
  - Information for families on facilitating sleep at home
  - Impact of food and water on alertness
  - Appropriate use of stimulants such as caffeine, NoDoz, and energy drinks.

Controls—Report incidents and accidents: For this component, you would develop an incident investigation process that includes fatigue-related questions such as
- Length of duty time prior to work task.
- Length of time awake since last "major" sleep period (>two hours) prior to work task.
- Length of last "major" sleep period (>two hours).
- Total length of any nap(s) since last "major" sleep period.
- Total amount of sleep in the 24 hours prior to the work task (including naps).
- Total amount of sleep in the 48 hours prior to the work task (including naps).
- Time periods worked two days prior to day of task.

Folkard (2003), a well-known sleep expert from Wales, wrote CAA Paper 2002/2006 entitled *Work Hours of Aircraft Maintenance Personnel*. In it, in addition to a lot of information on sleep and fatigue, he provides a suggestion as to what a Fatigue Policy might look like for an M&E organization. A brief description of this Policy is presented in the following.

Daily working hour limits
- No scheduled shift should exceed 12 hours.
- No shift should be extended beyond a total of 13 hours by overtime.
- A minimum rest period of 11 hours should be allowed between the end of a shift and the beginning of the next. This should not be compromised by overtime.

Weekly working hour limits
- Scheduled work hours should not exceed 48 hours in any period of seven successive days.
- Total work, including overtime, should not exceed 60 hours or seven successive work days before a period of rest days.
- A period of rest days should include a minimum of two successive rest days continuous with 11 hours off between shifts (i.e., a minimum of 59 hours off). This rest time should not be compromised by overtime.

Annual working hour limits
- Whenever possible, the aim should be for a total of 28 days of annual leave. This should not be reduced to less than 21 days of annual leave by overtime.

Breaks
- maximum of four hours of work before a break.
- A minimum break period of 10 minutes plus 5 minutes for each hour worked since the start of the work period or the last break.

Night shift limits
- A span of successive night shifts should be limited to six for shifts of up to eight hours long, four for shifts of over 8- to 10-hour long, and two for shifts of over 10 hours. These limits should not be exceeded by overtime.
- A span of night shifts should be immediately followed by a minimum of two successive rest days continuous with 11 hours off between shifts (i.e., a minimum of 59 hours off) and this should be increased to three successive rest days (i.e., 83 hours off) if the preceding span of night shifts exceeds three (or 36 hours of work). These limits should not be compromised by overtime.
- The finish time of the night shift should not be later than 0800 a.m.

Limits on morning/day shifts
- Should not start before 0600 a.m. Should start between 0700 and 0800 a.m.
- A span of successive morning or day shifts that start before 0700 a.m. should be limited to four, immediately following that there should be a minimum of two successive rest days continuous with the 11-hour off between shifts (i.e., a minimum of 59-hour off). This limit should not be compromised by overtime.
- Whenever possible, aircraft maintenance engineers should be given at least 28 days' notice of their work schedule.

Additional recommendations
- Employers should develop risk-management systems to control fatigue.
- Educational programs should be developed to increase the awareness of problems associated with shift work.
- Personnel should be required to report for duty adequately rested.
- Personnel should be discouraged or prevented from working a second job during the work week or working another job on days off.

Where to go for information
- TC
  - http://www.tc.gc.ca/eng/civilaviation/standards/sms-frms-menu-634.htm
- US FAA
  - *www.mxfatigue.com*
- American Academy of Sleep Medicine
  - http://www.aasmnet.org/
  - http://www.sleepeducation.com/

SAFETY CULTURE/JUST CULTURE

The concepts of safety culture and just culture became known in the industry following work by James Reason, Patrick Hudson, David Marx, and the FAA's Christopher Hart.

Professor James Reason (1997, 1998) of Manchester University in England published several books/articles in this area. Hudson of Leiden University in The Netherlands worked with Professor Reason in this area.

In the early 2000s, Christopher Hart of the U.S. FAA led an attempt, called the Global Aviation Information Network, to develop a globally integrated aviation safety database that included all safety-related data in any format and could be searched by then undeveloped search engines that could comb through mounds of disparate data looking for important safety information. The integrated database was never realized. However, the Global Aviation Information Network industry group published a document in 2004 entitled "Safety Culture—The Way Ahead?: Theory and Practical Principles" that helped spread the word on safety culture and just culture.

Finally, David Marx, who helped develop the Boeing MEDA process in the mid-1990s, started his own consulting business, now called Outcome Engenuity, in the area of just culture using a legal model of determining how to deal with bad outcomes due to errors, risk-taking behavior, and reckless behavior. This is a proprietary process, so we will not discuss it further in this chapter. However, David's Just Culture work has had an important impact on helping organizations (aircraft maintenance, hospitals, etc.) deal with bad outcomes in a just (fair) way. His work can easily be found on the internet.

As Professor James Reason had the most impact in this field from an aircraft maintenance perspective, we will use his theory of what a safety culture and what a just culture are. A safety culture is an organizational culture in which safety plays a major role (as in the airline industry). A safety culture is composed of five different elements:

1. *Informed culture*: A culture in which those who manage and operate the system have current knowledge about the human, technical, organizational, and environmental factors that determine the safety of the system as a whole.
2. *Reporting culture*: A culture in which people are willing to report errors and near misses.
3. *Learning culture*: A culture in which people have the willingness and competence to draw the right conclusions from its safety information system, and the will to implement major reforms when the need is indicated.
4. *Just culture*: A just culture is that in which not only an atmosphere of trust is present and people are encouraged or even rewarded for providing essential safety-related information, but there is also a *clear line between acceptable and unacceptable behavior.*
5. *Flexible culture*: A culture that has organizational flexibility typically characterized as shifting from the conventional hierarchical structure to a flatter professional structure.

As it is somewhat hard to *regulate* the informed, learning, and flexible culture aspects of a safety culture, the focus has been on the reporting culture and just culture aspects of the safety culture.

On the reporting side of this culture, because of SMS, there is a need for maintenance organizations to have hazard reporting systems for their employees to use. These systems are in use and were discussed in the *Internal Hazard Reporting System* section. The need for a Just Culture to support a Reporting Culture is now understood by many in the industry. At this time, EASA is moving toward regulating a Just Culture in the M&E world. The major issue in implementing a Just Culture focuses around determining *the clear line* between acceptable and unacceptable behavior.

FAA made great strides in error reporting and just culture with the ASAP (FAA, 2002) and with the FAA Compliance Philosophy (FAA, 2015), Order 8000.373. These programs promote voluntary reporting in an open and just manner. The Compliance Philosophy chooses traditional enforcement as the last measure to help achieve regulatory compliance.

RESEARCH FINDINGS REGARDING MAINTENANCE HUMAN FACTORS ISSUES

DOCUMENTATION AND FAILURE-TO-FOLLOW PROCEDURES

Although the main concept being addressed here is failure-to-follow procedures (FFP), one of the reasons proposed for FFP is that the procedures are not good enough to use. So, we will first discuss research on documentation followed by research on FFP.

Documentation

In the late 1990s, there was concern expressed by John Goglia, NTSB Board Member and former mechanic, who thought that aircraft maintenance manuals (AMMs) were not well written, contained errors, and had not been validated. This led the FAA Office of Aviation Research to give a contract to Alex Chaparro and Loren Groff of Wichita State University to study this issue. The study was carried out over three phases beginning in 2001 (see Chaparro and Groff, 2001a, 2001b, 2002).

Phase 1 examined aviation industry procedures for developing maintenance technical data. Phase 2 documented user problems with maintenance technical data. Phase 3 identified improvements to the development of maintenance technical data. In Phase 1, five aircraft manufacturers were surveyed regarding company policy, communication, data tracking, user feedback, and error reduction efforts. The five industry participants represented both regional and large commercial transport manufacturers. Phase 1 survey results found three maintenance technical data issues:

1. Inconsistent development process guidelines
2. Reactive rather than proactive response to user feedback
3. Inadequate assessment of errors involving usability as opposed to accuracy.

In Phase 2 of the research, Chaparro and Groff used a questionnaire to gather information from aircraft technicians about their opinions regarding errors in current AMMs, AMM usage rates, and general manual quality. On-site interviews of technicians were also conducted to gather feedback about the types of problems encountered with manuals, the associated impact, and suggestions for improving manuals. Technicians surveyed worked on a range of aircraft from smaller corporate jets to large Boeing aircraft. Chaparro and Groff found that, although user evaluations of the accuracy and quality of technical manuals are generally good, the technicians rated manuals as having poor usability. This finding supported the need for a higher level of user involvement during the document development process.

In Phase 3 of the research, Chaparro and Groff outlined a series of recommendations to address problem areas identified in Phase 1 and 2. They recommended that

1. Manufacturers and operators improve communication between technicians submitting change requests and technical writers to ensure prompt feedback of actions
2. That maintenance procedures be validated using standard HF techniques

3. That the industry cooperate in the development of a system akin to MSG-3 (Maintenance Steering Group—3, which is a logic process used by aircraft OEMs for determining what maintenance tasks need to be carried out and when in order to maintain aircraft airworthiness) for identifying maintenance procedures that should be systematically validated

4. That manufacturers maintain databases with a history of user reported errors, feedback to the user, and actions taken. By tracking the history of user error reports, manufacturers can target maintenance procedures for validation that have the greatest potential impact on safety or economics

5. Finally, they provided an example (using the MSG-3 process) of how the aforementioned recommendations could be implemented.

Some of the specific survey findings from these studies are presented next in the "Failure-to-Follow Procedures" section.

Failure-to-Follow Procedures

Failure-to-follow procedures (FFP) is one of the leading causal factors leading to maintenance-caused events. It is also the major cause of rule violations by Aircraft Maintenance Technicians (AMTs) that leads to an FAA Letter of Investigation. This section will summarize the findings on FFP as it relates to rule violations and maintenance-caused events. Then data on a survey regarding the quality of maintenance procedures will be presented. Then some data on the reasons why AMTs do not follow procedures will be presented.

(Note: In this section, *procedures* refers to work instructions used by an AMT to carry out a maintenance task. This would typically include the AMMs provided by the OEM/Design Approval Holders (OEM/DAHs), Component Maintenance Manuals provided by the OEM/DAHs and their vendors, and work instructions written by the M&E function of an airline (or, less likely, by their MRO organizations), including Task Cards and Engineering Orders (also referred to as Engineering Authorizations). The term procedure will not refer to M&E processes, such as a required process to do a tool inventory or clean up the area after completion of a task.)

Research on Failure-to-Follow Procedures

A 2002 report entitled "Root Cause Analysis of Rule Violations by Aviation Maintenance Technicians" (Patankar, 2002) for the FAA Office of Aviation Medicine analyzed 1,555 cases from FAA's records regarding rule violation cases that were closed between January, 1998, and December, 2000. The most common rule violations involved are as follows:

- § 43.13(a): Each person performing maintenance ... shall use the methods, techniques, and practices prescribed in the current manufacturer's maintenance manual .... He shall use the tools, equipment, and test apparatus necessary to assure completion of the work in accordance with accepted industry practices.
- § 43.13(b): Each person maintaining or altering, or performing preventive maintenance, shall do that work in such a manner and use materials of such a quality that the condition of the aircraft ... will be at least equal to its original or properly altered condition ....
- § 43.13(c): ... the methods, techniques, and practices contained in the maintenance manual ... constitute acceptable means of compliance with this section.
- § 43.9: (a) *Maintenance record entries* ... each person ... shall make an entry in the maintenance record of that equipment containing the following information:
- (1) A description ... of work performed.
- (2) The date of completion of the work performed.
- (3) The name of the person performing the work ....

- (4) If the work ... has been performed satisfactorily, the signature, certificate number, and kind of certificate held by the person approving the work.
- § 43.16—Each person performing an inspection or other maintenance specified in an Airworthiness Limitations section ... shall perform the inspection or other maintenance in accordance with that section ....

The findings from Patankar (2002) are show in Table 4.5. As can be seen, a majority of the violations were of (538 of 944 or 57%) were of § 43.13(a), which means that the AMT did not follow the procedures from the perspective of using the methods, techniques, and practices or did not use the tools, equipment, and test apparatus specified in the procedure.

In the mid-1990s, The Boeing Company, working with nine of its customer airlines, developed a process for investigating the causal factors related to maintenance-caused events called the MEDA. The MEDA philosophy is that CFs in the workplace lead to maintenance errors (e.g., installation errors, servicing errors, and inspection errors), which, in turn, lead to maintenance events (e.g., flight delays, flight cancellations, and aircraft damage). In 1998–2000, the FAA Office of Medicine funded a research study (Johnson and Watson, 2001) in which three Part 121 M&E organizations used the MEDA process and provided the MEDA findings for publication. In all, 125 MEDA investigations were carried out and provided to the researchers.

The study found that the most common category of CFs to maintenance error was *Information*, which refers to the documentation that AMTs use to carry out a maintenance task—48 (38%) of the 125 investigations found that information contributed to the error that caused the investigated event. In 24 (50%) of the investigations, the specific CF was that the information was *not used*, in 12 (25%) the specific CF was that the information was *unavailable*, in 11 (23%) the specific CF was that the information was *not understandable*, and in 6 (13%) the specific CF was that the information was *incorrect*.

A U.S. major airline provided data to a Flight Safety Foundation committee that the first author served on. They provided 191 of their MEDA investigations. In 47 (25%) of the 191 investigations, the category of information was causal. In 15 (32%) of the 47 investigations, the specific CF was *not used*, in 13 (28%) the specific CF was *inadequate*, in 3 (6%) the specific CF was *not understandable*, in 3 (6%) the specific CF was *incorrect*, in 2 (4%) the specific CF was *unavailable/inaccessible*, in 1 (2%) the specific CF was *too much/conflicting information*, and in 1 (2%) the specific CF was *uncontrolled*.

In 2004, the United Kingdom (UK) Flight Safety Committee published a report that listed the top 10 causes of maintenance mishaps. These causes were as follows (from most frequent to least frequent):

TABLE 4.5

Frequency of Aircraft Maintenance Technician Rule Violations Closed between January, 1998 and December, 2000, as a Function of Regulation and Type of Maintenance Organization

Regulation	121 Frequency	145 Frequency
§ 43.13(a)	538	44
§ 43.13(b)	249	22
§ 43.13(c)	58	0
§ 43.9(a)	44	8
§ 43.16	55	4
Total of Above	944	78
Total of All	944	86

---

1. Failure-to-follow published technical data or local instructions
2. Using unauthorized procedure not referenced in technical data
3. Supervisors accepting nonuse of technical data or failure-to-follow maintenance instructions
4. Failure-to-document maintenance properly in maintenance records or maintenance work packages
5. Inattention to detail/complacency
6. Incorrectly installed hardware on an aircraft/engine
7. Performing an unauthorized modification to the aircraft
8. Failure to conduct a tool inventory after completion of the task
9. Personnel not trained or certified to perform the task
10. Ground support equipment (GSE) improperly positioned for the task.

Aforementioned items 1, 2, and 6 basically state that the AMT did not follow the procedures. Item 3 implicates AMT supervisors for accepting the fact that the AMT did not use the technical instructions and/or did not follow the technical instructions.

Jim Burin of the Flight Safety Foundation carried out some research from an FAA database to develop some high-level points on *not following procedures*. He found that

- Procedures are not referenced (used) by the AMT because
  - Procedures are not available.
  - Procedures are available, but not used during the work process.
- Procedures exist but are not implemented correctly because
  - Procedures are misunderstood.
  - Procedures are understood but work process was not followed exactly as written.

The earlier data suggest some of the reasons that AMTs do not follow procedures—they simply do not get (print out or copy) the procedure and use it, the procedure is unavailable, there is a problem with the procedure (not understandable/incorrect/inadequate), and the supervisors accept the fact that the AMT is not using/following the procedure.

Chaparro and Groff (2001a, 2001b, 2002) presented their findings on a three-part evaluation of aviation technical manuals carried out for the FAA's Office of Aviation Research. As part of this research, they conducted a survey with AMTs. Approximately 296 surveys were filled out by AMTs who worked on large aircraft (Boeing, Airbus, and McDonnell Douglas) and approximately 81 were filled out by AMTs who worked on smaller aircraft (Cessna and Embraer). AMTs were asked

1. "How often have you found an error in text descriptions or procedures?" Responses were *Never* (12%), *Very Rarely* (42%), *Occasionally* (35%), *Often* (10%), and *Very Often* (1%)
2. "How often have you found an error in illustrations?" Responses were *Never* (14%), *Very Rarely* (49%), *Occasionally* (29%), *Often* (7%), and *Very Often* (1%)
3. "How often have you found an error in diagrams?" Responses were *Never* (17%), *Very Rarely* (49%), *Occasionally* (27%), *Often* (6%), and *Very Often* (1%).

A majority of AMTs Never or Rarely found errors in the technical manuals.

AMTs were also asked "How would you rate the following aspects of AMMs?" on a five-point scale from Very Good through Neutral to Very Poor. The questions referred to the following:

1. *Usefulness of information*: *Very Good* (17%), *Good* (57%), *Neutral* (19%), *Poor* (6%), *Very Poor* (1%)
2. *Quality of illustrations*: *Very Good* (13%), *Good* (43%), *Neutral* (27%), *Poor* (14%), and *Very Poor* (3%)

3. *Quality of diagrams*: *Very Good* (11%), *Good* (43%), *Neutral* (30%), *Poor* (12%), *Very Poor* (4%)
4. *Clarity of text descriptions*: *Very Good* (10%), *Good* (37%), *Neutral* (33%), *Poor* (16%), *Very Poor* (4%).

A plurality to majority of AMTs rated these four aspects of AMMs as *Very Good* or *Good*.

One area of rating of the maintenance instructions that did not score as positively had to do with whether the AMMs described the easiest way to do a procedure and whether the AMM author understands how to do maintenance. Using a five-point scale from Strongly Agree through Neutral to Strongly Disagree

1. *The maintenance manual describes the easiest way to conduct each procedure.*: *Strongly Agree* (2%), *Agree* (16%), *Neutral* (35%), *Disagree* (38%), *Strongly Disagree* (9%)
2. *The writers of the manual understand the way I do maintenance procedures.*: *Strongly Agree* (1%), *Agree* (12%), *Neutral* (33%), *Disagree* (35%), *Strongly Disagree* (19%)

Thus, although a plurality to majority of AMTs do not usually find errors in the manual and find the major aspects of the manuals to be good, a majority of the AMTs do not think that the manuals are written in a way that describes the easiest way to do the task, possibly because the manual authors do not understand how to carry out maintenance procedures in the way that AMTs would like to carry out the procedure.

There is one study regarding a survey of AMTs asking them why *people* do not follow procedures. It was carried out by Human Reliability Associates Ltd. from England and is cited in the United Kingdom Civil Aviation Authority publication CAP 716 (United Kingdom Civil Aviation Authority, 2003). AMTs from England were asked to agree or disagree with the statements such as *Procedures are not used because they are inaccurate*. Table 4.6 provides the results of this survey. (It is important to note that AMTs were not simply asked why they did not follow procedures.)

Issues such as "they are inaccurate," Practicality, Optimization, and Presentation are mostly related to how the procedure is written by the OEM/DAH, vendors, and M&E publications staff. Issues regarding "they are out-of-date," Accessibility, and Policy are related to M&E policies and processes regarding procedure use. The Usage issues seem more related to AMT attitude. The highest agreed to statements dealt with how long it would take if the procedures were followed *to the letter*, people preferring to rely on their own skills, and people assuming that they know what the procedure is (so they do not have to use the documentation).

Summary

FFP was the major reason for FAA Letters of Investigation regarding rule violations. The AMTs either did not use the methods, techniques, and practices prescribed in the current manufacturer's maintenance manual or the tools, equipment, and test apparatus called out in the maintenance manual. Documentation issues were also the most common contributing factors to AMT errors that lead to aircraft events. Most often the specific contributing factor was that the AMT did not use the procedure, but problems with procedures (unavailable/incorrect/inadequate). In 2004, a UK Flight Safety Committee found that FFP, using unauthorized procedures, and supervisors accepting this were the top three causes of maintenance mishaps. In the early 2000s, a survey of AMTs found that maintenance manuals did not often contain mistakes and that the major aspects of the manuals were good, the manuals were often not written in the way an AMT would carry out the task, possibly because the manual authors did not understand how AMTs carry out procedures. Finally, in a survey in the UK of AMTs regarding "why people don't follow procedures," issues regarding how the procedures are written, procedure accessibility and company policy on the use of procedures, and AMT attitudes about using procedures were the leading reasons given for not using procedures to carry out a maintenance task.

TABLE 4.6

Percentage of AMTs in England Agreeing with a Statement about Why "People Don't Follow Procedures"

Procedures Are Not Followed Because...		% Agreeing
Accuracy	... they are inaccurate	21
	... they are out-of-date	45
Practicality	... they are unworkable in practice	40
	... they make it more difficult to do the work	42
	... they are too restrictive	48
	... they are too time consuming	44
	... if they were followed "to the letter" the job couldn't be done in time	62
Optimization	... people usually find a better way of doing the job	42
	... they do not describe the best way to carry out the work	48
Presentation	... it is difficult to know which is the right procedure	32
	... they are too complex and difficult to use	42
	... it is difficult to find the information you need within the procedure	48
Accessibility	... it is difficult to locate the right procedure	50
	... people are not aware that a procedure exists for the job they are doing	57
Policy	... people do not understand why they are necessary	40
	... no clear policy on when they should be used	37
Usage	... experienced people don't need them	19
	... people resent being told how to do their job	34
	... people prefer to rely on their own skills and experience	72
	... people assume they know what is in the procedure	70

A Systems View of Dealing with Failure-to-Follow Procedures

Given the earlier findings about why AMTs do not always follow procedures, what can be done about it to increase procedure use? In the following, we briefly discuss a systems view on how to deal with the FFP at an M&E organization.

First, there should be a high-level General Maintenance Manual policy on the expectations regarding AMT use of procedures. These expectations could range from "do not need to refer to procedure to carry it out if you know how to do the task" through "must at least review the procedure before going out to do the task" to "must have a copy of the procedure and refer to it periodically while carrying out the task." Certainly, the "must have a copy ..." expectation is the one most likely to encourage following procedures. Some aircraft M&E organizations have also started using a Critical Behavior approach to procedure use. That is, what is expected of the AMT regarding procedures is spelled out explicitly in the Critical Behavior list, and AMTs are provided training on what this means. Moreover, the role of the supervisor/manager, and possibly the lead, should be explicitly stated with regard to them enforcing the policy on procedure use. That is, when they see that the policy on procedure use is not being followed, they should intervene and help the AMT move back to the expected behavior.

Second, the M&E organization should have processes in place for supporting AMTs in following the policy on procedure use. For example, adequate printers should be available for AMTs to print out procedures easily and quickly. Alternatively, if the procedures are on laptops, adequate laptops should be available. In addition, management needs to ensure that all of the tools, tooling, GSE, parts, consumables, and so on, to carry out the procedure need to be available to the AMT. Moreover, the organization should consider kitting remove and replace tasks, servicing tasks, and so on, such that all of the paperwork, tools, parts, consumables, and so on, are in a kit for the AMT to use.

Third, for airline M&E generated procedures (e.g., Engineering Orders and Task Cards) best practices should be used in the development of these documents. This includes the following:

- Use clear, precise writing style
- Use brief and to the point language—use modifiers sparingly
- Use Simplified Technical English
- Avoid using acronyms and abbreviations
- Use a column format, not narrative paragraphs
- Make each procedural step a single, least-operation step (use only one action verb per statement)
- List procedural steps in the proper sequence
- Use a simple numbering system
- Build in checklists where useful
- Use graphics/figures where helpful
- Use graphic aids with callouts, where helpful
- Where the arrow is pointing on a figure should be immediately obvious
- Be consistent in referring to items/actions—do not change names from step to step or from a step to a figure
- Write WARNING and CAUTION statements using current international standards (American National Standards Institute or the International Organization for Standardization)
- Have a second technical person check the procedure for technical accuracy (verification)
- Get a document user (mechanic/technician) to *try out* the procedure and give feedback about usability issues (validation)
- Is each step easily understood
- Are the steps in the order in which the user would carry out the task?
- Does the procedure bring about the desired result?
- Use a fourth-to-sixth grade reading level (which is automatically achieved if you use Simplified Technical English).

Fourth, the aircraft M&E organization should have a change process that requires/allows the AMT to file a *change request* if the technician knows that the procedure is incorrect or cannot be followed as written. If this was an M&E written procedure, then an M&E process would exist and be followed to fix the procedure problem. If the procedure was written by the OEM/DAH, then the organization should have a process in place for contacting the OEM/DAH to get the procedure fixed. To support the AMT in finishing the task, when there is a problem with the procedure that must be fixed, is to have a 24/7 Engineering Help Line that is staffed with engineers who have the authority to change/fix procedures. The Help Line should have some goal (e.g., 1 hour) in which to fix the procedure and get an approved, fixed version of the procedure to the AMT in order to finish the task. Moreover, Maintenance Tips provided by airline maintenance organization to mechanics/technicians via company intranet can also be used to warn AMTs of issues that can arise during specific maintenance tasks.

Summary

FFP is one of the most common causal factors in maintenance-related incidents and accidents. Therefore, the issue should be addressed. AMTs give many reasons for not following procedures, which need to be addressed systematically through company policy and processes. The policies would specify what is expected of the AMT with regard to procedure use, and then management would be responsible for making sure that the policy is followed.

Visual Inspections

As mentioned in the beginning of the current chapter, a 737 accident due to incorrectly performed visual inspections was a major impetus to begin research and application in the field of MHF. Much of the inspection research for the FAA was carried out by Dr. Colin Drury while he was at the State University of New York (SUNY), Buffalo. A culmination of his work (Drury 2005) is found in his *Handbook of System Reliability in Airframe and Engine Inspection*. This handbook discusses the general concept of an inspection and defines and discusses all of the major types of inspections done during aircraft maintenance (e.g., visual, borescope, magnetic particle, eddy current, etc.).

The following research summary will focus only on visual inspections. The definition of visual inspections and their importance are discussed. This information comes from Dr. Rankin's knowledge of commercial aircraft maintenance and inspection. This will be followed by a discussion on the inspection process. Finally, we will summarize the Drury Handbook research findings regarding the five major factors that affect visual inspection:

1. Task factors
2. Inspector (operator) factors
3. Equipment factors
4. Environment factors
5. Social factors.

Types of Visual Inspections

More than 80% of inspections on aircraft are visual inspections. Visual inspection is often the most economical and fastest way to find defects on an aircraft. Airframe manufacturers and airlines depend on regular visual inspections to ensure the continued airworthiness of their aircraft. Typical airframe defects found using visual inspection include cracks, corrosion, and disbonding. Other defects that can be found by visual inspection include system and component wear, accidental damage, and environmental damage from long-term storage, sunlight, saltwater, and so on.

Visual inspection means inspection using either or all of human senses such as vision, hearing, touch, and smell. Visual Inspection typically means inspection using raw human senses and/or any nonspecialized inspection equipment. There are three major types of inspections used during aircraft maintenance:

1. General visual inspection (GVI)
2. Detailed inspection (DET)
3. Special detailed inspection (SDI)—these inspections involve specialized inspection equipment (e.g., borescope and NDT) and will not be discussed.
   Definition of GVI: A visual examination of an interior or exterior area, installation or assembly to detect obvious damage, failure or irregularity. This level of inspection is *made from within touching distance* unless otherwise specified. A *mirror* may be necessary to enhance visual access to all exposed surfaces in the inspection area. This level of inspection is made under normally available lighting conditions such as daylight, hangar lighting, *flashlight* or *drop-light* and may require removal or opening of access panels or doors. Stands, ladders, or platforms may be required to gain proximity to the area being checked.
   Definition of DET: An intensive visual examination of a specific structural area, system, installation, or assembly to detect damage, failure, or irregularity. Available lighting is normally supplemented with a direct source of good lighting at an intensity deemed appropriate by the inspector. Inspection aids, such as mirrors, magnifying lenses, and so

on, may be used. Surface cleaning and elaborate access procedures may be required. So, the main way that the DET differs from GVI is that the inspector is told exactly where to go and what to look for and can use magnifying lenses, if appropriate.

Operating rules for GVI and DET: Normal cleanup procedures are to be used prior to conducting GVI or DET. Specified cleanup procedures are to be used for special detailed inspections, if necessary. Sealant and corrosion protection finishes should only be removed when specified and restored in accordance with the Corrosion Prevention Manual after the task is completed. Excessive dust, debris, or overspray of corrosion inhibiting compounds found during any inspection are considered to be an unsatisfactory condition possibly reducing the fire resistance of the airplane design. Cleanup of these materials should be a standard part of maintenance activity.

Inspection Process

Visual inspection involves both a *search* and a *decision* function. During the *search* function, the inspector must move their eyes around the item to be inspected. The area around the line of sight in which it is possible to achieve detection is called the *visual lobe.* The size of the visual lobe depends on the defect and background characteristics, illumination and the inspector's peripheral visual acuity. However, *the average visual lobe is around 0.9 inches (2.3 centimeter) vertical and 1.2 inches (3.0 centimeter) horizontal at 20 inches (50 centimeter) distance.* That means when an inspector is doing a GVI at 20 inches (50 centimeter), as defined, he/she can only achieve detection in an area of approximately 1 inches by 1 inches (2.5 by 2.5 centimeter) per eye fixation. Successive fixations of the visual lobe occur about three seconds. Thus, the size of the area being inspected directly affects the search time (limited time on a large area brings the speed/accuracy tradeoff into play). Systematic searches are always more efficient than random searches.

The second part of the inspection is the *decision* function. When the inspector sees an anomaly, he/she must make a decision, which has one of four possible outcomes as shown in Figure 4.17. Perfect inspection would mean that the inspector never misses a true defect and never calls a defect when it does not exist.

How do we measure how good an inspector is? We typically measure how good an inspector is by the Hit rate or the True Positive rate. "The inspector found 97% of the defective parts. He's a good inspector." But Hit rate by itself is not a good indicator of inspector performance. For example, what strategy could I take in inspection so that I NEVER missed a defect? (Answer: call everything a defect.) But, if I took that strategy, the Miss/False Negative rate would go up (and we would be *fixing* problems on the aircraft that do not exist, which is expensive).

	Actual defect?	
	Yes	No
"Defect"	Hit true positive	False alarm false positive
"No defect"	Miss false negative	Correct accept true negative

Inspector says..

FIGURE 4.17 Four possible outcomes from the decision part of a visual inspection.

So, to measure how good an inspector's performance is, we must look at both Hit rate and False alarm rate. If we look at both measures at the same time, then we can answer questions such as, which of the following two inspectors is a better inspector?

- One with a 90% Hit rate and a 40% False Alarm rate
- One with a 70% Hit rate and 10% False Alarm rate

This can be done using the Receiver Operating Characteristic (ROC) Curve, which was developed using signal detection theory.

Signal detection theory is a methodology that can be used to quantify the ability to discriminate between information-bearing energy patterns (called a stimulus in humans or a signal in machines) and random energy patterns that distract from the information (called noise). The noise comes from background stimuli, random activity of some system (such as *noise* in a phone line), or the nervous system of the human. This is illustrated in Figure 4.18. There are a number of determiners of how a detecting system (inspector) will detect a signal (defect), and where its threshold level for calling *defect* will be.

An example of how this works in an easy to understand example is provided in Figure 4.19. There is a Canadian football game happening, and the fans are coming through the turnstile. About half of the fans are female. Your job is to guess the gender of the fan in based on their height, which is given to you. In Canada the average height of women is 1.633 meter (5 feet 4 ½ inches), the average height of men is 1.760 meter (5 feet 9 inches), and the exact point of overlap of the two distributions is at 1.696 meter (5 feet 7 inches). It is, hopefully, easy to see that any person who is more than 1.696 meter is more likely male than female with the opposite for any person under 1.696 meter. The closer the person is to 1.696 meter, the more error there is in the call.

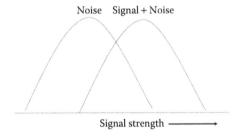

FIGURE 4.18   Signal detection theory showing the overlap between the noise distribution and the signal + noise distribution.

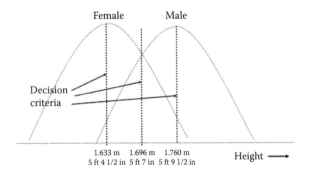

FIGURE 4.19   Which gender is coming through the turnstile based on height?

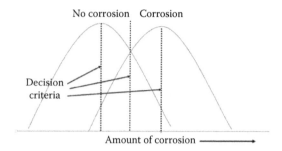

FIGURE 4.20 Overlap of the distributions for no corrosion and corrosion.

Now let us discuss how the decision applies in a more real life situation—such as doing a GVI and looking for corrosion. Suppose you have an inspector looking for corrosion. The corrosion is a bit hard to tell from no corrosion because of coloration and deterioration issues (Figure 4.20).

So, the inspector may be unwilling to call *corrosion* unless he/she is extremely certain that they are correct. As there is a penalty to pay if you call *corrosion* when none exists—additional work and may keep the aircraft out of service. In other words, the inspector's *decision criteria* may change on the basis of certain things, such as

- Training and feedback (knowledge of results and feed forward information)
- The *cost* of a missed defect
- The *cost* of calling something a defect when it is not a defect
- Pressure from management
- Pressure from working colleagues
- Whether you are inspecting your own aircraft/helicopter or a customer's.

The purpose of the aforementioned discussion is to point out two things. First, most inspection decisions are not so *clear cut* that the inspector is 100% certain that he/she is correct when they call *defect* or *no defect* because of the overlap of the *no corrosion* and *corrosion* visual distributions. Second, the decision criteria for when an inspector calls *defect* can be influenced by many different things. This also means the decision criteria can change from day to day depending on things such as management pressure (not to find defects because we have to get this plane out) and other factors.

So, how do we measure how good an inspector is? First, we have to have samples for him/her to inspect in which we know whether the sample does or does not contain a defect. Then they inspect all of the items and call *defect* or *no defect* for each item. Then we calculate their Hit or True Positive Rate and their False Alarm or False Positive Rate. Then we plot it on the ROC as illustrated in Figure 4.21. Note that the x axis is for the False Alarm Rate and the y axis is for the Hit Rate. The diagonal line runs from (0, 0) to (1, 1) is the random guess line. If you are below the line, you could do better by simply guessing. Inspector A is at point (0.27, 0.63) and Inspector C is at point (0.12, 0.77). Which one is a better inspector? We can determine this by drawing a line from (0, 0) to point A (B) and then from point A (B) on to (1, 1). Then we could determine the size of the triangles formed by the two lines and the *random guess* line. The bigger the size (maximum size is 0.5 units), the better the inspector. Note a perfect inspector is at point (0, 1) and the triangle for that would be 0.5 units. Therefore, in this example, C is a much better inspector than A.

Factors Affecting Visual Inspections

In the following, we discuss the effects of five major factors on visual inspection:

1. Task factors
2. Inspector (operator) factors

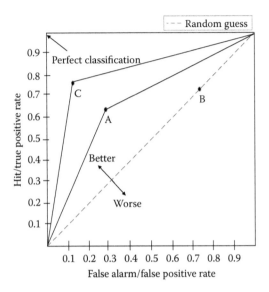

FIGURE 4.21 Receiver operating characteristic (ROC) curve.

3. Equipment factors
4. Environment factors
5. Social factors.

Task Factors

Size/complexity of the object searched: In general, search time is linearly related to either search field area or number of inspectable items in the field.

Number of different types of defects: The greater the number of types of defects, the slower the search performance and/or the lower the *hit* rate.

Defect/background contrast: Higher defect/background contrast produces faster and more accurate searches.

Defect size: Larger defects are generally found more accurately and faster than smaller defects.

Probability that the item contains a defect: The *a priori* probability that there will be a defect is positively related to hit rate and false alarm rate.

Timing/pacing: In general, a self-paced inspection task is more effective than an externally paced task.

Inspector Factors

Inspector demographics: Age, experience, and gender (no affect)

Training: Not the same as experience. Even experienced inspectors can improve their performance (often dramatically) with a well-designed training program, which is based on a task analysis and which provides Knowledge of Results to the trainees.

Visual performance: Not related to visual acuity as measured by a Snellen Chart (the typical eye chart used by your optometrist) if you can correct to 20/20. The size of the visual lobe is a predictor of inspector performance.

Cognitive performance: People with the ability to disembed objects from a complex/confusing background make better inspectors.

Inattentional blindness: The concept of inattentional blindness has been developed over the past 20 years. This helps one to explain how somebody does not see something that is *right in front of them* or in *plain sight*. It is also known as perceptual blindness. It is exploited by

illusionists and magicians. Most perceptual processing occurs outside of conscious awareness, as our senses are bombarded with a large amount of inputs (e.g., sights, sounds, and
smells). *Attention* acts as a filter to quickly examine sensory input and selects a small
percentage for full processing and conscious processing. Attention capture and awareness
capture occur at two different stages of processing.

The most famous example of inattentional blindness comes from the work of Daniel Simons and
Christopher Chabris (Chabris and Simons 2010). In a famous video there are three people in white
shirts passing a basketball to each other and three people in black shirts are passing a basketball to
each other (Figure 4.22). Your *job* is to count the number of times that people in the white shirt pass
the basketball. As you are counting, a person in a black gorilla suit casually strolls through the scene.
Around half of the subjects in these experiments do not see the gorilla even though it was *right in
front of them.*
What are the factors that influence inattentional blindness?

- *Conspicuity*: An object's ability to catch your attention.
  - *Sensory conspicuity*: Physical properties an object has, for example, bright colors,
    flashing lights, movement, and so on.
  - *Cognitive conspicuity*: Familiarity and meaning of an object to the perceiver.
- *Mental workload*: A person's cognitive resources, which interfere with processing of other
  stimuli
  - Either too high or too low (automatic processing) can be limiting
- *Expectation*: When a person expects certain things to happen, he/she tends to block out
  other possibilities
  - Experts are more prone to inattentional blindness than beginners
  - Confirmation bias strengthens expectancy effects
- *Capacity*: A measure of how much attention must be focused to complete a task
  - Can be lessened by drugs, alcohol, fatigue, and age.

Inattentional blindness and human errors/accidents. Someone performing a task simply fails to
see what should have been plainly visible and cannot explain the lapse afterward. Inattentional

FIGURE 4.22   Screen capture from Daniel Simons and Christopher Chabris study on inattentional blindness.

blindness causes accidents when attention mistakenly filters away important information, due to a combination of factors: low conspicuity, divided attention, high expectation, and low arousal. People are unaware of the blindness. Training mainly affects conscious, voluntary behavior. Inattentional blindness should be viewed as generally a good thing. It is the price we pay for the gift of attention. Without this ability to block out the irrelevant stimuli in our environment, we could not function.

Equipment Factors

Magnification: Increasing magnification may only change the speed-accuracy tradeoff with higher magnification improving target detection at the expense of speed. Not allowed in GVI.

Field integration: Using a *known perfect item* for comparison during the inspection. This improves inspection.

Environment Factors

Visual environment: Correct lighting is important both for physical inspection and to avoid glare in computer-based inspection.

Auditory environment: No necessarily clear affects, although some studies show that noise >90 Db worsened inspection performance.

Thermal environment: Some data to suggest that really hot and really cold environments have a negative effect on inspection.

Workplace comfort: No data. However, the myth that comfortable inspectors lose vigilance has been refuted.

Social Factors

Working period: Detection performance decreases rapidly over the first 20–30 minutes of a vigilance task (e.g., watching radar), although this phenomenon is hard to replicate in typical visual inspection tasks.

Job design: Rest periods have been shown to improve performance.

Supervision, instruction and other pressures: From signal detection theory, we know that criterion used by an inspector for reporting defects is influenced by the sum of all biases on the inspector. These biases are affected by a priori probability of a defect and also by the perceived costs of misses and false alarms, which can be affected by supervisory instructions/reprimands.

Information environment: Both feedback of inspection performance (knowledge of results) and feed forward information (where to look for a defect) have been shown to improve inspection performance.

Summary

More than 80% of inspections on aircraft are visual inspections. Visual inspection is often the most economical and fastest way to find defects on an aircraft. The ROC curve can be used to determine how good an inspector is. Inattentional blindness is a factor in visual inspection. There are five sets of factors that influence visual inspections: task factors, subject factors, equipment factors, environment factors, and social factors.

PROJECTIONS ON THE FUTURE OF MAINTENANCE HUMAN FACTORS

THE IMPACT OF SAFETY MANAGEMENT SYSTEMS

The SMS initiatives became formalized by ICAO (see ICAO 9859) in 2006 as a Standard and Recommended Practice regarding the implementation of an SMS in airports, air traffic control, and commercial aviation. Within commercial aviation, Flight Operations and Maintenance were initially expected to implement their portion of the SMS, although this has been extended to include dispatch, ground operations/cargo loading, and flight attendants. At the time of this publication, all NAAs have an SMS regulation. With an SMS, every organization must have an organized and documented means to identify hazards, to assess those hazards for risk, to mitigate unacceptable risk to acceptable risk, and to monitor progress in maintaining that acceptable risk. It is a new way of doing business and must promote an invigorated corporate safety culture. This has required commercial aviation maintenance organizations to implement reactive hazard investigation processes (such as MEDA), proactive internal hazard reporting processes, and, possibly, predictive hazard observation processes (such as M-LOSA).

In addition, the breadth of an SMS has recently been expanded to include a FRMS for maintenance and implementation of a Just Culture to enhance hazard reporting. So the immediate future for MHF is to implement predictive hazard identification processes (M-LOSA), a FRMS, and a Just Culture.

THE IMPACT OF HUMAN FACTORS ON UNMANNED AERIAL SYSTEMS

Although Unmanned Aerial Systems (UAS) is a critically important and significant topic, it does not alleviate the importance of the human in the system, especially with regard to maintenance. Today's small UASs are tomorrow's passenger carriers, without a doubt. Although some UAS will, no doubt, be able to fix themselves, the human maintainer will be a critical part of the aviation system. The MHF issues with regard to UAS maintenance will be part of the future of MHF.

REFERENCES

Avers, K. A. and Johnson, W. B. (2011). A Review of Federal Aviation Administration Fatigue Research: Transitioning Scientific Results to the Aviation Industry. *Aviation Psychology and Applied Human Factors.* Hogrefe Publishing. vol. 1, no. 2, 87–98. http://doi.org/10.1027/2192-0923/a000016

Aviation Safety Research Act of 1988, P.L. 100-591, 102 Stat. 3011.

Chabris, C. and Simons, D. (2010). *The Invisible Gorilla: How our Intuitions Deceive Us.* New York: Crown Publishers.

Chaparro, A. and Groff, L. (2001a). *Human Factors Survey of Aviation Technical Manuals Phase 1 Report: Manual Development Procedures.* Springfield, VA: National Technical Information Service (NTIS).

Chaparro, A. and Groff, L. (2001b). *Human Factors Survey of Aviation Technical Manuals Phase 2 Report: User Evaluation of Maintenance Documents.* DOT/FAA/AR-02/34, Springfield, VA: National Technical Information Service (NTIS).

Chaparro, A. and Groff, L. (2002). *Reliability Survey of Aviation Technical Manuals Phase 3 Report: Final Report and Recommendations.* DOT/FAA/AR-02/123, Springfield, VA: National Technical Information Service (NTIS).

Civil Aviation Safety Authority of Australia, CASA (2013). Safety behaviors: Human factors for engineers. A multi-media training system. Available at: safetyproducts@casa.gov.au.

Drury, C. G. (2005). *Handbook of System Reliability in Airframe and Engine Inspection.* Prepared for Federal Aviation Administration under Contract #2004-G-037. Available at: HFSkyway.faa.gov

Edwards, E. (1972). Man and machine: Systems for safety. *Proceedings of British Airline Pilots Association Technical Symposium* (pp. 21–36). London: British Airline Pilots Association.

Federal Aviation Administration (2000). *Maintenance Resource Management Training.* (Advisory Circular 120–72). Washington, DC: Federal Aviation Administration.

Federal Aviation Administration (2002). Aviation safety action program, advisory circular AC 120–66B, Washington, DC. Available at http://www.faa.gov/safety/programs_initiatives/aircraft_aviation/asap/policy/.

Federal Aviation Administration (2005). The operator's manual for human factors in aviation maintenance, Washington, DC. Available at www.humanfactorsinfo.com.

Federal Aviation Administration (2010). *Order 1110.110E Research, Engineering, and Development Advisory Committee.*

Federal Aviation Administration (2015). *The FAA compliance philosophy, FAA Order 8000.373.* June 26, 2015.

Folkard, S. (2003). CAA Paper 2002/06, Work Hours of Aircraft Maintenance Personnel. Available at www.caa.co.uk.

Hawkins, F. H. (1993). *Human Factors in Flight.* Aldershot, UK: Ashgate Publishing.

Helmreich, R. L., Butler, R. E., Taggard, W. R., and Wilhelm, J. A. (1995). Behavioral markers in accidents and incidents: Reference list. NASA/UT/FAA Technical Report 95-1. Austin, TX: The University of Texas.

Johnson, W. B. (1991). The national plan for aviation human factors: Maintenance research issues. *Proceedings of the 35th Annual Meeting of the Human Factors Society.* San Francisco, CA: The Human Factors Society, pp. 27–32.

Johnson, W. B. (1998). Past, present, and future of human factors in aviation. *Proceedings of the 12th Symposium on Human Factors in Aviation Maintenance.* Joint meeting of CAA, FAA, and Transport Canada at Gatwick Airport, UK, March 10–12, 1998.

Johnson, W. B. (Ed.). (2005). *The Operator's Manual for Human Factors in Aviation Maintenance.* Washington, DC: Federal Aviation Administration, Available at www.hf.faa.gov/opsmanual.

Johnson, W. B. (2015a). Human factors training: How FAA trains airworthiness inspectors. *Aircraft Maintenance Technology Magazine.* August, 2015, pp. 30–31.

Johnson, W. B. (2015b). Two countries apply trusted human factors model to manage safety. *FAA Aviation Mx Human Factors Quarterly Newsletter.* 3(1). March 2015. Available at www.humanfactorsinfo.com.

Johnson, W. B. and Avers, K. A. (Eds.). (2014). *The Operator's Manual for Human Factors in Aviation Maintenance.* Washington, DC: Federal Aviation Administration, Available at See www.humanfactorsinfo.com.

Johnson, W. B. and Maddox, M. E. (2007). A Model to Explain Human Factors in Aviation Maintenance. *Avionics News,* April.

Johnson, W. B. and Watson, J. (2001). *Installation Error in Airline Maintenance.* Washington, DC: Federal Aviation Administration Office of Aviation Medicine.

Maddox, M. (Ed.). (2007). *Human Factors Guide for Aviation Maintenance and Inspection.* Washington, DC. Available at www.humanfactorsinfo.com.

National Transportation Safety Board (1989). Aircraft accident report: Aloha airlines flight 243, April 28, 1988 (NTSB-AAR-89/03). Washington, DC: National Transportation Safety Board.

Patankar, M. (2002). *Root Cause Analysis of Rule Violations by Aviation Maintenance Technicians.* Washington, DC: FAA Office of Aviation Medicine.

Rankin, W., Hibit, R., Allen, J., Jr., and Sargent, R. (2000). Development and evaluation of the maintenance error decision aid (MEDA) process. *International Journal of Industrial Ergonomics,* 26, 261–276.

Rankin, W. L. (2007). MEDA investigation process. Boeing *Aero,* Quarter 02, pp. 1–8.

Rankin, W. L. and Allen, J. P., Jr. (1996). Boeing introduces MEDA: Maintenance error decision aid. Boeing *Airliner,* Quarter 02, pp. 14–21.

Rasmussen, J. (1983). Skills, rules, knowledge; Signals, signs, and symbols, and other distinctions in human performance models. *IEEE Transactions on Systems, Man and Cybernetics,* 13, 257–266.

Reason, J. (1998). Achieving a safe culture: Theory and practice. *Work & Stress,* 12(3), 293–306.

Reason, J. T. (1995). A system approach to organizational error. *Ergonomics,* 38, 1708–1721.

Reason, J. T. (1997). *Managing the Risks of Organizational Accidents.* New York: Ashgate Publishing.

Shappell, S. A. and Weigmann, D. A. (2000). The human factors analysis and classification System—HFACS. *FAA Civil Aeromedical Institute for the Office of Aviation Medicine, DOT/FAA/AM-00/7,* Springfield, VA: National Technical Information Service (NTIS).

United Kingdom Civil Aviation Authority (2003). CAP 716. *Aviation Maintenance Human Factors (EASA/ JAR 145 Approved Organizations) Guidance Material on the UK CAA Interpretation of Part-145 Human Factors and Error Management Requirements.* London: UK CAA.

Wiegmann, D. A. and Shappell, S. A. (2003). *A Human Error Approach to Aviation Accident Analysis: The Human Factors Analysis and Classification System.* Burlington, VT: Ashgate Publishing.

5 Distributed Work in the National Airspace System
Traffic Flow Management and Airline Operations Control

Philip J. Smith and Ellen J. Bass

CONTENTS

Human-Centered Design of a Distributed Work System 108
Allocation of Strategic versus Tactical Decisions... 108
Design for Coordination versus Collaboration 109
Incorporation of Overlapping Responsibilities to Provide Safety Nets....................................... 110
The Future of Airport Surface Management: A Case Study 110
Conclusion 112
References.. 113

The National Airspace System (NAS) relies heavily on distributed responsibilities across individuals in multiple roles in order to help ensure safety and efficiency (Smith et al., 2001, 2010). By distributing responsibilities, the workload and cognitive complexity for any individual are reduced to make tasks tractable. By carefully defining these responsibilities, partial redundancies are also introduced to provide safety nets (Smith et al., 2012a).

The other chapters in this book have discussed the roles of pilots and controllers. Those roles represent one important aspect of the NAS as a distributed work system (Hutchins and Klausen, 1996; Hinds and Kiesler, 2002; Salas et al., 2008; Harwood, 2010) and merit careful study from that perspective. However, for this chapter, we look at two other groups within the NAS, Federal Aviation Administration (FAA) traffic flow managers and airline aircraft dispatchers, who have a number of responsibilities that complement those assigned to pilots and controllers. Some of these responsibilities are tactical in nature. If, for example, a flight needs to divert to an alternate airport the airline dispatcher—who has shared responsibility for the operation of the flight with the pilot (14 CFR §121.395 [FAA, 2017a], 14 CFR §121.533 [FAA, 2017b])—is very much involved in determining the nearest *suitable* airport. Similarly, FAA traffic flow managers frequently make tactical flight amendments, changing the planned route that a dispatcher originally filed for a flight as it is taxiing out (FAA, 2016a, 2016b).

Many of the responsibilities of traffic managers and dispatchers, however, are more strategic in nature. From an FAA perspective, they involve identifying constraints in the NAS and implementing plans to safely maximize throughput given these constraints. For example, a traffic manager at the FAA's Air Traffic Control Systems Command Center (ATCSCC) may implement a ground delay program (FAA, 2016b) for flights scheduled to arrive at Philadelphia International Airport (PHL) over a six-hour period, reducing the number of arrivals per hour. From an airline operations control perspective, such strategic decisions involve managing the airline's schedule and the development of plans for implementing that schedule (such as planning the route for a specific flight).

HUMAN-CENTERED DESIGN OF A DISTRIBUTED WORK SYSTEM

Abstractly, the design of the NAS provides a classic example of a distributed work system. The architecture for distributing work can be categorized along several interacting dimensions:

- Allocation of strategic versus tactical decisions (Smith et al., 2003)
- Design for coordination versus collaboration (Lee and Bass, 2014)
- Incorporation of overlapping responsibilities to provide safety nets (Smith and Billings, 2009).

These different dimensions have important human factors implications, which are discussed briefly in the following.

ALLOCATION OF STRATEGIC VERSUS TACTICAL DECISIONS

One of the challenges in managing a stochastic process such as the NAS is the time-varying nature of uncertainty (Smith et al., 2003). As the time for an event approaches, there is often less uncertainty about what will actually happen. However, some of the alternate options for dealing with the event may no longer be available as time progresses (Bolton, Göknur, & Bass, 2013).

In the case of traffic flow management, an illustration would be the uncertainty about whether forecast convective weather in the terminal airspace around JFK will actually materialize from 2100 to 0200Z. If a decision is made at 1500Z to use a ground delay program to reduce arrivals into JFK from 2100 to 0200Z, the decision about what rate(s) to assign will be based on a forecast that is much less precise, potentially resulting in underdelivery or overdelivery of inventory (arrivals) relative to capacity, with an associated impact on delays. However, if this decision is delayed until 1800Z, it will be too late to delay the departures of flights from the West Coast, which means the short hauls into JFK could experience longer delays. Moreover, if the decision is delayed until individual flights have pushed back from the gate (using a call-for-release strategy), then the set of flights affected will again be quite different.

Thus, from a human performance perspective, the system designers need to consider how to design operational concepts, procedures, training, and supporting technologies to support making four decisions given the time-varying nature of the uncertainty:

- When should I make a decision?
- What alternative decisions should I consider?
- What alternative should I implement?
- Should I adapt my plan over time based on how the scenario is evolving?

Note that the fourth decision highlights the fact that strategic and tactical decisions are not independent. If the system is designed to support continuous adaptive planning, strategic decisions can explicitly include contingency plans or enable ad hoc replanning as necessary.

Note also that such a division of labor in terms of strategic versus tactical decisions is often consistent with the goal of decomposing the system into a set of nearly independent subtasks that can be assigned to different people (or in some systems, across some combination of people and automation) (Bass, et al., 2011). For instance, airline dispatchers have a primary role in developing the flight plan (route, cruise altitude, alternative airport, fueling, etc.) for a flight (a strategic decision), whereas the flight crews have a primary role in making small tactical adjustments while the aircraft is enroute in response to encounters with weather or requests from air traffic control (ATC) made in order to maintain separation (tactical decisions).

Although the strategic decisions made by the dispatcher have to anticipate potentially necessary tactical responses, to make these strategic decisions some of the data, expertise and cognitive processes used by the dispatcher are quite different from the requirements for pilots when making tactical adjustments to their trajectories while in flight. In addition, the focus of attention is quite different. The dispatch staff is looking at a bigger picture of the system (in terms of traffic flow management, weather, and airline scheduling) when developing flight plans, whereas the pilot is focused on more immediate impacts to his/her flight. Thus, a decomposition based on access to the necessary data, expertise, cognitive processing, and attentional requirements necessary to support these strategic versus tactical decisions can be very effective.

DESIGN FOR COORDINATION VERSUS COLLABORATION

The word *nearly* used earlier in discussion about system decomposition is critical in terms of designing a distributed system. As Figure 5.1 indicates, although coordinated activity is preferred, collaboration is necessary when scenarios arise that require the interaction of individuals fulfilling different roles.

By coordination, we mean a system in which each individual performs his/her task independently, simply providing the outputs of that task to the other individuals without any need to interact in a richer, unstructured manner (talking with one another, texting, emailing, etc.). An example of such coordination is provided by Ground Delay Programs as they operate today. When a traffic manager at the Systems Command Center sets an arrival rate at JFK 25% below the normal capacity for an airport, software translates this into a set of arrival time slots that are then assigned to different airlines based on a ration-by-schedule prioritization. The responsible airline ATC coordinator (a specialized role for a dispatcher) does not have to talk with or otherwise interact with that traffic manager in any rich sense, he/she just has to know which arrival time slots have been allocated to his/her airline. He/she can then use airline software to help in swapping flights among his/her airline's slots to adjust the airline's schedule, for example, giving priority to a flight that has a large number of passengers with international connections at JFK. In short, the traffic manager and dispatcher coordinate rather than collaborate.

In addition to supporting such coordination, however, enabling procedures, training and software also must be incorporated to ensure that the responsible individuals will detect the need to shift from coordinated work to collaborative work when necessary, and to ensure that they can effectively collaborate. Today, for example, if a flight that is included in a Ground Delay Program has an issue that requires special handling from an airline perspective, the dispatcher can submit a request to a website hosted by the Systems Command Center (the *TCA page*) asking for an exception, which is then evaluated by a traffic manager at the Systems Command Center in consultation with the relevant enroute facility or airport.

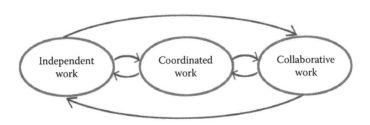

FIGURE 5.1 Forms of interaction: independence versus coordination versus collaboration.

INCORPORATION OF OVERLAPPING RESPONSIBILITIES
TO PROVIDE SAFETY NETS

The design of the NAS is very sensitive to the fact that people and technologies are fallible (Reason, 1990; Pidgeon and O'Leary, 2000; Dekker, 2003; Bolton and Bass, 2012). There are numerous safety nets in place so that if one *component* fails, another one will detect this and respond appropriately to ensure a safe operation. The manner in which responsibilities have been distributed plays an important role in providing these safety nets.

At a very high level, the safety net provided through the design of the NAS as a distributed work system is illustrated by the fact that, at some level of abstraction at a given point in time, two pilots, a controller, a dispatcher, multiple traffic managers, and various software functions are *looking* at a flight. As a result, any one of them may detect a problem with that flight and trigger a response to ensure its safety.

For instance, the pilots, a controller and a collision avoidance system are all *looking* at the flight relative to other aircraft in a very direct sense with a responsibility for ensuring separation from other aircraft. Moreover at a more abstract level, the dispatcher and several traffic managers are implicitly considering that flight when they ask the question: Has the weather developed in an unexpected way that requires some change in the routing for flights in the vicinity of that weather? In addition, the dispatcher has an alerting function that is watching a specific flight and indicates if it has deviated significantly from its planned route.

Thus, by assigning overlapping responsibilities to operational staff and designers (through their technologies) who have different primary tasks with different expertise and experiences, and who are looking at different (partially overlapping) data, slips, mistakes and hardware failures, are much more likely to be detected and dealt with.

THE FUTURE OF AIRPORT SURFACE MANAGEMENT: A CASE STUDY

The management of departures from an airport is a highly distributed subsystem within the NAS. Consider the task relative to the airport layout illustrated in Figure 5.2 for a future concept involving departure surface metering.

In this figure, the dark gray lines are runways, and the light ones are taxi-ways. For this discussion, the vertical runway in the center (18C) and the vertical runway to the right (18L) are being used exclusively for departures.

Figure 5.3 provides an oversimplified indication of the individuals who affect airport surface management directly or indirectly. An Air Route Traffic Control Center (ARTCC) is a regional air traffic center; a Terminal Radar Approach Control Facility (TRACON) is responsible for metroplex airspace within an ARTCC; an ATCT is an airport ATC tower.

Figure 5.4 shows the departure routes mapped to Runways 18C and 18L for this airport. The center five routes to the north, east, south, and west are the standard departure routes. The adjacent routes (located in what is normally arrival airspace) are available for the flexible use of arrival airspace for departures if conditions warrant.

To provide a sense of how they all interact and impact airport departure management, consider the following example based on the FAA's operational concept for the use of virtual queues to better manage surface traffic (Atkins et al., 2004; Doble et al., 2009; Smith et al., 2011, 2012b; Simaiakis et al., 2014).

- Convective weather is approaching the airport from the west, impacting some of the departure fixes.
- Two of the departure fixes to the west/northwest, WILEY and WICKR, are forecast to be impacted from 1700 to 1800Z, reducing departure throughput.
- At 1645Z, the TRACON traffic manager, in consultation with the surrounding ARTCC traffic manager (collaboration by voice with shared displays), decides to place a throughput (Miles-in-Trail or MIT) restriction on WILEY and WICKR (20 MIT as one).

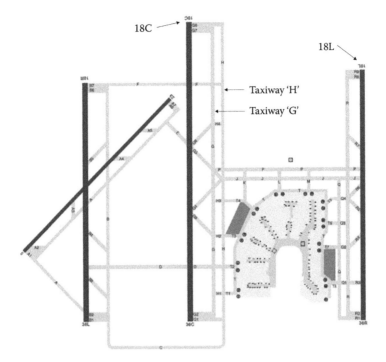

18C

18L

Taxiway 'H'

Taxiway 'G'

FIGURE 5.2 Sample airport layout.

- This is communicated to the Departure Reservoir Coordinator (DRC)—a new role for a traffic manager at the ATCT (coordination electronically).
- The DRC looks at the lineup for departures using 18C and finds that 30% of the flights scheduled to depart from 1,700 to 1,800Z are filed to depart via WICKR and WILEY and that, with the MIT restriction, throughput is predicted to be reduced from 52 to 34 flights from 1,700 to 1,800Z. The good news is that there are only four flights filed to WICKR and WILEY that are actively taxiing at this time, so there is an opportunity to hold a number of the WICKR and WILEY departures at their gates to change the lineup so that splitters (flights filed to depart via unimpacted departure fixes) can be placed between the WICKR and WILEY departures in order to maximize throughput.
- The DRC checks with the Tower Supervisor and learns that the ground controller plans to queue all departures filed to depart WILEY and WICKR on the taxiway H, and all other departures using 18C on taxiway G (coordination by voice).
- The DRC sets a target runway queue length for taxiway G of eight flights and a target queue length of eight for taxiway H in order to reduce taxi out times and queue lengths but still have enough inventory filed to the different fixes in case the weather develops differently than expected. The predicted departure throughput for 18C with this resultant change in the departure lineup increases to 47 flights from 1,700 to 1,800Z.
- The Departure Metering Program (with *virtual departure queues*) is put into effect and is communicated to the flight operators (coordination electronically), and the FAA sends target spot times (which are translated by the flight operators into target pushback times) to the flight operators.
- Airline ATC Coordinators (dispatchers with supervisory responsibilities) look at the departure delays assigned to particular flights and (with software support) swap some of their high priority flights with low priority flights to move up their pushback times.

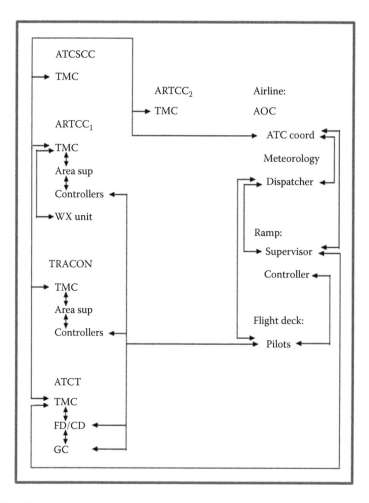

FIGURE 5.3 Distributed roles in the NAS.

- These changes in pushback times are communicated to the FAA for use by traffic managers at the ARTCC and TRACON and to the DRC and controllers in the ATCT (coordination electronically).
- The assigned pushback times are also communicated to airline ramp controllers, the gate, and pilots (coordination electronically).

As the aforementioned steps indicate, this future design for airport departure management focuses on a highly distributed work architecture, enabled by a number of technologies. The example itself somewhat oversimplifies the full richness of this system but serves to illustrate the interweaving of coordination and collaboration among individuals assigned to a number of different roles. (Issues such as interactions of Departure Metering Programs with Ground Delay Programs and the use of alternative strategies such as reroutes instead of the use of splitters add further interesting twists to the actual functioning of this operational concept.)

CONCLUSION

The goal of this discussion has been to identify conceptual issues that need to be addressed in designing the architecture for a distributed work system as well as to use the NAS as an example to illustrate these issues. Several important dimensions were highlighted, including the need to

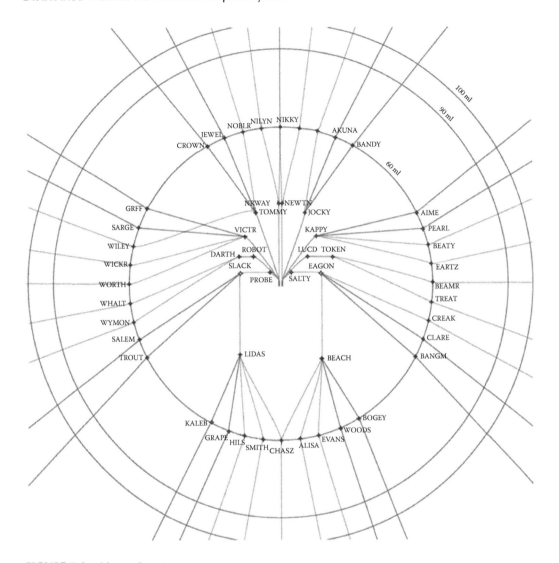

FIGURE 5.4 Airport departure routes.

- Integrate strategic and tactical decision making.
- Support both coordination and collaboration when the system can only be decomposed into a set of *nearly* independent subtasks and ensure effective transitions between these two forms of interaction.
- Decompose the system into subtasks in which the necessary data, expertise, and processing capabilities are available for the individuals assigned particular roles and responsibilities.
- Incorporate overlapping responsibilities in order to provide safety nets.

REFERENCES

Atkins, S., Brinton, C., and Rogowski, S. (2004). Surface management system field trial results. *Proceedings of 2004 AIAA 4th Aviation Technology,* Integration and Operations (ATIO) Forum, Chicago, IL.

Bass, E.J., Bolton, M.L., Feigh, K.M., Griffith, D., Gunter, E., Mansky, W., and Rushby, J. (2011). Toward a multi-method approach to formalizing human-automation interaction and human-human communications. *2011 IEEE International Conference on Systems, Man, and Cybernetics.* October 9–12, 2011, Anchorage, AK, pp. 1817–1824.

Bolton, M.L. and Bass, E.J. (2012). Using model checking to explore checklist-guided pilot behavior. *The International Journal of Aviation Psychology*, 22(4), 343–366.

Bolton, M.L. Göknur, S., and Bass, E.J. (2013). Framework to support scenario development for human-centered alerting system evaluation. *IEEE Transactions on Human-Machine Systems*, 43(6), 595–607.

Dekker, S. (2003). Failure to adapt or adaptations that fail: Contrasting models on procedures and safety. *Applied Ergonomics*, 34(3), 233–238.

Doble, N., Timmerman, J., Carniol, T., Klopfenstein, M., Tanino, M., and Sud, V. (2009). Linking traffic management to the airport surface: Departure flow management and beyond. *Proceedings of the Eighth USA/Europe Air Traffic Management Research and Development Seminar*, Napa, CA.

FAA (2016a). JO 7119.65. Change 2. Air traffic control. (Accessed on July 6, 2017) https://www.faa.gov/documentLibrary/media/Order/7110.65_ATC_Chg_2_dtd_11-10-16.pdf

FAA (2016b). JO 7210.3Z. Change 2. Facility operation and administration. Part 5 traffic management system. https://www.faa.gov/regulations_policies/orders_notices/index.cfm/go/document.information/documentID/1028577.

FAA (2017a). Electronic code of federal regulations. Title 14 Chapter I subchapter G Part 121 subpart M—Airman and crewmember requirements. §121.395 Aircraft dispatcher: Domestic and flag operations. http://www.ecfr.gov/cgi-bin/text-idx?SID=be19bf4b3efef3017473887aaed8c248&mc=true&node=se14.3.121_1395&rgn=div8.

FAA (2017b). Electronic code of federal regulations. Title 14 Chapter I subchapter G Part 121 Subpart T—Flight operations. §121.533 Responsibility for operational control: Domestic operations. http://www.ecfr.gov/cgi-bin/text-idx?SID=be19bf4b3efef3017473887aaed8c248&mc=true&node=se14.3.121_1533&rgn=div8.

Harwood, G. (2010). Design principles for successful virtual teams. In Nemiro, J., Beyerlein, M., Bradley, L., and Beyerlein, S. (Eds.), *The Handbook of High-Performance Virtual Teams: A Toolkit for Collaborating Across Boundaries*. San Francisco, CA: Jossey-Bass.

Hinds, P. and Kiesler, S. (Eds.). (2002). *Distributed Work*. Cambridge, MA: MIT Press.

Hutchins, E. and Klausen, T. (1996). Distributed cognition in an airline cockpit. In Engestom, Y. and Middleton, D. (Eds.) *Cognition and Communication at Work*. New York: Cambridge University Press. pp. 15–34.

Lee, D.L. and Bass, E.J. (2014). Delegation for authority and autonomy: An assignment and coordination model. *2014 IEEE International Conference on Systems, Man, and Cybernetics*. October 5–8, 2014, San Diego, CA, pp. 1759–1766.

Pidgeon, N. and O'Leary, M. (2000). Man-made disasters: Why technology and organizations (sometimes) fail. *Safety Science*, 34(1), 15–30.

Reason, J. (1990). *Human Error*. New York: Cambridge University Press.

Salas, E., Cooke, N., and Rosen, M. (2008). On teams, teamwork and team performance: Discoveries and developments. *Human Factors*, 50 (3), 540–547.

Simaiakis, I., Khadilkar, H., Balakrishnan, H., Reynolds, T., and Hansman, R.J. (2014). Demonstration of reduced airport congestion through pushback rate control. *Transportation Research Part A: Policy and Practice*, 66, 251–267.

Smith, P.J., Beatty, R., Hayes, C., Larson, A., Geddes, N., and Dorneich, M. (2012a). Human-centered design of decisions-support systems. In Jacko, J. (Ed.), *The Human-Computer Interaction Handbook: Fundamentals, Evolving Technologies, and Emerging Applications*, 3rd ed., pp. 589–621.

Smith, P.J., Beatty, R., Spencer, A., and Billings, C. (2003). Dealing with the challenges of distributed planning in a stochastic environment: Coordinated contingency planning. *Proceedings of the 2003 Annual Conference on Digital Avionics Systems*, Chicago, IL.

Smith, P.J. and Billings, C. (2009). Layered resilience. In Nemeth, C., Hollnagel, E., and Dekker, S. (Eds.), *Resilience Engineering Perspectives*, Volume 2. (pp. 413–430). Hampshire, UK: Ashgate.

Smith, P.J., Fernandes, A.B., Durham, K., Evans, M., Spencer, A., Beatty, R. Wiley, E., and Spencer, A. (2011). Airport surface management as a distributed supervisory control task. *Proceedings of the 2011 AIAA Digital Avionics Systems Conference*, Orlando, FL.

Smith, P.J., McCoy, E., and Orasanu, J. (2001). Distributed cooperative problem-solving in the air traffic management system. In Klein, G. and Salas, E (Eds.), *Naturalistic Decision Making*. Mahwah, NJ: Erlbaum, pp. 369–384.

Smith. P.J., Spencer, A.L., and Billings, C. (2010). The design of a distributed work system to support adaptive decision making across multiple organizations. In Kathleen, L. Mosier and Ute M. Fischer, (Eds.), *Informed by Knowledge: Expert Performance in Complex Situations*. New York: Taylor & Francis Group.

Smith, P.J., Weaver, K., Fernandes, A., Durham, K., Evans, M., Spencer, A., and Johnson, D. (2012b). Supporting distributed management of the airport surface. *Proceedings of the 2012 AIAA Digital Avionics Systems Conference*, Williamsburg, VA.

6 Accident Investigation

William J. Bramble, Jr.

CONTENTS

Introduction 115
Some History 116
Grappling with Human Factors 117
Building a Human Performance Investigation Capability at the NTSB 118
Increasing Sophistication and the Development of Investigative Guidance 124
The Rise of Safety Management and Data-Drive Safety Efforts 126
Theoretical Perspectives 127
Notable Successes 128
Continuing Challenges 128
References 129

INTRODUCTION

Accident investigation is a sense-making activity, a way of understanding events that frighten us, and, hopefully, finding ways to reduce the risk of those events in the future. It is a multidisciplinary field that requires collaboration among professionals of many backgrounds, including flight operations, air traffic control, maintenance, and cabin safety. The participation of these varied specialties is needed to understand the complexity of aviation work and to develop a sufficiently detailed record on which to base an analysis. Due to this disciplinary diversity, accident investigation is also a team activity, and the National Transportation Safety Board (NTSB) has developed an organizational framework based on these technical specialties to impose order on the process.

Investigative work is divided up by working groups led by an investigator from each technical specialty. Operations investigators examine flight deck operations and work in the cockpit. Survival factors investigators examine cabin operations and the work of flight attendants and airport rescue personnel. Air traffic control investigators examine air traffic operations in towers and centers. Maintenance investigators examine work in the maintenance hangar. Engineering investigators examine the design and functioning of aircraft structures, systems, and powerplants. The role of the human factors investigator is uniquely broad. Human factors investigators are charged with examining issues that may span these domains. This boundary crossing makes the work of the human factors investigator unique and sometimes leads to confusion about the investigators' role relative to the role of other investigative team members.

The role of the human factors investigator is spelled out in the International Civil Aviation Organization's (ICAO's) *Manual of Aircraft Accident and Incident Investigation* (International Civil Aviation Organization, 2014). According to this manual, the objectives of human factors investigation are to

- Determine how breakdowns in human performance may have caused or contributed to an occurrence.
- Identify safety hazards related to limitations in human performance.
- Identify ways to eliminate or reduce the consequences of faulty human actions or decisions.

Interpretation of these objectives depends on one's knowledge of human factors and one's preferred theoretical perspective on the nature of human error.

SOME HISTORY

Flying was risky in the early days of aviation. About half of U.S. Air Mail pilots died in crashes during the service's first two years of operation (U.S. Centennial of Flight Commission, 2017), underscoring the riskiness of early flight. Aircraft accidents of the 1920s were attributed to a range of factors, including defective equipment, faulty maintenance, and of course, human causes (Wilson, 1949). In 1926, the U.S. Congress established a Bureau of Air Commerce to establish and enforce new safety rules. The Bureau's Aeronautics Branch was responsible for investigating crashes. However, a series of high profile accidents led to the establishment of a new three-member Air Safety Board. In 1940, the Civil Aeronautics Board (CAB) was formed, and it assumed the functions of both regulating safety and investigating accidents.

By this time, airlines were ferrying passengers around the globe in aircraft with as many as 50 seats. Airline accidents were attributed to a variety of factors including structural failures, engine failures, terrain, weather, darkness, and human failures including *poor technique, defective judgment, carelessness* (Wilson, 1949, p. 443). During and after World War II, manufacturers adopted new technologies, including the gyropilot, turbo-supercharger control, propeller-feathering mechanism, and control surface compensator. Such technologies were seen as encroaching on the responsibilities of pilots, but the industry regarded the tools as *instrumental in eliminating human error*. After the war, airlines campaigned for accelerated deployment of ground and airborne navigation equipment, radar, and high-intensity approach lighting. Fatal accident rates began to drop (Figure 6.1).

A series of midair collisions in the 1950s, including the 1956 collision of a United Airlines DC-7 and a Trans World Airlines L-1049 over the Grand Canyon, spurred development of a more robust air traffic control system, and, in 1958, Congress established a new Federal Aviation Agency (later renamed the Federal Aviation Administration or FAA). The FAA took charge of safety regulation and enforcement, whereas the CAB retained responsibility for regulating air commerce and investigating accidents. The march of aviation technology continued. Manufacturers introduced the first jet airliner, the DeHavilland Comet. The Comet was plagued

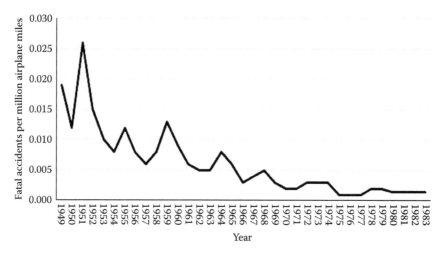

FIGURE 6.1 Fatal accidents per million airplane miles flown, scheduled operations, 1949–1983. (From Compilation of CAB and NTSB statistics. With Permission.)

with safety issues, but it heralded a new era in commercial aviation. The fatal accident rate for U.S. scheduled airlines continued its decline.

The 1960s saw a proliferation of narrow-body jet airliners. These airplanes carried more than 100 passengers, including the Douglas DC-8, DC-9, and DC-10, and the Boeing B-727 and B-737. These aircraft were equipped with more reliable engines and could fly at higher altitudes. This reduced the risk of engine failure and allowed flights to better avoid hazardous weather. The fatal accident rate reached an impressive new low of around 0.003 fatal accidents per million air-plane miles. In 1966, a multimodal NTSB was established within the new U.S. Department of Transportation (DOT), and the NTSB was given responsibility for investigating civil aircraft acci-dents. Originally, part of the DOT, NTSB was made independent in 1975 to enhance its ability to report objectively on the adequacy of DOT regulation and oversight.

The 1970s saw the introduction of wide-body jetliners carrying more than 200 passengers, such as the Boeing B-747, DC-10, Lockheed L-1011, and Airbus A-300. Now, an airline crash truly raised the prospect of mass fatalities, and this unfortunately did happen a few times in that decade. The worst of these accidents occurred in 1977 in Tenerife, Spanish Canary Islands, when two B-747 airplanes collided on a foggy runway (Comisión de Investigación de Accidentes e Incidentes de Aviación Civil, 1978). The decline in the fatal accident rate that had been occurring over the previ-ous decades began to stall, and the aviation safety community found itself searching for answers. The majority of major aircraft accidents were being attributed to pilot error, and this prompted increased interest in human factors.

GRAPPLING WITH HUMAN FACTORS

Aviation safety professionals have long recognized the need to look beyond pilot error to understand aviation safety issues. As McFarland (1946) stated

> Only in isolated cases is it possible to apportion the causes of the accident to specific faults rather than to an accumulation of contributing factors. Many accidents that have been attributed to pilot error may have resulted from excessive demands on the air crews. While training and selection procedures may be improved, it is unlikely that human limitations in operating aircraft can be appreciably altered. The aeronautical engineer, however may be able to simplify the duties of the pilot and thus reduce the likelihood of error. (p.561)

A human factors branch was first established at the CAB in 1959, but its scope was fairly limited. It was primarily concerned with identifying injury mechanisms and evaluating the crashworthiness of vehicles. In the 1960s, the division's responsibilities were expanded to include investigation of operators' medical, psychological, and physical fitness for duty (Doyle, 1968). By the 1970s, the division's efforts had expanded to include human performance, which was essentially an activity aimed at identifying reasons for unexpected deviations from desired crew performance. With this addition, the efforts of the branch's investigators were divided into three areas:

- Medical and crush injury factors
- Survival and other postcrash factors and
- Human performance factors

The Board approached the investigation of human factors as part of an *investigative triad* Miller (1971). This was a decompositional approach in which investigative activities were divided into three areas: *man, machine,* and *medium. Man* stood for the human element, *machine* for the aircraft, and *medium* for operational aspects. Investigative divisions were organized accordingly. They included an aircraft branch, a human factors branch, and an operational factors branch. The aircraft branch examined mechanical aspects of the aircraft. The human factors branch examined physiology, psychology, human engineering, and survivability. The operations branch explored *operational aspects*. NTSB managers recognized a significant amount of overlap between human factors and operational aspects, acknowledging that the division of labor on topics like flightcrew training was somewhat arbitrary (Miller, 1971).

Miller (1971), who directed the NTSB's Bureau of Aviation Safety in the early 1970s, considered investigation of human factors *the greatest single technological challenge* for investigators. He urged exploration of pilot skill, judgment, and personality and felt that the solutions to problems in these areas would involve education, enforcement, and engineering. He also saw the need for an investigative specialty that could tie together the disparate factors identified in an investigation, including organizational aspects. He felt this specialty should be systems safety, a new discipline in which he was a pioneering influence (Miller, 1954) that focused on "the integration of skills and resources, specifically organized to achieve accident prevention over the life cycle of an air vehicle system" (Miller, 1965). According to Miller (1971), the NTSB explored the hiring of systems safety specialists; however, systems safety never became a major area of specialization at the NTSB.

Danaher (1971) was chief of the NTSB's human factors branch in the 1970s. He observed that NTSB findings were too often, "objective summaries of what happened rather than statements of the true underlying cause of the accident." He felt that addressing these underlying causes was important and that it would require greater attention to human factors; however, he did not describe how the investigation of such factors should be carried out. In a 1974 paper, the National Air Transportation Associations (Macy, 1974) criticized the NTSB's existing approach to investigating human factors that included probing a pilot's mood, feelings, and habits. The organization argued that investigators lacked adequate research describing the factors affecting pilots' performance of operational tasks, and it advocated for the study of pilot behavior in realistic operational contexts.

Shortly thereafter, a group of NASA researchers began research examining crew performance, and they issued some recommendations for studying crew behavior and investigating human factors in aircraft accidents and incidents (Barnhart et al., 1975; Cooper et al., 1980). The NASA group recommended the creation of a timeline of significant behavioral events, the making of inferences about how these events affected the broader event sequence, which they called a function analysis. They also urged the application of information processing models to gain deeper insight into the context of flightcrew errors. The approach reflected trends in psychology at the time, which emphasized the study of cognition and compared the functioning of the brain with a computer.

Schleede (1979), a senior NTSB air safety investigator, tried this approach in his own investigations and found it helpful. He encouraged other investigators to use of it but pointed out that doing so would require a change of mindset. Schleede observed that accident investigators, many of whom were engineers, were most comfortable applying deductive reasoning to their investigations. He argued that the use of deductive reasoning was not possible when analyzing human performance as it was when analyzing mechanical failures because some of the factors influencing human performance (such as cognition) were not directly observable, and because the influence of various factors on behavior was probabilistic. The shift to a more inductive approach was essential, he reasoned, if the air safety community was to make significant progress addressing safety issues related to pilot error.

BUILDING A HUMAN PERFORMANCE INVESTIGATION CAPABILITY AT THE NTSB

In 1977, the NTSB hired its first engineering psychologist to investigate human performance issues. This investigator, Alan Diehl, participated in at least two significant investigations during his time at the board. The first was an accident involving United Airlines Flight 173, which crashed after running out of fuel in Portland, Oregon (National Transportation Safety Board, 1979). As Kayten (1993) wrote, this accident occurred soon after NASA's safety researchers had developed a new vocabulary and framework for analyzing crew performance called cockpit resource management (CRM). Diehl identified CRM-related deficiencies in the crew's performance and drafted the Board's first recommendation urging the provision of CRM training to airline pilots emphasizing "the merits of participative management for captains and assertiveness training for other cockpit crewmembers." Subsequently, Diehl participated in the investigation of a commuter airline accident in Rockland, Maine. He also identified CRM issues in that accident, and he uncovered management pressures encouraging risky pilot decisions (National Transportation Safety Board, 1980).

These investigations, and sustained industry pressure to do a better job, addressing the factors underlying pilot error prompted the NTSB to develop a systematic framework for investigating human performance. This frame work was created following a review of investigative protocols used by various organizations, including the U.S. Air Force, Army, and Navy, and by British, Canadian, and the Air Line Pilots Association (National Transportation Safety Board, Unpublished) by the human factors branch. After reviewing these models, the branch evaluated the utility of the various models for discovering human performance issues that had been documented in recent accidents investigated by the NTSB (Walhout, 1981). The resulting framework included six focus areas: behavioral, medical, operational, task, equipment, and environmental factors (Figure 6.2).

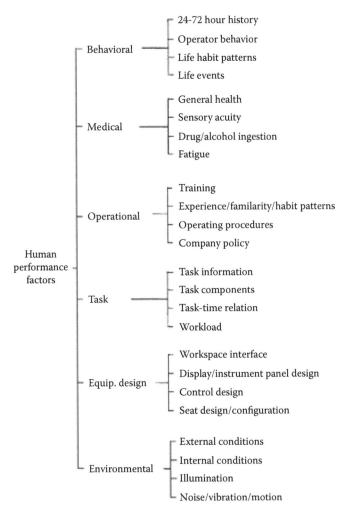

FIGURE 6.2 NTSB human performance investigation model. (From Walhout, G., *ISASI Forum*, 14, 21–22, 1981. With Permission.)

The model did not include a separate category for organization factors, which is surprising, given the success of the Rockland, Maine, investigation, which was touted as a significant accomplishment (Walhout, 1981). The lack of a separate focus area for organizational factors probably reflects the absence at that time of a comprehensive theoretical model addressing the role of organizational factors in accident development.

The human factors branch tested its new human performance model during the investigation of a 1981 air taxi accident near Spokane, Washington (National Transportation Safety Board, 1981). This first formal human performance investigation group was led by an engineering psychologist named Janis Stoklosa, who had filled the position left vacant by Alan Diehl. The group consisted of investigators with a variety of backgrounds. In addition to engineering psychology, medicine and flight operations specialists were included. By using the new human performance investigation model as a guide, this group identified shortcomings in the design of the approach charts and navigation equipment (Stoklosa, 1983). The approach the crew was cleared to conduct required them to tune in two consecutive distance measuring equipment (DME) stations. The crew tuned in the first but not the second DME station, causing them to descend prematurely. Stoklosa and colleagues linked this error to the DME, which could utilize a DME station other than the one specified in its display window if a certain button was pressed.

Walhout (1981) reported that the first formal human performance investigative group encountered difficulty gaining acceptance from other NTSB investigators. Few understood the group's purpose or role. This prompted "an intense educational effort … concerning the purpose of this new investigative technique." Unpublished NTSB training materials dated the following year explain, "In response to the directions of the Office of the Managing Director, has developed a program to investigate the human performance issues in accidents in order to better define the 'why' of accident causes. This program is multi-modal; it has far-reaching potential in accident prevention through the identification of the often complex interrelationships of factors which can affect a person's ability to perform his or her duties." The materials went on to explain that the NTSB's human performance approach offered a "systematic and standardized method to gather, analyze and integrate the medical, operational, environmental, vehicle design, and psycho-physiological factors obtained during accident investigation" (National Transportation Safety Board, 1982, p. 1).

The new human performance model was also used during the investigation of an accident involving Air Florida Flight 90 that crashed during takeoff from Washington National Airport (National Transportation Safety Board, 1982). When exploring the *behavior* element, investigators noted that the first officer in this accident had noticed something wrong with thrust during the takeoff roll and they concluded that he was insufficiently assertive, and the captain was insufficiently receptive to his comments (Stoklosa, 1983). As a result, the NTSB reiterated the recommendation from the Portland accident that urged CRM training. Stoklosa also described the use of a functional flow chart method when examining task-related factors. Consequently of this method, the Board concluded that time pressure likely played a role in the crew's difficulty recognizing and diagnosing the problem with the thrust. Consequently, the agency recommended the installation of runway distance markers at certain airports to help crews evaluate takeoff performance and described the potential benefit of an automated takeoff performance monitoring system.

In 1983, NTSB split the human factors branch into two and created a new, stand-alone human performance division (another new division was created to investigate survival factors). The agency cited three recent investigations when justifying the need for the new division (Kayten, 1993): Allegheny Airlines Flight 485 (National Transportation Safety Board, 1972); Eastern Airlines Flight 401 (National Transportation Safety Board, 1973); and Air Florida Flight 90 (National Transportation Safety Board, 1982). The new human performance division employed six investigators and supported NTSB investigations in all modes of transportation. Ron Schleede, the first chief of the NTSB human performance division, concentrated the division's activities on direct support of investigations of significant interest, educating other (still skeptical) NTSB investigators about the "significance of human performance factors in all accident investigations," and cataloguing human performance data to support future aggregate analyses (Bradley, 1984).

The late 1980s was a very busy period for the new division. Major aircraft accidents happened somewhat frequently, and the division provided direct support in the following cases:

- The 1985 crash of a Delta Lockheed L-1011 that encountered wind shear on approach to Dallas Fort Worth International Airport (National Transportation Safety Board, 1986)
- The 1985 stall and crash of a Galaxy Airlines Lockheed L-188 shortly after takeoff from Reno, Nevada (National Transportation Safety Board, 1986)
- The 1986 midair collision and crash of an Aeronaves de Mexico DC-9 and a general aviation airplane near Cerritos, California (National Transportation Safety Board, 1987)
- The 1987 loss of control and crash of a Continental Airlines DC-9 on takeoff from Denver, Colorado (National Transportation Safety Board, 1988)
- The 1987 crash of a Northwest Airlines MD-82 during takeoff from Detroit Metropolitan Airport (National Transportation Safety Board, 1988)
- The 1988 crash of a Delta Airlines crash Boeing B-727 during an attempted takeoff at Dallas, Texas (National Transportation Safety Board, 1989)
- The 1989 crash of a United Airlines DC-10 in Sioux City, Iowa after an uncontained engine failure (National Transportation Safety Board, 1990)

Human performance issues explored in these investigations included pilot decision-making, procedure design, training, and pilot control (National Transportation Safety Board, 1986); procedural compliance, stress, workload, and training (National Transportation Safety Board, 1986); visual performance and attention (National Transportation Safety Board, 1987); preemployment screening and crew pairing practices (National Transportation Safety Board, 1988); checklist design and procedural compliance (National Transportation Safety Board, 1988) (National Transportation Safety Board, 1989); and maintenance inspection and quality control (National Transportation Safety Board, 1990). Investigator training was spotty. One former investigator was sent out on a major accident before receiving any formal training in the role (Brenner, personal communication, 2017).

One particularly noteworthy accident in the 1980s was the crash of United Airlines Flight 232 (National Transportation Safety Board, 1990). After experiencing a catastrophic engine failure that damaged the airplane's major flight control systems, the crew improvised a control strategy using differential thrust and managed a partially controlled impact that saved the lives of many passengers. NTSB concluded that the crew's performance, which involved efficient distribution of communication and maximum utilization of a fourth crewmember (Predmore, 1991), was "highly commendable and greatly exceeded reasonable expectations." Across the airline industry, this was taken as prime evidence of "the value of cockpit resource management training" which had been provided at some airlines for nearly a decade. During these investigations, human performance investigators supported the development of numerous recommendations. Some of these advocated for the continued development of CRM training programs. Others addressed different aspects of training, and some addressed the design of standard operating procedures and the design and reliability of certain equipment, including collision avoidance and configuration warning devices.

In 1990, the NTSB's human performance function was again reorganized. Half of its investigators were assigned to a new aviation human performance division, others to a new sister division in the NTSB Office of Surface Transportation. Thereafter, the agency's aviation human performance investigators focused their efforts on the unique aspects of this complex mode of transportation. The new human performance division employed three investigators, two of whom had PhDs and aviation technical experience (Brenner, personal communication, 2017). This staffing level has been maintained, more or less, to the present day. Beginning in the early 1990s, however, all aviation human performance investigators have been equipped with PhDs in psychology and at least some aviation training and experience (usually a private pilot's license). After plateauing in the

1980s, airline accident rates increased and the frequency of accidents roughly doubled in the 1990s (National Transportation Safety Board, 2002b). The NTSB's aviation human performance investigators remained busy supporting many additional major investigations. Investigations supported during the early 1990s included the following:

- The 1990 runway collision between a Northwest Airlines DC-9 and a Northwest Airlines B-727 in Detroit Metropolitan Airport (National Transportation Safety Board, 1991)
- The 1991 runway collision between an Eastern Airlines B-727 and a Beechcraft King Air A100 in Atlanta Hartsfield International Airport (National Transportation Safety Board, 1991)
- The 1991 runway collision between a USAir B-737 and a Fairchild Metroliner at Los Angeles International Airport (National Transportation Safety Board, 1991)
- The 1991 in-flight structural breakup and crash of a Continental Express EMB-120 near Eagle Lake, Texas (National Transportation Safety Board, 1992)
- The 1992 aerodynamic stall and crash after liftoff of a USAir Fokker F-28 at LaGuardia International Airport (National Transportation Safety Board, 1993)
- The 1993 crash during approach of an Express Airlines II Jetstream 31 near Hibbing Minnesota (National Transportation Safety Board, 1994b)
- The 1994 aerodynamic stall and crash during approach of an American International Airlines DC-8 at Guantanamo Bay, Cuba (National Transportation Safety Board, 1994)
- The 1994 aerodynamic stall and crash during approach of an Atlantic Coast Airlines Jetstream 41 near Columbus, Ohio (National Transportation Safety Board, 1994)
- The 1994 crash during attempted go-around of a USAir DC-9 at Charlotte/Douglas International Airport (National Transportation Safety Board, 1995)
- The 1994 uncontrolled descent and crash during approach of a USAir B-737 near Alaquippa, Pennsylvania (National Transportation Safety Board, 1999)

Specific issues explored during the early 1990s included leadership, decision-making, communication, airport signage, and procedures for air traffic control (National Transportation Safety Board, 1991); visual performance, airport lighting, distraction, and procedural compliance (National Transportation Safety Board, 1991); visual performance, vigilance, workload, and supervision (National Transportation Safety Board, 1991); organizational culture, procedural compliance, and regulatory oversight (National Transportation Safety Board, 1992); visual performance (National Transportation Safety Board, 1993); communication, monitoring, and oversight (National Transportation Safety Board, 1994); decision-making, monitoring, scheduling, and fatigue (National Transportation Safety Board, 1994); planning, communication, monitoring, distraction, and response warnings (National Transportation Safety Board, 1994) (National Transportation Safety Board, 1995); and pilot response to unexpected malfunctions (National Transportation Safety Board, 1999).

In 1994, the NTSB Office of Research and Engineering published a study of 37 flightcrew-involved major accidents that occurred between 1978 and 1990 (National Transportation Safety Board, 1994). This study, which drew heavily upon human performance and operational information contained in NTSB accident reports, has been widely cited in the human factors literature. It contained a number of interesting findings including the following:

- The captain was the flying pilot in more than 80% of the accidents studied.
- Procedural, tactical decision, and monitoring/challenging errors were quite common.
- Almost half of the accidents occurred on the first duty day of a flightcrew pairing.
- Half of the accident-involved captains had been awake for over 12 hours.

The study's focus on flightcrew coordination, use of procedures, decision-making, monitoring, and time since waking reflects the strong emphasis on the concept of CRM that influenced safety work during the 1980s, and the agency's growing interest in flightcrew fatigue as a safety issue.

Investigations supported by the aviation human performance division during the late 1990s included the following:

- The 1996 wheels-up landing of a Continental Airlines DC-9 at Houston Intercontinental Airport (National Transportation Safety Board, 1997)
- The 1996 in-flight fire and crash of a Valujet DC-9 near Miami, Florida (National Transportation Safety Board, 1997)
- The 1996 in-flight loss of control and crash of an Airborne Express DC-8 near Narrows, Virginia (National Transportation Safety Board, 1997)
- The 1996 uncontained engine failure of a Delta MD-88 at Pensacola Regional Airport (National Transportation Safety Board, 1998)
- The 1996 in-flight breakup and crash of a TWA B-747 near East Moriches, New York (National Transportation Safety Board, 2000c)
- The 1997 crash during approach of a Korean Air B-747 near Agana, Guam (National Transportation Safety Board, 2000a)
- The 1997 crash during landing of a FedEx MD-11 at Newark International Airport (National Transportation Safety Board, 2000)
- The 1999 crash during landing of an American Airlines MD-82 at Little Rock National Airport (National Transportation Safety Board, 2001)
- The 1999 crash of an Egypt Air B-767 near Nantucket, Massachusetts (National Transportation Safety Board, 2002)

Issues explored included checklist design, procedural compliance, flightcrew training, and regulatory oversight (National Transportation Safety Board, 1997); equipment, flightcrew training, flightcrew procedures, maintenance procedures, and contract maintenance oversight (National Transportation Safety Board, 1997); stall warning systems, simulator fidelity, and stall recovery procedures (National Transportation Safety Board, 1997); maintenance inspection procedures (National Transportation Safety Board, 1998); eyewitness testimony (National Transportation Safety Board, 2000); briefing, monitoring, fatigue, and alerting systems (National Transportation Safety Board, 2000); stabilized approach procedures and aircraft handling during bounce recovery (National Transportation Safety Board, 2000); decision-making, fatigue, stress, and procedural noncompliance (National Transportation Safety Board, 2001); and mental health and a crash resulting from intentional pilot action (National Transportation Safety Board, 2002).

In 1996, the NTSB added an in-house medical officer to its staff (Garber, 1999). Previously, human performance and survival factors investigators were responsible for documenting and analyzing relevant medical issues with assistance from other U.S. government agencies (such as the Armed Forces Institute of Pathology or Civil Aerospace Medical Institute) as needed when complex issues arose. The new medical officer was not classified as an investigator, so the aviation human performance investigators remained responsible for collecting relevant medical information in aircraft accident investigations. The medical officer was also located in a different agency office and was quite busy, because he supported investigations in all modes of transportation. However, he worked closely with human performance investigators to analyze medical factors, and this gave the agency's human performance investigators more time to devote to the exploration of nonmedical factors underlying human error.

INCREASING SOPHISTICATION AND THE DEVELOPMENT
OF INVESTIGATIVE GUIDANCE

Although CRM remained a paradigm for evaluating crew performance and was extended by various academic and industry safety specialists (Helmreich, 1996) (Helmreich, Merritt, & Wilhelm, 1999) (Klinect, 2005), the work of Reason (1990, 1997) became increasingly influential during the 1990s. Building on the work of Rasmussen (1983), Reason linked error types and cognitive control strategies to redundant layers of safety defenses embedded in a hierarchical organizational structure to create his now ubiquitous *Swiss Cheese* model. According to this model, all productive systems have redundant layers of defenses at the managerial, supervisory, and workplace level that are designed to guard against hazards. Hazards only lead to accidents when multiple layers of this defense structure are breached. Reason's framework eased the categorization of factors shaping human performance, linked them to organizational functions, and promoted consideration of a variety of accident countermeasures. It shifted the focus of prevention efforts off front-line operation and onto the organizational system, prompting increased consideration of organizational influences on behavior, including supervision, management, and organizational culture.

Reason's works profoundly influenced safety thinking for the next two decades. It helped shape the content of the first international guidance for investigating human factors in aircraft accidents, released in 1993 by the ICAO (International Civil Aviation Organization, 1993). ICAO, a UN agency established at the 1944 Convention on International Civil Aviation in Chicago, develops standards and recommended practices for aviation around the world. Guidelines spelled out in Annex 13 to the Convention on International Civil Aviation (International Civil Aviation Organization, 2016) are agreed upon by most of the world's nations.[*] ICAO's initial guidance was the product of an ICAO Flight Safety and Human Factors Study Group that was active from approximately 1989 to 2005 (Maurino, 2017). The group, which included representatives from the United States, Canada, Australia, United Kingdom, France, Japan, and Russia, sought ways to promote and facilitate investigation of human factors in aircraft accidents. After identifying a lack of international guidance on the topic, a decision was made to develop and release written guidance. The Australian Bureau of Air Safety Investigation and the Transportation Safety Board of Canada played a highly influential role in this process.

The resulting guidance material described the need for investigation of human factors, provided a methodology for conducting such investigations, and outlined how findings should be reported. The document's philosophical approach urged an *all-encompassing view* of the accident event. Investigators were urged to identify active and latent failures, and the defenses that failed to prevent them from propagating through the various layers of a productive system (including upper management, supervision, and the activities of front-line employees). For additional inspiration, investigators were referred to the work of Reason (1990), as well as the software–hardware–environment–liveware model developed by Edwards (1972) and refined by Hawkins (1987), which emphasizes the search for mismatches in the interface between people (liveware) and other major elements of an engineered system (software, hardware, environment, and other liveware). In contrast to earlier investigators who argued that "considerable training specialized training is required to effectively unravel and understand the mechanisms motivating human behavior" (Cierbelj, 1970), the circular argued that a human factors investigator needed only a sound aviation knowledge and some human factors training. Specifically, it asserted that

> The measure of the good human factors investigator is not his or her professional qualifications in behavioral sciences, but rather the ability to determine, with the help of specialists if necessary, what information is relevant, to ask the right questions, to listen to the answers and to analyze the information gathered in a logical and practical way. (International Civil Aviation Organization, 1993, p. 17)

---

[*] Minor differences are spelled out in statements of difference submitted to ICAO by individual member states.

The circular described an inductive reasoning process for analyzing human factors evidence and establishing cause–effect relationships. Investigators were encouraged to identify possible links using human factors research and couch their conclusions in terms of probabilities and likelihoods. An approach developed by the Australian Bureau of Air Safety Investigation was highlighted. It involved three steps: (1) testing for existence of a performance-shaping condition, (2) testing for influence of the condition on human performance, and (3) testing for the validity or relevance of the hypothesized relationship to the goal of accident prevention. The provision of this analysis structure addressed the previously-described gap in the NTSB's 1981 model, which did not specify an analysis method.

Strauch (1999) served as a chief of the aviation human performance division during the 1990s and was also heavily involved with the training of NTSB investigators. Strauch was influenced by the error taxonomies of Senders and Moray (1991) and the work of Reason (1990). Strauch also felt that human performance investigation could be improved by the adoption of an investigative protocol. He outlined his personal approach in a 2002 book (Strauch, 2002). In this book, Strauch defined human performance investigation as the application of a systematic study of the *relationships between antecedents and errors* and between errors and accidents (p. 170). Strauch focused on the need to cross-reference various forms of evidence to establish what happened and to draw on the available research literature to evaluate possible causal links. His writing was highly focused on the nuts and bolts of investigation, including the conduct of interviews and the interpretation of recorded flight data and other forms of investigative information. In line with Reason's and ICAO's thinking, he was also concerned with the exploration of company, regulatory, and cultural influences and played a leading role in the organization of a 2013 forum on the topic of safety culture.

Another engineering psychologist who briefly worked for NTSB in the 1990s, Doug Wiegmann, also went on to coauthor a book about investigating human factors (Wiegmann and Shappell, 2003). Wiegmann argued that the *Swiss Cheese* model could be applied even more systematically than had been advocated by ICAO. Wiegmann and Shappell's Human Factors Analysis and Classification system hewed closely to Reason's (1990) model of active and latent failures. It specified four layers of factors that could affect safety in a complex system: (1) unsafe acts of operators, (2) preconditions for unsafe acts, (3) unsafe supervision, and (4) organizational influences. Wiegmann and Shappell (2003) demonstrated high inter-rater reliability when coding the findings of completed accident investigations. However, these codings were applied retroactively for gaining insight into accident investigations that had already been completed (Wiegmann and Shappell, 2001). The model did not specify precisely how human performance investigation was to be conducted, or how linkages between cause and effect should be tested and established. This made the model less useful for conducting an investigation.

Strauch (2015) opined that human performance investigation at the NTSB remained largely unwedded to any particular theory or method of human factors and that accident analysis at the agency was "legalistic, based on logic and internal consistency and the preponderance of evidence" and dependent on the research literature available at the time. After reviewing the works of Heinrich (1941), Perrow (1999), Dekker (2002), Leveson (2004), and Griffin et al. (2010), he argued that "no model has been proposed that appears to have replaced the methodology currently employed for their effectiveness in explaining operator performance in accident investigation scenarios." Strauch argued that a failure to coalescence around any single theoretical model or approach stemmed from the fact that theoretical models could never truly replicate the complexity of accident causation, and that the highly varied underpinnings of causation meant that the theoretical model that was best suited to analyzing a particular accident would naturally vary. He predicted that researchers would continue to apply models of accident causation after investigations had already been completed, with the relevant issues identified by skilled investigators operating from expert knowledge and experience.

Evan Byrne, who joined the NTSB in 1996 and served as the chief of the NTSB's aviation human performance division from 2001 to the present, inherited from Strauch the responsibility for training NTSB investigators in the investigation of human factors. He revised the 1981 human performance investigation model categories in 2005, condensing them from six to the following four: (1) physiological, (2) behavioral, (3) cognitive, and (4) workplace environment factors. Byrne advocated careful study of standards for performance; examination of the characteristics, recent activities, and actions of key personnel; a close look at the workplace environment (both physical and social) in which the performance was occurring; and the identification of limitations of the human cognitive system that could help explain any errors that were documented. Like Strauch, Byrne urged consideration of team, organizational and cultural factors on behavior, leaving it to individual human performance investigators to seek guidance from the published works of authors such as Reason (1997) and Vaughan (1997). However, the human performance investigation guidance contained in the NTSB's Major Team Investigations Manual (2002) has not been revised to reflect these changes.

In 2014, ICAO released the aforementioned updated guidance on the investigation of human performance in aircraft accident investigation. (International Civil Aviation Organization, 2014)

In addition to specifying the objectives of human performance investigation. ICAO urged human factors investigations to be systematic, fully integrated into the broader investigation, and based on a systems-oriented approach. Additional guidance is available in the *Manual of Aircraft Accident and Incident Investigation* (International Civil Aviation Organization, 2014), *Human Factors Training Manual* (International Civil Aviation Organization, 1998), and Circular 240 (International Civil Aviation Organization, 1993).

THE RISE OF SAFETY MANAGEMENT AND DATA-DRIVE SAFETY EFFORTS

An important national development in the investigation of aviation safety issues, including those related to human factors, was the formation of a White House Commission on Aviation Safety and Security in 1996, a month after the accident involving TWA flight 800 (National Transportation Safety Board, 2000). This commission was chaired by U.S. Vice President Al Gore and included participation from numerous other public officials, including the chairman of the NTSB (White House Commission on Aviation Safety and Security, 1997). Facing a predicted tripling in the number of airline passengers over the next decade, the Commission expressed concerned about a potential increase in the frequency of accidents. It advocated aggressive action to reduce the already low accident rate by 80%. To accomplish this goal, the commission urged several actions. Some of these actions were technical, such as advocating more widespread installation of enhanced ground proximity warning systems. Others were more abstract, such as urging the formation of a government-industry partnership to improve and integrate research, standards, regulations, procedures, and infrastructure. A National Civil Aviation Review Commission report published months later reiterated this and urged the FAA and industry to develop a comprehensive integrated safety plan that would establish priorities based on "objective, quantitative analysis of safety information and data" (National Civil Aviation Review Commission, 1997).

The Commercial Aviation Safety Team (CAST), and FAA-industry partnership, was established as a result of these recommendations. Formed in 1998, CAST's mission was to "to significantly increase public safety by adopting an integrated, data-driven strategy to reduce the fatality risk in commercial air travel" (Commercial Aviation Safety Team, 2017). In the two decades since its founding, CAST has pioneered methods for systematically analyzing safety-related information and converting its findings into tangible safety improvements undertaken by the aviation industry. By leveraging increasingly large repositories of confidential operational safety data, accident reports, and other sources of data, CAST has completed analysis projects on major areas of aviation risk with human factors components, ranging from controlled flight into terrain accidents (Commercial Aviation Safety Team, 1998), to approach and landing accidents (Commercial Aviation Safety Team, 1999), to in-flight loss of control accidents. CAST studies produced a litany of related safety recommendations on which the industry has actively worked. CAST has been a major success story for aviation safety improvement and government-industry cooperation.

CAST's efforts have been complementary to the rise of the safety management system, a paradigm that involves managing safety by measuring safety indicators so that it can be controlled like other business processes. Since CAST was formed, standards for safety management systems (SMS) development and implementation have been created for use by individual operators and national regulatory agencies (Federal Aviation Administration, 2010; International Civil Aviation Organization, 2012). The rise of SMS is an important development in the history of aviation safety because it has move the locus of control from reactive accident investigations and centralized government control to proactive analysis and decentralized control. This shift, in turn, resulted from two earlier changes, a move toward the deregulation of airlines that began in the 1980s, and the development of increasingly sophisticated technologies for monitoring and recording the progress of airline flights and flagging safety-related incidents or deviations from a desirable range of operating parameters. The FAA has been intimately involved as a partner in such efforts, such as the Aviation Safety Information Analysis and Sharing system. Until recently, the NTSB was excluded from direct participation due to concerns about the confidentiality of such sensitive data and an unwillingness to involve a transparent public agency in such analysis.

THEORETICAL PERSPECTIVES

Views on human performance evolved considerably over the last 100 years. The late nineteenth and early twentieth century saw much attention paid to the identification of ideal methods for performing work and the selection of workers who were best-suited for particular tasks. Rapid advances in technology and the massive mobilization of personnel that occurred during World War II, however, caused the emphasis to shift toward examination of the "psychological capacities and limitations of individuals" and the design of equipment that was more compatible with their needs (Fitts, 1947). The impact of this shift on perspectives in aviation safety is evident in the writing of McFarland (1946), who argued that although studies of air transport accidents indicated that 75%–85% could be attributed to human error, "neither the operating characteristics of the plane nor the performance of the pilot can be considered as completely separate variables."

This change in the approach to development of technological systems gave rise to the human factors profession (Sanders & McCormick, 1993), and it was accompanied by a shift in perspectives on human error. Reason (2000) has made the distinction between *person* and *system* views of error. According to Reason, the person view of error focuses on the personal accountability of individuals and blames them for forgetfulness, inattention, or moral weakness, whereas the system view sees error as part of the human condition and emphasizes the role of system design in preventing and mitigating error. Similarly, Dekker (2014) has written about the shift from an *old view* to a *new view* of error. According to Dekker, the *old view* regards human error as the cause of accidents, whereas the *new view* sees it as the consequence of upstream factors involving the use of technology and the design of tools and tasks. Although Dekker describes this shift as the natural progression of safety thinking, Reason (2015) argues that the person view is the oldest and still the most commonly held view of error.

System-oriented views of human error are widely embraced in the human factors community, but they are not universally accepted. The person model is particularly appealing in cultures that emphasize personal responsibility and freedom of choice. It is also reinforced by various human cognitive biases. These include the hindsight bias (a tendency to see outcomes as more predictable after they have occurred) and the fundamental attribution error (a tendency to attribute others' failings to personal characteristics while de-emphasizing contextual factors). Additional biases include the *just world* hypothesis, which is the tendency to believe that bad things happen to bad people, and outcome bias, which is a tendency to discount the idea that bad outcomes can result from good decisions. From an organizational standpoint, placing blame on individual operators is less complex and consumes fewer investigative resources. It also compartmentalizes responsibility, reducing the likelihood that powerful interests will be threatened, and it may provide reassurance to an anxious public concerned about finding quick solutions to avoid the potential for recurrence.

The person model has not been particularly useful from the standpoint of learning about safety-related risks and preventing accidents, however. It leads to premature closure of investigations that stop short of identifying systemic issues underlying recurring patterns of accidents and, therefore, to less learning and fewer, less useful safety recommendations. For this reason, the charters of many safety organizations include a provision specifying that the purpose of a safety investigation is the identification of safety hazards and the development of recommendations for improvement of safety, rather than allocation of blame. Still, accident investigators are human. Many have had little formal training in psychology or human factors, and all organizations are subject to the pressures and biases described earlier. For these reasons, the person model survives—even among members of the professional accident investigation community.

NOTABLE SUCCESSES

One outcome of decades of professional accident investigation has been the encouragement of new safety technologies. Enhanced ground proximity warning systems have reduced the risk of controlled flight into terrain accidents, and traffic collision alerting systems have reduced the risk of midair collisions, and ground radar surveillance systems such as ASDE-X have improved controller awareness of airport surface traffic. Although the evolution of aviation technology has introduced new human factors challenges, such as how to keep humans in the loop when more and more of their cognitive work is being shared with machines, it has also contributed to extremely high levels of safety that exist in the airline industry today. In fact, the system has become so safe that several recent years have passed with no fatal accidents among major U.S. airlines, despite the carriage of hundreds of millions of passengers.

CONTINUING CHALLENGES

Human factors, in general, is concerned with "understanding of interactions among humans and other elements of a system" and the application of "theory, principles, data, and other methods to design in order to optimize human well-being and overall system performance" (International Ergonomics Association, 2017). Moreover, human, organizational, and systemic factors are ubiquitous causal factors for accidents in hazardous industries, meaning human factors expertise is critical to understanding accidents. However, most accident investigators are technical or operational specialists (Reason, 2002), which sometimes leads to a lack of familiarity with human factors.

In the 1940s, the division of labor between humans and machines was relatively straightforward. Humans and machines shared responsibility for simple control tasks, and humans had exclusive domain over complex, higher order tasks. Human factors specialists of the day devoted attention to determining the optimal arrangements and physical properties of controls and displays to minimize pilot workload and errors (Fitts, 1947). The introduction of microcomputers led to more automation and more complex technological systems and new pathways to failure, and the issue of mismatches between humans and technological systems has become more complex.

ICAO also asserts that the "sole objective of the investigation of an accident or incident shall be the prevention of accidents and incidents .... It is not the purpose of this activity to apportion blame or liability." In this investigator's personal experience, however, it appears that the public, journalists, and even some aviation industry professionals can have difficulty separating the search for safety-related explanations from allocation of blame or responsibility.

Although safety investigation is focused on learning from failure, blame fulfills needs that are not necessarily aligned with this goal. The inclination toward a blame-oriented perspective may stem from cultural views about free will and professional responsibility, cognitive biases, organizational resource limitations, or from reasons of administrative convenience (Maurino, Reason et al. 1995). As a result, there is sometimes a significant gap between the views of the human factors investigator and the views of other constituencies to an accident investigation. Successfully navigating this situation requires a sophisticated understanding of human factors and a well-developed interpersonal skillset. As few accident investigation agencies are led by human factors experts, most human factors investigators cannot rely on positional authority to exert influence.

REFERENCES

Barnhart, W., Billings, C., Cooper, G., Gilstrap, R., Lauber, J., Orlady, H., ... Stephens, W. (1975). A method for the study of human factors in aircraft operations. *Report No. NASA TM X-62,472*. Moffet Field, CA: U.S. Government Printing Office.

Billings, C. E., Lauber, J. K., Funkhouser, H., Lyman, E. G., and Huff, E. M. (1976). *NASA Aviation Safety Reporting System Quarterly Report Number 76-1, April 15, 1976 through July 14, 1976. Report No. NASA TM X-3445*. Washington, DC: U.S. Government Printing Office.

Billings, C. R. (1984). Human factors in aircraft accidents: Results of a 7-year study. *Aviation, Space, and Environmental Medicine*, 5(10):960–965.

Bradley, P. (1984, September). What did the pilot do and why? A new department at the NTSB searches on the causes of "pilot error." *Flying*, 111(9):65–66.

Cierbelj, A. (1970). Conduct of an investigation, Part 2. *First Annual Meeting of the Society of Air Safety Investigators*, (p. 111). Washington, DC.

Civil Aeronautics Board. (1961). *Statistical Review: U.S. Air Carrier Accidents Calendar Year 1959*. Washington, DC: Author.

Comisión de Investigación de Accidentes e Incidentes de Aviación Civil. (1978). *Collision Aeronaves Boeing 747 PH-BUF DE K.L.M. Y Boeing 747 N 736 PA de Panam en Los Rodeos (Tenerife) el 27 de Marzo de 1977*. Madrid, Spain.

Commercial Aviation Safety Team. (1998). *Controlled flight into terrain (CFIT), Joint Safety Analysis Team (JSAT), Results and analysis, September 1, 1998, Revision A, November 20, 1998*. Available at http://cast-safety.org/pdf/jsat_cfit.pdf: Author.

Commercial Aviation Safety Team. (1999). *Approach and Landing, Joint Safety Analysis Team (JSAT), Results and Analysis, September 10, 1999*. Available at http://cast-safety.org/pdf/jsat_approach-landing.pdf: Author.

Commercial Aviation Safety Team. (2008). *Safety Enhancement 30, Revision 5, August, 2008, Mode Awareness and Energy State Management, Final Report*. Available at http://cast-safety.org/pdf/cast_automation_aug08.pdf.: Author.

Commercial Aviation Safety Team. (2017). *Background*. Available at http://www.cast-safety.org/apex/f?p=18 0:1:13716458632650::NO::P1_X:background, Retrieved February 10, 2017.

Cooper, G. E., White, M. D., and Lauber, J. K. (1980). Resource management on the flight deck. *NASA Conference Publication 2120*. San Francisco, CA: U.S. Government Printing Office: National Air and Space Administration.

Danaher, J. (1971). The recording of human factors information in aircraft accidents. *Proceedings of the Second Annual Meeting of the Society of Air Safety Investigators*, Los Angeles, CA, pp. 122–129.

Dekker, S. (2002). Reconstructing human contributions to accidents: The new view on error and performance. *Journal of Safety Research*, 33(3):371–385.

Dekker, S. (2014). *Safety Differently: Human Factors for a New Era* (2nd ed.), (p. 312). Boca Raton, FL: CRC Press.

Doyle, B. (1968). Aeromedical investigations of civil aircraft accidents. In W. Reals (Ed.), *Medical Investigation of Aviation Accidents* (pp. 38–70). Chicago, IL: College of American Pathologists.

Edwards, E. (1972). Man and machine: Systems for safety. *Proceedings of the British Airline Pilots Association Technical Symposium,* (pp. 21–36). London, UK: British Airline Pilots Association.

Federal Aviation Administration. (2010). *Safety management systems for aviation service providers. AC No. 120-92A.* Washington, DC: U.S. Government Printing Office.

Fitts, P. (1947). Psychological research on equipment design in the AAF. *American Psychologist,* 2(3):93–s98.

Garber, M. (1999). Aeromedical accident investigation at the National Transportation Safety Board: A view from the inside. *Aviation, Space, and Environmental Medicine,* 70(4):407.

Griffin, T. G., Young, M. S., and Stanton, N. A. (2010). Investigating accident causation through information network modeling. *Ergonomics,* 53(2):198–210.

Hawkins, F. (1987). *Human Factors in Flight.* Aldershot, UK: Gower Technical Press.

Heinrich, H. W. (1941). *Industrial Accident Prevention* (2nd ed.). New York: McGraw-Hill.

Helmreich, R. (1996). The evolution of crew resource management. *Paper Presented at the IATA Human Factors Seminar,* Warsaw, Poland, October 31, 1996.

Helmreich, R., Merritt, A. C., and Wilhelm, J. A. (1999). The evolution of crew resource management training in commercial aviation. *International Journal of Aviation Psychology,* 9, 19–32.

International Civil Aviation Organization. (1993). *Human Factors Digest No. 7: Investigation of Human Factors in Accidents and Incidents. ICAO Circular 240-AN/144.* Montréal, Canada: Author.

International Civil Aviation Organization. (1998). *Human Factors Training Manual (Doc 9683).* Montréal, Canada: Author.

International Civil Aviation Organization. (2012). *Safety Management Manual* (3rd ed.). *Doc 9859, AN/474.* Montreal: Author.

International Civil Aviation Organization. (2014). *Manual of Aircraft Accident and Incident Investigation* (Doc 9756, 2nd ed.) Montréal, Canada: Author.

International Civil Aviation Organization. (2016). *Annex 13 to the Convention on International Civil Aviation* (11th ed.) Montréal, Canada: Author.

International Ergonomics Association. (2017). *Definitions and Domains of Ergonomics.* http://www.iea.cc/whats/index.html. Retrieved February 24, 2017.

Kayten, P. (1993). The accident investigator's perspective. In E. K. Wiener (Ed.), *Cockpit Resource Management* (pp. 283–314). San Diego, CA: Academic Press.

Klinect, J. (2005). *Line Operations Safety Audit: A Cockpit Observation Methodology for Monitoring Commercial Airline Safety Performance.* Doctoral dissertation. Austin, TX: University of Texas at Austin.

Leveson, N. (2004). A new accident model for engineering safer systems. *Safety Science,* 42(4):237–270.

Macy, A. (1974). Comments of A. Martin Macy, Vice President—Operations, national air transportation associations. *Proceedings of the Society of Air Safety Investigators Fifth Annual International Seminar,* October 1–3, 1974. Washington, DC.

Maurino, D. (2017). *Personal communication with Daniel Maurino, former Coordinator of the ICAO Flight Safety and Human Factors Study Programme,* February 10, 2017.

Maurino, D. E., Reason, J., Johnston, N., and Lee, R. B. (1995). *Beyond Aviation Human Factors.* New York: Routledge.

McFarland, R. (1946). *Human Factors in Air Transport Design.* New York: McGraw-Hill.

Miller, C. (1954). Applying lessons learned from accident investigations to design through a systems safety concept. *Paper Presented at the Flight Safety Foundation Seminar,* Santa Fe, New Mexico.

Miller, C. (1965). *Safety and Semantics.* Los Angeles, CA: University of Southern California, Aerospace Safety Division.

Miller, C. (1971). The bureau of aviation safety views human factors in aviation accident investigation. *Society of Air Safety Investigators Second Annual Air Safety Seminar,* October 26–28, 1971, (pp. 18–22). Los Angeles, CA.

National Civil Aviation Review Commission. (1997). *Avoiding Aviation Gridlock and Reducing the Accident Rate.* Washington, DC: U.S. Government Printing Office.

National Transportation Safety Board. (1971). *Annual Review of Aircraft Accident Data: U.S. Air Carrier Operations Calendar Year 1969.* Washington, DC: U.S. Department of Commerce, National Technical Information Service.

National Transportation Safety Board. (1972). Aircraft accident report, Allegheny Airlines, Inc., *Allison Prop Jet Convair 340/440, N5832, New Haven, Connecticut, June 7, 1971.* (Report No. NTSB-AAR-72-20). Washington, DC: U.S. Government Printing Office.

National Transportation Safety Board. (1973). Aircraft accident report: Eastern Air Lines, Inc., *L-101, N310EA, Miami, Florida, December 29, 1972* (Report No. NTSB-AAR-73-14). Washington, DC: U.S. Government Printing Office.

National Transportation Safety Board. (1975a). Aircraft accident report, Eastern Air Lines, Inc., *Douglas DC-9-31, N8984E, Charlotte, North Carolina, September 11, 1974.* (Report No. NTSB AAR/75-9). Washington, DC: U.S. Government Printing Office.

National Transportation Safety Board. (1975b). Aircraft accident report, Trans World Airlines, Inc., *Boeing 727-231, N54328, Berryville, Virginia, December 1, 1974.* (Report No. NTSB AAR/75-16). Washington, DC: U.S. Government Printing Office.

National Transportation Safety Board. (1975c). Aircraft accident report, Western Air Lines, Inc., *Boeing 737-200, N4527W, Casper, Wyoming, March 31, 1975.* (Report No. NTSB/AAR-75-15). Washington, DC: U.S. Government Printing Office.

National Transportation Safety Board. (1976). Aircraft accident report, Alaska Airlines, Inc., *Boeing 727-81, N124AS, Ketchikan, Alaska, April 5, 1976.* (Report No. AAR/76-20). Washington, DC: U.S. Government Printing Office.

National Transportation Safety Board. (1979). Aircraft accident report, United Airlines, Inc., *McDonnell-Douglas, DC-8-61, N8082U, Portland, Oregon, December 28, 1978.* (Report No. NTSB-AAR-79-7). Washington, DC: U.S. Government Printing Office.

National Transportation Safety Board. (1980). Aircraft accident report: Downeast Airlines, Inc., *DeHavilland DHC-6-200, N68DE, Rockland, Maine, May 30, 1979.* (Report No. NTSB-AAR-80-5). Washington, DC: U.S. Government Printing Office.

National Transportation Safety Board. (1981). Aircraft accident report: Cascade Airways, Inc., *Beechcraft 99A, N390CA, Spokane, Washington, January 20, 1981.* (Report No. NTSB-AAR-81-11). Washington, DC: U.S. Government Printing Office.

National Transportation Safety Board. (1982a). *Air Florida, Inc., Boeing 737-222, N62AF, Collision with 14th street bridge, Near Washington National Airport, Washington, D.C., January 13, 1982.* (Report No. NTSB-AAR-82-8). Washington, DC: U.S. Government Printing Office.

National Transportation Safety Board. (1982b). *Human Performance Program.* Washington, DC: Unpublished.

National Transportation Safety Board. (1982c). *NTSB Aviation Accident Investigation Protocol for Human Performance.* Washington, DC: NTSB.

National Transportation Safety Board. (1986a). Aircraft accident report, Delta Airlines, Inc., *Lockheed L-1011-385-1, N726DA, Dallas/Fort Worth International Airport, Texas, August, 2, 1985.* (Report No. NTSB/AAR-86/05). Washington, DC: U.S. Government Printing Office.

National Transportation Safety Board. (1986b). Aircraft accident report, Galaxy Airlines, Inc., *Lockheed Electra-L-188C, N5532, Reno, Nevada, January 21, 1985.* (Report No. NTSB/AAR-86/01). Washington, DC: U.S. Government Printing Office.

National Transportation Safety Board. (1987a). *Aircraft accident report, Collision of Aeronaves de Mexico, S.A., McDonnell Douglas DC-9-32, XA-JED and Piper PA-28-181, N4891F, Cerritos, California, August 31, 1986.* (Report No. NTSB/AAR-87/07). Washington, DC: U.S. Government Printing Office.

National Transportation Safety Board. (1987b). *Annual Review of Aircraft Accident Data: U.S. Air Carrier Operations, Calendar Year 1983.* Washington, DC: U.S. Government Printing Office.

National Transportation Safety Board. (1988a). *Aircraft Accident Report, Continental Airlines, Inc., Flight 1713, McDonnell Douglas DC-9-14, N626TX, Stapleton International Airport, Denver, Colorado, November 15, 1987.* Washington, DC: U.S. Government Printing Office.

National Transportation Safety Board. (1988b). *Aircraft Accident Report, Northwest Airlines, Inc., McDonnell Douglas DC-9-82, N312RC, Detroit Metropolitan Wayne County Airport, Romulus, Michigan, August 16, 1987.* (Report No. NTSB/AAR-88/05). Washington, DC: U.S. Government Printing Office.

National Transportation Safety Board. (1989). *Aircraft accident report, Delta Air Lines, Inc., Boeing 727-232, N473DA, Dallas-Fort Worth International Airport, Texas, August 31, 1988.* (Report No. NTSB/AAR-89/04). Washington, DC: U.S. Government Printing Office.

National Transportation Safety Board. (1990). *Aircraft accident report, United Airlines Flight 232, McDonnell Douglas DC-10-10, Sioux Gateway Airport, Sioux City Iowa, July 19, 1989.* Washington, DC: U.S. Government Printing Office.

National Transportation Safety Board. (1991a). *Aircraft accident report, Northwest Airlines, Inc., Flights 1482 and 299, Runway incursion and collision, Detroit Metropolitan/Wayne County Airport, Romulus, Michigan, December 3, 1990.* (Report No. NTSB/AAR-91/05). Washington, DC: U.S. Government Printing Office.

National Transportation Safety Board. (1991b). *Aircraft accident report, Runway collision of USAir Flight 1493, Boeing 737 and Skywest Flight 5569 Fairchild Metroliner, Los Angeles International Airport, Los Angeles, California, February 1, 1991.* (Report No. NTSB/AAR-91/08). Washington, DC: U.S. Government Printing Office.

National Transportation Safety Board. (1991c). *Runway Collision of Eastern Airlines Boeing 727, Flight 111 and Epps Air Service Beechcraft King Air A100, Atlanta Hartsfield International Airport, Atlanta, Georgia, January 18, 1990.* (NTSB Report No. NTSB/AAR-91/03). Washington, DC: U.S. Government Printing Office.

National Transportation Safety Board. (1992). *Aircraft Accident Report, Britt Airways, Inc., DBA Continental Express Flight 2574, In-flight structural breakup, EMB-120RT, N33701, Eagle Lake, Texas, September 11, 1991.* (Report No. NTSB/AAR-92/04). Washington, DC: U.S. Government Printing Office.

National Transportation Safety Board. (1993a). *Aircraft Accident Report, Takeoff Stall in Icing Conditions, USAir flight 405, Fokker F-28, N485US, Laguardia Airport, Flushing, New York, March 22, 1992.* (Report No. NTSB/AAR-93/02). Washington, DC: U.S. Government Printing Office.

National Transportation Safety Board. (1993b). *Annual Review of Aircraft Accident Data, U.S. Air Carrier Operations, Calendar Year 1989.* (Report No. NTSB/ARC-93/01). Washington, DC: U.S. Government Printing Office.

National Transportation Safety Board. (1994a). *A Review of Flightcrew-involved Major Accidents of U.S. Air Carriers, 1978 Through 1990.* (Report No. NTSB/SS-94/01). Washington, DC: U.S. Government Printing Office.

National Transportation Safety Board. (1994b). *Aircraft accident report, Controlled collision with terrain, Express II Airlines, Inc./Northwest Airlink Flight 5719, Jetstream BA-3100, N334PX, Hibbing, Minnesota, December 1, 1993.* (Report No. NTSB/AAR-94/05). Washington, DC: U.S. Government Printing Office.

National Transportation Safety Board. (1994c). *Aircraft accident report, Stall and loss of control on final approach, Atlantic Coast Airlines, Inc., United Express Flight 6291, Jetstream 4101, N304UE, Columbus, Ohio, January 7, 1994.* (Report No. NTSB/AAR-94/07). Washington, DC: U.S. Government Printing Office.

National Transportation Safety Board. (1994d). *Aircraft accident report, Uncontrolled collision with terrain, American International Airways Flight 808, Douglas DC-8-61, N814CK, U.S. Naval Air Station, Guantanamo Bay, Cuba, August 18, 1993.* (Report No. NTSB/AAR-94/04). Washington, DC: U.S. Government Printing Office.

National Transportation Safety Board. (1995). *Aircraft accident report, Flight into terrain during missed approach, USAir Flight 1016, DC-9-31, N954VJ, Charlotte/Douglas International Airport, Charlotte, North Carolina, July 2, 1994.* (Report No. NTSB/AAR-95/03). Washington, DC: U.S. Government Printing Office.

National Transportation Safety Board. (1997a). *Aircraft accident report, In-flight fire and impact with terrain, Valujet Airlines Flight 592, DC-9-32, N904VJ, Everglades, Near Miami, Florida, May 11, 1996.* (Report No. NTSB/AAR-97/06). Washington, DC: U.S. Government Printing Office.

National Transportation Safety Board. (1997b). *Aircraft accident report, Uncontrolled flight into terrain, ABX Air (Airborne Express), Douglas DC-8-63, N827AX, Narrows, Virginia, December 22, 1996.* (Report No. NTSB/AAR-97/05). Washington, DC: U.S. Government Printing Office.

National Transportation Safety Board. (1997c). *Aircraft accident report, Wheels-up landing, Continental Airlines Flight 1943, Douglas DC-9 N10556, Houston, Texas, February 19, 1996.* (Report No. NTSB/AAR-97/01). Washington, DC: U.S. Government Printing Office.

National Transportation Safety Board. (1998). *Aircraft accident report, Uncontained engine failure, Delta Air Lines Flight 1288, McDonnell Douglas MD-88, N927DA, Pensacola, Florida, July 6, 1996.* (Report No. NTSB/AAR-98/01). Washington, DC: U.S. Government Printing Office.

National Transportation Safety Board. (1999). *Aircraft accident report, Uncontrolled descent and collision with terrain, USAir Flight 427, Boeing 737-300, N513AU, Near Aliquippa, Pennsylvania, September 8, 1994.* (Report No. NTSB/AAR-99/01). Washington, DC: U.S. Government Printing Office.

National Transportation Safety Board. (2000a). *Aircraft accident report, Controlled flight into terrain, Korean Air Flight 801, Boeing 747-300, HL7468, Nimitz Hill, Guam, August 6, 1997.* (Report No. NTSB/AAR-00/01). Washington, DC: U.S. Government Printing Office.

National Transportation Safety Board. (2000b). *Aircraft accident report, Crash during landing, Federal Express, Inc., McDonnell Douglas MD-11, N611FE, Newark International Airport, Newark, New Jersey, July 31, 1997.* (Report No. NTSB/AAR-00/02). Washington, DC: U.S. Government Printing Office.

National Transportation Safety Board. (2000c). *Aircraft accident report, In-flight breakup over the Atlantic Ocean, Trans World Airlines Flight 800, Boeing 747-131, N93119, Near East Moriches, New York, July 17, 1996.* (Report No. NTSB/AAR-00/03). Washington, DC: U.S. Government Printing Office.

National Transportation Safety Board. (2001). *Aircraft accident report, Runway overrun during landing, American Airlines Flight 1420, McDonnell Douglas MD-82, N215AA, Little Rock, Arkansas, June 1, 1999.* (Report No. NTSB/AAR-01/02). Washington, DC: U.S. Government Printing Office.

National Transportation Safety Board. (2002a). *Aircraft accident brief, EgyptAir Flight 990, Boeing 767-366ER, SU-GAP, 60 miles south of Nantucket, Massachusetts, October 31, 1999.* (Report No. NTSB/AAB-02/01). Washington, DC: U.S. Government Printing Office.

National Transportation Safety Board. (2002b). *Annual review of aircraft accident data: U.S. air carrier operations, Calendar year 1999.* (Report No. NTSB/ARC-02/03). Washington, DC: U.S. Government Printing Office.

National Transportation Safety Board. (2002c). *Aviation Investigation Manual: Major Team Investigations.* Washington, DC: Author.

National Transportation Safety Board. (2017). *History of the National Transportation Safety Board.* https://www.ntsb.gov/about/history/pages/default.aspx. Retrieved February 22, 2017.

Perrow, C. (1999). *Normal Accidents.* Princeton, NJ: Princeton University Press.

Predmore, S. (1991). Microcoding of communications in accident investigation: Crew coordination in United 811 and United 232. *Paper Presented at the 6th International Symposium on Aviation Psychology.* Columbus, OH: The Ohio State University.

Rasmussen, J. (1983). Skills, rules, knowledge: Signals, signs and symbols and other distinctions in human performance models. *IEEE Transactions: Systems, Man & Cybernetics,* SMC-13(3):257–267.

Reason, J. (1990). *Human Error.* New York: Cambridge University Press.

Reason, J. (1997). *Managing the Risks of Organizational Accidents.* Burlington, VT: Ashgate Publishing.

Reason, J. (2000). Human error: Models and management. *British Medical Journal,* 320(7237):768–770.

Reason, J. (2002). Forward. In B. Strauch (Ed.), *Investigating Human Error: Incidents, Accidents and Complex Systems* (pp. xi–xvi). Aldershot, UK: Ashgate Publishing.

Reason, J. (2015). *A Life in Error: From Little Slips to Big Disasters.* Burlington, VT: Ashgate Publishing.

Sanders, M. and McCormick, E. J. (1993). *Human Factors in Engineering and Design* (7th ed.). New York: McGraw-Hill.

Schleede, R. (1979). Application of a decision-making model to the investigation of human error in aircraft accident. *Proceedings of the Tenth International Seminar of the International Society of Air Safety Investigators,* Montreal, Canada, September 24–27, 1979, (pp. 66–82).

Senders, J. W. and Moray, N. P. (1991). *Human Error: Cause, Prediction, and Reduction.* Hillsdale, NJ: Lawrence Erlbaum Associates.

Stoklosa, J. (1983). Accident investigation of human performance factors. *Proceedings of the Second Symposium on Aviation Psychology.* Columbus, OH: The Ohio State University.

Strauch, B. (1999). Managing aviation human factors aircraft accident investigations: Lessons learned. *Proceedings of the Human Factors and Ergonomics Society 43rd Annual Meeting* (pp. 957–961). Houston, TX: Human Factors and Ergonomics Society.

Strauch, B. (2002). *Investigating Human Error: Incidents, Accidents, and Complex Systems.* Aldershot, UK: Ashgate Publishing.

Strauch, B. (2015). The use of investigation models to explain accident causation and operator performance. *Proceedings of the Human Factors and Ergonomics Society 59th Annual Meeting* (pp. 961–965). Los Angeles, CA: Human Factors and Ergonomics Society.

U.S. Centennial of Flight Commission. (2017). *The Post Office Flies the Mail, 1918–1924.* Article retrieved February 22, 2017, http://www.centennialofflight.net/essay/Government_Role/1918-1924/POL3.htm.

Vaughan, D. (1997). *The Challenger Launch Decision: Risky Technology, Culture, and Deviance at NASA.* Chicago, IL: University of Chicago Press.

Walhout, G. (1981). Proceedings of the Twelfth International Seminar of the International Society of Air Safety Investigators Hyatt-Regency Hotel. *ISASI Forum,* 14(4):21–22.

White House Commission on Aviation Safety and Security. (1997). *Final Report to President Clinton,* February 12, 1997. Washington, DC: U.S. Government Printing Office.

Wiegmann, D. A., and Shappell, S. A. (2001). Human error analysis of commercial aviation accidents: Application of the human factors analysis and classification system (HFACS). *Aviation, Space, and Environmental Medicine,* 72(11):1006–1016.

Wiegmann, D. A., and Shappell, S. A. (2003). *A Human Error Approach to Aviation Accident Analysis: The Human Factors Analysis and Classification System.* Aldershot, UK: Ashgate Publishing.

Wilson, G. A. (1949). *Air Transportation.* New York: Prentice Hall.

Section II

Regulations and Certification

7 An Overview of Flight Deck Human Factors in the U.S. Aviation Regulatory System

Kathy Abbott

CONTENTS

Introduction...137
Aviation Regulation—A Brief Primer 137
Regulatory Requirements and Guidance for Flight Deck...140
Airworthiness Requirements for Equipment Design—Examples 141
    14 CFR Part 25 Section 25.1329 Flight Guidance Systems....................................141
    14 CFR 25.1302 Installed Systems and Equipment for Use by the Flight Crew and
    European Aviation Safety Agency Certification Specification CS 25.1302.............142
Other Aircraft Certification Standards 142
    Flight-Crew Training and Operational Requirements...142
Concluding Remarks..144
References...144

INTRODUCTION

The current chapter presents a perspective of the U.S. regulatory system for aviation, administered by the Federal Aviation Administration (FAA). The chapter begins with a brief description of the underlying philosophy of aviation regulation as a form of risk management. It then described the history and philosophical basis of some of the key aviation regulations in the U.S. system.

The chapter will then focus on how and where flight-deck human factors (HF) are integrated in the regulations with an emphasis on flight deck–related applications. In particular, this chapter will discuss how HF applies within the regulatory structure for equipment design, flight-crew training, flight-crew procedures, and flight-crew operations.

AVIATION REGULATION—A BRIEF PRIMER

The legal origin for the FAA's regulatory activities is founded in the U.S. Constitution and is generally considered to have begun with the Air Commerce Act, enacted in 1926.[*] This act commissioned the Secretary of the Department of Commerce to be responsible for fostering air commerce, issuing and enforcing air traffic rules, certifying pilots and aircraft, and operating and maintaining air navigation aids (NAVAID). Birnbach and Longridge (1993) provided a historical perspective on the FAA and provide a more detailed history of the evolution of the legal structure.

---

[*] The material in this section is based on the regulatory primer perspective from the RTCA Task Force on Certification (RTCA 1999). In particular, the contributions of several members of RTCA Task Force 4 on Certification were instrumental in developing the material for the regulatory primer, including John Ackland, Tony Broderick, and Tom Imrich.

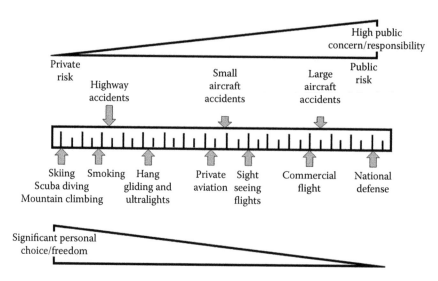

FIGURE 7.1 *Personal* versus *Public* risk assumption.

One way to look at regulation is to consider it as a form of safety risk management, and to consider where a society places certain activities on a notional continuum of safety risk. Figure 7.1 shows such a continuum, which describes private risk at one end and public risk at the other. It depicts different activities and where they may be placed by a society. In the continuum shown in the figure, the left end of the continuum represents private risk, with activities such as scuba diving, mountain climbing, and skiing as examples of items that a society might place on this end. These activities at this end correspond to significant personal choice and freedom, with low public concern and responsibility.

Activities considered to be of high public concern and responsibility are represented on the right end of the continuum, and examples include commercial flights and national defense. In the United States (and many other societies), large aircraft accidents are placed to the right end of the continuum, because they are considered to be of high public concern and responsibility. Highway accidents and small airplane accidents are further to the left on the scale, as this is the choice of the society.

Another society may choose to place these activities and safety-related events at different points on the continuum. For example, another society may choose to place private and commercial aviation at the same *risk* point (and therefore the same level of public concern).

Society also determines the role of the government in managing or mitigation of safety risk. Consider the continuum shown in Figure 7.2, which illustrates where the potential government role may be. The leftmost end is intended to represent activities or concerns that are primarily personal and commercial. At this end, the society chooses to limit the government's role—possibly in a role to simply enable the activity. At the rightmost end, one can find activities or concerns that are inherently governmental, such as national defense. At this end, the government actually conducts or controls the activity.

The location of any particular activity is driven by the will of the society for which the government works. In the United States, for example, private aviation is considered to be less of a public concern, and therefore less of a public responsibility by the government. Commercial aviation is much more a public concern and is expected to have the highest level of safety. Therefore, legislation requires that government oversight and standards are more stringent for commercial aviation than for private aviation.

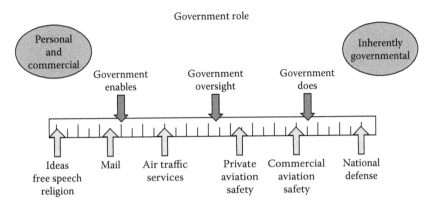

FIGURE 7.2   Continuum depicting potential governmental role.

The aviation regulations were initiated through the Air Commerce Act of 1926, which assigned responsibility and regulatory authority for aviation to the Aeronautics Branch of the Department of Commerce. This branch had the following objectives:

1. Establish airworthiness standards and associated system of aircraft registration.
2. Administer examination and licensing procedures for aviation personnel and facilities.
3. Establish uniform rules for air navigation.
4. Establish new airports.
5. Encourage the development of civil aviation.

Basically, the fundamental governmental responsibilities are to

1. Assure that aircraft do not fall on the public (assure the airworthiness of the aircraft);
2. Assure the *highest level of safety* for public transportation;
3. 3 Assure at least a basic level of safety for other *certificated aircraft* passengers;
4. Assure that aircraft can satisfy safety-related inter-aircraft responsibilities for mutual separation (e.g., the requirement for altitude-encoding transponders in certain airspace)

The means that are used to accomplish these responsibilities are as follows:

- Certifying air vehicles and supporting ground elements—if, and as necessary
- Establishing operating rules—*rules of the road*
- Providing or empowering certain capabilities (e.g., certain services, facilities, or capabilities agreed to by the aviation system users, or by the public)

In part, these functions are accomplished via some type of *certification*. Here, certification means the approval and authorization for aircraft; personnel (e.g., pilots), operations, procedures, facilities, and equipment.

The current regulations that the FAA administers are contained in the U.S. Code, specifically, Title 14—Code of Federal Regulations (CFR), Aeronautics and Space Chapter 1—FAA, Department of Transportation. Three subchapters of particular interest to this discussion are Subchapter C, Aircraft; Subchapter F—Air Traffic and General Operating Rules; and Subchapter G, Air Carriers and Operators for Compensation or Hire: Certification and Operations. Subchapter C includes the Airworthiness Standards for various categories of aircraft (including Part 25 for Transport Category Airplanes). Subchapter F contains the general operating and flight rules (Part 91) and Subchapter G contains Part 121 Operating Requirements: Domestic, Flag, and Supplemental Operations and

Part 135—Operating Requirements: Commuter and On-Demand Operations and Rules Governing Persons On Board Such Aircraft.

Part 25 and several other Parts in Subchapter C contain airworthiness standards for aircraft. These requirements are considered to be *point in time* regulations, because once compliance is found with one of these regulations (such as issuance of a Type Certificate for an airplane type design), it is not revisited unless the type design of the airplane changes (e.g., adding equipment or system not part of the original type design). Therefore, any change in any of the regulations in Part 25 does not result in the change of existing certificated aircraft type designs.

In contrast, the operating rules (Parts 91, 121, 135, etc.) are *continuous applicability* rules, and therefore, when a regulation is changed, the operators certificated under that operating rule must comply according to the date from when the rule is effective. Pilot training and qualification requirements fall under the operating regulations, allowing changes to be made continuously as needed.

The aforementioned discussion describes the underlying philosophy of the regulations, which represent requirements. Regulatory material can have one or more motivations, as discussed in the following:

- *Minimum standards*: The regulations might describe the minimum standards for a required characteristic, such as the performance of a system on an aircraft.
- *Protection, such as 14 CFR Part 193*: This part describes when and how the FAA protects from disclosure safety and security information that is submitted voluntarily to the FAA.
- *Incentives for equipage by giving operational credit*: For example, aircraft with autoland capability (and corresponding pilot qualification) have the potential to fly to lower visibilities than aircraft without such capability.

The FAA publishes several other types of documents, in addition to regulations. One such type of document is an advisory circular (AC), which provides guidance from the FAA to the external community. An AC may contain guidance on means of compliance with particular regulations or may provide other information of interest to the aviation community (e.g., one AC lists all the published ACs).

Advisory circulars are numbered using a system that corresponds to the regulations for which it provides information. For example, AC 25.1329 Approval of Flight Guidance Systems (FAA 2006b) provides a means of compliance with 14 CFR 25.1329 (FAA 2006a). Another example of an AC is AC 120-76A for Approval of Electronic Flight Bags (FAA 2003b). This includes both airworthiness and operational approval guidance for approving a technology or type of system, rather than a specific regulation.

The following material discusses both regulations and advisory circulars related to HF in airworthiness of equipment design, flight-crew training, and operational approval.

REGULATORY REQUIREMENTS AND GUIDANCE FOR FLIGHT DECK

Congress recently directed the FAA to use the following definition of human factors: "human factors means a multidisciplinary field that generates and compiles information about human capabilities and limitations and applies it to design, development, and evaluation of equipment, systems, facilities, procedures, jobs, environments, staffing, organizations, and personnel management for safe, efficient, and effective human performance, including people's use of technology."[*] This is a very broad definition that is intended to cover all the ways that HF is applied by the FAA, whether air traffic applications, aircraft/flight-deck operations, maintenance, commercial space, or other.

[*] S.2658 Federal Aviation Administration Reauthorization Act of 2016. https://www.congress.gov/bill/114th-congress/senate-bill/2658/text

The current chapter will describe how HF is related to regulatory material for equipment design, flight-crew training, and flight-crew operations. In each area, regulatory material is discussed together with some illustrative HF aspects.

AIRWORTHINESS REQUIREMENTS FOR EQUIPMENT DESIGN—EXAMPLES

Chapter 9 of this book describes the certification of aircraft and aircraft systems.[*] HF is related to many of the airworthiness regulations. A thorough review of the HF considerations and where they can be found in aircraft certification standard and guidance material may be found in Yeh et al. (2016). A few of the Part 25 regulations will be discussed next to illustrate in more detail how HF considerations are incorporated into the airworthiness requirements for equipment design.

14 CFR Part 25 Section 25.1329 Flight Guidance Systems

This regulation describes the airworthiness requirements for Flight Guidance Systems (FGS), including autopilots, autothrust systems, flight directors, and associated flight-crew interfaces (FAA 2006a). Operational experience showed that flight-crew errors and confusion were occurring when operating the FGS and its subsystems (FAA 1996), including vulnerabilities that can be mitigated in the equipment design. Therefore, the airworthiness requirements were updated to address these issues, and to address changes in technology and capabilities of FGS.

As one example of how the equipment design requirements were updated to support crew coordination, paragraph (j) requires that the alert for autopilot disengagement must be done in a way to assure that the information is available to each pilot:

(j) Following disengagement of the autopilot, a warning (visual and auditory) must be provided to each pilot and be timely and distinct from all other cockpit warnings. (FAA 2006a)

The Advisory Circular for this regulation makes it clear that the intent is that the alert associated with disengagement of the autopilot(s) must be implemented in a way to support flight-crew coordination:

> "It should sound long enough to ensure that it is heard and recognized by the pilot and other flight-crew members, but not so long that it adversely affects communication between crew members or is a distraction." (FAA 2006b)

Paragraph (i) explicitly addresses the need to support error management through preventing errors, and through the equipment design providing feedback on current modes of operation:

(i) The flight guidance system functions, controls, indications, and alerts must be designed to minimize flight-crew errors and confusion concerning the behavior and operation of the flight guidance system. (FAA 2006a)

Paragraph (i) also says: "Means must be provided to indicate the current mode of operation, including any armed modes, transitions, and reversions. Selector switch position is not an acceptable means of indication. The controls and indications must be grouped and presented in a logical and consistent manner. The indications must be visible to each pilot under all expected lighting conditions." (FAA 2006a). This portion of the regulation supports the requirement for equipment design to support crew monitoring of the status of the FGS.

One point worth of note is that this regulatory material addresses systems but does not use the term *automation*. The FAA does not apply aircraft certification requirements any differently for so-called automation as a separate type of system than for other complex systems. Instead, the systems are evaluated for compliance with the regulations based on their function.

---

[*] Chapter 9 Certification of Aircraft and Aircraft Systems, this book.

14 CFR 25.1302 Installed Systems and Equipment for Use by the Flight Crew and European Aviation Safety Agency Certification Specification CS 25.1302

Another airworthiness regulation was developed jointly by the FAA, the Joint Aviation Authorities (JAA), the European Aviation Safety Agency (EASA), North and South American industry, and European industry to address the need for the equipment design to support error management by the pilots (EASA 2007b). This regulation was written to require the equipment design to have characteristics that are known to avoid error. Specifically, the equipment must provide the information and controls necessary for the pilots to do the tasks associated with the intended function of the equipment, and the controls and information must be in a usable form.

In addition, the regulation was written on the basis of the understanding that even well-qualified pilots using well-designed systems will make errors. Therefore, the equipment design must support detection and recovery aspects of error management. The first sentence of paragraph (d) explicitly addresses this:

(d) To the extent practicable, installed equipment must enable the flight crew to manage errors resulting from the kinds of flight crew interactions with the equipment that can be reasonably expected in service, assuming the flight crew is acting in good faith (EASA 2007; FAA 2013).[*][†]

This regulation requires that the equipment design provide the information needed to perform the tasks associated with the intended function of the equipment—and this includes monitoring of the equipment—and that the equipment provide information about its operationally relevant behavior. The FAA and EASA regulations are harmonized, which results in the United States and Europe having consistent requirements for this aspect of equipment design.

The airworthiness regulations described earlier are *point in time* regulations, as discussed earlier. Therefore, any new airplane type must meet the requirements, but existing airplanes do not. As there are many aircraft that received their aircraft certification approval before these regulations were implemented, such aircraft do not necessarily meet the requirements for the equipment design to support crew coordination, error management, and crew monitoring. Thus, the mitigations required in the flight-crew training and procedures are especially important for such aircraft.

OTHER AIRCRAFT CERTIFICATION STANDARDS

Many other aircraft certification requirements have HF-related aspects. Examples include 14 CFR 25.1322 Aircraft Alerting Systems;

Flight-Crew Training and Operational Requirements

Flight-crew training and qualification requirements include many aspects that affect pilot performance and, therefore, involve the application of human factors. One major area (but by no means the only one) is that of Crew Resource Management (CRM). The U.S. regulations include a requirement for training of CRM principles and topics for pilots and dispatchers. These requirements for CRM are codified into 14 CFR Part 121 Section 121.404, Compliance dates: Crew and dispatcher resource management training, which states

> After March 19, 1998, no certificate holder may use a person as a flight crewmember, and after March 19, 1999, no certificate holder may use a person as a flight attendant or aircraft dispatcher unless that person has completed approved crew resource management (CRM) or dispatcher resource management (DRM) initial training, as applicable, with that certificate holder or with another certificate holder. (FAA 1996)

---

[*] 14 CFR 25.1302 Installed Systems and Equipment for Use by the Flightcrew. http://www.airweb.faa.gov/Regulatory_ and_Guidance_Library/rgFAR.nsf/0/FE0119F0EBD3BAC086257B9C0051C97C?OpenDocument&Highlight=25.1302

[†] As with all regulations, the regulatory material should be read and considered in its entirety, together with other applicable regulations.

The requirement extends beyond air carriers operating under Part 121. On December 20, 1995, the FAA published Air Carrier and Commercial Operator Training Programs. This final rule required CRM training for crew members in their training programs by certificate holders conducting operations under part 135 that are required to comply with part 121 training and qualification requirements, such as those certificate holders that conduct commuter operations with airplanes for which two pilots are required by aircraft certification rules, and those that conduct commuter operations with airplanes having a passenger seating configuration of 10 seats or more.

The regulation itself does not specify the content of the training, but Advisory Circular (AC) 120-51E (FAA 2004) provides guidance for the content of U.S. operator training programs to address CRM. Subjects such as crew coordination and communication, error management, and flight-crew monitoring are specifically described in the AC. This AC also discusses the importance of pre- and post-training session briefings, and ways to evaluate the pilots' performance as a result of the training, among other topics. As an example of related guidance outside the United States, CAA UK (2006) provides guidance for the content of CRM training.

Moreover, as with all the operating regulations, the requirement for CRM training for flight crews is a continuous applicability requirement. Thus, improvements to the guidance can be made and applied as more is learned about effective implementation of training for CRM.

The FAA also recognizes that human factors is an important element of effective flight-crew procedures. The design of procedures should embody that coordination. Degani and Weiner (1994) describe that there are several aspects to design of the procedures that can promote crew coordination:

1. *Reduced variance*: The procedure triggers a predetermined and expected set of actions.
2. *Feedback*: Procedures specify expected feedback to other crew members (e.g., callouts). This feedback can detail (1) the current, and/or expected system state; (2) the actions that are currently being conducted; (3) the system outcome; and (4) an indication of task completion. There are several ways in which this feedback is provided: (1) verbally (callouts, callback, etc.); (2) nonverbally (gestures, manual operation—such as pulling down the gear lever); (3) via the interface (when the configuration of the system is significantly changed, for example, all displays are momentarily blank when power is switched from APU to engine-driven generators, this provides clear feedback to the other pilot); and (4) via the operating environment (when slats/flaps are extended during approach, there is a clear aerodynamic feedback—pitch change).
3. *Information transfer*: Procedures convey, or transfer, information from one agent to others (e.g., *the after takeoff checklist is complete*).

Another area that is important to consider for crew coordination is the delineation of duties among the Pilot Flying (PF), the Pilot Not Flying (PNF)/Pilot Monitoring (PM), and the Flight Engineer (if present). This includes the identification of who does which tasks, which pilot calls for particular procedures, which pilot reads them and which one responds.

The FAA has published an advisory circular that provides guidance for implementation of Standard Operating Procedures (SOPs) that address these areas,* each task, and that mental model is founded on SOPs. The procedures templates included in the AC provide recommended steps for crew coordination and communication tasks.

This AC specifically highlights the need for pilots to perform monitoring tasks, based on recognition of the role that inadequate monitoring played in previous accidents. For example, the National

---

* AC 120-71B Standard Operating Procedures and Pilot Monitoring Duties for Flight Deck Crewmembers. http://rgl.faa.gov/Regulatory_and_Guidance_Library/rgAdvisoryCircular.nsf/0/CF313985D7C52500862580A500692715?OpenDocument&Highlight=120-71b

Transportation Safety Board has identified that inadequate crew monitoring or challenging was involved in 84% of crew-involved accidents (NTSB 1994, Sumwalt 2004).

The AC also reflects the conversion of the term *Pilot Not Flying* to *Pilot Monitoring* to reflect what the pilot *is* doing, rather than what the pilot is not doing. In addition, it helps one to emphasize the importance of the monitoring task. This emphasis must be part of the philosophy that forms the foundation of the SOPs.

Another area that can involve application of HF is the evaluation of operational approvals. The general process of approval or acceptance of certain operations, programs, documents, procedures, methods, or systems is an orderly method used by FAA inspectors to ensure that such items meet regulatory standards in the operating rules and provide for safe operating practices.[*] This can vary from evaluation of a single system (such as an Electronic Flight Bag) or operations that involve equipment, flight crew training, and flight-crew procedures (such as Performance-Based Navigation operations [Barhydt and Adams 2006]).

CONCLUDING REMARKS

The current chapter has presented a regulatory perspective on HF, given primarily from a U.S. point of view. It is not comprehensive, as HF considerations can be found throughout the regulatory material related to flight-deck airworthiness, flight-crew training, and flight-crew operations and procedures—among other application areas.

The FAA and other regulatory authorities around the world recognize the importance of HF as an important contributor to aviation safety. Although most of this chapter has focused on the pilots and flight decks, the regulatory system is recognizing the importance of resource management for other personnel in the aviation system, including dispatchers (as evidenced by the requirement for dispatcher resource management training), stated earlier, and guidance for training of maintenance personnel (FAA 2000).

This propagation of resource management considerations throughout the regulatory material reflects the growing understanding of the importance of this area—but it needs to be even more widespread. Application of CRM concepts to air traffic personnel, to the communication and coordination between pilots and ATS, among pilots, maintenance, dispatchers, cabin crew, and others remain important and can be improved even further.

REFERENCES

Barhydt, R. and Adams, C. (2006). *Human Factors Considerations for Performance-Based Navigation.* National Aeronautics and Space Administration Technical Memorandum 2006-214531, Hampton, VA.

Birnbach, R. and Longridge, T. (1993). The regulatory perspective, in *Cockpit Resource Management*, Wiener, E., Kanki, B., and Helmreich, R., (Eds.), Academic Press, San Diego, CA.

Civil Aviation Authority United Kingdom. (2006). *Crew Resource Management (CRM) Training: Guidance for Flight Crew, CRM Instructors (CRMIS) and CRM Instructor-Examiners (CRMIES).* (CAP 737). UK Civil Aviation Authority, Gatwick Airport South, West Sussex, UK.

Degani, A. and Wiener, E. (1994). On the design of flight-deck procedures. *NASA Contractor Report.* National Aeronautics and Space Administration, Moffet Field, CA. June 1994.

European Aviation Safety Agency. (2007a). *Certification Specifications for Large Aeroplanes CS-25 Amendment-3, Book 1, Airworthiness Code, Subpart F Equipment, General,* CS-25.1302 Installed systems and equipment for use by the flight crew. September 19, 2007.

European Aviation Safety Agency. (2007b). *Certification Specifications for Large Aeroplanes CS-25 Amendment-3, Book 2, Acceptable Means of Compliance, Subpart F Equipment,* General, AMC-25.1302 Installed systems and equipment for use by the flight crew. September 19, 2007.

---

[*] Order 8900.1 CHG 266, Volume 3 General Technical Administration, Chapter 1 The General Process for Approval or Acceptance of Air Operator Applications. See fsims.faa.gov

Federal Aviation Administration. (1996). *Human Factors Team Report on: The Interfaces Between Flightcrews and Modern Flight Deck Systems*. Federal Aviation Administration, Washington, DC. June 18, 1996.

Federal Aviation Administration. (2000). *Maintenance Resource Management Training* (FAA Advisory Circular AC 120-72). Department of Transportation, Washington, DC. September 28, 2000.

Federal Aviation Administration. (2003a). *Standard Operating Procedures for Flight Deck Crewmembers and Pilot Monitoring Duties* (FAA Advisory Circular AC 120-71B). Department of Transportation, Washington, DC. January, 2003.

Federal Aviation Administration. (2003b). *Guidelines for the Certification, Airworthiness, and Operational Approval of Electronic Flight Bag Computing Devices* (FAA Advisory Circular 120-76A). Department of Transportation, Washington, DC. March 17, 2003.

Federal Aviation Administration. (2003c). *Crew Resource Management Training* (FAA Advisory Circular 120-51E). Department of Transportation, Washington, DC. January 22, 2004.

Federal Aviation Administration (2004). Advisory Circular AC 120-51E, *Crew Resource Management Training*. Department of Transportation, Washington, DC. January 22, 2004.

Federal Aviation Administration. (2006a). *Title 14 United States Code of Federal Regulations Part 25*, Section 25.1329. (*Flight Guidance Systems*). Department of Transportation, Washington, DC. July 2006.

Federal Aviation Administration. (2006b). Advisory circular 25.1329-1B, *Approval of Flight Guidance Systems*, Washington, DC. July 17, 2006.

Federal Aviation Administration. (2013). Advisory Circular AC 25.1302-1, *Installed Systems and Equipment for Use by the Flightcrew*. Department of Transportation, Washington, DC. May 3, 2013.

National Transportation Safety Board. (1994). *Safety Study: A Review of Flightcrew-Involved, Major Accidents of U.S. Air Carriers, 1978 through 1990*. Report no. NTSB/SS-94/01. NTSB, Washington, DC, 1994.

RTCA. (1999). Final report of task force 4, *Certification*. Washington, DC.

Sumwalt, R. L. (1995). Enhancing flight crew monitoring skills can reduce altitude deviations, in *Flight Safety Digest*. Flight Safety Foundation, Alexandria, VA. December 1995.

Sumwalt, R. L. (2004). Enhancing flight crew monitoring skills can increase corporate aviation safety. *49th Corporate Aviation Safety Seminar*, Flight Safety Foundation. Tucson, AZ. April 27–29, 2004.

Yeh, M., Swider, C., Jo, Y. J., and Donovan, C. (2016). *Human Factors Considerations in the Design and Evaluation of Flight Deck Displays and Controls*—Version 2, (DOT/FAA/TC-16/56; DOT-VNTSC-FAA-17-02). U.S. DOT Volpe National Transportation Systems Center, Cambridge, MA.

8 Certification of Aircraft and Aircraft Systems

Robert E. Joslin

CONTENTS

Regulations, Guidance, and Policy 148
Standards... 149
Certification Basis 149
Regulators 151
Project Specific Certification Plan 151
Intended Human Operator... 186
Human Factors Certification Plan 187
  Introduction.. 187
  System Description 187
Certification Requirements 188
Methods of Compliance.. 188
Other Considerations 190
System Safety Assessments 190
Operational Considerations... 191
Certification Documentation... 191
Certification Schedule... 191
Use of Designees and Identification of Individual Designated Engineering
Representatives/Designated Airworthiness Representatives.. 192
Orders.. 192
Advisory Circulars.. 192
FAA Reports 193
Policy Statements.. 193

Aircraft[*] and aircraft systems that operate in the National Airspace System[†] are required to be airworthy,[‡] defined as being built to an approved design and in a safe condition for flight. The certification of an approved design for new or modified aircraft and aircraft systems is determined through a type certification[§] process administered by the Federal Aviation Administration (FAA) via a *show of compliance* to the applicable Federal Aviation Regulations (FAR) under United States Code (U.S.C.) Title 14 Code of Federal Regulations (14CFR). Although there is not one specific regulation for the certification of human factors, the requirements to accommodate the

---

[*] 14CFR §1.1—Aircraft means a device that is used or intended to be used for flight in the air.

[†] The National Airspace System (NAS) is the common network of U.S. airspace; air navigation facilities, equipment, and services, airports or landing areas; aeronautical charts, information, and services; rules, regulations, and procedures, technical information, and manpower and material. Included are system components shared jointly with the military. FAA Aeronautical Information Manual (AIM).

[‡] 14CFR §3.5(a)—Airworthy means the aircraft conforms to its type design and is in a condition for safe operation.

[§] FAA Order 8110.4_ Type Certification.

human element in aircraft design are threaded throughout the FAR, hence, are a significant part of the type certification process.

The FAA issues type certificates (TC) for a complete aircraft, engine, or propeller, whereas supplemental TC (STC) are issued for aircraft systems or components, as required under 14CFR Part 21—certification procedures for products, articles, and parts. The term *type certificate*, or *TC*, applies to the original TC, supplemental TCs, and amended TCs, unless otherwise specified. Although a TC establishes that an aircraft or aircraft system was built to design, it does not constitute approval to operate the aircraft or aircraft system in the National Airspace System. An operational certification or approval must be obtained through the issuance of an appropriate airworthiness (A/W) certificate, also under 14CFR Part 21.

Other aircraft certification–related processes can be supported by, or conducted in conjunction with the TC process, such as technical standard order[*] that provides the minimum FAA approved performance standard for specified materials, parts, processes, and appliances used on civil aircraft, or type validation[†] that is a special form of certification used to establish the compliance of an imported product to the importing state's applicable airworthiness standards.

The current chapter will focus on the FAA regulations, guidance, and policies for human factors considerations in the certification of approved designs for aircraft flight decks (cockpits) in new or modified aircraft and aircraft systems. The examples are for Part 25-Transport Category Airplanes; however, the content is applicable, in part or in toto, to other category aircraft (i.e., Part 27/29 for normal/transport category rotorcraft, and Part 23 for normal, utility, acrobatic, and commuter category airplanes).

REGULATIONS, GUIDANCE, AND POLICY

The human factors regulatory requirements that must be met are mandated by law under the FAR, which encompass objective criteria (e.g., ".... right rudder for nose right") and subjective criteria (e.g., "... may not require exceptional piloting skill or alertness"). Within the FAR the terms *shall* or *must* are used to indicate mandatory requirements, and the terms *should* and *may* are used when something is recommended but not required. However, the language in the FAR may not always provide sufficient detail in describing how a *show of compliance* can be fulfilled. In those cases, the FAA develops Advisory Circulars[‡] (AC) and Policy (Public) Statements[§] (PS) to provide additional guidance and information. The contents of ACs and PSs are not mandatory or regulatory unless incorporated by reference into a regulation or the *certification basis*.

An AC describes an acceptable means, but not the only means, for complying with one or more related regulations and normally focuses on a specific system or function (e.g., controls, alerting). The means (or method) of compliance (MOC) can be by analysis, test (e.g., flight, ground, simulator, laboratory, bench), design similarity, equivalent level of safety[¶,**], special conditions,[††] or exemption.[‡‡] The applicant can also propose their own MOC, which must be approved by the FAA. PS normally just provide general guidance on a single topic or issue, or acceptable practices on how

---

[*] FAA Order 8150.1_ Technical Standard Order Program.
[†] FAA Order 8110.52_ Type Validation and Post-type Validation Procedures.
[‡] FAA Order 1320.46_ FAA Advisory Circular System.
[§] FAA Order IR 8100.16_ Aircraft Certification Service Policy Statement, Policy Memorandum, and Deviation Memorandum Systems.
[¶] A level of safety at least equal to that provided by the rule from which relief is sought.
[**] FAA Order 8110.112_ Standardized Procedures for Usage of Issue Papers and Development of Equivalent Levels of Safety Memorandums.
[††] Special conditions are issued only if the existing applicable airworthiness standards do not contain adequate or appropriate safety standards for the aircraft, aircraft engine, or propeller due to novel or unusual design features of the product to be type certificated.
[‡‡] A petition for exemption is a request to FAA by an individual or entity asking for relief from the requirements of a current regulation.

to find compliance with a specific FAR section or paragraph. The FAA also promulgates Orders, Notices, and internal Quality Management System documents with policy, instructions, and work information that provide direction to FAA personnel and FAA designees on how they are expected to carry out their certification responsibilities.

STANDARDS

The certification of aircraft systems requires compliance with technical and operational performance standards as referenced in FAA AC and Technical Standard Orders. However, the FAA does not unilaterally develop standards; rather they solicit the development of government/industry consensus standards through various globally recognized organizations, such as the Radio Technical Commission for Aeronautics[*] (RTCA), Society of Automotive Engineers[†] (SAE), and the American Society for Testing and Materials. Within many of these standards are specific quantitative/objective (e.g., font/typeface) and qualitative/subjective (e.g., fatigue) human factors guidelines and considerations.

CERTIFICATION BASIS

Determining what regulations are applicable and what guidance and policies should be followed, the *certification basis*,[‡] is a fundamental part of the type certification process when considering human factors in the design approval of new or modified aircraft and aircraft systems. FAR are periodically added, amended, or deleted; hence, each regulation has an effective date, or amendment level. The amendment level in effect on the date that an applicant submits their application to their servicing geographical FAA Aircraft Certification Office (ACO) to initiate a project for the type design certification of a new or modified aircraft or aircraft system establishes what regulations are binding for their *show of compliance*. The FAA will then issue a Certification Program Notification[§] letter that includes the original date of application, which then formally establishes the effectivity date for the *certification basis*. The human factors–related regulations that form the *certification basis* for the design approval are tied to the tasks performed by the pilot, such as seeing and reading aircraft instruments, moving the aircraft flight controls, manipulating switches and knobs, ensuring sufficient clearance with aircraft structure, and viewing the world out the cockpit window. These tasks are considered in the context of workload and the performance of the required tasks, according to the phase of flight, type of operation, and the urgency of the situation. The overarching requirement is that the aircraft and aircraft systems must perform their intended function, comply with the applicable regulations, and be safe for flight. There are also other design areas with human factors considerations (e.g., egress, crashworthiness, maintenance/inspection, crew sleeping quarters/rest facilities); however, they involve other occupants of the aircraft and not just the cockpit/cockpit crew, hence, are not completely addressed in this chapter.

In some cases, although infrequently, a FAR that became effective after the *certification basis* was established may be imposed if the FAA Administrator deems it a safety critical item, whereas in other cases a modification to an existing aircraft or aircraft system may be able to retain part or all of their original *certification basis* and not step up to the FAR in effect on their date of application. The applicant may also voluntarily elect to include any FAR in their certification basis. The *certification basis* can include guidance material, such as AC and PS. Table 8.1 shows a notional Human Factors Certification Compliance Checklist for a Part 25-Transport Category Airplane.

---

[*] FAA Order 1110.77Q RTCA, Inc. (Utilized as an Advisory Committee).
[†] FAA Order 8110.116 Procedures for Requesting Development of SAE Standards.
[‡] FAA Order 8110.48 How to Establish the Certification Basis for Changed Aeronautical Products.
[§] FAA Order 8110.115 Certification Project Initiation and Certification Project Notification.

TABLE 8.1

Human Factors Certification Plan-Outline

1. INTRODUCTION
2. SYSTEM DESCRIPTION
   a. Intended Function
   b. Flight Deck Layout Drawings
   c. Underlying Principles for Automation Logic
   d. Underlying Principles for Crew Procedures
   e. Assumed Pilot Characteristics
3. CERTIFICATION REQUIREMENTS
4. METHODS OF COMPLIANCE (MOC)
   a. Drawings
   b. Configuration Description
   c. Statement of Similarity
   d. Evaluations, Assessments, Analyses
   e. Engineering Evaluations or Analyses
   f. Mock-up Evaluations
   g. Part-Task Evaluations
   h. Simulator Evaluations
   i. In-Flight Evaluations
   j. Demonstrations
   k. Inspection
   l. Tests
      • Bench Tests
      • Ground Tests
      • Simulator Tests
      • Flight Tests
   m. Other Considerations
      • Degree of Integration/Independence
      • Novelty/Past Experience
      • Complexity/Level of Automation
      • Criticality
      • Dynamics
      • Level of Training Required
      • Subjectivity of Acceptance Criteria
5. SYSTEM SAFETY ASSESSMENTS (SSA)
6. OPERATIONAL CONSIDERATIONS
   a. Airplane Flight Manual (AFM)
   b. Master Minimum Equipment List (MMEL)
   c. Flightcrew Operating Manual (FCOM)
   d. Quick Reference Handbook (QRH)
7. CERTIFICATION DOCUMENTATION
8. CERTIFICATION SCHEDULE
   a. Certification Plan Submittals
   b. Flight Deck Reviews, Early Prototype Reviews, Simulator Reviews, and Flight Test Demonstrations
   c. Coordination meetings
9. USE OF DESIGNEES AND IDENTIFICATION OF INDIVIDUAL DESIGNATED ENGINEERING/ AIRWORTHINESS REPRESENTATIVE (DER/DAR)

REGULATORS[*]

Flight Test Pilots, Flight Test Engineers, Aircraft Evaluation Pilots, and Human Factors Specialists from the FAA Aircraft Certification Service, Flight Standards Service, and Aircraft Evaluation Groups are responsible for *finding compliance* with the human factors associated with the design approval and certification of new or modified aircraft and aircraft systems. The applicant is encouraged to have the involvement of these FAA personnel throughout the design process through technical reviews, bench/laboratory and simulator demonstrations, and research and development flights. The aforementioned activities are not mandated by regulation; however, the actual *finding of compliance*, or Type Inspection Authorization[†] (TIA), is required to be conducted by FAA personnel or FAA designees.[‡,§] The TIA is prepared by the FAA ACO or Organization Designation Authorization (ODA) and is used to authorize official conformity, airworthiness inspections, and flight tests necessary to fulfill certain requirements for TC and amended TC certification. The TIA is issued when the FAA examination of the applicant's technical data required for type certification is completed or has reached a point in which it appears that the aircraft or component being examined will meet the applicable regulations. The ground and flight tests specified in the TIA are then conducted by FAA personnel, who may use their discretionary authority to accept any portion of the company tests without further inspection. The FAA may also delegate some or all of the TIA issuance and TIA tests to FAA designees.

PROJECT SPECIFIC CERTIFICATION PLAN

The *certification basis* forms the foundation of the Project Specific Certification Plan (PSCP), which is the overall plan proposed by the entity (*applicant*) seeking a TC. The PSCP in turn points to individual certification plans to accomplish specific tasks for the *show of compliance*, such as ground test plan, aircraft performance and flying qualities flight test plan, electromagnetic interference test plan, and others. Although the guidance for the basic contents of a PSCP do not explicitly call for a stand-alone certification plan that consolidates all human factors issues, FAA PS[¶,**] encourage its use when certifying an entire new aircraft or multiple complex integrated systems, rather than scattering human factors compliance throughout the various system specific test plans. Table 8.2 shows a notional outline of a Human Factors Certification Plan.

---

[*] FAA Order 8100.5_ Aircraft Certification Service Mission, Responsibilities, Relationships, and Programs.
[†] FAA Order 8110.4_ Type Certification.
[‡] FAA Order 8100.8_ *Designee Management Handbook*; FAA Order 8100.15_ Organization Designation Authorization Procedures.
[§] FAA Order 8110.37_ *Designated Engineering Representative* (DER) Handbook.
[¶] Policy Statement Number PS-ANM111-1999-99-2, Guidance for Reviewing Certification Plans to Address Human Factors for Certification of Transport Airplane Flight Decks, September 29, 1999.
[**] Policy Statement Number PS-ACE100-2001-004, Guidance for Reviewing Certification Plans to Address Human Factors for Certification of Part 23 Small Airplanes, August 29, 2002.

TABLE 8.2
Compliance Matrix for Part 25 Regulations Related to Flightcrew Human Factors[a]

Regulation Title 14 CFR Part 25	Human Factors (HF) Requirements	Applicable Amendment	Method(s) of Compliance	Person or Entity Finding Compliance	Applicable Guidance
Example: §25.777(a)	Each cockpit control must be located to provide convenient operation and to prevent confusion and inadvertent operation.	Amdt??	DE, FT	J.Doe	AC 20-175 Controls for Flight-Deck Systems; AC 25.1302-1, Installed Systems and Equipment for Use by the Flightcrew; AC 25-11B Electronic Flight Displays
Performance					
25.101(h)-General	**Subpart B—Flight**				
	"The procedures established [for accelerate-stop distances, takeoff, landing, changes in the airplane's configuration, speed, power, and thrust, balked landings, and missed approaches] must—				
	(1) Be able to be consistently executed in service by crews of average skill;				
	(2) Use methods or devices that are safe and reliable; and				
	(3) Include allowance for any time delays, in the execution of the procedures, that may reasonably be expected in service.				
25.105(b)-Takeoff	"No takeoff made to determine the data required by this section may require exceptional piloting skill or alertness."				
25.109(e)(3)-Accelerate-stop distance	" … means other than wheel brakes may be used to determine the accelerate-stop distance if that means is such that exceptional skill is not required to control the airplane."				
25.125(b)(5)-Landing	"The landings may not require exceptional piloting skill or alertness."				
25.125(c)(3)-Landing	" … means of deceleration other than wheel brakes may be used if that means is such that exceptional skill is not required to control the airplane."				

(Continued)

TABLE 8.2 (Continued)
Compliance Matrix for Part 25 Regulations Related to Flightcrew Human Factors[a]

Regulation Title 14 CFR Part 25	Human Factors (HF) Requirements	Applicable Amendment	Method(s) of Compliance	Person or Entity Finding Compliance	Applicable Guidance
Controllability and Maneuverability					
	Subpart B—Flight				
25.143(b)-General	"It must be possible to make a smooth transition from one flight condition to any other flight condition without exceptional piloting skill, alertness or strength, and without danger of exceeding the airplane limit load factor under any probable operating condition, including —"				
	(1) The sudden failure of the critical engine.				
	(2) For airplanes with three or more engines, the sudden failure of the second critical engine when the airplane is in the en route, approach, or landing configuration and is trimmed with the critical engine inoperative.				
	(3) "Configuration changes, including deployment or retraction of deceleration devices."				
25.145(c)-Longitudinal Control	"It must be possible, without exceptional piloting skill, to prevent loss of altitude when complete retraction of the high lift devices from any position is begun during steady, straight, level flight ..."				
25.149(d)-Minimum Control Speed	"During recovery, the airplane may not assume any dangerous attitude or require exceptional piloting skill, alertness, or strength to prevent a heading change of more than 20 degrees."				
25.149(e)-Minimum Control Speed	"... enable the takeoff to be continued using normal piloting skill."				
25.149(h)-Minimum Control Speed	"The airplane may not exhibit hazardous flight characteristics or require exceptional piloting skill, alertness, or strength."				

(Continued)

TABLE 8.2 (Continued)
Compliance Matrix for Part 25 Regulations Related to Flightcrew Human Factors[a]

Regulation Title 14 CFR Part 25	Human Factors (HF) Requirements	Applicable Amendment	Method(s) of Compliance	Person or Entity Finding Compliance	Applicable Guidance
Stability	**Subpart B—Flight**				
25.173(d)-Static Longitudinal Stability	"… stabilize on speeds above and below the desired trim speeds if exceptional attention on the part of the pilot is not required to return to and maintain the desired trim speed and altitude."				
25.177(b)-Static Lateral-Directional Stability	… the static lateral stability may not be negative unless the divergence is "gradual, easily recognized, and easily controlled by the pilot."				
25.181(b)-Dynamic stability	"… … the Dutch roll "must be controllable with normal use of the primary controls without requiring exceptional pilot skill."				
Stalls	**Subpart B—Flight**				
25.203(c)-Stall characteristics	"For turning flight stalls, the action of the airplane after the stall may not be so violent or extreme as to make it difficult, with normal piloting skill, to effect a prompt recovery and to regain control of the airplane."				
25.207(a)-Stall warning	"Stall warning … must be clear and distinctive to the pilot."				
25.207 (b)-Stall warning	"… the warning … must give clearly distinguishable indications. …. However, a visual stall warning device that requires the attention of the crew within the cockpit is not acceptable by itself."				
Ground and Water Handling Characteristics	**Subpart B—Flight**				
25.233(b)-Directional Stability and Control	"Landplanes must be satisfactorily controllable, without exceptional piloting skill or alertness in power-off. landings … "				

(Continued)

TABLE 8.2 (Continued)
Compliance Matrix for Part 25 Regulations Related to Flightcrew Human Factors[a]

Regulation Title 14 CFR Part 25	Human Factors (HF) Requirements	Applicable Amendment	Method(s) of Compliance	Person or Entity Finding Compliance	Applicable Guidance
Miscellaneous Flight Requirements					
	Subpart B—Flight				
25.251(c)-Vibration and Buffeting	"… there may be no buffeting condition, in normal flight, including configuration changes during cruise, severe enough to interfere with control of the airplane, to cause excessive fatigue to the crew …"				
25.253(a)(2)-High Speed Characteristics	"Allowing for pilot reaction time after effective inherent or artificial speed warning occurs, it must be shown that the airplane can be recovered to a normal attitude and its speed reduced to VMO/MMO without—				
	(i) "Exceptional piloting strength or skill …"				
	(iii) Buffeting that would impair the pilot's ability to read the instruments of control the airplane for recovery.				
General	**Subpart D—Design and Construction**				
25.601-General	"The airplane may not have design features or details that experience has shown to be hazardous or unreliable. The suitability of each questionable design detail and part must be established by tests."				
Control Systems	**Subpart D—Design and Construction**				
25.671-General	(a) "Each control and control system must operate with the ease, smoothness, and positiveness appropriate to its function."				
	(c) "The airplane must be shown by analysis, tests, or both, to be capable of continued safe flight and landing after any of the following failures or jamming in the flight control system… without requiring exceptional piloting skill or strength. Probable malfunctions must have only minor effects on control system operation and must be capable of being readily counteracted by the pilot."				

(Continued)

TABLE 8.2 (Continued)
Compliance Matrix for Part 25 Regulations Related to Flightcrew Human Factors[a]

Regulation Title 14 CFR Part 25	Human Factors (HF) Requirements	Applicable Amendment	Method(s) of Compliance	Person or Entity Finding Compliance	Applicable Guidance
25.672-Stability augmentation and automatic and power-operated systems	(a) "A warning which is clearly distinguishable to the pilot under expected flight conditions without requiring his attention must be provided for any failure in the stability augmentation system or in any other automatic or power-operated system which could result in an unsafe condition if the pilot were not aware of the failure. Warning systems must activate the control systems."				
	(b) "The design of the stability augmentation system or of any other automatic or power-operated system must permit initial counteraction of failures of the type specified in § 25.671(c) without requiring exceptional pilot skill or strength, by either the deactivation of the system, or a failed portion thereof, or by overriding the failure by movement of the flight controls in the normal sense."				
25.677-Trim systems	(a) Trim controls must be designed to prevent inadvertent or abrupt operation and to operate in the plane, and with the sense of motion, of the airplane.				
	(b) "There must be means adjacent to the trim control to indicate the direction of the control movement relative to the airplane motion. In addition, there must be clearly visible means to indicate the position of the trim device with respect to the range of adjustment."				
25.679-Control System Gust Locks	(a) "If the device when engaged, prevents normal operation of the control surfaces by the pilot, it must (1) Automatically disengage when the pilot operates the primary flight controls in a normal manner or (2) limit operation of the airplane so that the pilot receives unmistakable warning at the start of takeoff."				

(Continued)

TABLE 8.2 (*Continued*)

Compliance Matrix for Part 25 Regulations Related to Flightcrew Human Factors[a]

Regulation Title 14 CFR Part 25	Human Factors (HF) Requirements	Applicable Amendment	Method(s) of Compliance	Person or Entity Finding Compliance	Applicable Guidance
25.685-Control system details	(a) "Each detail of each control system must be designed and installed to prevent jamming, chafing, and interference from cargo passengers, loose objects, or the freezing of moisture."				
	(b) "There must be a means in the cockpit to prevent the entry of foreign objects into places where they would jam the system."				
25.697-Lift and Drag Devices, Controls	(a) "Each lift device control must be designed so that the pilots can place the device in any takeoff, en route, approach, or landing position . . . Lift and drag devices must maintain the selected positions, except for movement produced by an automatic positioning or load limiting device, without further attention by the pilots."				
	(b) Each lift and drag device control must be designed and located to make inadvertent operation improbable. Lift and drag devices intended for ground operation only must have means to prevent the inadvertent operation of their controls in flight if that operation could be hazardous.				
	(c) "The rate of motion of the surfaces in response to the operation of the control and the characteristics of the automatic positioning or load limiting device must give satisfactory flight and performance characteristics under steady or changing conditions of airspeed, engine power, and airplane attitude."				

(*Continued*)

TABLE 8.2 (*Continued*)

Compliance Matrix for Part 25 Regulations Related to Flightcrew Human Factors[a]

Regulation Title 14 CFR Part 25	Human Factors (HF) Requirements	Applicable Amendment	Method(s) of Compliance	Person or Entity Finding Compliance	Applicable Guidance
25.699-Lift and Drag Device Indicator	(a) There must be means to indicate to the pilots the position of each lift or drag device having a separate control in the cockpit to adjust its position. In addition, an indication of unsymmetrical operation or other malfunction in the lift or drag device systems must be provided when such indication is necessary to enable the pilots to prevent or counteract an unsafe flight or ground condition, considering the effects on flight characteristics and performance.				
	(b) There must be means to indicate to the pilots the takeoff, en route, approach, and landing lift device positions.				
	(c) "If any extension of the lift and drag devices beyond the landing position is possible, the controls must be clearly marked to identify this range of extension."				
25.703-Takeoff Warning System	"A takeoff warning system must be installed and must meet the following requirements:				
	(a) The system must provide to the pilots an aural warning that is automatically activated during the initial portion of the takeoff roll if the airplane is in a configuration ... that would not allow a safe takeoff ..."				
Landing Gear	**Subpart D—Design and Construction**				
25.729-Retracting mechanism	(e) Position indicator and warning device. If a retractable landing gear is used, there must be a landing gear position indicator easily visible to the pilot or to the appropriate crew members (as well as necessary devices to actuate the indicator) to indicate without ambiguity that the retractable units and their associated doors are secured in the extended (or retracted) position. The means must be designed as follows:				

(*Continued*)

TABLE 8.2 (*Continued*)

Compliance Matrix for Part 25 Regulations Related to Flightcrew Human Factors[a]

Human Factors (HF) Requirements	Applicable Amendment	Method(s) of Compliance	Person or Entity Finding Compliance	Applicable Guidance
Regulation Title 14 CFR				
Part 25				
1. If switches are used, they must be located and coupled to the landing gear mechanical systems in a manner that prevents an erroneous indication of "down and locked" if the landing gear is not in a fully extended position, or of "up and locked" if the landing gear is not in the fully retracted position. The switches may be located where they are operated by the actual landing gear locking latch or device.				
2. The flight crew must be given an aural warning that functions continuously, or is periodically repeated, if a landing is attempted when the landing gear is not locked down.				
3. The warning must be given in sufficient time to allow the landing gear to be locked down or a go-around to be made.				
4. There must not be a manual shut-off means readily available to the flight crew for the warning required by paragraph (e)(2) of this section such that it could be operated instinctively, inadvertently, or by habitual reflexive action.				
5. The system used to generate the aural warning must be designed to minimize false or inappropriate alerts.				
6. Failures of systems used to inhibit the landing gear aural warning, that would prevent the warning system from operating, must be improbable.				
7. A flightcrew alert must be provided whenever the landing gear position is not consistent with the landing gear selector lever position.				

(*Continued*)

TABLE 8.2 (Continued)
Compliance Matrix for Part 25 Regulations Related to Flightcrew Human Factors[a]

Regulation Title 14 CFR Part 25	Human Factors (HF) Requirements	Applicable Amendment	Method(s) of Compliance	Person or Entity Finding Compliance	Applicable Guidance
Personnel and Cargo Accommodations	**Subpart D—Design and Construction**				
25.771-Pilot Compartment	(a) Each pilot compartment and its equipment must allow the minimum flight crew … to perform their duties without unreasonable concentration or fatigue …				
	(c) If provisions is made for a second pilot, the airplane must be controllable with equal safety from either pilot seat.				
	(d) "The pilot compartment must be constructed so that, when flying in rain or snow, it will not leak in a manner that will distract the crew …"				
	(e) "Vibration and noise characteristics of cockpit equipment may not interfere with safe operation of the airplane."				
25.773-Pilot Compartment View	(a) Nonprecipitation conditions. For non-precipitation conditions, the following apply:				
	(1) Each pilot compartment must be arranged to give the pilots a sufficiently extensive, clear, and undistorted view, to enable them to safely perform any maneuvers within the operating limitations of the airplane, including taxiing, takeoff, approach, and landing.				
	(2) Each pilot compartment must be free of glare and reflection that could interfere with the normal duties of the minimum flight crew (established under Sec. 25.1523). This must be shown in day and night flight tests under nonprecipitation conditions.				

(Continued)

TABLE 8.2 (*Continued*)
Compliance Matrix for Part 25 Regulations Related to Flightcrew Human Factors[a]

Regulation Title 14 CFR Part 25	Human Factors (HF) Requirements	Applicable Amendment	Method(s) of Compliance	Person or Entity Finding Compliance	Applicable Guidance
	(b) Precipitation conditions. For precipitation conditions, the following apply:				
	(1) The airplane must have a means to maintain a clear portion of the windshield, during precipitation conditions, sufficient for both pilots to have a sufficiently extensive view along the flight path in normal flight attitudes of the airplane. This means must be designed to function, without continuous attention on the part of the crew, in [specified precipitation conditions].				
	(2) No single failure of the systems used to provide the view required by paragraph (b)(1) of this section must cause the loss of that view by both pilots in the specified precipitation conditions				
	(i) A window that is openable under the conditions prescribed in paragraph (b)(1) of this section when the cabin is not pressurized, provides the view specified in that paragraph, and gives sufficient protection from the elements against impairment of the pilot's vision; or				
	(ii) An alternate means to maintain a clear view under the conditions specified in paragraph (b)(1) of this section, considering the probable damage due to a severe hail encounter.				
	(3) The first pilot must have a window that—				
	(i) Is openable under the conditions prescribed in paragraph (b)(1) of this section when the cabin is not pressurized,				
	(ii) Provides the view specified in paragraph (b)(1) of this section,				
	(iii) Provides sufficient protection from the elements against impairment of the pilot's vision,				

(Continued)

TABLE 8.2 (Continued)
Compliance Matrix for Part 25 Regulations Related to Flightcrew Human Factors[a]

Regulation Title 14 CFR Part 25	Human Factors (HF) Requirements	Applicable Amendment	Method(s) of Compliance	Person or Entity Finding Compliance	Applicable Guidance
	(4) The openable window specified in paragraph (b)(3) of this section need not be provided if it is shown that an area of the transparent surface will remain clear sufficient for at least one pilot to land the airplane safely in the event of—				
	(i) Any system failure or combination of failures which is not extremely improbable, in accordance with Sec. 25.1309, under the precipitation conditions specified in paragraph				
	(b) (1) of this section.				
	(ii) An encounter with severe hail, birds, or insects.				
	(c) Internal windshield and window fogging. The airplane must have a means to prevent fogging of the internal portions of the windshield and window panels over an area which would provide the visibility specified in paragraph (a) of this section under all internal and external ambient conditions, including precipitation conditions, in which the airplane is intended to be operated.				
	(d) Fixed markers or other guides must be installed at each pilot station to enable the pilots to position themselves in their seats for an optimum combination of outside visibility and instrument scan. If lighted markers or guides are used they must comply with the requirements specified in Sec. 25.1381.				
25.777–Cockpit Controls	(a) Each cockpit control must be located to provide convenient operation and to prevent confusion and inadvertent operation.				
	(b) The direction of movement of cockpit controls must meet the requirements of Sec. 25.779. Wherever practicable, the sense of motion involved in the operation of other controls must correspond to the sense of the effect of the operation upon the airplane or upon the part operated. Controls of a variable nature using a rotary motion must move clockwise from the off position, through an increasing range, to the full on position.				

(Continued)

TABLE 8.2 (*Continued*)

Compliance Matrix for Part 25 Regulations Related to Flightcrew Human Factors[a]

Regulation Title 14 CFR Part 25	Human Factors (HF) Requirements	Applicable Amendment	Method(s) of Compliance	Person or Entity Finding Compliance	Applicable Guidance
	(c) The controls must be located and arranged, with respect to the pilots' seats, so that there is full and unrestricted movement of each control without interference from the cockpit structure or the clothing of the minimum flight crew (established under Sec. 25.1523) when any member of this flight crew, from 5′2″ to (6′3″) in height, is seated with the seat (belt and shoulder harness (if provided)) fastened …				
	(d) Identical powerplant controls for each engine must be located to prevent confusion as to the engines they control.				
	(e) Wing flap controls and other auxiliary lift device controls must be located on top of the pedestal, aft of the throttles, centrally or to the right of the pedestal centerline, and not less than 10 inches aft of the landing gear control.				
	(f) The landing gear control must be located forward of the throttles and must be operable by each pilot when seated with seat [belt and shoulder harness (if provided)] fastened.				
	(g) Control knobs must be shaped in accordance with Sec. 25.781. In addition, the knobs must be of the same color, and this color must contrast with the color of the control knobs for other purposes and the surrounding cockpit.				
	(h) If a flight engineer is required as part of the minimum flight crew (established under Sec. 25.1523), the airplane must have a flight engineer station located and arranged so that the flight crewmembers can perform their functions efficiently and without interfering with each other.				

(Continued)

TABLE 8.2 (Continued)
Compliance Matrix for Part 25 Regulations Related to Flightcrew Human Factors[a]

Regulation Title 14 CFR Part 25	Human Factors (HF) Requirements	Applicable Amendment	Method(s) of Compliance	Person or Entity Finding Compliance	Applicable Guidance
25.779-Motion and Effect of Cockpit Controls	**(1) Primary Controls**				
	Motion and effect				
	Aileron ---------- Right (clockwise) for right wing down				
	Elevator ---------- Rearward for nose up				
	Rudder ---------- Right pedal forward for nose right				
	(2) Secondary Controls				
	Motion and effect				
	Flaps (or auxiliary lift devices). Forward for flaps up: rearward for flaps down				
	Trim tabs (or equivalent). Rotate to produce similar rotation of the airplane about an axis parallel to the axis of the control				
	(b) Powerplant and auxiliary controls:				
	(1) Powerplant Controls				
	Motion and effect				
	[Power or thrust ----------] Forward to increase forward thrust and rearward to increase rearward thrust				
	Propellers ---------- Forward to increase rpm				
	Mixture ---------- Forward or upward for rich				
	Carburetor air heat ---------- Forward or upward for cold				
	Super-charger ---------- Forward or upward for low blower. For turbosuperchargers, forward, upward, or clockwise, to increase pressure				
	(2) Auxiliary Controls				
	Motion and effect				
	Landing gear ---------- Down to extend				

(Continued)

TABLE 8.2 (Continued)

Compliance Matrix for Part 25 Regulations Related to Flightcrew Human Factors[a]

Regulation Title 14 CFR Part 25	Human Factors (HF) Requirements	Applicable Amendment	Method(s) of Compliance	Person or Entity Finding Compliance	Applicable Guidance
25.781-Cockpit Control Knob Shape	Cockpit control knobs must conform to the general shapes (but not necessarily the exact sizes or specific proportions) in the following figure:				

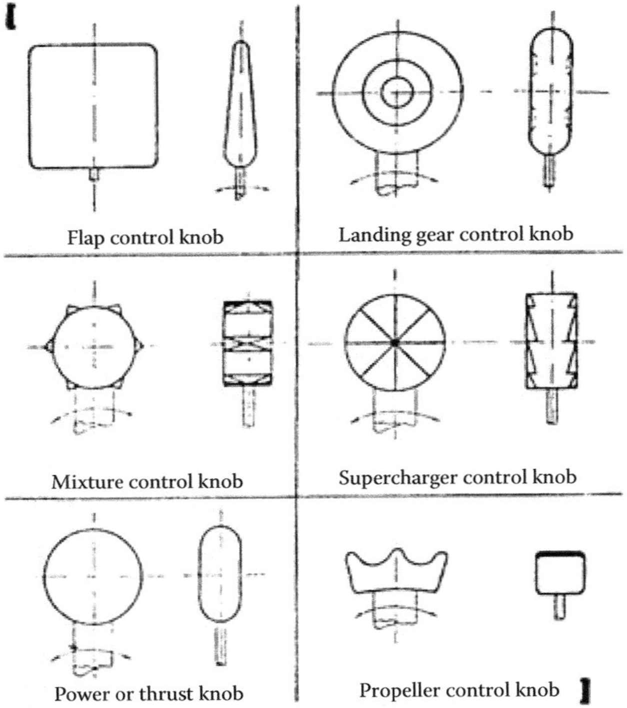

Flap control knob

Landing gear control knob

Mixture control knob

Supercharger control knob

Power or thrust knob

Propeller control knob

(Continued)

TABLE 8.2 (Continued)
Compliance Matrix for Part 25 Regulations Related to Flightcrew Human Factors[a]

Regulation Title 14 CFR Part 25	Human Factors (HF) Requirements	Applicable Amendment	Method(s) of Compliance	Person or Entity Finding Compliance	Applicable Guidance
25.785-Seat, berths, safety belts, and harnesses	(g) Each seat at a flight deck station must have a restraint system consisting of a combined safety belt and shoulder harness with a single-point release that permits the flight deck occupant, when seated with the restraint system fastened, to perform all of the occupant's necessary flight deck functions. There must be a means to secure each combined restraint system when not in use to prevent interference with the operation of the airplane and with rapid egress in an emergency.				
	(k) "Each projecting object that would injure persons seated or moving about the airplane in normal flight must be padded."				
25.809-Emergency Exit Arrangement	(a) Each emergency exit, including each flightcrew emergency exit, must be a movable door or hatch in the external walls of the fuselage, allowing an unobstructed opening to the outside. In addition, each emergency exit must have means to permit viewing of the conditions outside the exit when the exit is closed. The viewing means may be on or adjacent to the exit provided no obstructions exist between the exit and the viewing means. Means must also be provided to permit viewing of the likely areas of evacuee ground contact. The likely areas of evacuee ground contact must be viewable during all lighting conditions with the landing gear extended as well as in all conditions of landing gear collapse.				

(Continued)

TABLE 8.2 (*Continued*)

Compliance Matrix for Part 25 Regulations Related to Flightcrew Human Factors[a]

Regulation Title 14 CFR Part 25 — Human Factors (HF) Requirements	Applicable Amendment	Method(s) of Compliance	Person or Entity Finding Compliance	Applicable Guidance
(b) "Each emergency exit must be openable from the inside and the outside except that sliding window emergency exits in the flight crew area need not be openable from the outside if other approved exits are convenient and readily accessible to the flight crew area. Each emergency exit must be capable of being opened, when there is no fuselage deformation."				
(c) "The means of opening emergency exits must be simple and obvious; may not require exceptional effort; and must be arranged and marked so that it can be readily located and operated, even in darkness. Internal exit-opening means involving sequence operations (such as operation of two handles or latches, or the release of safety catches) may be used for flightcrew emergency exits if it can be reasonably established that these means are simple and obvious to crewmembers trained in their use."				
(h) When required by the operating rules for any large passenger-carrying turbojet-powered airplane, each ventral exit and tailcone exit must be— (1) Designed and constructed so that it cannot be opened during flight and (2) Marked with a placard readable from a distance of 30 in and installed at a conspicuous location near the means of opening the exit, stating that the exit has been designed and constructed so that it cannot be opened during flight.				

(*Continued*)

TABLE 8.2 (Continued)

Compliance Matrix for Part 25 Regulations Related to Flightcrew Human Factors[a]

Regulation Title 14 CFR Part 25	Human Factors (HF) Requirements	Applicable Amendment	Method(s) of Compliance	Person or Entity Finding Compliance	Applicable Guidance
General	**Subpart E—Powerplant**				
25.941-Inlet, Engine, and Exhaust Compatibility	(b) The dynamic effects of the operation of these (including consideration of probable malfunctions) upon the aerodynamic control of the airplane may not result in any condition that would require exceptional skill, alertness, or strength on the part of the pilot to avoid exceeding an operational or structural limitation of the airplane and				
	(c) In showing compliance with paragraph (b) of this section, the pilot strength required may not exceed the limits set forth in § 25.143(d), subject to the conditions set forth in paragraphs (e) and (f) of § 25.143.				
Powerplant Controls and Accessories	**Subpart E—Powerplant**				
25.1141-Powerplant Controls-General	Each powerplant control must be located, arranged, and designed under Secs. 25.777 through 25.781 and marked under Sec. 25.1555. In addition, it must meet the following requirements:				
	(a) Each control must be located so that it cannot be inadvertently operated by persons entering, leaving, or moving normally in, the cockpit.				
	(b) Each flexible control must be approved or must be shown to be suitable for the particular application.				
	(d) Each control must be able to maintain any set position without constant attention by flight crewmembers and without creep due to control loads or vibration.				

(Continued)

TABLE 8.2 (*Continued*)

Compliance Matrix for Part 25 Regulations Related to Flightcrew Human Factors[a]

Regulation Title 14 CFR Part 25	Human Factors (HF) Requirements	Applicable Amendment	Method(s) of Compliance	Person or Entity Finding Compliance	Applicable Guidance
	(f) For powerplant valve controls located in the flight deck there must be a means:				
	(1) For the flight crew to select each intended position or function of the valve.				
	(2) To indicate to the flightcrew:				
	(i) The selected position or function of the valve.				
	(ii) When the valve has not responded as intended to the selected position or function.				
25.1142-Auxiliary Power Unit Controls	Means must be provided on the flight deck for starting, stopping, and emergency shutdown of each installed auxiliary power unit				
25.1143-Engine Controls	(b) Power and thrust controls must be arranged to allow—				
	(1) Separate control of each engine.				
	(2) Simultaneous control of all engines.				
	(c) Each power and thrust control must provide a positive and immediately responsive means of controlling its engine.				
	(e) If a power or thrust control incorporates a fuel shutoff feature, the control must have a means to prevent the inadvertent movement of the control into the shutoff position. The means must—				
	(1) Have a positive lock or stop at the idle position and				
	(2) Require a separate and distinct operation to place the control in the shutoff position.				

(Continued)

TABLE 8.2 (Continued)
Compliance Matrix for Part 25 Regulations Related to Flightcrew Human Factors[a]

Regulation Title 14 CFR Part 25	Human Factors (HF) Requirements	Applicable Amendment	Method(s) of Compliance	Person or Entity Finding Compliance	Applicable Guidance
25.1145-Ignition Switches	(b) There must be means to quickly shut off all ignition by the grouping of switches or by a master ignition control.				
	(c) Each group of ignition switches, except ignition switches for turbine engines for which continuous ignition is not required, and each master ignition control must have a means to prevent its inadvertent operation.				
25.1147-Mixture Controls	(a) If there are mixture controls, each engine must have a separate control. The controls must be grouped and arranged to allow—				
	(1) Separate control of each engine and				
	(2) Simultaneous control of all engines.				
	(b) Each intermediate position of the mixture controls that corresponds to a normal operating setting must be identifiable by feel and sight.				
	(c) The mixture controls must be accessible to both pilots. However, if there is a separate flight engineer station with a control panel, the controls need be accessible only to the flight engineer.				
25.1149-Propeller Speed and Pitch Controls	(a) There must be a separate propeller speed and pitch control for each propeller.				
	(b) The controls must be grouped and arranged to allow—				
	(1) Separate control of each propeller.				
	(2) Simultaneous control of all propellers.				
	(d) The propeller speed and pitch controls must be to the right of, and at least one inch below, the pilot's throttle controls.				

(Continued)

TABLE 8.2 (Continued)
Compliance Matrix for Part 25 Regulations Related to Flightcrew Human Factors[a]

Regulation Title 14 CFR Part 25	Human Factors (HF) Requirements	Applicable Amendment	Method(s) of Compliance	Person or Entity Finding Compliance	Applicable Guidance
25.1153-Propeller Feathering Controls	(a) There must be a separate propeller feathering control for each propeller. The control must have means to prevent its inadvertent operation. (b) If feathering is accomplished by movement of the propeller pitch or speed control lever, there must be means to prevent the inadvertent movement of this lever to the feathering position during normal operation.				
25.1155-Reverse Thrust and Propeller Pitch Settings below the Flight Regime	Each control for reverse thrust and for propeller pitch settings below the flight regime must have means to prevent its inadvertent operation. The means must have a positive lock or stop at the flight idle position and must require a separate and distinct operation by the crew to displace the control from the flight regime (forward thrust regime for turbojet powered airplanes).				
25.1159-Supercharger Controls	Each supercharger control must be accessible to the pilots or, if there is a separate flight engineer station with a control panel, to the flight engineer.				
25.1161-Fuel Jettisoning System Controls	Each fuel jettisoning system control must have guards to prevent inadvertent operation. No control may be near any fire extinguisher control or other control used to combat fire.				
General	**Subpart F—Equipment**				
25.1301-Function and Installation	(a) Each item of installed equipment must— (1) Be of a kind and design appropriate to its intended function. (2) Be labeled as to its identification, function, or operating limitations, or any applicable combination of these factors. (3) Be installed according to limitations specified for that equipment. (4) Function properly when installed …				

(Continued)

TABLE 8.2 (*Continued*)
Compliance Matrix for Part 25 Regulations Related to Flightcrew Human Factors[a]

Regulation Title 14 CFR Part 25	Human Factors (HF) Requirements	Applicable Amendment	Method(s) of Compliance	Person or Entity Finding Compliance	Applicable Guidance
25.1302- Installed systems and equipment for use by the flightcrew.	This section applies to installed systems and equipment intended for flightcrew members' use in operating the airplane from their normally seated positions on the flight deck. The applicant must show that these systems and installed equipment, individually and in combination with other such systems and equipment, are designed so that qualified flightcrew members trained in their use can safely perform all of the tasks associated with the systems' and equipment's intended functions. Such installed equipment and systems must meet the following requirements:				
	(a) Flight-deck controls must be installed to allow accomplishment of all the tasks required to safely perform the equipment's intended function, and information must be provided to the flightcrew that is necessary to accomplish the defined tasks.				
	(b) Flight deck controls and information intended for the flight crew's use must:				
	(1) Be provided in a clear and unambiguous manner at a resolution and precision appropriate to the task.				
	(2) Be accessible and usable by the flightcrew in a manner consistent with the urgency, frequency, and duration of their tasks.				
	(3) Enable flightcrew awareness, if awareness is required for safe operation, of the effects on the airplane or systems resulting from flightcrew actions.				
	(c) Operationally-relevant behavior of the installed equipment must be:				
	(1) Predictable and unambiguous.				
	(2) Designed to enable the flight crew to intervene in a manner appropriate to the task.				

(*Continued*)

TABLE 8.2 (*Continued*)

Compliance Matrix for Part 25 Regulations Related to Flightcrew Human Factors[a]

Regulation Title 14 CFR Part 25	Human Factors (HF) Requirements	Applicable Amendment	Method(s) of Compliance	Person or Entity Finding Compliance	Applicable Guidance
	(d) To the extent practicable, installed equipment must incorporate means to enable the flight crew to manage errors resulting from the kinds of flightcrew interactions with the equipment that can be reasonably expected in service. This paragraph does not apply to any of the following:				
	(1) Skill-related errors associated with manual control of the airplane.				
	(2) Errors that result from decisions, actions, or omissions committed with malicious intent.				
	(3) Errors arising from a crewmember's reckless decisions, actions, or omissions reflecting a substantial disregard for safety.				
	(4) Errors resulting from acts or threats of violence, including actions taken under duress.				
25.1303-Flight and Navigation Instruments	(a) The following flight and navigation instruments must be installed so that the instrument is visible from each pilot station:				
	(1) A free air temperature indicator or an air-temperature indicator which provides indications that are convertible to free-air temperature.				
	(2) A clock displaying hours, minutes, and seconds with a sweep-second pointer or digital presentation.				
	(3) A direction indicator (nonstabilized magnetic compass).				
	(b) The following flight and navigation instruments must be installed at each pilot station:				
	(1) An airspeed indicator. If airspeed limitations vary with altitude, the indicator must have a maximum allowable airspeed indicator showing the variation of V_{MO} with altitude.				
	(2) An altimeter (sensitive).				
	(3) A rate-of-climb indicator (vertical speed).				

(Continued)

TABLE 8.2 (*Continued*)

Compliance Matrix for Part 25 Regulations Related to Flightcrew Human Factors[a]

Regulation Title 14 CFR Part 25	Human Factors (HF) Requirements	Applicable Amendment	Method(s) of Compliance	Person or Entity Finding Compliance	Applicable Guidance
	(4) A gyroscopic rate-of-turn indicator combined with an integral slip-skid indicator (turn–and-bank indicator) except that only a slip-skid indicator is required on large airplanes with a third attitude instrument system usable through flight attitudes of 360° of pitch and roll and installed in accordance with (Sec. 121.305[k]) of this title.				
	(5) A bank and pitch indicator (gyroscopically stabilized).				
	(6) A direction indicator (gyroscopically stabilized, magnetic or nonmagnetic).				
	(c) The following flight and navigation instruments are required as prescribed in this paragraph:				
	(1) A speed warning device is required for turbine engine powered airplanes and for airplanes with V_{MO}/M_{MO} greater than 0.8 V_{DF}/M_{DF} or 0.8 V_D/M_D. The speed warning device must give effective aural warning (differing distinctively from aural warnings used for other purposes) to the pilots, whenever the speed exceeds V_{MO} plus 6 knots or M_{MO} +0.01. The upper limit of the production tolerance for the warning device may not exceed the prescribed warning speed.				
	(2) A machmeter is required at each pilot station for airplanes with compressibility limitations not otherwise indicated to the pilot by the airspeed indicating system required under paragraph (b) (1) of this section.				
25.1305-Powerplant Instruments	... prescribes the powerplant instruments that are required.				

(Continued)

TABLE 8.2 (Continued)

Compliance Matrix for Part 25 Regulations Related to Flightcrew Human Factors[a]

Regulation Title 14 CFR Part 25	Human Factors (HF) Requirements	Applicable Amendment	Method(s) of Compliance	Person or Entity Finding Compliance	Applicable Guidance
25.1307-Miscellaneous Equipment	(d) Two systems for two-way radio communications, with controls for each accessible from each pilot station, ...				
	(e) Two systems for radio navigation, with controls for each accessible from each pilot station, ...				
25.1309-Equipment, systems, and installations	(a) The equipment, systems, and installations ... must be designed to ensure that they perform their intended functions under any foreseeable operating condition.				
	(b) The airplane systems and associated components ... must be designed so that ... the occurrence of ... failure conditions which would reduce the .... ability of the crew to cope with adverse operating conditions is improbable.				
	(c) Warning information must be provided to alert the crew to unsafe system operating conditions, and to enable them to take appropriate corrective action. Systems, controls, and associated monitoring and warning means must be designed to minimize crew errors which could create additional hazards.				
	(d) " ... The [compliance] analysis must consider ... the crew warning cues, corrective action required, and the capability of detecting faults."				
25.1321-Arrangement and Visibility	(a) Each flight, navigation, and powerplant instrument for use by any pilot must be plainly visible to him from his station with the minimum practicable deviation from his normal position and line of vision when he is looking forward along the flight path.				

(Continued)

TABLE 8.2 (Continued)
Compliance Matrix for Part 25 Regulations Related to Flightcrew Human Factors[a]

Regulation Title 14 CFR Part 25 / Human Factors (HF) Requirements	Applicable Amendment	Method(s) of Compliance	Person or Entity Finding Compliance	Applicable Guidance
(b) The flight instruments required by Sec. 25.1303 must be grouped on the instrument panel and centered as nearly as practicable about the vertical plane of the pilot's forward vision. In addition—				
(1) The instrument that most effectively indicates attitude must be on the panel in the top center position.				
(2) The instrument that most effectively indicates airspeed must be adjacent to and directly to the left of the instrument in the top center position.				
(3) The instrument that most effectively indicates altitude must be adjacent to and directly to the right of the instrument in the top center position.				
(4) The instrument that most effectively indicates direction of flight must be adjacent to and directly below the instrument in the top center position.				
(c) Required powerplant instruments must be closely grouped on the instrument panel. In addition—				
(1) The location of identical powerplant instruments for the engines must prevent confusion as to which engine each instrument relates and				
(2) Powerplant instruments vital to the safe operation of the airplane must be plainly visible to the appropriate crewmembers.				
(d) Instrument panel vibration may not damage or impair the accuracy of any instrument.				
(e) If a visual indicator is provided to indicate malfunction of an instrument, it must be effective under all probable cockpit lighting conditions.				

(Continued)

TABLE 8.2 (*Continued*)

Compliance Matrix for Part 25 Regulations Related to Flightcrew Human Factors[a]

Regulation Title 14 CFR Part 25	Human Factors (HF) Requirements	Applicable Amendment	Method(s) of Compliance	Person or Entity Finding Compliance	Applicable Guidance
25.1322-Flightcrew Alerting	(a) Flightcrew alerts must:				
	(1) Provide the flight crew with the information needed to				
	(i) Identify non-normal operation or airplane system conditions and				
	(ii) Determine the appropriate actions, if any.				
	(2) Be readily and easily detectable and intelligible by the flight crew under all foreseeable operating conditions, including conditions where multiple alerts are provided.				
	(3) Be removed when the alerting condition no longer exists.				
	(b) Alerts must conform to the following prioritization hierarchy based on the urgency of flightcrew awareness and response:				
	(1) *Warning*: For conditions that require immediate flightcrew awareness and immediate flightcrew response.				
	(2) *Caution*: For conditions that require immediate flightcrew awareness and subsequent flightcrew response.				
	(3) *Advisory*: For conditions that require flightcrew awareness and may require subsequent flightcrew response.				
	(c) Warning and caution alerts must:				
	(1) Be prioritized within each category, when necessary.				
	(2) Provide timely attention-getting cues through at least two different senses by a combination of aural, visual, or tactile indications.				
	(3) Permit each occurrence of the attention-getting cues required by paragraph (c)(2) of this section to be acknowledged and suppressed, unless they are required to be continuous.				

(Continued)

TABLE 8.2 (Continued)
Compliance Matrix for Part 25 Regulations Related to Flightcrew Human Factors[a]

Regulation Title 14 CFR Part 25 Human Factors (HF) Requirements	Applicable Amendment	Method(s) of Compliance	Person or Entity Finding Compliance	Applicable Guidance
(d) The alert function must be designed to minimize the effects of false and nuisance alerts. In particular, it must be designed to:				
(1) Prevent the presentation of an alert that is inappropriate or unnecessary.				
(2) Provide a means to suppress an attention-getting component of an alert caused by a failure of the alerting function that interferes with the flightcrew's ability to safely operate the airplane. This means must not be readily available to the flightcrew so that it could be operated inadvertently or by habitual reflexive action. When an alert is suppressed, there must be a clear and unmistakable annunciation to the flightcrew that the alert has been suppressed.				
(e) Visual alert indications must:				
(1) Conform to the following color convention:				
(i) Red for warning alert indications,				
(ii) Amber or yellow for caution alert indications and				
(iii) Any color except red or green for advisory alert indications.				
(2) Use visual coding techniques, together with other alerting function elements on the flight deck, to distinguish between warning, caution, and advisory alert indications, if they are presented on monochromatic displays that are not capable of conforming to the color convention in paragraph (e)(1) of this section.				
(f) Use of the colors red, amber, and yellow on the flight deck for functions other than flightcrew alerting must be limited and must not adversely affect flightcrew alerting.				

(Continued)

TABLE 8.2 (*Continued*)
Compliance Matrix for Part 25 Regulations Related to Flightcrew Human Factors[a]

Regulation Title 14 CFR Part 25	Human Factors (HF) Requirements	Applicable Amendment	Method(s) of Compliance	Person or Entity Finding Compliance	Applicable Guidance
25.1326-Pitot Heat Indication Systems	If a flight instrument pitot heating system is installed, an indication system must be provided to indicate to the flight crew when that pitot heating system is not operating. The indication system must comply with the following requirements: (a) The indication provided must incorporate an amber light that is in clear view of a flight crewmember. (b) The indication provided must be designed to alert the flight crew if either of the following conditions exist: (1) The pitot heating system is switched *off*. (2) The pitot heating system is switched on and any pitot tube heating element is inoperative.				
25.1329-Flight Guidance Systems	(a) Quick disengagement controls for the autopilot and autothrust functions must be provided for each pilot. The autopilot quick disengagement controls must be located on both control wheels (or equivalent). The autothrust quick disengagement controls must be located on the thrust control levers. Quick disengagement controls must be readily accessible to each pilot while operating the control wheel (or equivalent) and thrust control levers. (c) Engagement or switching of the flight guidance system, a mode, or a sensor may not cause a transient response of the airplane's flight path any greater than a minor transient, as defined in paragraph (n)(1) of this section. (d) Under normal conditions, the disengagement of any automatic control function of a flight guidance system may not cause a transient response of the airplane's flight path any greater than a minor transient.				

(Continued)

TABLE 8.2 (Continued)
Compliance Matrix for Part 25 Regulations Related to Flightcrew Human Factors[a]

Regulation Title 14 CFR Part 25 — Human Factors (HF) Requirements	Applicable Amendment	Method(s) of Compliance	Person or Entity Finding Compliance	Applicable Guidance
(e) Under rare normal and non-normal conditions, disengagement of any automatic control function of a flight guidance system may not result in a transient any greater than a significant transient, as defined in paragraph (n)(2) of this section.				
(f) The function and direction of motion of each command reference control, such as heading select or vertical speed, must be plainly indicated on, or adjacent to, each control if necessary to prevent inappropriate use or confusion.				
(g) Under any condition of flight appropriate to its use, the flight guidance system may not produce hazardous loads on the airplane, nor create hazardous deviations in the flight path. This applies to both fault-free operation and in the event of a malfunction, and assumes that the pilot begins corrective action within a reasonable period of time.				
(h) When the flight guidance system is in use, a means must be provided to avoid excursions beyond an acceptable margin from the speed range of the normal flight envelope. If the airplane experiences an excursion outside this range, a means must be provided to prevent the flight guidance system from providing guidance or control to an unsafe speed.				
(i) The flight guidance system functions, controls, indications, and alerts must be designed to minimize flightcrew errors and confusion concerning the behavior and operation of the flight guidance system. Means must be provided to indicate the current mode of operation, including any armed modes, transitions, and reversions. Selector switch position is not an acceptable means of indication. The controls and indications must be grouped and presented in a logical and consistent manner. The indications must be visible to each pilot under all expected lighting conditions.				

(Continued)

TABLE 8.2 (Continued)
Compliance Matrix for Part 25 Regulations Related to Flightcrew Human Factors[a]

Regulation Title 14 CFR Part 25	Human Factors (HF) Requirements	Applicable Amendment	Method(s) of Compliance	Person or Entity Finding Compliance	Applicable Guidance
	(j) Following disengagement of the autopilot, a warning (visual and auditory) must be provided to each pilot and be timely and distinct from all other cockpit warnings.				
	(k) Following disengagement of the autothrust function, a caution must be provided to each pilot.				
	(l) The autopilot may not create a potential hazard when the flightcrew applies an override force to the flight controls.				
	(m) During autothrust operation, it must be possible for the flightcrew to move the thrust levers without requiring excessive force. The autothrust may not create a potential hazard when the flightcrew applies an override force to the thrust levers.				
	(n) For purposes of this section, a transient is a disturbance in the control or flight path of the airplane that is not consistent with response to flightcrew inputs or environmental conditions.				
	(1) A minor transient would not significantly reduce safety margins and would involve flightcrew actions that are well within their capabilities. A minor transient may involve a slight increase in flightcrew workload or some physical discomfort to passengers or cabin crew.				
	(2) A significant transient may lead to a significant reduction in safety margins, an increase in flightcrew workload, discomfort to the flightcrew, or physical distress to the passengers or cabin crew, possibly including non-fatal injuries. Significant transients do not require, in order to remain within or recover to the normal flight envelope, any of the following: (i) Exceptional piloting skill, alertness, or strength.				

(Continued)

TABLE 8.2 (Continued)
Compliance Matrix for Part 25 Regulations Related to Flightcrew Human Factors[a]

Regulation Title 14 CFR Part 25	Human Factors (HF) Requirements	Applicable Amendment	Method(s) of Compliance	Person or Entity Finding Compliance	Applicable Guidance
	(ii) Forces applied by the pilot which are greater than those specified in § 25.143(c).				
	(iii) Accelerations or attitudes in the airplane that might result in further hazard to secured or non-secured occupants.				
	(b) Unless undimmed instrument lights are satisfactory under each expected flight condition, there must be a means to control the intensity of illumination.				
25.1447-Equipment standards for oxygen dispensing units	(b) If certification for operation up to and including 25,000 feet is requested, an oxygen supply terminal and unit of oxygen dispensing equipment for the immediate use of oxygen by each crewmember must be within easy reach of that crewmember. For any other occupants, the supply terminals and dispensing equipment must be located to allow the use of oxygen as required by the operating rules in this chapter.				
Operating Limitations	**Subpart G—Operating Limitations and Information**				
25.1523-Minimum Flight Crew	The minimum flight crew must be established so that it is sufficient for safe operation, considering—				
	(a) The workload on individual crewmembers.				
	(b) The accessibility and ease of operation of necessary controls by the appropriate crewmember.				
	(c) The kinds of operation authorized under § 25.1525.				
	The criteria used in making the determinations required by this section are set forth in Appendix D.	7			

(Continued)

TABLE 8.2 (*Continued*)
Compliance Matrix for Part 25 Regulations Related to Flightcrew Human Factors[a]

Regulation Title 14 CFR Part 25	Human Factors (HF) Requirements	Applicable Amendment	Method(s) of Compliance	Person or Entity Finding Compliance	Applicable Guidance
25.1331(a)-Instruments Using a Power Supply	(1) Each instrument must have a visual means integral with, the instrument, to indicate when power adequate to sustain proper instrument performance is not being supplied. The power must be measured at or near the point where it enters the instruments. For electric instruments, the power is considered to be adequate when the voltage is within approved limits …				
	(3) "If an instrument presenting navigation data receives information from sources external to that instrument and loss of that information would render the presented data unreliable, the instrument must incorporate a visual means to warn the crew, when such loss of information occurs, that the presented data should not be relied upon."				
25.1357-Circuit Protection Devices	(d) If the ability to reset a circuit breaker or replace a fuse is essential to safety in flight, that circuit breaker or fuse must be located and identified so that it can be readily reset or replaced in flight. Where fuses are used, there must be spare fuses for use in flight equal to at least 50% of the number of fuses of each rating required for complete circuit protection.				
	(f) For airplane systems for which the ability to remove or reset power during normal operations is necessary, the system must be designed so that circuit breakers are not the primary means to remove or reset system power unless specifically designed for use as a switch."				
25.1381-Instrument Lights	(a) The instrument lights must—				
	(1) Provide sufficient illumination to make each instrument, switch and other device necessary for safe operation easily readable unless sufficient illumination is available from another source and				
	(2) Be installed so that—				
	(i) Their direct rays are shielded from the pilot's eyes and				
	(ii) No objectionable reflections are visible to the pilot				

(*Continued*)

TABLE 8.2 (*Continued*)

Compliance Matrix for Part 25 Regulations Related to Flightcrew Human Factors[a]

Regulation Title 14 CFR Part 25	Human Factors (HF) Requirements	Applicable Amendment	Method(s) of Compliance	Person or Entity Finding Compliance	Applicable Guidance
Markings and Placards	**Subpart G-Operating Limitations and Information**				
25.1541-General	(b) Each marking and placard prescribed in paragraph (a) of this section—				
	(1) Must be displayed in a conspicuous place.				
	(2) May not be easily erased, disfigured, or obscured.				
25.1543-Instrument markings-General	"For each instrument—				
	When markings are on the cover glass of the instrument, there must be means to maintain the correct alignment of the glass cover with the face of the dial and				
	(b) Each instrument marking must be clearly visible to the appropriate crewmember."				
25.1545-Airspeed Limitation Information	"The airspeed limitations required by § 25.1583(a) must be easily read and understood by the flight crew."				
25.1555-Control Markings	(a) Each cockpit control, other than primary flight controls and controls whose function is obvious, must be plainly marked as to its function and method of operation.				
	(b) Each aerodynamic control must be marked under the requirements of Secs. 25.677 and 25.699.				
	(c) For powerplant fuel controls—				
	(1) Each fuel tank selector control must be marked to indicate the position corresponding to each tank and to each existing cross feed position.				
	(2) If safe operation requires the use of any tanks in a specific sequence, that sequence must be marked on, or adjacent to, the selector for those tanks.				
	(3) Each valve control for each engine must be marked to indicate the position corresponding to each engine controlled.				

(*Continued*)

TABLE 8.2 (Continued)
Compliance Matrix for Part 25 Regulations Related to Flightcrew Human Factors[a]

Regulation Title 14 CFR Part 25	Human Factors (HF) Requirements	Applicable Amendment	Method(s) of Compliance	Person or Entity Finding Compliance	Applicable Guidance
	(d) For accessory, auxiliary, and emergency controls—				
	(1) Each emergency control (including each fuel jettisoning and fluid shutoff control) must be colored red.				
	(2) Each visual indicator required by Sec. 25.729(e) must be marked so that the pilot can determine at any time when the wheels are locked in either extreme position, if retractable landing gear is used.				
25.1561-Safety Equipment	(a) Each safety equipment control to be operated by the crew in emergency, such as controls for automatic liferaft releases, must be plainly marked as to its method of operation.				
	(b) Each location, such as a locker or compartment, that carries any fire extinguishing, signaling, or other lifesaving equipment must be marked accordingly.				
	(c) Stowage provisions for required emergency equipment must be conspicuously marked to identify the contents and facilitate (the easy) removal of the equipment.				
	(d) Each liferaft must have obviously marked operating instructions.				
	(e) Approved survival equipment must be marked for identification and method of operation.				
25.1563-Airspeed Placard	"A placard showing the maximum airspeeds for flap extension for the takeoff, approach, and landing positions must be installed in clear view of each pilot."				

[a] Methods of Compliance: FT = Flight Test; GT = Ground Test; AN = Analysis; DE = Design; SI = Similarity; ELOS = Equivalent Level of Safety Finding; PE$_{xmpt}$ = Petition for Exemption.

INTENDED HUMAN OPERATOR[*]

The first requirement for the evaluation of human factors in any discipline is defining the intended *human* user for which the system is being designed. The FAR only states that aircraft cockpits are to be designed for flight crewmembers between the heights of 5'2" and 6'3" for transport category airplanes (14CFR Part 25), and 5'2" to 6'0" for transport (14CFR Part 29) and normal (14CFR Part 27) category helicopters, without prescribing a specific anthropometric database or any other body dimensions other than stature. Furthermore, no stature dimension(s) are provided at all for normal category (14CFR Part 23) airplanes. The FAA has recognized the importance of extending anthropometric design beyond just stature, and over the years has issued guidance, in the form of PS[†] and AC[‡] that expand on the intent of the regulation with regard to defining the human operator beyond just stature to include other body dimensions, such as sitting height, sitting eye height, functional reach, arm length, hand size, which can have significant effects on the geometric acceptability of the flight deck for pilots within the specified height range. Furthermore, the FAA guidance asserts that other dimensions do not necessarily correlate well with height or with each other. The FAA and the aviation industry have accepted the aircraft design requirement for a 5th percentile female to a 95th percentile male, which was based on historical military standards and reaffirmed as the center 90% of the target user population using a multivariate approach. The intended user is assumed to wear a variety of apparel (e.g., jacket, gloves, headset) and be in the normal seated position in the flight deck at the design eye (reference) point[§] (DEP) with seat/shoulder belts fastened, if required and so equipped.

In addition to standing height requirements, U.S.C. Title 14 of the CFR has both quantitative and qualitative force, skill, and workload requirements for manipulation of flight controls and performance of tasks. For example, throughout the 14 CFR design and certification requirements for aircraft performance and flying qualities, the statement is made that the aircraft should be designed so that it does not require *exceptional piloting skill, alertness*, or *strength*. However, no definition or quantitative measures are provided for *exceptional skill, alertness*, or *strength*; hence, it is subjectively evaluated. The FAA does not prescribe any specific subjective workload rating scale; however, the legacy scales used by industry and the military, such as Bedford Workload Scale,[¶] the Subjective Workload Assessment Technique[**] (SWAT), and NASA Task Load Index[††] (NASA-TLX), have been generally accepted by the FAA along with some aircraft specific scales developed by manufacturers. There is also no prescribed anthropometric database for body dimensions or strength/control forces specified in the FAR or any other FAA regulatory guidance; hence it is left to the discretion of each aircraft designer to negotiate with the FAA Aircraft Certification Service to determine an acceptable database that is representative of the intended users.

---

[*] Joslin, R. E. (2014, September). Examination of Anthropometric Databases for Aircraft Design. In Proceedings of the Human Factors and Ergonomics Society Annual Meeting (Vol. 58, No. 1, pp. 1–5). SAGE Publications.

[†] Policy Statement Number PS-ANM100-01-03A, Factors to Consider When Reviewing an Applicant's Proposed Human Factors Methods for Compliance for Flight Deck Certification, February 7, 2003.

[‡] FAA Advisory Circular 25.1302-1_ Installed Systems and Equipment for Use by the Flightcrew.

[§] The position at each pilot's station from which a seated pilot achieves the required combination of outside visibility and instrument scan. It is normally a point fixed in relation to the aircraft structure (neutral seat reference point) at which the midpoint of the pilot's eyes should be located when seated at the normal position. The DEP is the principal dimensional reference point for the location of flight-deck panels, controls, displays, and external vision.

[¶] Roscoe, A. H., & Ellis, G. A. (1990). A subjective rating scale for assessing pilot workload in flight: A decade of practical use (No. RAE-TR-90019). Royal Aerospace Establishment Farnborough (United Kingdom).

[**] Rubio, S., Díaz, E., Martín, J., & Puente, J. M. (2004). Evaluation of subjective mental workload: A comparison of SWAT, NASA-TLX, and workload profile methods. Applied Psychology, 53(1), 61–86.

[††] Hart, S. G., & Staveland, L. E. (1988). Development of NASA-TLX (Task Load Index): Results of empirical and theoretical research. Advances in Psychology, 52, 139–183.

HUMAN FACTORS CERTIFICATION PLAN

The following paragraphs outline the general contents of a Human Factors Certification Plan as specified in FAA PS.*

INTRODUCTION

Provides a short overview of the certification project, the certification program in general, and the purpose of the Human Factors Certification Plan specifically.

SYSTEM DESCRIPTION

Describes the general features of the flight-deck, system, and/or component being presented as part of a certification project, and the ways in which they interact (e.g., crew procedures and the basic design concepts as well as the principles and operational assumptions that underlie the design of the flight crew interfaces). The description is intended to cover both the design of the systems (hardware, software) and the tasks (physical, cognitive, perceptual, and procedural) the pilots will perform when using the systems in the context of their overall responsibilities. Special attention should be given to any new or unique features or functions and how the flight crew will use them, such as the following topic areas:

Intended function: The following items, if appropriate, should be identified focusing on new or unique features that affect the crew interface or the allocation of tasks between the pilot(s) and the airplane systems:
  - The intended function of the system from the pilot's perspective
  - The role of the pilot relative to the system
  - The procedures (e.g., type of approach procedures) expected to be flown
  - The assumed airplane capabilities (e.g., communication, navigation, and surveillance)

Flight deck layout drawings: Drawings of the flight-deck layout, even if they are only preliminary, can be very beneficial for providing an understanding of the intended overall flight-deck arrangement (controls, displays, sample display screens, seating, stowage, etc.). Scheduled updates to the drawings should be provided, so that the Certification Team's knowledge of the layout progresses as the design matures. Special attention should be given to any of the following that are novel or unique:
  - Arrangements of the controls, displays, or other flight-deck features or equipment
  - Controls, such as a cursor control device, or new applications of existing control technologies
  - Display hardware technology

For the items identified earlier, sketches of the crew interfaces for the specific systems can be helpful in providing an early understanding of the features that may have certification issues. Descriptions of interface, button, knob function, anticipated system response, alerting mechanism, mode annunciation, and so on, should be included with the drawing so that the documentation adequately covers each component or system that the pilot must interact with.

---

* Public Statement Number PS-ANM111-1999-99-2, Guidance for Reviewing Certification Plans to Address Human Factors for Certification of Transport Airplane Flight Decks, September 29, 1999.

Underlying principles for automation logic: For designs that involve significant automation, the way the automation operates and communicates that operation to the pilot can have significant effects on safety. Key topics could include the following:
- Operating modes
- Principles underlying mode transitions
- Mode annunciation scheme
- Automation engagement/disengagement principles
- Preliminary logic diagrams, if available

Underlying principles for crew procedures: As the design of the systems and the development of the associated procedures are interrelated, it is useful to describe the underlying guidelines or principles that form the basis for the crew procedures. Key topics could include the following:
- The expected use of memorized procedures with confirmation checklists versus read-and-do procedures/checklists
- Crew interactions during procedure/checklist accomplishment
- Automated support for procedures/checklists, if available

Assumed pilot characteristics: In addition to the anthropometric considerations described earlier in this chapter, the plan should include a description of the pilot group that the manufacturer expects will use the flight-deck design. This description could include assumptions about the following:
- Previous flying experience (e.g., ratings, flying hours)
- Experience with similar or dissimilar flight-deck designs and features, including automation
- Expected training that the pilots will receive on this flight-deck design, or assumptions regarding expected training

CERTIFICATION REQUIREMENTS

The current section should list and describe the human factors–related regulations and other requirements and may include a compliance checklist matrix with specific references to the detailed subparagraphs that will be covered by the Human Factors Certification Plan.

METHODS OF COMPLIANCE

The Certification Team should request the detailed plans for showing compliance as the plans evolve with the program. It is recommended that coordination meetings be held several times during the certification program to review the compliance checklist in detail and the associated test plans, as they are developed. This will help all parties reach agreement on how the tests, demonstrations, and other data-gathering efforts will be sufficient to show compliance. Of special importance is ensuring that the methods proposed will provide enough fidelity to identify human factors issues early enough to avoid adversely affecting the certification schedule.

In this section, the methods that will be used to demonstrate compliance with the relevant regulations should be delineated. The review and discussion of the MOC is an opportunity for the FAA and the applicant to work together to identify potential human factors issues early in the certification program. The MOC are not mutually exclusive. All of the methods should be described in enough detail to give the Certification Team confidence that the results of the chosen method will provide the necessary information for finding compliance. Examples of methods to demonstrate compliance are as follows:

Drawings: Layout drawings and/or engineering drawings that show the geometric arrangement of hardware or display graphics.

Configuration description: A description of the layout, arrangement, direction of movement, and so on or a reference to similar documentation.

Statement of similarity: A description of the system to be approved and a previously approved system, which details their physical, logical, and operational similarities, with respect to compliance with the regulations.

Evaluations, assessments, analyses: Evaluations conducted by the applicant or others (not the FAA or a designee), who provides a report to the FAA. These include the following:

Engineering evaluations or analyses: These assessments can involve a number of techniques, including such things as procedure evaluations (complexity, number of steps, nomenclature, etc.); reach analysis via computer modeling; time-line analysis for assessing task demands and workload; or other methods, depending on the issue being considered.

Mock-up evaluations: These types of evaluations use physical mockups of the flight deck and/or components, typically to assess of functional reach and structure clearance.

Part-task evaluations: These types of evaluations use devices that emulate (using flight hardware, simulated systems, or combinations) the crew interfaces for a single system or a related group of systems. Typically, these evaluations are limited by the extent to which acceptability may be affected by other flight deck tasks.

Simulator evaluations: These types of evaluations use devices that present an integrated emulation (using flight hardware, simulated systems, or combinations) of the flight deck and the operational environment. They can also be *flown*, with response characteristics that replicate, to some extent, the responses of the airplane. Typically, these evaluations are limited by the extent to which the simulation is a realistic, high fidelity representation of the airplane, the flight deck, the external environment, and crew operations. The types of pilots (test, instructor, and airline) used in the evaluations and the training they receive may significantly affect the results and their utility.

In-flight evaluations: These types of evaluations use the actual airplane. Typically, these evaluations are limited by the extent to which the flight conditions of particular interest (e.g., weather, failures, and unusual attitudes) can be located/generated and then safely evaluated in flight. The types of pilots (test, instructor, and airline) used in the evaluations and the training they receive may significantly affect the results and their utility.

Demonstrations: Similar to evaluations (described earlier), but conducted by the applicant with participation by the FAA or its designee. The applicant provides a report, requesting FAA concurrence on the findings. Examples of demonstrations include the following:
- Mock-up demonstrations
- Part-task demonstration
- Simulator demonstration

Inspection: A review by the FAA or its designee, who will be making the compliance finding.

Tests: Evaluations conducted by the FAA or a designee, which may encompass the following:
- *Bench tests*: These are tests of components in a laboratory environment. This type of testing is usually confined to showing that the components perform as designed. Typical bench testing may include measuring physical characteristics (e.g., forces, luminance, and format) or logical/dynamic responses to inputs, either from the user or from other systems (real or simulated).
- *Ground tests*: These are tests conducted in the actual airplane, while it is stationary on the ground. In some cases, specialized test equipment may be employed to allow the airplane systems to behave as though the airplane was airborne.
- *Simulator tests*: See simulator evaluations earlier.
- *Flight tests*: These are tests conducted in the actual airplane. The on-ground portions of the test (e.g., preflight, engine start, and taxi) are typically considered flight test rather than ground test. The methods identified previously cover a wide spectrum from documents that simply describe the product, to partial approximations, to methods

that replicate the actual airplane and its operation with great accuracy. Features of the product being certified and the types of human factors issues to be evaluated are key considerations when selecting which method is to be used. The characteristics described in the following can be used to help in coming to agreement regarding what constitutes the minimum acceptable MOC for any individual requirement. When a product needs to meet multiple requirements, some requirements may demand more complex testing whereas others can be handled using simple descriptive measures. It is important to note that the following characteristics are only general principles. They are intended to form the basis for discussions regarding acceptable MOC for a specific product with respect to a requirement.

OTHER CONSIDERATIONS

- *Degree of integration/independence*: If the product to be approved is a stand-alone piece of equipment that does not interact with other aspects of the crew interface, less-integrated MOC may be acceptable. However, if the product is tightly tied to other systems in the flight deck, either directly or by the ways crews use them, it may be necessary to use methods that allow the testing of those interactions.
- *Novelty/past experience*: If the technology is mature and well understood, less-rigorous methods may be appropriate. More rigorous methods may be called for if the technology is new, is used in some new application, is new for the particular applicant, or is unfamiliar to the certification personnel.
- *Complexity/level of automation*: More complex and automated systems typically require test methods that will reveal how that complexity will manifest itself to the pilot, in normal and backup or reversionary modes of operation.
- *Criticality*: Systems that are central to the interface design may require testing in the most realistic environments (high-quality simulation or flight test), because any problems are likely to have serious consequences.
- *Dynamics*: If the control and display features of the product are highly dynamic, the compliance methods should be capable of replicating those dynamic conditions.
- *Level of training required*: If the product is likely to require a significant amount of training to operate, the interfaces may need to be evaluated in an environment that replicates the full spectrum of activities in which the pilot may be involved.
- *Subjectivity of acceptance criteria*: Requirements that have specific, objectively measurable criteria can often employ simpler methods for demonstrating compliance. As the acceptance criteria become more subjective, more integrated test methods are needed, so that the evaluations take into account the aspects of the integrated flight deck that may affect those evaluations. The main objective is to carefully match the method to the product and the underlying human factors issues. It is also important for the Certification Team to recognize that several methods may be acceptable for any given requirement and applicants should be allowed to select among the acceptable methods, choosing the ones that best fit their compliance strategy, schedule, and cost considerations.

SYSTEM SAFETY ASSESSMENTS*

Typically, system safety assessments (i.e., Functional Hazard Assessment, Failure Modes and Effects Analysis, Fault Tree Analysis, etc.) are accomplished by the applicant's engineering group that is responsible for each system. However, for each assessment planned, the applicant should

---

* FAA Advisory Circular 25.1309-1_ System Design and Analysis.

describe how any human factors elements will be addressed (such as crew responses to failure conditions) and other assumptions that must be made about crew behavior. These assumptions should be reviewed by the full Certification Team to ensure that no assumptions are being made that will require the flight crew to compensate for failures beyond their expected capabilities. These human factors considerations can be documented in the individual system safety assessments, or the applicant may elect to describe them in the Human Factors Certification Plan, with references to the associated system safety assessments.

OPERATIONAL CONSIDERATIONS

The applicant may have specific goals associated with the operational certification of the airplane or system that could influence the design and its evaluation. In this section, the applicant will typically describe how these operational considerations will be integrated into the part 25 aspects of the certification project. It would be useful to identify operational requirements that have been factored into the type design. This section of the Certification Plan also may include how the operational certification, as captured in the following documents, will influence the MOC:

- Airplane Flight Manual
- Master Minimum Equipment List
- Flight Crew Operating Manual
- Quick Reference Handbook

The Aircraft Evaluation Groups, Flight Standards Operations representatives, and Human Factors Specialists on the Certification Team should be involved in the review of this section of the Human Factors Certification Plan.

CERTIFICATION DOCUMENTATION

The plan should indicate the types of documentation that will be submitted to show compliance or otherwise document the progress of the certification program. This section may list the specific documentation (test report number, analysis report number, etc.) that will be used to support compliance with the subject regulation. They may also be indicated in the compliance matrix.

CERTIFICATION SCHEDULE

Include the major milestones of the certification program, such as the following:

Certification plan submittals: The Certification Team should expect periodic updates to the Human Factors Certification Plan as the certification program progresses. The applicant should be encouraged to submit the first Human Factors Certification Plan as soon as possible after the start of the program. The applicant should be reassured that draft, preliminary information is acceptable and appropriate, provided that it is updated and finalized in a timely manner (as documented in the schedule and agreed to jointly by the FAA and the applicant).

Flight deck reviews, early prototype reviews, simulator reviews, and flight-test demonstrations: The Human Factors Certification Plan can document planned design reviews. Even in cases in which the reviews are not directly associated with finding compliance, such as Multiple Pilot Suitability and Usability Evaluations, they can be very helpful in the following ways:
- Providing the Certification Team with an accurate and early understanding of the crew interface tradeoffs and design proposals

- Allow the certification team to provide the applicant with early feedback on any potential certification issues
- Support cooperative teaming between the applicant and the certification team, in a manner consistent with the Certification Process Improvement initiative

Coordination meetings: Coordination meetings with other certification authorities, or meetings with other FAA ACOs on components of the same certification project or related projects, should be documented in the schedule. The Certification Team can use the information in the schedule to determine if sufficient coordination and resources are planned for the certification program.

USE OF DESIGNEES AND IDENTIFICATION OF INDIVIDUAL DESIGNATED ENGINEERING REPRESENTATIVES/ DESIGNATED AIRWORTHINESS REPRESENTATIVES

Describe how the applicant will make use of Designated Engineering Representatives, Designated Airworthiness Representatives, or other designees during the certification program[*,†]

ORDERS

FAA Order 8100.5_ Aircraft Certification Service Mission, Responsibilities, Relationships, and Programs
FAA Order8100.8_ Designee Management Handbook
FAA Order 8100.15_ Organization Designation Authorization Procedures
FAA Order 8110.4_ Type Certification
FAA Order 8110.37_ Designated Engineering Representative (DER) Handbook
FAA Order 8110.48 How to Establish the Certification Basis for Changed Aeronautical Products
FAA Order 8110.52_ Type Validation and Post-type Validation Procedures
FAA Order 8110.112_ Standardized Procedures for Usage of Issue Papers and Development of Equivalent Levels of Safety Memorandums
FAA Order 8110.115 Certification Project Initiation and Certification Project Notification

ADVISORY CIRCULARS

FAA Advisory Circular (AC) 20-175—Controls for Flight Deck Systems
FAA Advisory Circular (AC) 21-40—Application Guide for Obtaining a Supplemental TC
FAA Advisory Circular AC 23-8—Flight Test Guide for Certification of Part 23 Airplanes
FAA Advisory Circular (AC) 23-21—Airworthiness Compliance Checklists Used to Substantiate Major Alterations for Small Airplanes
FAA Advisory Circular (AC) 23-24—Airworthiness Compliance Checklists for Common Part 23 Supplemental TC (STC) Projects
FAA Advisory Circular AC 23-15—Small Airplane Certification Compliance Program
FAA Advisory Circular AC 23-17—Systems and Equipment Guide for Certification of Part 23 Airplanes and Airships
FAA Advisory Circular (AC) 23.1309-1—System Safety Analysis and Assessment for Part 23 Airplanes
FAA Advisory Circular (AC) 23.1311-1—Installation of Electronic Display in Part 23 Airplanes
FAA Advisory Circular AC 25-7—Flight Test Guide for Certification of Transport Category Airplanes
FAA Advisory Circular (AC) 25-11—Electronic Flight Deck Displays
FAA Advisory Circular (AC) 25.773-1—Pilot Compartment View Design Considerations
FAA Advisory Circular (AC) 25.1302-1 - Installed Systems and Equipment for Use by the Flightcrew
FAA Advisory Circular (AC) 25.1309-1—System Design and Analysis

---

[*] FAA Order8100.8_ *Designee Management Handbook*; FAA Order 8100.15_ Organization Designation Authorization Procedures.
[†] FAA Order 8110.37_ *Designated Engineering Representative* (DER) Handbook.

FAA Advisory Circular (AC) 25.1322-1—Flightcrew Alerting
FAA Advisory Circular (AC) 25.1523-1—Minimum Flightcrew
FAA Advisory Circular (AC) 27-1—Certification of Normal Category Rotorcraft
FAA Advisory Circular (AC) 29-2—Certification of Transport Category Rotorcraft

FAA REPORTS

Yeh, M., Young, J.J., Donovan, C., & Gabree S. (2013). *Human Factors Considerations in the Design and Evaluation of Flight Deck Displays and Controls: Version 1.0 (No. DOT-VNTSC-FAA-13-09,)*. Office of Aviation Research.
Gabree, S., Chase, S., & Cardosi K. (2014). *Use of Color on Airport Moving Maps and Cockpit Displays of Traffic Information (CDTIs) (No. DOT-VNTSC-FAA-14-13,)*. Office of Aviation Research.

POLICY STATEMENTS

Policy Statement Number PS-ANM111-1999-99-2, Guidance for Reviewing Certification Plans to Address Human Factors for Certification of Transport Airplane Flight Decks, September 29, 1999.
Policy Statement Number PS-ACE100-2001-004, Guidance for Reviewing Certification Plans to Address Human Factors for Certification of Part 23 Small Airplanes, August 29, 2002.
Policy Statement Number PS-ANM100-01-03A, Factors to Consider When Reviewing an Applicant's Proposed Human Factors Methods for Compliance for Flight Deck Certification, February 7, 2003.

9 Safety Systems and Practices

Alan Stolzer

John is a student at a large university in the Midwest and is pursuing a degree in Aeronautical Science. One of his curricular requirements is to take a course on safety management systems (SMS) in the aviation industry. Having no background in safety, he is a little nervous about the subject matter, so he arranges to meet a friend of his father who works in the safety department at a major air carrier. John explains that he wants to do well in the course and, thus, would like to have a better understanding of SMS before the class begins in the fall. Bill is the Flight Operations Quality Assurance (FOQA) manager for the airline and is well acquainted with SMS, so the two meet at a local coffee house. After some pleasantries, Bill decides to start at the end with his explanation of SMS. He begins:

An organization that has embraced and incorporated SMS into the way it operates is markedly different than one that has not, and you can tell the difference by talking with people and examining their operation. In addition, this is true of any organization, whether it is an aviation company, a school, a movie theater, or a nuclear power plant facility. If I were asked to determine the state of SMS in an organization, I would want answers to these questions (Stolzer and Goglia 2015, pp. 39–43). Employees would be asked the following questions:

- What are your areas of greatest risk or threats to the operation?
- How do you mitigate those risks or threats?
- How do you know that? An SMS organization has established an effective means of communicating this information to its employees, and not just generic information, but information that is relevant to the employee's work.
- Do you feel empowered to suggest changes to improve safety? Can you give me examples of when that has occurred and the outcomes?
- What happens when someone makes a mistake? What if a theater worker mistakenly leaves a rear door unlatched during a movie show? What if a mechanic leaves a fluid spill on the floor in the hangar area? Are they punished? Is a system in place for reporting such mistakes?
- How do you know management is committed to safety? SMS simply will not be effective unless the top person in the organization is fully invested in it, so can you show me examples and evidence?

After questioning employees, I would examine the organization's documents and records to look for clear policy statements and process descriptions with means to measure performance of those processes. These means are usually called *key performance indicators,* and effective process descriptions will have them. Documents will also reveal how it is determined whether the processes are meeting their safety targets. It will also describe the decision-making processes used for management review. The record keeping must always be up-to-date and as extensive as necessary to satisfy its purpose. It must also be organized, accessible, and manageable—a tall order, but one the organization takes very seriously. For the management team, I have a different set of questions:

- Are your safety programs reactive or proactive?
- How much money is invested in these programs and how is their effectiveness measured?

- How frequently are employees trained on these programs?
- How is change managed in the organization? SMS must lead to change, so I want to know about the change management system, and I want to see it in policy and process descriptions and understand how change management is monitored. The benefits of SMS are lost if change is not institutionalized.

John is taking notes and appears to be getting the gist of it and wants to know more. Bill is happy to oblige.

Okay, let's talk about what SMS is and where it came from, and then we can delve into some specifics. The International Civil Aviation Organization (ICAO) is a specialized agency of the United Nations and is charged with ensuring the safe and orderly growth of international civil aviation throughout the world. As such, ICAO has been an active proponent of SMS and has incorporated SMS requirements into its standards and recommended procedures. ICAO defines SMS as, "a systematic approach to managing safety, including the necessary organizational structures, accountabilities, policies and procedure" (ICAO 2013, p. 12). The Federal Aviation Administration (FAA), which regulates civil aviation in the United States, defines it slightly differently: "SMS means the formal, top-down business approach to managing safety risk, which includes a systematic approach to managing safety, including the necessary organizational structures, accountabilities, policies and procedures" (FAA Order VS 8000.367). Notice the differences between the two definitions, such as FAA's inclusion of *formal*, *top down*, and *business approach* terms.

For most people, the beginning of SMS was in June 2006 when the FAA released its first SMS Advisory Circular (AC), 120-92, entitled, *Introduction to Safety Management System for Air Operators*. That AC laid out the concept of SMS and provided guidance for it. SMS, however, is not something that was drawn out on a chalkboard; it has evolved from a combination of many different systems and theories dating back decades, including system safety, quality management systems (QMSs) (or earlier, total quality management), environmental management systems, occupational health and management, and systems and theories of management in general. These are concepts and systems that many industries, including aviation, have been incorporating into its operations over time. Thus, SMS has been the result of an evolutionary process that has brought ideas from other scientific and management domains to bear on the problem of managing safety in the field of aviation (Stolzer and Goglia 2015, pp. 38–39).

John seems impressed with SMS and asks Bill if SMS is responsible for the improvements in aviation safety. Bill smiles and offers the following.

Aviation is so much safer now than it was years ago, but for many reasons. Improved technology in the form of better, more reliable engines, as well as advances in avionics, automation, satellite-based navigation systems, weather radar, automatic dependent surveillance-broadcast systems, ground proximity warning systems, and high fidelity aircraft simulators, just to name a few, have had a profound effect on aviation safety.

Pilot training is far better than it used to be, partly because of the better technology employed and partly because we are no longer training pilots the way the Wright brothers did. We do a better job of developing effective curricula, structuring complex simulation sessions with cascading failures, and placing emphasis on professionalism and crew resource management. Training systems are well designed, data driven, and they enable flight crews to remediate areas of deficiency efficiently and effectively.

Approaches to safety have also improved. In the past, our mindset was what has been called the *fly-crash-fix-fly* approach; that is, we would fly airplanes, experience the occasional airplane crash, and then attempt to determine the cause(s) and introduce a solution to prevent it from happening again. This approach is highly reactive and, although important even today, it is insufficient to lead to the improvements we want to make.

John relates that his father has mentioned the four components of SMS, and asks Bill to explain them.

The four components of SMS are safety policy, safety risk management (SRM), safety assurance (SA), and safety promotion, I'll explain each one.

Within *Safety policy*, the organization establishes not only policies but also the procedures and organizational structures used to achieve its safety performance. The safety policy defines responsibility, authority, accountability, and expectations, so all involved know what the organization desires to achieve regarding safety and how it will do so. This is far more than a simple policy statement; it is an important document of all safety policies for the benefit of all organization stakeholders.

Top management's involvement in the safety activities of the organization must be stipulated in the safety policy, and the policy should recognize that the ultimate responsibility for safety cannot be delegated. One of the ways top management can convey the importance of SMS is to include a regular management review of the SMS at the CEO level.

Policies also stipulate requirements for all operational departments to establish their own procedures, controls, training, process measurements, and change management systems.

SRM is the heart of SMS. SRM has three primary components: hazard identification, risk analysis and assessment, and control of risk. There are also some other concepts that need to be understood with respect to SRM, including system and task analysis, residual and substitute risk, and system operation.

First, let us distinguish hazards from risks. A *hazard*, according to ICAO, is "a condition or an object with the potential to cause death, injuries to personnel, damage to equipment or structures, loss of material, or reduction of ability to perform a prescribed function" (ICAO 2013, pp. 2–24). A key word in this definition is *potential*, referring to a condition that has the capability of causing death, injury, and so on.

Risk, by contrast, is "the projected likelihood and severity of the consequence or outcome from an existing hazard or situation" (ICAO 2013, pp. 2–27). So, based on that definition, it is apparent that a risk is a mathematical calculation of the consequence of something occurring.

Examples of hazards are a crosswind, a pool of hydraulic fluid on the hangar floor near a walk path, a wrench left in the engine cowling after an aircraft inspection, a fatigued pilot, and a thunderstorm. Remember, with each of these conditions, there is the potential for something bad to happen.

As the objective of SMS is to mitigate and control risk, it is vital to find hazards wherever they exist. There are fundamentally two approaches to identify hazards: one is through *operational observations,* and the other is through *process analysis*.

Operational observations comprise the most common method of identifying hazards. These entail using data collected from the operation, whether that information comes from programs such as Aviation Safety Action Program (ASAP), FOQA, Line Operations Safety Audit, or any of our other safety programs. The information might also come from the safety investigator's analysis of the environment, the aircraft, the pilots, and so on, following an incident or accident.

Process analysis is what is most often discussed by ICAO and the FAA, and it is essentially using subject matter experts (SMEs) to analyze a process, determine the hazards existing in that process, and further break down each of the hazards into specific components. In doing so, risk can be enumerated and addressed. There are tools to aid in this activity, such as the hazard/risk process decision program chart (see Figure 9.1) (Stolzer and Goglia 2015, p. 134). This activity requires a high degree of familiarity with the activity being assessed; thus, the selection of SMEs is critical.

No.	Type of operation or activity	State the generic hazard (hazard statement)	Risk(s) description	Current measures to reduce risk, and risk index	Further action to reduce risk, and resulting risk index

FIGURE 9.1 Hazard/risk process decision program chart.

Let us now briefly discuss risks. Whenever hazards are present, and they always are, those hazards can manifest into risks. Risks must be assessed, and normally, both the severity and probability are assessed against a predetermined standard. A couple of points to keep in mind are as follows:

- To understand the risks associated with an activity, one needs to understand the attendant hazards.
- Determination of the severity and probability is inherently estimated. As there is a degree of uncertainty, the best estimates are based on a distribution rather than single point values. In fact, the same could be said of the risk controls.

Let's talk about the risk matrix. The risk matrix is used throughout the industry to assess risk, and used properly, it is a reasonably effective tool at driving decision-making—its purpose. The risk matrix has two components—a table and a chart. As we have to categorize the severity and the likelihood, we will use a table to inform the analyst about what the category choices are and the meaning or definition of each category (see Tables 9.1 and 9.2). ICAO provides examples of these definitions, but an organization can use whatever system it desires. The calculation of the risk index is simple: R (risk) = S (severity) X L (likelihood).

ICAO's risk matrix is shown in Figure 9.2. The categories for severity and likelihood comprise the top and left-side axes of the chart, and the cells are the product of the two categories. The cells are divided into color-coded regions; usually, green indicates acceptable risk, yellow indicates tolerable, and red intolerable, or as shown in this risk matrix, the regions are labeled.

TABLE 9.1
Safety Risk Probability Table

Likelihood	Meaning	Value
Frequent	Likely to occur many times (has occurred frequently)	5
Occasional	Likely to occur sometimes (has occurred infrequently)	4
Remote	Unlikely to occur, but possible (has occurred rarely)	3
Improbable	Very unlikely to occur (not known to have occurred)	2
Extremely improbable	Almost inconceivable that the event will occur	1

Source: International Civil Aviation Organization (ICAO), Safety Management Manual (SMM), 3rd ed., Doc 9859 AN/474, Montreal, Canada, ICAO, 2013.

TABLE 9.2
Safety Risk Severity Table

Severity	Meaning	Value
Catastrophic	• Equipment destroyed • Multiple deaths	5
Hazardous	• A large reduction in safety margins, physical distress, or a workload such that the operators cannot be relied upon to perform their tasks accurately or completely • Serious injury • Major equipment damage	4
Major	• A significant reduction in safety margins, a reduction in the ability of the operators to cope with adverse operating conditions as a result of an increase in workload or as a result of conditions impairing their efficiency • Serious incident • Injury to persons	3
Minor	• Nuisance • Operating limitations • Use of emergency procedures • Minor incident	2
Negligible	• Few consequences	1

Source: International Civil Aviation Organization (ICAO), Safety Management Manual (SMM), 3rd ed., Doc 9859 AN/474, Montreal, Canada, ICAO, 2013.

Risk probability	Risk severity				
	Catastrophic A	Hazardous B	Major C	Minor D	Negligible E
Frequent 5	5A	5B	5C	5D	5E
Occasional 4	4A	4B	4C	4D	4E
Remote 3	3A	3B	3C	3D	3E
Improbable 2	2A	2B	2C	2D	2E
Extremely improbable 1	1A	1B	1C	1D	1E

FIGURE 9.2   Risk matrix.

Most operators use a management guidance document that describes what action must be taken should the risk index fall in the yellow or red regions, and in what timeframes. Remember, the goal is to mitigate risk in as expeditious a manner as possible.

The risk matrix is necessarily backward looking or forensic in nature; after all, how are the determinations made of severity and likelihood levels? These should not simply be guesses; rather, we often derive these values based on actual events (e.g., hard landings as determined by FOQA and ASAP) using a measure of exposure as the denominator (e.g., total number of landings).

To stress this point, the risk matrix is used throughout the aviation industry and in a host of applications from daily operations and strategic planning to system design and others.

One of the things to keep in mind is that we assess risk for a purpose, and that is to mitigate it, and the way we generally do that is to put controls in place once we fully understand the risk. However, one has to be very careful at this stage. Two concepts you will learn about include *residual risk* and *substitute risk*. *Residual risk* is the risk remaining after a control has been implemented, and that risk is determined by the SMEs, once again redoing the risk assessment after the control has been put in place. A simple equation for this concept is:

$$Residual\ Risk = Hazard - Control\ Effectiveness$$

Substitute risk refers to the introduction of new, unintended risks into the process as a result of implementing controls for the previously identified risk; so in effect, you are substituting one risk for another. Again, this situation is identified by performing subsequent risk analyses using SMEs—a very important point to remember.

> This concept of controls in the SRM process seems important to John, and he wants to know more. Bill does not hesitate to dive deeper.

After identifying the hazards and assessing risk, we now choose one or more controls to mitigate the risk. There are essentially four steps involved in hazard control (Stolzer and Goglia 2015, p. 234):

1. Ensure that the hazard is recognized and identified
2. Determine and select hazard control(s)
3. Assign responsibility for implementing control(s)
4. Measure and monitor the effectiveness of the control(s)

We have already discussed some of this, so let me briefly explain the last two steps. Once a hazard control is selected, the responsibility for implementing that control must be assigned, and there should be no doubt in anyone's mind about which party is responsible. The risk control should be clearly described and documented.

SA is all about monitoring the system and measuring the effectiveness of the controls, and the best way to do this is to follow quality management principles. A good QMS will provide the structure to manage the myriad sources of data for SA, and will provide confidence that the monitoring task is being managed effectively.

The field of system safety gives us a prioritization method for evaluating hazard controls, and the SMS practitioner should be familiar with it. According to Brauer (2006, pp. 99–103), the hierarchical order of control effectiveness is as follows:

- Eliminate the hazard (obviously, not always possible)
- Reduce the hazard level (either the severity or the likelihood)
- Provide safety devices (examples are switch guards, handholds, and nonslip walking surfaces)
- Provide warnings or advisories; examples are stall warning horns, stick shakers, fire alarms (Human action must be taken to prevent risk of harm to persons or damage to equipment.)
- Provide safety procedures (such as emergency actions when a safety problem occurs)

Let's take a look at a very simplistic example. I am a notoriously poor suitcase packer, and I arrive at a destination hotel room in need of reironing my shirts. I look in the closet for a clothes iron and none is there, so I call the front desk and am informed that for safety reasons hotel policy precludes having irons in guest rooms. What did the hotel do? They eliminated the hazard.

Suppose, the hotel considers changing its policy to allow clothes irons in guest rooms, what other controls could they put in place? Well, they could use only irons that have an auto shut-off feature to *reduce the hazard level*. There are not too many *safety devices* for irons, unfortunately, but the incorporation of coiled cords not extending across traffic paths is one, and perhaps antidrip features that stop water from flowing when the iron cools below a certain temperature is another. Of course, *warnings* could include an audible noise, if the iron is turned on too long or gets too hot, or in worst cases scenarios, a fire alarm that sounds if the iron catches something on fire. Safety procedures might include a list of steps for hotel guests to follow to ensure iron safety. Although it is hard to argue against eliminating the hazard (the more conservative approach), my shirts would have been far more presentable had they opted for one of these other controls.

The hazard control process is not an easy one due to the complexity of the aviation industry. There are interrelationships among people, technical systems, organizations, and the environment. To aid the analyst in visualizing these interfaces, several models from the discipline of system safety are used. Two examples are the SHELL model and the 5-M model.

The SHELL model is named for its components: software, hardware, environment, liveware, and liveware (again). As depicted in Figure 9.3, liveware is at the center of the diagram, and represents the persons affecting the system. Liveware relates with all other elements of the system. Think of software as comprising the nonphysical components of the system, and hardware as comprising the machines and all of their components. Environment includes a host of factors, such as temperature, oxygen, altitude, sleep deprivation, spatial disorientation, illusions, political and economic considerations, and others. The other liveware represents all the other people and groups we interact with in the aviation system.

Another model for visualizing the system is the 5-M model (see Figure 9.4); the 5 Ms are mission, man, machine, management, and media. In this model, the mission is at the center of the system. Man represents the human element of the system, whereas machine represents the hardware and software, as they were described in the previous model. Management represents the policies, rules, regulations, and structure in which we operate, and media refers to the environment.

Bill pauses to assess his student's understanding and to let him catch up on his note-taking. Bill proceeds to discuss the next topic on his list.

SA is the third component of SMS. Thus far, we have put in place the policies, procedures, and expectations regarding safety performance, performance measures, and change management system—all required in safety policy, and we have identified hazards, assessed risk, determined

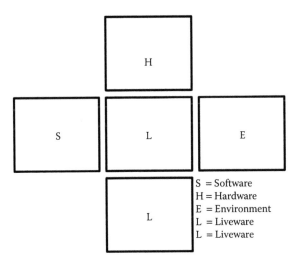

FIGURE 9.3   Shell model.

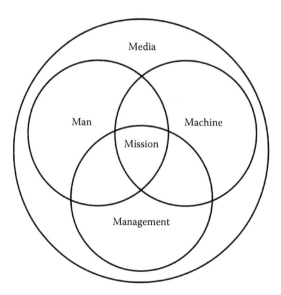

FIGURE 9.4 5-M model.

controls, and reassessed risk—required in SRM, so now it is time to monitor the system and ensure that the controls we implemented are having the intended effect on safety performance. This is accomplished through the diligent efforts of the operators of the system, safety professionals, and management through regular reviews.

SMS is data driven, and SA is no exception. The system will be monitored through all available data sources, including: audits—both internal and external, investigations of incidents and accidents, key process indicators, and input from employee reporting systems such as ASAP. This information, as appropriate, should funnel into a system for management review. The challenge is to use the correct information that paints the clearest picture of the situation, not to overwhelm management with nonrelevant information. I like to say that we separate the signal from the noise and present the former.

A point to make in this discussion is that much of the information for SA comes from front-line personnel. They are typically the process owners and have the best information about how the process works. These people are key to the process and are tasked with monitoring and reporting safety performance, typically through an internal auditing program.

Something that should be stressed in any discussion of SMS is change management, and I will mention it again with respect to SA. Without an effective change management processes in place, an organization will typically revert back to its original state, even after making improvements. Change management enables an organization to move from its current state to a desired state without such regression, that is, the change is *institutionalized*. An effective change management system does this without negative consequences or adverse reaction from the organization's workers.

The fourth component is *Safety Promotion*. An SMS continuously promotes safety as a core value through communication and awareness, as well as through training and qualification programs. Safety promotion also includes encouraging a positive safety culture, so let's discuss what that is.

First and most importantly, *safety culture* begins at the top of the organization through policies and procedures designed to ensure that every person in the organization understands his or her role in maintaining a safe operation and actively participates in identifying hazards and mitigating risk. A *reporting culture* is characterized by an organization that enables information to move through all levels and feed into a system empowered to solve problems. Simply stated, people are not afraid to report issues they observe because they understand when problems are revealed safety will be continuously improved.

A *just culture* is one in which employees are held accountable for their actions, but they are not blamed for honest mistakes. One can easily imagine how an employee's willingness to report a safety issue would be influenced by his or her fear of being punished. Thus, the organization has a responsibility to make clear what is acceptable and unacceptable behavior and encourage reporting of safety issues so it can both learn of and address them. Reporting is essential, and treating people fairly is how to encourage reporting. In summary, safety culture is critical to a successful SMS.

> John continues to write feverishly and asks Bill what else he needs to know. "There are some things I'd like to expound on if you can handle a bit more information." John nods and Bill begins.

First, let's talk about *process descriptions*; these are essential for process-based SRM and SA. Although it might seem a little extensive for our discussion, I want to emphasize how important it is. These critical and complex process descriptions include the following elements (Stolzer and Goglia 2015, pp. 202–211):

1. Purpose—aids in fully understanding the process with a clear and concise statement
2. Scope—defines the limits, boundaries, and restrictions for the process
3. Interfaces—include all of the inputs to and outputs of a process
4. Stakeholders—identify the people or parties who accomplish the work or otherwise interface with the process
5. Responsibility—identifies the person(s) responsible for the process
6. Authority—identifies the person who manages the process and is accountable for its success
7. Procedure description/flowchart—provides a graphical depiction of the flow of activities in a procedure
8. Materials/equipment/supplies—create a listing of these items, useful for understanding the process
9. Hazards—recognize problem areas in a mature SMS as an integral part of process descriptions
10. Controls—identify those used so SMS practitioners can be intimately familiar with controls in the organization's processes
11. Process measures—signify what changes are needed throughout the workflow, particularly, at the output of the process; SA cannot exist without them
12. Targets—measure the effectiveness of controls against predetermined milestones and objectives as part of SA
13. Assessment—determines whether or not the targets have been met and the procedures that differ a bit between SRM and SA, but this will require a lengthier discussion
14. Change management—assures coordination among stakeholders that is imperative, and that agreed-upon change is incorporated into process descriptions
15. Qualifications—required for participants in process—supplies needed training for persons with responsibility for the conduct of any part of the process
16. Documentation/record keeping—determines what documents are needed to perform the process and what records should be maintained

Any student of SMS should practice writing process descriptions by incorporating all of these facets in complete detail until they become second nature.

Next, let's talk about *system descriptions*. An overlooked aspect of SMS is the second "S" in the term. To develop SMS as a system and not a program, one must first be able to describe the system. The system descriptions should always describe its boundaries. For example, if the analyst conducting hazard identification does not have a complete understanding of the system, he or she may overlook certain hazards that live in those unexamined components. But importantly, the intent of a system description is to bring together the disparate elements of the organization into an overarching view of the system as a whole, thus, enabling overall system analysis.

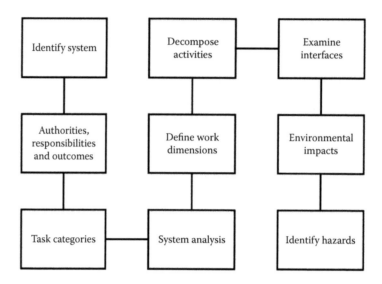

FIGURE 9.5 System description. (From Stolzer, A. and Goglia, J., Safety *Management Systems in Aviation*, 2nd ed., Ashgate Publications, Aldershot, England, 2015.)

A system description can be thought of as an account of structures, procedures, processes, people, equipment, and facilities used to accomplish the organization's mission. It provides a detailed view of an organization sufficient to understand how hazards are related to parts of the description. The system description provides the foundation upon which proactive hazard identification can occur.* Its various components include the items shown in Figure 9.5.

> Bill could sense John has been sitting on a question, and he is correct. Earlier in the conversation, Bill stressed the importance and limitations of risk matrices in analyzing risk, so John asks if there are any reasonable alternatives. Bill is happy to expound on his earlier comments.

Largely due to its simplicity and intuitive framework, the risk matrix approach is common in aviation, but it is not without its detractors. As so much stock is put in the use of matrices for risk analysis in assessing risk, setting priorities, management decision-making, and guiding resource allocation, I would encourage every SMS practitioner to read two publications: (1) a journal article by Tony Cox (2008) entitled, *What's Wrong with Risk Matrices?* and (2) a book by Douglas Hubbard (2009) entitled *The Failure of Risk Management: Why It's Broken and How to Fix It.* These two publications will enlighten the reader about how these matrices work and their inherent limitations.

As described by Hubbard, the risk matrix is fundamentally a scoring method, and even though we use SMEs to assign values, there are weaknesses with this approach that even the best SMEs cannot overcome. According to Hubbard (2009, pp. 122–123), there are three problems with scoring methods: (1) They do not consider issues about perception of risks and uncertainties; (2) descriptions of likelihood are interpreted differently by different SMEs, even after attempts to standardize the meanings; and 3) the methods themselves introduce their own sources of error.

Cox (2008) points out that very little empirical or theoretical study has been conducted on how well the risk matrix approach succeeds in improved decision-making. It is remarkable that a technique relied upon so heavily in this industry has never been validated as accurate. Cox's research reveals that when probability (likelihood) and consequence (severity) are negatively correlated, the information provided from the risk matrix can be *worse than useless* and *worse than random* (p. 500), and negative correlation may be common in practice. Cox asserts that, due to these issues and others, risk matrices should be used with caution and with a thorough understanding of their limitations.

---

* A procedure for developing a system description is discussed in *Safety Management Systems in Aviation*, Ashgate, 2015.

Hubbard offers three improvements to risk management:

1. Adopt an approach appropriate for modeling uncertain systems. Use Monte Carlo models with calibrated probabilities.
2. Be a scientist. Consider history and forecasts, and compare forecasts with predictions once the events occur. Use empirical observations when possible.
3. Build a community focused on expressing analysis in probabilistic terms.

Hubbard believes the solution lies in using more quantitative approaches such as Monte Carlo. To the cynic who offers the objection that the method would not work in his or her industry or application, Hubbard (2009, p. 162) states that, however unique you think your measurement problem is, it is been done before; you have more data than you think; you need less data than you think; getting more data does not cost as much as you might think; and you probably need different data than you think.

John surmises that perhaps the aviation industry needs to begin a transition to more quantitative methods. Bill agrees and explains that NASA transitioned to probabilistic risk analysis—including Monte Carlo—after the space shuttle challenger accident. Although NASA has the resources and skills sets to do that, the airline industry, by and large, does not. We need highly trained experts in safety and risk management in our safety departments who have skills in Probabilistic Risk Analysis/Monte Carlo, statistics, and data analysis to raise the level of rigor and science used in airline safety.

Bill has two more topics in his mind to discuss with John. One is quality management and the other is SMS effectiveness.

In 2008, shortly after the FAA released its first AC on SMS, then FAA's Associate Administrator for Aviation Safety, Nick Sabatini, wrote the following in an FAA FAASTeam news release.[*]

"The FAA's Aviation Safety organization is well on its way to adopting an SMS. Adopting a QMS and achieving International Organization for Standardization 9001 registration in 2006 was a critical first step. In fact, aviation safety was the first federal organization to achieve certification to the International Organization for Standardization 9001:2000 quality management standard as a single corporate management system."

When effectively combined, the SMS and QMS ensure safety and quality products. Stated simply: SMS needs a QMS, because SMS requires documented, repeatable processes. The QMS provides the strong foundation upon which an SMS is built.

In AC 120-92 (FAA 2006, p. 2), The FAA states that "an SMS is essentially a quality management approach to controlling risk," "The SMS incorporates internal evaluation and quality assurance concepts that can result in more structured management and continuous improvement of operational processes" (p. 2), and "...quality management techniques can be used to provide a structured process for ensuring that they achieve their intended objectives and, where they fall short, to improve them. Safety management can, therefore, be thought of as a quality management of safety related operational and support processes to achieve safety goals" (p. 3).

On the contrary, in the past few years, the FAA seems to have moved away from stressing quality management as an underpinning of SMS, but let there be no doubt that its importance has not diminished. The quality guru Kaoru Ishikawa once said that 95% of all quality problems in the factory could be solved with the application of fundamental quantitative tools (Ishikawa 1985). The tools known as basic quality tools are listed in the following, and these should be in wide use in aviation safety departments:

- Flowcharts—provide graphical representation of a process using standard symbology
- Pareto charts—provide graphical representation of the factors having the greatest cumulative effect on the issue under study

---

[*] https://www.faasafety.gov/hottopics.aspx?id=58.

- Cause-and-effect diagrams—enable an analyst to understand underlying causes and effects
- Control charts—offer output charts from Statistical Process Control, a quantitative problem-solving process
- Check sheets—provide for the recording and tracking of event occurrences
- Scatter diagrams—graphically depict the relationship between two variables
- Histograms—graphically depict frequency distributions

In addition, many tools used in QMSs are very applicable to safety management, including strategic planning, strengths, weaknesses, opportunities, and threats (SWOT) analysis, benchmarking, performance measurement, balanced scorecards, and dashboard performance systems. There is a plethora of resources for learning about these tools and methods, and all safety/SMS practitioners would be well served by gaining expertise in the use of the ones that can assist their organizations in performing the SRM and SA functions.

The final topic for discussion is *SMS effectiveness*. Sometimes critics of the FAA lament that the agency seems to introduce initiatives without a full understanding of their effectiveness, and then there is little follow-up, after-the-fact, to determine whether the program produced the desired results. Applied to our current topic, one might ask how we know that SMS is having the intended effect in improving aviation safety.

This is a complicated question and one not easily answered. Although it's tempting to look at pre-and post-implementation safety statistics, determining improvement (if there is) and proclaiming success for one initiative or another can be misleading.

At the risk of offending some of my colleagues, let's look at a recent example. As a result of a White House Commission on Aviation Safety and Security, the Commercial Aviation Safety Team (CAST) was formed in 1998 with the goal of reducing the commercial aviation fatality rate in the U.S. by 80% within 10 years.[*] CAST built a structure, formed committees, and recruited some of the best safety minds in the industry to establish what appears to be excellent processes. The processes include scoring methods, discussed earlier, for examining accidents and incidents and developing interventions to improve safety. From 1998 to 2008, aviation safety did indeed improve dramatically. The following is a quote from the CAST website:

> The CAST model has been extremely successful in the United States. Safety experts report that by implementing the most promising safety enhancements, the fatality rate of commercial air travel has been reduced in the United States by 83 percent over the last 10 years.

The implication in this statement and many others issued in 2008 and later is that CAST initiatives produced that improvement (though some of the news releases qualified it by including terms such as *in part*, *largely due to CAST*, or similar).

Let me be crystal clear: *It is highly likely that CAST initiatives played a very large role in the safety improvement realized during that period!* And I applaud all of the dedicated people who work tirelessly in service of CAST. (Note: CAST won the prestigious Collier Trophy in 2008 for achievements in aeronautics and astronautics.)

But how do we know? Is it possible the aggregate total of hundreds of other efforts happening concurrently produced that improvement in safety, and that CAST initiatives made a relatively small contribution to the success? How can we isolate CAST's contribution from all the others, and have any efforts been made to do so? In statistics, variables unaccounted for are called *confounding variables*, and they often render a research project worthless. It is no small task to: (a) account for all of the variables in a complex problem and (b) isolate the effects of each of the independent variables on the dependent variable; that is the parallel we are discussing here.

---

[*] http://www.cast-safety.org/about_background.cfm.

Given the resource expenditures required by many of these FAA initiatives, we should be making more of an effort to measure the success or effectiveness of them to guide future decision-making.

Fortunately, there are efforts underway to measure the effectiveness of SMS. One promising research project is examining the use of data envelopment analysis (DEA) to model the inputs and outputs to the process. DEA is a mathematical programing technique and uses a multifactor analysis model to measure the relative efficiencies of decision making units (DMUs), or organizations, based on ratios of specified inputs and outputs. The performance of the DMUs are measured against the best performing DMU(s) in the dataset; the best performing DMU is termed the *efficient frontier*. Note: efficiency and effectiveness are considered synonymous.

Using this approach, DEA models are being developed for each of the four SMS components discussed earlier. The models are based on inputs and outputs collected via a survey from a wide range of aviation organizations.

One of the advantages of utilizing DEA in this context is that it not only identifies organizations in which SMS is not as effective as it is in others, it also indicates how the organization can become more *SMS effective*. For example, Figure 9.6 shows the distribution of scores for the safety policy component in a trial study.

There are 11 columns in the histogram; one for each 10% range, and the last one indicating how many efficient/effective organizations were there. In addition, the software reveals which specific organizations are included in these various outputs, that is, organizations numbered 1, 4, 8, 22, 28, and 32 were determined to be efficient. As indicated, there were six organizations deemed effective, one deemed marginally effective, and 17 determined to be at various degrees of inefficiency.

Figure 9.7 depicts the improvements that could be made to improve the efficiency/effectiveness score. For safety policy, the large green portion of the pie (SPO_03) indicates that the most opportunity for improvement may result from dedicating a larger overall budget allocation to system safety; the remaining improvements (SPO_02 and SPO_01) could be made by increasing the number of employees with safety responsibilities.

This is just an example of the efforts being made to address the important issue of determining the effectiveness of SMS in the aviation industry. Much more refinement of this model needs to be done, but it is definitely a step in the right direction.

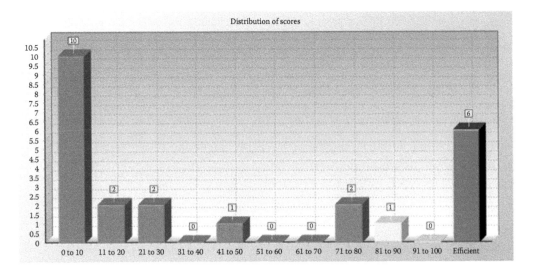

FIGURE 9.6   Distribution of scores for the safety policy component as determined by DEA.

Total potential improvements

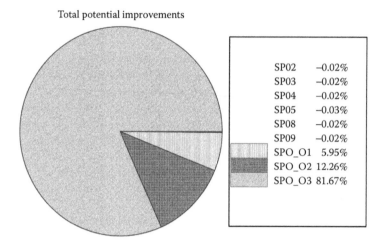

SP02	−0.02%
SP03	−0.02%
SP04	−0.02%
SP05	−0.03%
SP08	−0.02%
SP09	−0.02%
SPO_O1	5.95%
SPO_O2	12.26%
SPO_O3	81.67%

FIGURE 9.7   Potential improvements to safety policy component as determined by DEA.

John thanks Bill for his time and sharing his expertise about aviation safety and SMS. At this point, Bill asks John if he knows who the instructor will be for SMS course in the fall semester. John replies that he does not, and Bill discloses that it is him. He explains to John that collegiate aviation programs sometimes make use of the expertise of airline personnel to teach certain subject matter, and Bill has been hired as an adjunct professor by John's university—a coincidence to be sure. John tells Bill he should do well in the course as he now knows the material, to which Bill says that he will definitely get an "A" for the first day or two, but there is much more to SMS than what was covered in their brief conversation. John sighs, again thanks Professor Bill for his time, and heads off to learn more about SMS.

REFERENCES

Brauer, R. (2006). *Safety and Health for Engineers*. Hoboken, NJ: John Wiley & Sons.
Cox, L. (2008). What's wrong with risk matrices? *Risk Analysis,* 28(2), 497–512.
Federal Aviation Administration (FAA) (2006). Introduction to safety management systems for air operators. Advisory circular 120-92. Retrieved June 15, 2016 from http://rgl.faa.gov/Regulatory_and_Guidance_Library/rgAdvisoryCircular.nsf/list/AC%20120-92/$FILE/AC%20120-92.pdf.
Federal Aviation Administration (FAA) (2008). Aviation safety (AVS) safety management system requirements. Order VS 8000.367. Retrieved June 15, 2016 from https://www.faa.gov/documentLibrary/media/Order/VS%208000.367.pdf.
Hubbard, D. (2009). *The Failure of Risk Management: Why It's Broken and How to Fix It*. Hoboken, NJ: John Wiley & Sons.
International Civil Aviation Organization (ICAO) (2013). *Safety Management Manual (SMM)*, 3rd ed. (Doc 9859 AN/474). Montreal, Canada: ICAO.
Ishikawa, K. (1985). *What Is Total Quality Control?* Englewood Cliffs, NJ: Prentice Hall.
Stolzer, A., Goglia, J. (2015). *Safety Management Systems in Aviation*, 2nd ed. Aldershot, UK: Ashgate Publishing.

10 Security

Douglas Harris

CONTENTS

Introduction 210
Background 210
   A Brief History of Aviation Security 210
   Threats to Aviation Security 211
   The Roles of Human Operators 212
   Factors Influencing Operator Performance 213
Assessments of Aviation Security 215
   Heritage Foundation Assessment 215
   RAND Corporation Assessment 215
   GAO Assessments of 2012, 2013, and 2014 216
   Reason Foundation Assessment 217
   The Aviation Security International Assessment 217
Research on Operator Capabilities and Performance 218
   Factors Affecting Visual Search and Detection 218
      Two-Component Model of Visual Search and Detection 218
   Object Recognition 218
      Target Prevalence 219
      Correctable Target-Detection Decisions 220
      Sleep Deprivation 220
      Image-Based Factors 221
      Single versus Multiple Target Search 221
   Assignment of Security Tasks 222
   Human Factors in Layers of Defense in Airport Security 223
   Human Factors in Cargo Screening 225
   Knowledge-Based and Image-Based Factors 225
   Addition of Color to X-Ray Images 225
   Computer-Based Training for Cargo Screening 226
   Detection of Deception 226
   Performance of Full Body X-Ray Scanners 227
Looking Ahead 228
   Solving Continuing Problems 228
      Air Line Pilots Association Perspective 228
      Department of Homeland Security Perspective 228
      Security Topics Identified at the 2014 Aviation Security Summit 229
   Addressing the Changing Mix of Future Threats 229
      Insider Threats 230
      Threats from Explosive Devices 230
      Threats against Airline Facilities and Airports 230
      Threats from Ranged Weapons 230

      Threats from Thermite-Based Incendiary Devices..230
      Threats in the Form of Cyber-Attacks...231
      Threats from Drones...231
    Capitalizing on New Procedures and Technologies 231
      Checkpoint of the Future..231
      Credential Authentication Technology 231
      Paperless Boarding Pass 232
      Biometric Identification..232
      Bottled Liquids Scanners 232
      Explosives Trace Detection 232
      Airport Scanner Game as a Research Tool...232
Summary and Conclusions 233
References..234

INTRODUCTION

The current chapter addresses the ever more critical human factors in aviation security, a subject in which human factors have played and will continue play a significant role. A historical background of human factors issues is presented, emphasizing research approaches and findings, and their resulting operational applications. In addition, the findings of periodic critical assessments of aviation security are summarized and their implications for human factors discussed. The chapter also looks ahead to the continuing problems that remain to be solved, predicting the changing mix of threats that will need to be addressed and anticipating the types of new procedures and technologies that will need to be assessed.

As we address the methods and concerns of aviation security, it may be helpful to keep in mind the magnitude and scope of the enterprise. To this end, some helpful statistics have been compiled by the Department of Homeland Security (DHS), the Transportation Security Administration (TSA), the General Accountability Office (GAO), and other organizations. For example, more than 650 million passengers and 700 million items of baggage undergo security screening each year at 450 airports in the United States. The TSA employs the full-time equivalent of 55,600 personnel to conduct passenger and baggage screening at these airports. Moreover, the total cost of these aviation security efforts is an estimated $8 billion per year.

Aviation security was addressed by Gibb and Lofaro in their chapter of the Handbook of *Aviation Human Factors*, second edition, which was published in 2009. It is the intent of this chapter to focus on findings and developments that have taken place since the publication of this earlier handbook, with the exception of the background information that follows. We start, then, with a brief history of aviation security, an accounting and description of threats to aviation security, the roles that have been assigned to human operators, and a description and assessment of factors that influence operator performance.

BACKGROUND

A Brief History of Aviation Security

Threat-detection systems were first implemented at airports in the early 1970s in response to increased hijackings of commercial aircraft. Since the introduction of these systems, human operators have played a key role in the detection of threats, such as guns, explosives, and possible terrorists, by examining X-ray images, resolving metal detector alarms, conducting body scans with metal detection wands, observing passenger's behavior, conducting physical searches of baggage, and maintaining order at screening checkpoints.

Ensuring the safety of air travel was soon recognized as a monumental task, and the effectiveness of the systems implemented to detect and deter threats was soon called into question and has been of concern ever since. In 1994, for example, at the request of the Federal Aviation Administration (FAA), the National Research Council formed a panel that assessed and reported on airline passenger security screening (NRC, 1996). The panel determined that the analytical and decision-making performance of security personnel was critical to the successful implementation of any security system, and that the allocation of functions between machine and operator is likely to have significant influence on the effectiveness of future systems. The panel stated that an ideal passenger screening system "would be capable of detecting both metallic and nonmetallic threat items in less than six seconds, with a high degree of accuracy (including a high detection rate and a low false-alarm rate). In addition, an ideal system would give the operator enough information, in an appropriate format, to allow for the speedy and accurate resolution of alarms."

The panel also determined that the screening systems could be made more effective by integrating the human operator completely into the system. The inability to maintain the needed level of operator performance was considered a major weakness of existing screening systems and a potential weakness of future systems. Moreover, improving operator performance should be considered as critical as technical improvements for enhancing current systems and ensuring the success of new systems.

Concerns for security were then greatly intensified after the terrorist attacks of September 11, 2001 in which four commercial aircraft were hijacked and used as weapons to destroy the World Trade Center twin towers and inflict severe damage to the Pentagon; moreover, these concerns have not diminished as evidence of worldwide terrorism continues to accumulate. In his review and assessment of aviation security at that time, Harris (2002) proposed a much broader and more flexible perspective for addressing the problem than was then in place. He proposed that instead of applying the same procedure to every passenger and item of baggage, the system should be adaptable so as to meet the full range of anticipated security threats with timely and appropriate responses based on the nature and extent of the potential threat. "Such a system would consist of the following integrated and coordinated components:

- Security teams of selected, cross-trained, motivated personnel performing tasks appropriate to the threats, with continuing measurement and feedback of their performance;
- An arsenal of technologies and procedures that can be quickly reconfigured and deployed to meet probable threats; and
- Timely intelligence continually disseminated to security teams on probable adversaries and threat scenarios."

Others also proposed alternative approaches to replace that of screening all passengers in the same manner. For example, Poole and Passantino (2003) suggested employing multiple levels of security involving the screening of passengers and baggage in accordance to assessments of their perceived risk.

As reported by the Congressional Research Service (Elias, 2014), the TSA, after having applied relatively inflexible methods to aviation security over the years, is now shifting away from this approach to what is being called a *risk-based* approach. Thus far, the development of this approach has focused on identification of high-risk passengers and expediting the processing of low-risk passengers. Issues yet to be tested are the efficacy of the specific program elements and the extent to which they complement each other and enhance the effectiveness of the overall security process.

THREATS TO AVIATION SECURITY

The Aviation and Transportation Security Act of 2001 established TSA as the federal agency with primary responsibility for securing the nation's civil aviation system including the screening of all passengers and properties transported by commercial passenger aircraft. TSA relies upon multiple

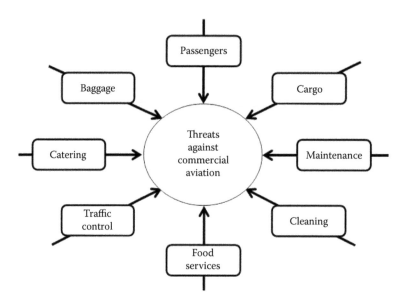

FIGURE 10.1 Potential avenues for threats against commercial aviation.

layers of security to deter, detect, and disrupt persons posing a potential risk to aviation security. The subsequent Act of 2007 further mandates the screening of 100% of cargo transported on passenger aircraft to determine whether cargo poses a threat to transportation security.

Although passengers, baggage, and cargo are the primary focus of security efforts, they are not the only sources of threats to aviation security (see Figure 10.1). Threats can also come from many processes that support airport operations: catering, maintenance, cleaning, baggage handling, air traffic control, retail, food services, and others. For example, members of the cleaning staff stashed guns and grenades in the plane's washroom to support the hijacking of TWA Flight 847, which led to 17 days of terror in 1985 (Gladwell, 2001). Thus, even perfect threat detection applied only to passengers, and their baggage would not necessarily result in acceptable levels of security.

According to Richard W. Bloom (2011), Chair of the Transportation Research Board Aviation Security and Emergency Management Committee, the most appropriate and comprehensive approach to assessing the sources of threat would be a comparative analysis of threats from people (passengers) versus things (cargo). His assessment is that the security threat from passengers or baggage or cargo changes from moment to moment depending on the continuous coupling of threat with vulnerability, qualified by the impact and probability of a successful terrorist attack. In addition, over time, the nature and extent of threats from passengers and cargo will change as the world changes.

The Roles of Human Operators

Aviation security, particularly checkpoint screening, places extensive reliance on human perception, decision-making, and judgment to detect and resolve potential threats. Assuring and enhancing the performance of human operators in their assigned roles will continue to be one of the major challenges in future efforts to incorporate improved security technologies and procedures (Elias, 2009).

For most of the history of aviation security operations, human operators have been called upon to perform tasks for which humans are poorly equipped—such as the monitoring and detection of rarely occurring, low-signal-to-noise-ratio signals embedded in the context of varying background configurations. This has been principally by default because technology has not been available to perform these tasks. An extensive body of research on human vigilance, for example, has led to the unavoidable conclusion that humans are poor monitors and that a variety of interventions and countermeasures directed toward improving detection performance in different settings and tasks over many years have demonstrated little benefit (Davies & Parsuraman, 1982; Mackie, 1987). Consequently, a primary objective in the design of security systems should be to remove the operator from this monitoring role. Emphasis should be given to developing and implementing screening technology designed to monitor, detect, and alert operators to potential anomalies and to the development of procedures to enable operators to determine whether or not the anomaly is a threat. That is, the monitoring/detection/alarm function should be allocated to machines, and the alarm resolution function to human operators.

Human operators have special unique capabilities needed for alarm resolution, such as pattern recognition, abstract reasoning, spatial visualization, and cognitive flexibility. As a consequence, an appropriate (even necessary) role for human operators is in the resolution of detected signals. However, even in this role, the combined effects of the design of the specific operator tasks, the environment in which the tasks are performed, and the selection and training of operators are critical to the realization of effective performance. Using human operators appropriately in threat-detection systems requires systematic efforts to match their capabilities with the requirements of the system and to minimize the effects of human limitations (Parasuraman, Molloy, Mouloua, & Hilburn, 1996; Rasmussen, 1986; Reason, 1990; Wiener, 1988).

Paradoxically, as threat-detection systems become more automated, human integration issues are likely to become even more important to their successful implementation. Human operators will be performing the more difficult and complex tasks that defy automation, such as alarm resolution. Moreover, the increased automation will introduce new performance issues that will need to be addressed. For example, an operator directly involved in X-ray signal detection necessarily monitors the performance of the X-ray machine. Degradation of images, easily detected by the operator, leads to appropriate equipment calibration or maintenance. If detection is automatic, on the other hand, the operator will need to have some other means of monitoring and ensuring that the technology is operating properly. Such problems introduced by automation can be resolved if they are recognized and addressed during system design and development.

In 2007, the TSA implemented the screening of passengers by observation techniques (SPOT) program that utilized operators in a very different manner from that previously discussed. The program employed behavior detection officers (BDOs) to observe passenger behavior to detect potential high-risk travelers. The assumption of the SPOT program was that cues observed from passenger behaviors might be indicative of the stress, fear, or deception likely to accompany destructive actions. According to TSA, as of 2012, more than 3,000 BDOs were authorized for deployment to 176 U.S. airports (DHS, 2012a).

BDOs, working in pairs, primarily conduct behavioral observations at airport screening checkpoints and by having brief verbal exchanges with passengers while they are waiting in line. A BDO identifies passengers for additional screening based on an evaluation system of previously identified behaviors; the BDO may involve a law enforcement officer to determine if legal intervention is needed. If determined to be a possible threat, the passenger may not be permitted to board the aircraft. According to TSA, SPOT referrals made during the year from October 2011 to September 2012 resulted in 199 arrests for having outstanding warrants, suspected transport of drugs, and being illegal aliens. The passenger throughput during that year was 657,000,000.

FACTORS INFLUENCING OPERATOR PERFORMANCE

Operator performance is a continuing concern as covert testing results have repeatedly demonstrated existing weaknesses in screening procedures and capabilities, weaknesses that could be exploited by terrorists or criminals seeking to attack the aviation system. These weaknesses can stem from policies, procedures, technology, and/or operator performance. But according to the Congressional Research Service (Elias, 2009), weaknesses in operator performance are the principal concern. A wide range of human factors considerations pertaining to operator performance were identified in this assessment as having potentially significant effects on operator performance, including procedures, training, fatigue, perception, detection, judgment, and decision-making.

The humans who operate the system have long been regarded as a potentially highly fallible and vulnerable element of the aviation security system. This should not be construed as a reflection on the dedication, commitment, or motivation of individual operators in performing their critical job functions. Rather, it reflects the combination of complex challenges faced by operators, such as limitations in the required capabilities, resourceful adversaries employing artful concealment methods, and time pressures to perform well while maintaining an efficient flow of passengers through security checkpoints.

In such a system, measurement and feedback of operator performance would seem essential to the development and maintenance of proficiency. Consider, for example, how little people would improve their bowling performance, and how soon they would stop bowling altogether if there was no system of keeping score and no feedback on how many and which pins were knocked down with each roll of the ball. The huge sports industry—players, teams, leagues, fans, tournaments, records, statistics, and magazines—is based on measuring performance and providing feedback (scores, standings, winners, and champions). The power of this process to enhance performance in military and industrial jobs has been known for decades; when it is absent, performance deteriorates (Harris & Chaney, 1969; Howell & Cooke, 1989; Ilgen & Klein, 1988; Swezey & Salas, 1992).

Airport security operators cannot be expected to maintain high levels of performance day after day, month after month without consistent, regular, realistic, accurate measurement, and feedback of their performance. In addition to providing a continuing basis for motivating and improving operators, performance measurement and feedback can also serve to enhance proficiency of the overall threat-detection system by the following:

- Diagnosing performance weaknesses and tailoring performance-based training to overcome the weaknesses identified
- Assessing the impact of new detection technologies and procedures and
- Providing criteria for validating the design and organization of the security processes

According to the TSA (2014b), threat image projection technology is now being employed on a daily basis to measure and provide feedback on operator proficiency in the X-ray detection of weapons and explosives. Potential threats, including guns and explosives, are projected onto X-ray images of carry-on bags passing through screening, thus permitting the testing of detection capabilities. These data permit the evaluation and feedback of individual detection performance and, also, help one formulate operator training programs. As the X-ray equipment is part of a network, every airport and X-ray monitor is able to receive automatic image updates from the TSA technology lab based on the latest intelligence on potential threats.

The ultimate success of on-line performance measurement for improving operator performance will have as much to do with the manner in which systems are introduced and employed as with the effectiveness of the systems themselves. Critical considerations include the detailed procedures needed for collecting and analyzing data, the extent to which the measurement system intrudes into operator screening tasks and the degree of understanding by operators of how the measures are to be obtained and used. Rarely, for example, should the results obtained from any on-line performance measurement be employed as the basis for punitive actions, such as fines, penalties, reprimands, or other punishments. Punitive actions based on performance measures will inevitably lead to subversion of the measurement process, contamination of the results obtained, and resistance to remedial measures that might be introduced to enhance performance.

ASSESSMENTS OF AVIATION SECURITY

Over the past 10 years, systematic, independent assessments of aviation security have been made by oversight organizations such as the Heritage Foundation, the GAO, the RAND Corporation, the Reason Foundation, and the 2015 Aviation Security International Summit. The results of these assessments serve to provide a varied perspective of the issues confronting aviation security and to identify a host of human factors problems and issues that need to be addressed.

HERITAGE FOUNDATION ASSESSMENT

The Heritage Foundation is an American conservative think tank based in Washington, D.C. The foundation's stated mission is to "formulate and promote conservative public policies based on the principles of free enterprise, limited government, individual freedom, traditional American values, and a strong national defense."

In its assessment of aviation security, the Heritage Foundation (2006) stated that "although well-intentioned, much of the effort to enhance aviation security since September 11, 2001, has done little to make the skies significantly safer." The assessment is critical of the Congress having vested both regulatory and operational responsibilities in the same agency, the TSA. According to the assessment, TSA's dual role of being both regulator and operator of baggage and passenger screening creates a serious conflict of interest.

The assessment points out that beginning in the 1980s, European airports began to develop a performance-contracting model under which the government set and enforced high performance standards, which airports then carried out by hiring security companies or occasionally using their own staff. Now, most countries in the world use this model, leaving the United States pretty much alone in having the national government actually operate its aviation security system. As a result, the Heritage Foundation report states that the prospects for any significant improvement in passenger and baggage security are grim.

RAND CORPORATION ASSESSMENT

The 2012 assessment by the RAND Corporation (Jenkins, 2012) starts with a historical perspective, taking the reader down the 40-year path of terrorist threats and aviation security responses and concluding that, over the long run, increasing security has made terrorist operations more difficult and pointing out that the effectiveness of security should be measured not in the number of weapons or explosive devices discovered at airport checkpoints but in the steadily declining number of attempted terrorist hijackings and sabotage attempts. The reason for this positive trend is in both the decline in the number of terrorist groups focusing on aviation and in the improvement in security. In addition, although some critics think airline security is a joke, terrorists take it very seriously because risks of betrayal and failure are just too great.

The RAND assessment observes that airport security personnel is under increasing stress because each terrorist innovation has added another security procedure and each added procedure complicates the search, slows down the screening process, and further stretches human resources. At the same time, passenger loads are increasing, whereas security budgets appear likely to decline. If the same number of operators is expected to perform more procedures on more passengers without letting the lines back up at checkpoints, performance can be expected to suffer. Meanwhile, public tolerance and cooperation are likely to fray.

Americans have come to hold unreasonable expectations that government should provide 100% security, and they quail when there is any failure. At the same time, they have little tolerance for inconvenience and react with outrage to intrusions into their privacy, and the safer people feel, the less tolerance they have for what they see as increasingly intrusive security.

Thus, the review concludes that aviation security is costly, controversial, and contentious; moreover, no other security measures directly affect such a large portion of the country's population. On account of the nature of the threat, aviation security is the most intrusive form of security and, as a consequence, is often at conflict with issues of civil liberties. But, the threat is real and not likely to go away soon; terrorists remain obsessed with attacking airplanes.

The RAND assessment proposed a sweeping review of aviation security and recommended that nongovernment research institutions undertake the task, and that each reviewer independently design an optimal aviation security system beginning with a clean slate. The competing models would be reviewed, and the best ideas, or combination of ideas, would be put forward. RAND justifies this effort by stating that even if the results resemble what is already in place, the process would offer some comfort that we are close to the best that we can do.

GAO Assessments of 2012, 2013, and 2014

The GAO is an independent, nonpartisan agency that supports the U.S. Congress in meeting its constitutional responsibilities by helping to improve the performance and ensure the accountability of the federal government. The stated objective of GAO evaluations is to provide Congress with timely information that is objective, fact based, nonpartisan, nonideological, fair, and balanced.

In 2012, the GAO assessed TSA's SPOT program and TSA's efforts to enhance cargo security on inbound aircraft. In its report (GAO, 2012), the GAO acknowledged that the TSA had not only taken actions to validate the science underlying its behavior-based passenger screening (SPOT program) but also stated that more work remains. GAO had reported earlier, in 2010, that TSA had deployed SPOT before first determining whether there was any scientifically valid basis for using behavior and appearance indicators to reliably identify passengers who might pose a risk. The GAO also had reported then that it was unlikely that the SPOT program had ever resulted in the arrest of anyone who is a terrorist, or who was planning to engage in terrorist related activity, even though there was evidence that terrorists had transited through SPOT airports.

The GAO also reported that the TSA has taken actions to enhance the security of cargo on inbound aircraft, but that challenges remain. In its June 2010 assessment, GAO had recommended that TSA develop a mechanism to verify the accuracy of all screening data. TSA concurred in part and required air carriers to report inbound cargo screening data but has not yet fully addressed the recommendation. In June 2012, TSA required air carriers to screen 100% of inbound air cargo transported on passenger aircraft by December 3, 2012. However, air carriers and TSA have faced challenges in implementing this requirement and in providing reasonable assurance that screening is being conducted at reported levels.

The 2013 assessment (GAO, 2013) concluded that the SPOT program does not meet its objectives and recommended that future funding for behavioral detection activities be limited. According to this assessment, available evidence does not support the use of behavioral indicators, such as those employed in the TSA's SPOT program to identify persons who may pose a risk to aviation security. GAO reviewed four meta-analyses of over 400 studies from the past 60 years and found that the human ability to accurately identify deceptive behavior based on behavioral indicators is the same as or only slightly better than chance. The GAO warned that until TSA can provide scientifically validated evidence demonstrating that behavioral indicators can be used to identify passengers who may pose a threat to aviation security, the agency risks funding an aviation security approach that has not been determined to be effective.

The efforts of TSA to introduce expedited passenger screening were addressed by the GAO in its 2014 assessment (GAO, 2014). This program, sometimes referred to by the TSA as *Managed Inclusion*, was implemented in 2011 and involves selecting passengers for expedited screening based on layers of risk assessments of all passengers. Testing of the system is scheduled to be completed by mid-2016. The number of passengers receiving expedited screening grew slowly at first but has increased more rapidly since late 2013 when the TSA expanded its employment of automated risk assessments of all passengers, in which a risk score was assigned to each passenger based upon available information.

TSA determines a passenger's risk score, and thus eligibility for expedited screening at the airport, using one of three risk assessment criteria: inclusion on a TSA precheck list of preapproved low-risk travelers, identification of passengers as low risk by TSA's risk assessment algorithm, or a real-time threat assessment at the airport using the managed inclusion process. To be on the TSA precheck list of approved travelers requires the completion and approval of an application and payment of a processing fee.

The GAO had previously examined several of the layers of passenger risk assessment employed in the managed inclusion process, raising concerns regarding its effectiveness, and recommending actions to TSA to strengthen them. In late 2014, TSA planned to begin testing managed inclusion as an overall system but could not provide specifics or a plan or documentation showing how the testing was to be conducted, the locations where it is to occur, how these locations are to be selected, or the timeframes for conducting testing at each location. GAO emphasized the need to employ established methodological practices for evaluation such as adequate sample size and random selection of participants to ensure the generalizability of results.

REASON FOUNDATION ASSESSMENT

In their 2013 report, "Overhauling U.S. Airport Security Screening," the Reason Foundation addressed what they see as a conflict of interest of the TSA (Poole & Ybarra, 2013). That is, as specified by the Aviation and Transportation Security Act of 2001, the TSA establishes security policy and regulates those that provide airport security while, at the same time, is itself the operator of the largest component of airport security: passenger and baggage screening. The recommendation is for the airport and not TSA to select the security contractor and to manage the contract under TSA regulatory oversight. This position is essentially the same as that taken by the Heritage Foundation in 2006 that was discussed earlier in this section of the chapter.

In their assessment, the Reason Foundation pointed out that separation of aviation security regulation from the provision of security services is called for by the International Civil Aviation Organization (ICAO), to which the United States (along with 189 other countries) is a signatory. Under the Chicago Convention that created ICAO, "contracting states are required to notify ICAO of any differences between their national regulations and practices" and ICAO's international standards. The United States has failed to notify ICAO that it does not comply.

The Foundation report further specified that under a performance contracting approach, with screening devolved to the airport, TSA would continue to certify screening companies that met its requirements (e.g., security experience, financial strength, screener qualifications, training, etc.). It would also spell out the screening performance measures (outcomes) that companies or airports would be required to meet. Airports would be free to either provide screening themselves or to competitively contract for a TSA-certified screening company. Companies bidding in response to the airport's RFP would propose their approach to meet the performance requirements, in terms of staff, procedures, and technology. This could include, for example, cross-training screeners to carry out other airport security duties, such as access and perimeter control. The airport would select the proposal that offered the best value, subject to TSA approval. TSA, in its role as regulator, would oversee all aspects of the airport's security operations, including adherence to federal laws and screening.

THE AVIATION SECURITY INTERNATIONAL ASSESSMENT

In their report, "Security Screening: Improving Human Performance for Greater Reliability," Aviation Security International (2015) stated that they were encouraged by the innovative designs and security solutions tested recently at larger airports. Specifically mentioned were the improved capabilities of detection equipment resulting from the introduction of new technologies. On the other hand, the assessment concluded that innovation in detection capabilities must extend beyond the introduction of new technologies alone. It must encompass improvements in capability, which can result from the effective alignment of people, procedures, and equipment, and where the role of technology is one of assisting human operators to better perform their detection tasks.

The Aviation Security International assessment references the results of research on airport security estimating that about 80% of all error events can be attributed to deficient human performance and, of these, the majority stem from organizational weaknesses. Their assessment states that the findings from analyses such as these can provide the basis for enhanced operator performance by sharing the results with operators. Feedback of this type, providing an opportunity to learn from performance results, has long been known and practiced by top-performing organizations.

RESEARCH ON OPERATOR CAPABILITIES AND PERFORMANCE

FACTORS AFFECTING VISUAL SEARCH AND DETECTION

Research has been conducted for decades to assess visual skills in search and detection for a variety of applications and, more recently, as they relate to performance in airport-security screening. In this section, we identify, describe, and assess those factors that have been identified as affecting the visual search and detection tasks employed to meet aviation security objectives.

Two-Component Model of Visual Search and Detection

Elements of the aviation security inspection task can be seen as being similar to inspection tasks found generally in other applications. They rely on the component of repetitive visual search of multiple target items on multiple images, and on the component of detection—the classification of a target either as acceptable or requiring further investigation. However, aviation security inspection has the added challenge of requiring operators to recognize each of the changing catalogue of potential threats whose general nature is known (i.e., guns, knives, etc.), but whose actual size, shape, and other characteristics are likely to change over time. In addition, threats are likely to be deliberately concealed, to rarely occur, and to lead to severe consequences if undetected (Schwaninger & Hofer, 2004).

In a study employing 416 experienced X-ray screeners, Ghylin, Drury, and Schwaninger (2006) tested the applicability of this two-component model as the basis for computer-based training of operators in aviation security X-ray screening. Application of the two-component inspection model was shown to result in large increases in detection performance and substantial reductions in response times. It also allowed the individual parts of the inspection process to be further analyzed and modeled to provide a better idea of what steps within the inspection processes have been changed by the training and why. The authors conclude that by having knowledge of the individual components within inspection, the technologies, methods, and procedures that current security inspectors utilize can be better honed to enhance the strengths of the human–system interface and correct their weaknesses.

OBJECT RECOGNITION

The detection of prohibited objects in passenger's bags using X-ray screening equipment requires the matching of X-ray images with object representations stored in visual memory. Visual knowledge of which objects are forbidden and what they look like is therefore critical to their successful detection. In studying the visual abilities and visual knowledge required of screeners to perform this task, Schwaninger, Hardmeier, and Hofer (2005) developed and applied tests of these capabilities.

The prohibited items test was developed to measure the extent to which a screener has acquired the necessary visual knowledge. The test contains a sample of different forbidden items from the international prohibited items lists placed in X-ray images of passenger bags to clearly reveal object shapes. As each image in the test can be inspected for 10 seconds, failing to recognize a threat item can be mainly attributed to a lack of visual knowledge.

The object recognition test was developed to measure visual processing and encoding. Three image-based factors can be distinguished that challenge different visual processing abilities. First, depending on the rotation within a bag, an object can be more or less difficult to recognize (effect of viewpoint). Second, other objects can be superimposed over prohibited items, which can impair detection performance (effect of superposition). Third, the number and type of other objects in a bag can challenge visual search and processing capacity (effect of bag complexity).

To examine the role of object recognition in the detection of threat objects in X-ray images of passenger baggage, a total of 134 aviation security screeners and 134 novices participated in a study. The criticality of the three image-based factors measured by the object recognition test was validated. The effect of view, superposition, and bag complexity was all highly significant. Regarding the effect of visual knowledge, as measured by the prohibited items test, detection performance of screeners was significantly superior to that of novices, consistent with the assumption that visual knowledge is also important to screener performance. Reliability was found to be high for both participant groups and tests, indicating that they can be used for measuring performance on an individual basis. The results confirm that X-ray detection performance relies on visual abilities necessary for coping with image-based effects such as view perspective, bag complexity, and superposition. Visual experience and training are necessary to know which items are prohibited, and what they look like in X-ray images of passenger bags.

Target Prevalence

Target prevalence is calculated as the ratio of the number of target events (target present) to the number of nontarget events (target absent) over a period of time. For example, if the number of opportunities for a target to be present were 100 and only one target appeared during this time, the target prevalence would be 1%.

Visual search for infrequently appearing targets (low-target prevalence) is characterized by a marked elevation of miss errors accompanied by a corresponding decrease in reaction times (Wolfe, Horowitz, & Kenner, 2005). A likely explanation might be that observers simply become faster and more careless when targets seldom appear. However, follow-up research by Wolfe and his associates (Wolfe et al., 2007) demonstrated that this target prevalence effect is more complex than just a simple speed-accuracy tradeoff. When two observers view the same sequence of stimuli, if prevalence errors were simply careless lapses, they should occur randomly and not be correlated between observers. However, their research showed that errors were strongly correlated. Rather than the product of carelessness, then, errors appear to be caused by shifts in decision criteria that lead operators to miss most of the same targets; when targets rarely appear, operators shift their criteria to a strongly conservative position resulting in the nondetection of infrequently appearing targets.

This pattern of results has been well known for years in other settings (Green & Swets, 1967; Mackworth, 1970; Maddox, 2002). Mackworth stated that one of the most important findings in vigilance research was the discovery that the probability that a signal will be detected is considerably reduced when the rate of target events is very small relative to the rate of nontarget events (often referred to as the background rate). The effect of reducing target prevalence is to make the decision criterion more conservative. The operator performing X-ray screening of passenger baggage will be less likely to call something a target if the *a priori* probability of the target is low.

Target prevalence, then, powerfully influences the results of visual search behavior. In most visual search experiments, in which targets are rare (such as in medical or airport screening), observers shift response criteria, leading to elevated target miss rates. Operators also speed target-absent responses and may make more motor errors. This could result from a simple speed/accuracy tradeoff with fast, frequent absent responses producing more miss errors. However, research completed by Wolfe and Van Wert (2010) showed that although very high target prevalence (98%)

shifted response criteria in the opposite direction, leading to elevated false alarms in a simulated baggage search, the very frequent target-present responses were not speeded and the rare target-absent responses were greatly slowed. This effect was further studied by varying target prevalence systematically more than 1,000 trials; detection accuracy and reaction times were measured for each trial. Observers' criterion and target absent response times closely tracked prevalence. These results support a model in which prevalence influences two parameters: the decision criterion governing the detection decision for each attended item, and the quitting threshold that governs the timing of target-absent responses.

Correctable Target-Detection Decisions

Fleck and Mitroff (2007) also addressed the problem of low detection rates for infrequently appearing targets in baggage screening by examining the mechanisms that hinder visual search and detection for these targets. As shown in the previous research, when targets are rarely present, observers respond more quickly, resulting in higher miss rates. Their research showed that when searching arrays similar to those viewed by airport baggage screeners, observers missed only 7% of the targets when target frequency was high (target present on 50% of trials) but missed an alarming 30% when target frequency was low (target present on 1% of trials).

They hypothesized that when targets are rarely present, observers adapt by responding too quickly, resulting in reporting errors and high miss rates. They proposed that misses are often due to response-execution errors rather than perceptual or identification errors. That is, airport screeners might actually have sufficient information to determine that a target is present but just respond too quickly indicating it is not. The investigators further hypothesized that when provided with the opportunity to correct their response, the operator would often be able to catch and correct the mistake. If so, low-target prevalence may not be a generalizable cause of high miss rates in visual search.

They tested this hypothesis with 1,400 target-detection trials divided into three blocks determined by the frequency with which targets were present. The experimental design was similar to that of Wolfe, Horowitz, and Kenner (2005); the critical difference was that half of the participants were given the opportunity to correct each response. The high-target-prevalence block consisted of 200 trials, 50% of which contained a target; the medium-prevalence block consisted of 200 trials, 10% of which contained a target: and the low-prevalence block consisted of 1,000 trials, 2% of which contained a target. Presentation of the blocks was counterbalanced among the 20 participants in the research and no order effects were noted. Half of the operators, the correction condition, were given the opportunity to correct their response before going on to the next trial.

Results for the no-correction condition replicated the prevalence effects of the Wolfe, Horowiz, and Kenner study. Miss rates were 10%, 19%, and 31% for the high-, medium-, and low-target-prevalence blocks, respectively. In contrast, the correction condition showed no significant effect of prevalence, with miss rates of 4%, 10%, and 10% for the high-, medium-, and low-prevalence blocks.

In summary, target prevalence did not influence the error rate in the correctable searches in this study. The option to correct mistakes appeared to parse out response-execution errors, thus minimizing the rise in miss rates previously found in searches for rare targets. Ultimately, improving real-world search and detection performance might be served by separately addressing errors of action and errors of perception. However, these conclusions need to be verified by studies conducted in actual aviation security screening operations.

Sleep Deprivation

Fatigue may be caused by a variety of factors, including intrinsic and extrinsic sleep disorders, as well as work and lifestyle-related changes in sleep schedules. Impairments from sleep loss have been found, in general, to produce increased variability in alertness and compensatory effort, resulting in increased rates of errors of omission (i.e., lapses) and errors of commission (i.e., incorrect responses). To what extent do night work and sleep loss impair target detection performance in aviation security tasks? Is the impact on visual detection sooner and more dramatically than other cognitive functions?

In a study to determine the effects on screener performance of fatigue induced by night work, sleep loss, repeated performance shifts, and time-on-task, screener performance was assessed by measuring the speed and accuracy of threat detection during a simulated luggage screening task (Basner et al., 2008). The study tested the implications of previous research that night work and sleep loss would increase both errors of omission (missed threats) and errors of commission (false alarms) in threat-detection performance.

The results confirmed that night work and sleep loss adversely affect performance of tasks performed by airport baggage screeners. Thus, fatigue induced by sleep deprivation in luggage screening personnel could potentially pose a threat for air traffic safety. In the future, methods could be developed to predict signal detection performance based on brief fitness-for-duty tests or objective monitoring of screener alertness during the screening task, and countermeasures developed and deployed in an effort to assure high levels of vigilance and detection performance.

Image-Based Factors

Among the factors that are likely to affect X-ray screener performance in the detection of threat objects in passenger baggage are view difficulty and threat item superposition. The nature and extent of image interpretation training received by screeners is also likely to interact with these image-based factors. The objective of an experimental study completed by Bolfing, Halbherr, and Schwaninger (2008) was to determine the relative importance of these image factors and the possible influences of screener training on them. The study employed 90 professional aviation security X-ray screening operators and 2,048 test images created automatically employing a set of image measurement algorithms.

Bivariate correlations between detection performance and image factors were analyzed to estimate the isolated impact of each single factor independently of any other. Multiple linear regression analyses were applied for estimating the overall impact of both image-based factors and computer-based training, respectively. Analyses of covariance were applied in to check for possible interaction effects between all variables. All analyses were applied separately for the four threat item categories: guns, knives, improvised explosive devices (IED), and others.

The findings from these analyses led to the conclusion that view difficulty can be addressed effectively with computer-based training of X-ray screening operators; training improved detection performance significantly, particularly for the more difficult views. The results for image superposition, however, were not so promising under the limitations of the screening technology employed in which only one image was presented per bag. The solution for superimposed images is likely to require more recent technology capable of providing multiple views for each bag.

This study showed, once again, that in addition to stable human abilities, there are also trainable skills that play a very important role in mediating detection performance. Knowledge-based factors such as knowing which objects are prohibited and what they look like in X-ray images can be effectively learned by screeners through computer-based training, a powerful tool to improve X-ray image interpretation competency of screeners.

Single versus Multiple Target Search

Research on industrial inspection tasks long ago determined that visual search for one type of possible defect (target) at a time, employing multiple searches, was more effective than searching for all possible defects (targets) at a time during a single search. When applied to aviation security searches of passenger bags, this finding would indicate that searching an X-ray image of a bag for one threat item (knife, gun, and bomb) at a time would result in a higher detection rate than searching for all three items at the same time. In industrial inspection, defect-by-defect searches of complex items of electronic equipment containing many types of potential defects were found to be about twice as effective as area-by-area searches of the items (Harris & Chaney, 1969).

Recent research has confirmed that the superiority of single-target searches also applies to X-ray screening of passenger baggage for purposes of aviation security. Menneer, Donnelly, Godwin, and Cave (2010) investigated the relationship between the search for one and the search for two targets and found that performance was lower in dual-target search compared with the combined performance for two independent single-target searches, and that the added response time required by the two searches disappeared with practice, but the reduced accuracy of single searches remained. Thus, the two searches, one for each type of target, produced better detection performance and, ultimately, did not require more search time.

Sensitivity was lower and the decision criterion more conservative in dual-target searches than in single-target searches, suggesting that the mental representation of the target was less effective in dual-target search than in single-target search. In addition, target-present responses were excessive under high target prevalence, and target-absent responses were excessive under low-target prevalence. These findings are important for aviation security screening tasks in which targets appear rarely and can differ from each other. For example, the low-target prevalence in X-ray security searches may magnify the dual-target deficiencies implicated in previous research with X-ray images. Such a result would increase the need for security personnel to consider alternatives to dual-target search, such as specialization in detecting one target type or training to encourage independent searches for each target.

ASSIGNMENT OF SECURITY TASKS

Depending on the country in which the airport is located, there are typically several parties responsible for aviation security activities at an airport: the government department responsible for civil aviation, the police, the airport operator, and aircraft operators who have contracts with security companies. There are also several categories of security tasks at airports such as passenger security screening, checked baggage security control, access control to restricted areas, cargo and mail security, and crisis management. An important human factors consideration in aviation security, and the object of some study, is the rationale and manner in which responsibility for each security task is assigned.

The ICAO, a specialized agency of the United Nations, provides the standards for defining and allocating aviation security tasks for its 190 member states that include the United States. It specifies that each state shall require the appropriate authority to define and allocate tasks and coordinate activities between the departments, agencies, and other organizations of the state, the airport and aircraft operators, and of the entities assigned with the responsibility for implementation of the national aviation security program.

The structure and assignment of security tasks has been the subject of study by several researchers. Feng (2003) reviewed the aviation security structures in various countries such as the United States of America, Canada, the United Kingdom, Australia, New Zealand, Singapore, and Japan. He found that only the United States had created a new national organization, the TSA, charged with overseeing transportation security under the DHS. The other countries place their security organizations under their respective ministries of transport. Yoo and Lee (2004) studied the responsibility structure of security tasks at international airports in several countries and compared the advantages and disadvantages of each system. They pointed out that systems emphasizing a governmental role, such as those in the United States of America, have better security performance, whereas systems that place the responsibility for security tasks on the airport operator have an advantage in maintaining the efficiency of overall airport operations. Askew (2004) researched passenger screening and argued that significant improvements in security screening outcomes and check point performance result from the implementation of comprehensive recruitment and training with recurrent training to remedy deficiencies, along with continuing assessment and modification of processes and procedures.

Yoo (2009) surveyed a sample of experienced aviation security practitioners to assess the relative importance of five principal aviation security tasks. Passenger screening was considered to be the most important among the five security tasks and the most in need of enhancement. Checked baggage control was assessed as being the second most important followed, in order, by access control, cargo security, and crisis management. According to Yoo, these results probably reflected respondents' awareness that there have been numerous major security incidents at their facilities, such as hijackings and sabotages caused by the failures of passenger or baggage security control.

HUMAN FACTORS IN LAYERS OF DEFENSE IN AIRPORT SECURITY

Airport security systems have been built up over the years out of layers of defense based on the security-in-depth model. As discussed in this section, the TSA has now defined a staggering 20 layers of defense to control security risks. This means that many others have security responsibilities beyond the more visible security personnel that we see at the airport when catching a flight. Human factors play a significant role in most of these layers of defense and in possible future improvements.

According to the TSA (2015), multiple layers of security are employed to ensure that the traveling public and the nation's transportation systems are adequately protected. Security is most often associated with the airport checkpoints operated by transportation security officers; however, checkpoints represent just one layer of security among the many in place to protect the various forms of transportation. There are different measures of security utilized within the aviation system including intelligence gathering and analysis, cross-checking passenger manifests against watch lists, random canine team screening at airports, and federal air marshals and federal flight deck officers that provide armed protection to the aircraft. According to the TSA, there are also additional security measures undisclosed to the public.

According to the TSA, each layer serves as a counterterrorism measure. In combination, these layers provide enhanced security creating a much stronger and protected transportation system to deter, detect, and prevent an attack from happening. Each of the 20 layers is described briefly in the following.

1. Intelligence: Intelligence officers work with government and international partners to identify threats and to determine current and future issues affecting aviation security.
2. Customs and Border Protection: As one of the largest and most complex security operations, its objective is to keep terrorists and their weapons out of the United States while also facilitating trading and passenger travel.
3. Joint Terrorism Task Force: The Joint Terrorism Task Force consists of personnel from local and national agencies brought together to protect regional airports by assuring that the security measures taken are up to national standards.
4. No Fly List and Passenger Prescreening: The no-fly list is developed by means of passenger prescreening and monitored to ban people considered potential threats from traveling in or out the United States on any commercial aircraft.
5. Crew Vetting: In the process of assuming flight duties, airline crew members submit identification details such as name, country of residence, date of birth, and, in the case of pilots and engineers, license numbers.
6. Visible Intermodal Prevention and Response Teams: Visible Intermodal Prevention and Response teams work with regional security and police departments to maximize the capabilities for the detection of terrorism plans.
7. Canine Teams: The TSA has an estimated 500 canine teams working with the local law enforcement agencies and operating in 70 airports and 14 mass transit systems to detect possible explosive substances at the airports perimeter and interior.

8. Behavior Detection Officers: Behavior detection officers operate at security checkpoints in an attempt to detect passengers who exhibit suspicious behavior.
9. Travel Document Checker: Passengers are required to hand or show their boarding pass and identification to a TSA document checker when entering the departure areas.
10. Checkpoint Transportation Security Officers: TSA agents are stationed at airport security checkpoints to assure that passengers are properly screened before entering the aircraft.
11. Checked Luggage Screening: All checked luggage that will go into the cargo compartment is required to undergo X-ray and explosive-trace screening.
12. Transportation security inspectors: Transportation security inspectors observe security operations to assess the TSA personnel, the jobs they are performing, and the effectiveness of TSA procedures.
13. Random Employee Screening: Any individual with access to the airport's secure areas is subject to random screening and thorough background checking.
14. Bomb Appraisal Officers: These officers provide guidance and education to other security personnel to enhance their capabilities at checkpoints. They typically have had operational experience in military or law enforcement bomb squads.
15. Federal Air Marshal Service: Air marshals typically work undercover and armed in passenger aircraft to address any problems that might occur in flight.
16. Federal Flight Deck officers: Selected pilots and other approved flight crews are, after being trained, permitted by the TSA to carry a firearm during flight operations.
17. Trained Flight Crew: All cabin crews are encouraged to complete self-defense training offered by the TSA to deal with assaults that might occur during flight.
18. Law Enforcement Officers: Armed law enforcement officers may be called upon to assist with security on an airplane or at an airport by arresting or suppressing individuals exhibiting problem behavior.
19. Hardened Cockpit Door: The three aircraft cockpit doors are hardened to prevent entry, thus providing an extra layer of threat prevention.
20. Passengers: Passengers are required to undergo a number of screening processes as they pass through the security checkpoint, possibly including X ray, metal detector, trace explosive detector, and a pat-down search under suspicious circumstances.

In their study of human factors in layers of defense in airport security (Andriessen, Gulijk, and Ale 2012), 177 security personnel and 180 passengers completed survey questionnaires. The objective of the study was to obtain the perspectives of both security personnel and passengers toward the various elements that constitute the layers of defense in aviation security. The results identified several themes that the authors considered important to understand the effectiveness of the layers of defense model.

Security staff feel their responsibility most strongly when their responsibilities are compared with those of other airport employees. The type and amount of training received appears to contribute to this effect.

Frequent flyers are the most informed about security procedures but also feel the least responsibility for the security of other passengers.

A passenger's level of education was found to influence his or her perception of security; passengers with higher levels of education were found to know more about the security elements and processes. Among airport security staff, on the other hand, systematic differences were not found. It was hypothesized that this was a result of the common elements in the training that security staff had received.

The technology employed in airport security was found to be little trusted by both passengers and staff. This is a remarkable finding considering that technology plays such a large role in many of the layers of defense employed by the TSA. It seems that security staff, although feeling responsible for doing their job right, cannot trust the tools they use to do this job.

Airport security staff considered their airport to be less safe than did the passengers. Whether this is caused by the specific knowledge of the process that they possess or because of their distrust of the technology employed was not clear from the results of this study.

HUMAN FACTORS IN CARGO SCREENING

Air cargo varies in content, how the content is packaged and situated, and the configuration and other characteristics of the aircraft. Content may be categorized by density, weight, size, economic value, signatures of explosives, weapons, and weapons components. The associated screening challenges include the following: possible electrostatic discharge, physical damage related to the method of screening, levels of specificity and sensitivity related to the cargo content, and terrorist knowledge of screening methods, which can lead to the development of countermeasures or to other means of exploitation. The determination as to whether or not a cargo container is acceptable is difficult because the inspected containers can be very large, whereas the prohibited items can be comparably small (e.g., an IED or contraband goods). Taking human factors into account is therefore of utmost importance for assuring effective cargo security screening.

According to the results of a review of human factors in cargo screening (Mendes, Michel, & Schwaninger, 2013), security measures have been strengthened substantially in recent years for passenger screening, but not so much for cargo screening. Most airports now use X-ray security screening for containers and unit load devices (a pallet or container used to load cargo on aircraft, permitting a large quantity of cargo to be bundled into a single unit) providing an image of the contents without the need for physical interference. Yet it is still the human operator (security officer) who needs to identify prohibited items in the X-ray image and make the decision on whether the inspected unit can be regarded as harmless or not. Consequently, methods must continue to be explored for enhancing operators' ability to identify threats.

KNOWLEDGE-BASED AND IMAGE-BASED FACTORS

From previous research results and observations of the performance of security personnel, both knowledge-based and image-based factors appear to have a significant impact on human detection performance (Michel, Mendes, de Ruiter, Koomen, & Schwaninger, 2014). Knowledge-based factors relate to knowing which items are prohibited, what they look like in X-ray images, and how they can be distinguished visually. They are particularly relevant for objects, such as IEDs, that are rarely seen in everyday life and that look quite different in an X-ray image. Image-based factors are characteristics of the X-ray images of the objects. For example, objects are more difficult to recognize if shown from an unusual viewpoint or when superimposed by other objects. Recognition is also affected by the complexity of the image, which depends on the number and types of other objects.

ADDITION OF COLOR TO X-RAY IMAGES

Recent X-ray technologies have material discrimination capability that employs color to aid the threat-detection process. Material types (such as organic, inorganic, and metal) are rendered in different colors that overlay the original gray image. According to Michel et al. (2014), although this technology holds the potential to improve the screening process, additional human factors evaluations are necessary because the introduction of color images has not demonstrated an improvement in comparison with the grayscale images in experimental tests to date. For example, colors introduced into the cognition and perception processes might serve as distractors during target detection and resolution.

Computer-Based Training for Cargo Screening

Research conducted by Mendes, Michel, and Schwaninger (2013) demonstrated that customized computer-based training can significantly improve operator performance in cargo X-ray image interpretation. During a three month study, one group of experienced cargo screeners received a minimum of 20 minutes of training per week, whereas a control group from the same population received no training over the same period. Both groups were tested before and after the training with a computer-administered test consisting of 240 X-ray images, half of which contained prohibited items relevant for customs concerns (e.g., drugs, weapons) as well as for security screening (e.g., explosives, IEDs). The X-ray images were provided in both grayscale and with color enhancement. The task of the screeners was to distinguish between clear images (nothing of concern) and images containing prohibited items (something to report). The results of the study revealed significant increases in detection performance for screeners having received the weekly training, whereas the control group showed no change. The effect of training was particularly high for the detection of IEDs. Moreover, the larger the number of training hours received by a screener, the better the test performance.

During training, the screener received an immediate response feedback as to whether the response was correct or not. When a prohibited item was contained in the image, information on this item was also provided. Test performance of individual participants correlated positively with the number of training hours they received; these results were similar for displays presented in grayscale and those with color enhancement. Further, a general decrease in the inspection time per image was found for the group that received training. It is noteworthy that the baseline detection performance of both groups measured before training was very low for certain prohibited items, further underlining the significance of training for cargo X-ray screening.

Detection of Deception

The majority of published research on detecting deception has set out to identify indicators of human behavior that can discriminate deceivers from truth-tellers. Behavioral indicators of deception relate to observable demeanor and/or actions such as nervousness, aggression, eye contact, fidgeting, and verbal hesitations. Although the effectiveness of behavioral indicator approaches has never been tested in a large-scale field trial, a meta-analysis of laboratory studies that used behavioral indicators to discriminate deceivers from truth-tellers revealed a mean rate for correct identification at only slightly above chance (Bond & DePaulo, 2006), inspiring little confidence that the approach would be effective for detecting a deceiver from among the large numbers of truth-tellers among airline passengers. Even so, as discussed earlier, the TSA initiated the use of behavioral indicators in 2007 in their SPOT program through the employment of behavioral detection officers. That program eventually came under considerable criticism for its cost and ineffectiveness and, as discussed earlier, in 2013, the GAO recommended that future funding for this program be limited.

Ormerod and Dando (2014) have proposed veracity testing for detecting deception, focusing on the verbal exchange between the sender (the individual attempting to deceive) and the receiver (the individual attempting to detect deception). (This approach, based on verbal interactions, differs from the deception indicators employed in the SPOT program that were based on observation of the nonverbal behavioral indicators discussed earlier.) Ormerod and Dando reviewed a number of research studies and identified six aspects of dyadic verbal exchanges that appear to have the potential for discriminating between deceivers and truth-tellers. These are summarized as follows:

1. Interviews that use tactical and strategic information-gathering methods. The receiver first explores the sender's accounts, guided by information known to them before the interview, but without revealing to the sender what is known. The sender's responses are then compared with the known information and challenged when inconsistencies arise by revealing the information (Dando, Bull, Ormerod, & Sandham, 2015).

2. Questioning styles that elicit rich verbal accounts, such as open questions that do not constrain responses, require expansive answers and commit passengers to an account of the truth concerning issues such as identity, background, and previous, current, or future activities. (Oxburgh, Myklebust, & Grant, 2010).
3. Tests of expected knowledge, which compare the content of what someone says with information already known (Blair, Levine, & Shaw, 2010).
4. Interviewing methods that restrict the verbal maneuvering, the strategic manipulation by deceivers of verbal content, and delivery to control a conversation to avoid detection (Taylor et al., 2013).
5. Procedures that raise the cognitive load faced by an interviewee, such asking a question unlikely to be anticipated (Walczyk, Igou, Dixon, & Tcholakian, 2013).
6. The length of responses—truth-tellers speak longer, say more, and use more unique words than deceivers (Morgan, Rabinowitz, Hilts, Weller, & Coric, 2013).

Ormerod and Dando (2014) developed a method for applying the veracity concept to deception detection and conducted a double-blind randomized-control study involving 165 security agents and 204 mock passengers at international airports. Each mock passenger was provided with a valid ticket, an itinerary of flights, and a deceptive cover story to be maintained during the security interview and queued with other passengers to pass through the security process. Security agents detected 66% of the deceptive passengers using the veracity testing methodology. This compared with the detection of less than 5% using behavioral indicator recognition methodology described earlier in this section.

PERFORMANCE OF FULL BODY X-RAY SCANNERS

Full-body X-ray scanners started replacing metal detectors at airports in many countries in 2007. A full-body scanner is a device that detects both metallic and nonmetallic objects on a person's body for security screening purposes, without the need to physically remove clothes or make physical contact. Depending on the specific technology, the operator may see an image of the person's naked body, or just a cartoon-like representation of the person, with indicators showing the location of any suspicious items. Full body X-ray scanners were faster than previous scanning methods (taking only about 15 seconds) and did not require passengers to physically remove clothing. In the United States, pursuant to the FAA Modernization and Reform Act of 2012, all full-body scanners operated in airports by the TSA now must use automated target recognition software, which replaces the picture of a nude body with the cartoon-like representation to show the screener the location of suspicious items. According to the TSA, there are now, in 2015, as many as 740 of these units employing the advanced imaging technology at 160 airports nationwide.

At this time, there have been no evaluation studies completed on the effectiveness of the recently employed full-body X-ray scanners. On the other hand, a number of studies were completed on the Rapiscan Secure 1000, the full-body scanner that was deployed at U.S. airports between 2009 and 2013. Most of these studies revealed vulnerabilities in passenger screening employing these scanners. A representative sample of this research was that completed by a team of researchers from the University of California, San Diego, the University of Michigan, and John Hopkins University (Mowery et al., 2014) that claimed theirs' was the first analysis and laboratory testing that was independent of the device's manufacturer and its customers, and the first to consider software as well as hardware.

The researchers concluded that the system was ineffective in reliably detecting concealed knives, guns, and explosives against an adaptive adversary who has access to a device for study and to use for testing and refining attacks. They also concluded that the flaws identified could be partly remediated through changes to procedures such as performing side scans in addition to front and back scans, and by screening subjects with magnetometers as well as a backscatter scanner. But they also concluded that these procedural changes would significantly lengthen screening times, thus eliminating one of the key advantages of the system.

Despite the flaws identified in the Rapiscan Secure 1000, the researchers did not categorically reject TSA's claim that advanced image technology represents the best available tradeoff for airport passenger screening. The millimeter-wave scanners that are currently deployed to airports are likely to behave differently from the system that was studied. However, it was recommended that these full-body scanners be subjected to independent, adversarial testing, and that this testing specifically consider software security.

LOOKING AHEAD

SOLVING CONTINUING PROBLEMS

Air Line Pilots Association Perspective

As the representative of more than 53,000 pilots who fly for airlines in the United States and Canada and with its 80-year history, Air Line Pilots Association provides a unique and important perspective for analyzing the current state of aviation security and identifying the continuing problems that need to be solved. It provides regular analyses of issues of importance to air line pilots, employees, and passengers on the internet at www.alpa.org. In a recent analysis, it identified and reported the following proposed improvements in aviation security that are most in need of additional research and development with attention to human factors considerations.

- Implementation of Threat-Based Security. Although TSA has publicly committed to pursuing a threat-based approach to aviation security and some steps have been taken, threat-based security should be adopted across the board as a foundational philosophy and as a plan of action to address today's threats. A threat-based approach will ultimately enhance passenger privacy, create a more efficient and effective screening system, and make better use of limited screening resources.
- Enhanced Security of All-Cargo Flight Operations. Continuing efforts need to be expended in closing the gaps in security requirements for all-cargo flight operations. In particular, improvements are needed in the background vetting of individuals with unescorted access to cargo aircraft and cargo, hardened flight deck door requirements, and perimeter security protocols that clearly define entry and exit procedures and dictate specific identification display and ramp security procedures.
- Introduction of Threatened Airspace Management. A prioritized plan for control of the national airspace system during a major security event is needed to provide pilots in command of airborne aircraft, or those about to take off, with real-time notification of significant and ongoing security concerns.
- Aircraft Protection from Laser Attacks. FAA statistics show that the threat to aviation safety from laser attacks is growing and real, with an exponential increase in reported laser attacks against aviation. Efforts are needed to develop and test possible countermeasures that will minimize both the likelihood of these attacks and their possible impact.

Department of Homeland Security Perspective

The long-range research and development test and evaluation plan prepared by DHS (2012b) provided strategic objectives for the continuing development and enhancement of aviation security. A brief summary of each of the objectives that have implication for future human factors research and development is provided here.

- Enhanced detection performance of security systems. The principal areas of research identified for meeting this goal are developing more reliable methods for characterizing the signatures and indicators of emerging threats, developing tools and procedures for more effective alarm resolution, and increasing the efficiency and certainty of the testing conducted to evaluate the implementation of new technologies.
- Improved passenger experience in the screening process. Research conducted to help meet this goal would focus on solutions, technological, and procedural, which might decrease the physical invasiveness of screening modes and increase the efficiency with which passengers proceed through the screening processes. Possible solutions include the development and evaluation of procedures that do not involve direct contact with or actions required by passengers.
- Enabled risk-based and intelligence-driven screening processes. It is hypothesized that risk-based screening would allow the TSA to focus resources on high-risk individuals or threat objects, increasing the overall effectiveness and efficiency of passenger aviation screening. Research would address the development and evaluation of systems that would provide for a rapid adjustment of risk posture, enhancement of behavior-based targeting, and improved data processing and decision-making.
- Enhanced threat response capability with more flexible security solutions. The concern is that present security structures and procedures are not sufficiently flexible to respond to changing or emerging threats. Research is required for the development security systems that are more easily and quickly adaptable to changing threats.
- Other desired improvements in technology and procedures. A variety of other desired improvements must be derived from the application of science to specific problems identified by the DHS. They include improvements in the following: blast mitigation capabilities, behavior detection and biometric identification, anomaly detection, high throughput threat detection, system resilience and recovery, and freight tamper prevention and detection.

Security Topics Identified at the 2014 Aviation Security Summit

The 14th Annual Security Summit was hosted on November 17, 2014 by the American Association of Airport Executives (AAAE) in cooperation with the DHS and TSA. A product of the summit was identification of the principal aviation security concerns of those in attendance, as summarized in the following.

Risk-Based Security: Constantly assessing the operating environment and being able to discern new threats as they present themselves is a capability being developed in the industry. Learning how to adapt under different conditions and to apply the appropriate security measures is an essential component of a risk-based security program. TSA is working hard to adapt its security protocols to this philosophy, finding that it allows for greater flexibility for both its own personnel and all those affected.

TSA Precheck Program: Approximately 30% of the flying public is now taking advantage of the TSA's Precheck program with efforts in place working to expand the program to include aviation employees such as flight crew members, fueling personnel, and ramp workers who have already undergone background checks to obtain airport access. New innovations in the program could allow better matching between program participants and their luggage should an associated security protocol be activated in baggage handling.

Addressing the Changing Mix of Future Threats

Ideally, new threats are met with fast, nonintrusive, and flawless passenger screening that is 100% effective against the latest terrorist devices and concealment methods, while anticipating and thwarting new threats. The system must effectively screen every passenger and every piece of checked and hand-carried luggage nearly 800 million times a year. At the same time, passenger screening should be discerning but democratic. Intelligence must keep terrorist suspects off flights but without errors that affect innocent travelers, and it must accomplish this without government-held databases that appear to threaten civil liberties. Although attaining all of these objectives would be considered by most involved in aviation security to be impossible, it is clear that knowing and understanding the changing mix of threats faced by aviation security is critical and worthy of an increasing level of effort.

The Combatting Terrorism Center (CTC) at the United States Military Academy in West Point, New York, in their 2011 assessment of terrorist threats to commercial aviation (Brandt, 2011) antici-pated the critical future threats to be addressed by aviation security. The CTC focused on insider threats, threats from ranged weapons, evolving threats from explosive devices, and threats against airline facilities and airports. The reports from the TSA and others added threats from thermite-based incendiary devices, threats in the form of cyber-attacks, and threats from drones.

Insider Threats

Examples of insider threats that have been discovered during the past 10 years include a man seek-ing to become a flight attendant to stage a suicide attack and attempts by terrorists to recruit airport employees such as baggage handlers, security staff, and mechanics to participate in various plots. Although TSA and airport investigators currently conduct criminal and terrorist database checks on potential airport, airline, and vendor employees who are to be granted access to secure areas, there are apparently vulnerabilities in their approach. For example, according to the CTC, street gangs have gained employment and carried out criminal activities such as narcotics trafficking, baggage theft, and prostitution at airports nationwide. The magnitude of this vulnerability is compounded because most airport employees working in secure areas do not undergo security screening prior to entering their workspace due to practical constraints

Threats from Explosive Devices

IEDs are likely to remain a significant threat to commercial aviation due to limitations in cur-rent screening technology. According to the CTC, even the latest enhancements in bomb detection technology can be defeated by surgically implanting a device or secreting it within a body cavity. Moreover, IEDs concealed within complex electronic devices are likely to defeat all but the most thorough visual inspection.

Threats against Airline Facilities and Airports

Airport facilities, such as fuel lines, arrival halls, and curbside drop-off points, are likely to con-tinue to be tempting targets, as are aircraft that can be reached by breaching the airport perimeter fencing to assault aircraft on runways, taxiing areas, and at gates. Airport facilities are likely to be more vulnerable in the future, particularly if terrorists continue their emphasis on making smaller scale attacks on targets of opportunity.

Threats from Ranged Weapons

Ranged weapons are small portable weapons such as rocket-propelled grenades and small sur-face-to-air missiles that can be launched at a target from a distance. These weapons are becoming more accessible to terrorist groups through thefts from military depots and through gifts from sympathizers.

Threats from Thermite-Based Incendiary Devices

In a bulletin released to its intelligence customers, the TSA (2014c) warned of a thermite-based incendiary device, the ignition of which on an aircraft at altitude could result in catastrophic dam-age and the death of every person on board. The bulletin stated that the devices are easily assembled and concealable, and that current TSA screening procedures would likely not recognize thermite-based mixtures. Thermite is made of rust and aluminum powder (both substances are easily obtainable) and, when ignited, burns violently at extremely high temperatures and is capable of spraying molten metal in all directions.

Threats in the Form of Cyber-Attacks

Cyber security was a principal concern raised at the November 17, 2014 Aviation Security Summit hosted by the AAAE in cooperation with the DHS and TSA. Cyber security is growing in importance because aircraft systems are typically not encrypted and therefore vulnerable to hackers. Central to this concern of cyber security is the human factor. In many cases, cyber security faults are caused by staff that click on a malware link or unknowingly breach security in hundreds of small ways.

Threats from Drones

Recent reports from pilots of commercial airliners have identified drones as a potential threat to aviation security. As reported in the Washington Post (Whitlock, 2014), two pilots reported sightings on the same day of large drones operating at high altitudes near their aircraft. One sighting over New York City from a pilot landing at LaGuardia Airport was of a black drone with an estimated 15-foot wingspan flying at about 5,500 feet above Lower Manhattan. The other sighting from two airliners approaching Los Angeles International Airport was of a remote-controlled aircraft the size of a trash can at an altitude of 6,500 feet. A NASA database contains 50 reports over the past 10 years, and the FAA database has 15 reports over just the past two years of close calls or improper flight operations involving drones.

CAPITALIZING ON NEW PROCEDURES AND TECHNOLOGIES

Risk-based approaches to aviation security have been proposed as part of a comprehensive, multilayered approach to aviation security rather than as an alternative approach. Risk-based programs are envisioned to closely interact with physical screening checkpoint measures to allow TSA to focus physical screening resources on unknown and elevated risk passengers. They also inform the protocols that TSA utilizes to modify security postures based on known or perceived threats. The TSA and others have proposed the development and evaluation of specific procedures and technologies to support the implementation of risk-based approaches to aviation security (Elias, 2014; Pistole, 2014; TSA, 2014a). These are presented and discussed in this final section of the chapter; each is subject to human factors considerations, concerns, and implications for research and testing.

Checkpoint of the Future

In its Press Release Number 35 of June 7, 2011, the International Air Transport Association discussed its vision for the *checkpoint of the future*, a series of neon-lit tunnels, each equipped with an array of eye-scanners, X-ray machines, and metal and liquid detectors. Heralding an end to *one size fits all screening*, the association says that passengers will be assigned a *travel profile* and ushered into one of three corridors accordingly. Thermal lie detection has been proposed to be implemented with a camera that can detect variations in facial temperature in response to questioning. Human factors are likely to play a significant role in the design of the tunnels and the passenger interaction with the various screening technologies and equipment.

Credential Authentication Technology

The TSA envisions the development of technology that would automatically authenticate identity documents presented to TSA screeners by passengers during the security checkpoint screening process. Assuming an optimal interaction between passengers, screeners, and the authentication technology, the system would both enhance security and increase screening efficiency by automatically verifying passenger's identification and obtaining their vetting status.

Paperless Boarding Pass

The paperless boarding pass, an encrypted bar code along with passenger identification and flight information, enables passengers to download their boarding pass to their cell phones or personal digital assistants. The objective of this approach is to streamline the boarding process at the checkpoint while also increasing the ability to detect fraudulent boarding passes. Passengers would continue to be required to show photo identification so officers can validate that the name on the boarding pass matches that of the passenger's identification. Critical issues to assuring the successful employment of paperless boarding passes are the user friendly design of the cell phone software, the handling of passengers without cell phones, and the expeditious handling of problem cases.

Biometric Identification

Biometric identification is being explored because it permits verification of a person's identity by employing his or her own unique set of biological identifiers—fingerprints, iris scans, or a combination of the two. The TSA is testing this technology at airports across the country. The success of this approach would seem to depend on the efficient, minimally intrusive, and error-proof manner in which passengers provide their fingerprints and iris scans.

Bottled Liquids Scanners

Bottled liquids screening systems are being considered by TSA to detect potential liquid or gel threats, which may be contained in a passenger's property. The technology employed must be capable of differentiating liquid explosives from other liquids in amounts that are permitted. Potentially applicable technology includes laser, infrared, and electromagnetic resonance. A critical consideration would appear to be the manner in which this screening component is incorporated into the overall passenger screening process to maintain adequate passenger flow.

Explosives Trace Detection

Explosives trace detection (ETD) is technology used at security checkpoints around the country to screen baggage and passengers for traces of explosives. Officers may swab a piece of carry-on or checked baggage or a passenger's hands and then place the swab inside the ETD unit to analyze it for the presence of potential explosive residue. This is an awkward and relatively time-consuming process for passengers; one that seems ripe for technological innovation. A recent problem, somewhat similar to this one that has been solved by technology, is that of measuring body temperature. The most common approach, employed for years, has required shaking a small thermometer, after it has been sanitized with alcohol, to lower the liquid below the level of the expected body temperature, inserting the bulb end of the thermometer in an oral cavity for approximately three minutes, and then reading the liquid level to determine the temperature. The new temporal artery thermometer simply requires a swipe across the forehead.

Airport Scanner Game as a Research Tool

Data collected from the video game, airport scanner, and analyzed by researchers at the Duke Institute for Brain Sciences, indicate the possible potential of such games as research tools. In the Airport Scanner game, players earn points by tapping on illegal items as simulated X-ray images of airport passenger bags roll across their screens. The game collects anonymous data on how players perform, providing researchers the results from about a million trials a day, far more information than could possibly be obtained in a more traditional laboratory setting. By analyzing these gameplay data, researchers have been able to identify some of the common mistakes the human brain makes while searching for things (Mitroff et al., 2015).

SUMMARY AND CONCLUSIONS

As stated in the introduction, the objective of this chapter is to address the ever more critical and significant role that human factors play in aviation security. A historical background of important security events and the associated human factors issues was presented, focusing mainly on those dating from publication of the previous *Handbook of Human Factors in Aviation* published in 2009. The chapter reviewed human factors research approaches and findings, and their resulting operational applications to elements of aviation security. An important part of this review was the findings of the periodic critical assessments of aviation security by the GAO and others, which were summarized and their implications for human factors research and development discussed. The chapter also looked ahead to the continuing human factors problems that remain to be solved, predicting the changing mix of threats that will need to be addressed by human factors research, and anticipating the types of new procedures and technologies for which human factors issues will need to be identified and resolved.

This handbook is being published at the time of a major turning point in aviation security. The TSA, after having applied relatively inflexible methods to aviation security since its inception, is now shifting away from this *one size fits all* approach to what is being called a *risk-based* approach. Thus far, the development of this approach has emphasized identifying and focusing on high-risk passengers, while expediting the processing of passengers predetermined to be of low risk. Issues yet to be tested are the efficacy of the specific program elements and the extent to which they complement each other and enhance the effectiveness of the overall security process. Discussion of the many human factors issues to be addressed, as elements of risk-based approaches are introduced and tested, constitutes an important part of this chapter.

Operator capabilities and performance are critical to the effectiveness of systems developed and installed to assure aviation security. Human visual search and detection skills, in particular, have played a critical and central role since the initiation of passenger and baggage screening at airports. The application of operator skills requires a repetitive visual search of multiple target items on multiple images, and on the detection and classification of a target as either acceptable or requiring further investigation. The results of research studies have shown that successful visual search and detection requires the application of a number of specific human skills and technologies while also overcoming factors likely to degrade performance. Human factors were found to play a significant role in aviation security beyond that of visual search and detection during passenger screening including the assignment of security tasks, establishment of layers of defense in airport security, screening of air cargo, detection of passenger deception, and the operation and performance of screening with full-body X-ray scanners.

Looking ahead, the chapter identified and discussed the following as continuing problems that will require human factors research and development from the perspective of the organizations most concerned with aviation security. From the perspective of the Air Line Pilots Association, the major continuing problems are the implementation of threat-based security, enhancement of the security of all-cargo flight operations, institution of airspace management, and aircraft protection from laser attack. From the perspective of the DHS, the following continuing problems should be given priority: enhancing the detection performance of security systems, improving the passenger experience in screening process, enabling risk-based and intelligence-driven screening processes, and enhancing threat response capability with more flexible screening solutions. The AAAE at their 2014 Annual Security Summit focused on risk-based security and enhancement and expansion of the TSA Precheck Program as their highest priorities for further development.

The changing mix of threats likely to be faced by aviation security in the future was addressed by the CTC. The CTC identified insider threats, threats from ranged weapons (small portable weapons such as rocket-propelled grenades and small surface-to-air missiles), evolving threats from explosive devices, and threats against airline facilities and airports. Reports from the TSA and others added threats from thermite-based incendiary devices, threats in the form of cyber-attacks, and threats from drones.

Risk-based approaches to aviation security have been proposed as part of a comprehensive, multilayered approach to aviation security rather than as an alternative approach. Risk-based programs are envisioned to closely interact with physical screening checkpoint measures to allow TSA to focus physical screening resources on unknown and elevated risk passengers. The anticipated new procedures and technologies to be employed in support of this approach are as follows: checkpoint reconfiguration, credential authentication technology, paperless boarding passes, biometric identification, bottled liquids scanners, enhanced explosives trace detection, enhanced explosives trace detection, and video games as research and training tools.

REFERENCES

Andriessen, H., Gulijk, C., & Ale, B. (2012). Human factors in layers of defense in airport security. In *11th International Probabilistic Safety Assessment and Management Conference and the Annual European Safety and Reliability Conference* (pp. 4325–4332). Red Hook, NY: Curran Associates.

Askew, G. (2004, November). *Who is the screening boss? Does it really matter?* Aviation Security (AVSEC) 2004 World Conference Proceedings, Vancouver, British Columbia, Canada.

Aviation Security International (2015, January). Security screening: Improving human performance for greater reliability. Retrieved March 8, 2015, from: http://www.asi-mag.com/security-screening-improving-human-performance-greater-reliability/

Basner, M., Rubinstein, J., Fomberstein, K. M., Coble, M.C, Ecker, A., Avinas, D., & Dinges, D.F. (2008). Effects of night work, sleep loss and time on task on simulated threat detection performance. *Sleep, 19,* 1251–1259.

Blair, J. P., Levine, T. R., & Shaw, A. S. (2010). Content in context improves deception detection accuracy. *Human Communication Research, 36*(3), 423–442.

Bloom, R. W. (2011, July–August). Airport security: Which poses the greatest threat—passengers or air cargo? *TR News,* pp. 21–27.

Bolfing, A., Halbherr, T., & Schwaninger, A. (2008). How image based factors and human factors contribute to threat detection performance in X-ray aviation security screening. *HCI and Usability for Education and Work, Lecture Notes in Computer Science, 5298,* 419–438.

Bond, C. F. Jr., & DePaulo, B. M. (2006). Accuracy of deception judgements. *Personality and Social Psychology Review, 10*(3), 214–234.

Brandt, B. (2011, November). Terrorist threats to commercial aviation: A contemporary assessment. *Combating Terrorism Center at West Point Military Academy.* Retrieved March 9, 2015, from: https://www.ctc.usma.edu/posts/terrorist-threats-to-commercial-aviation-a-contemporary-assessment

Dando, C. J., Bull, R., Ormerod, T. C., & Sandham, A. L. (2015). Helping to sort the liars from the truth-tellers: The gradual revelation of information during investigative interviews. *Legal and Criminological Psychology, 20*(1), 114–128.

Davies, D., & Parasuraman, R. (1982). *The psychology of vigilance.* London, UK: Academic Press.

Department of Homeland Security (DHS), Office of Inspector General (2013a, May). *Transportation Security Administration's Screening of Passengers by Observation Techniques.* (Report OIG-13-91.)

Department of Homeland Security (DHS) (2013b, August). Strategic plan for fiscal years (FY) 2012–2016. Retrieved March 9, 2015, from: http://www.dhs.gov/strategic-plan-fiscal-years-fy-2012-2016

Elias, B. (2009). *Airport passenger screening: Background and issues for Congress.* Congressional Research Service Report R40543.

Elias, B. (2014). *Risk-based approaches to airline passenger screening.* Congressional Research Service Report R43456.

Feng, C. (2003). A review of aviation security organization. In L. W. Lan (Ed.), *The new challenge of international transportation security* (pp. 73–84). Institute of Traffic and Transportation, National Chiao Tung University, Taiwan.

Fleck, M. S., & Mitroff, S. R. (2007). Rare targets rarely missed in correctable search. *Psychological Science, 18*(11), 943–947.

Ghylin, K., Drury, C., & Schwaninger, A. (2006, July). Two-component model of security inspection: Application and findings. *16th World Congress of Ergonomics, IEA 2006,* Maastricht, the Netherlands, July, 10–14, 2006.

Gibb, G., & Lofaro, R. (2009). Human factors in civil aviation security. In J. A., Wise, V. D. Hopkin, & D. J. Garland (Eds.), *Handbook of aviation human factors* (2nd Ed.), 27-1–27-30. Boca Raton, FL: CRC Press.

Gladwell, M. (2001). Safety in the skies. *New Yorker*, 50–53, October 1, 2001.

Government Accountability Office (GAO) (2012, September). *9/11 Anniversary observations on TSA's progress and challenges in strengthening aviation security*. General Accountability Office Publication GAO-12-1024T.

Government Accountability Office (GAO) (2013, November). *Aviation security: TSA should limit future funding for behavior detection activities*. General Accountability Office Publication GAO-14-159.

Government Accountability Office (GAO) (2014, December). *Rapid growth in expedited passenger screening highlights needs to plan effective security assessments*. General Accountability Office Publication GAO-15-150.

Green D. M., & Swets, J. A. (1967). *Signal detection theory and psychophysics*. New York: John Wiley & Sons.

Harris, D. H. (2002). How to really improve airport security. *Ergonomics in Design, 10*, 17–22.

Harris, D. H., & Chaney, F. B. (1969). *Human factors in quality assurance*. New York: John Wiley & Sons.

Howell, W. C., & Cooke, N. J. (1989). Training the human information processor: A review of cognitive models. In I. Goldstein (Ed.), *Training and development in organizations* (pp. 121–182). San Francisco, CA: Jossey-Bass.

Ilgen, D. R., & Klein, H. J. (1988). Individual motivation and performance: Cognitive influences on effort and choice. In J. P. Campbell & R. J. Campbell (Eds.), *Productivity in organizations* (pp. 143–176). San Francisco, CA: Jossey-Bass.

Jenkins, B. M. (2012). *Aviation security: After four decades, it's time for a fundamental review*. Homeland Security and Defense Center Occasional Paper, RAND Corporation, Santa Monica, CA.

Mackie, R. (1987). Vigilance research –Are we ready for countermeasures? *Human Factors, 29*, 707–724.

Mackworth, J. (1970). *Vigilance and attention*. Harmondsworth, UK: Penguin Books.

Maddox, W. T. (2002). Toward a unified theory of decision criterion learning in perceptual categorization. *Journal of the Experimental Analysis of Behavior, 78*, 567–595.

Mendes, M., Michel, S., & Schwaninger, A. (2013). Cargo screening: Enhancement of human factors. *Aviation Security International*, 29-30, October 2013.

Menneer, T., Donnelly, N., Godwin, H. J., & Cave, K. R. (2010). High or low target prevalence increases the dual-target cost in visual search. *Journal of Experimental Psychology: Applied, 16*(2), 133–144. doi:10.1037/a0019569

Michel, S., Mendes, M., de Ruiter, J., Koomen, G., & Schwaninger, A. (2014). Increasing X-ray image interpretation competency of cargo security screeners. *International Journal of Industrial Ergonomics, 44*, 551–560.

Mitroff, S. R., Biggs, A. T., Adamo, S. H., Dowd, E. W., Winkle, J., & Clark, K. (2015). What can 1 billion trials tell us about visual search? *Journal of Experimental Psychology: Human Perception & Performance, 41*(1), 1–5. doi:10.1037/xhp0000012

Morgan, C. A., Rabinowitz, R. G., Hilts, D., Weller, C. E., & Coric, V. (2013). Efficacy of modified cognitive interviewing compared to human judgments in detecting deception related to bio-threat activities. *Journal of Strategic Security, 6*, 100–119. doi:10.5038/1944-0472.6.3.9

Mowery, K., Wustrow, E., Wypych, T., Singleton, C., Comfort, C., Rescorla, E., Checkoway, S., Halderman, A., & Shacham, H. (2014, August), Security analysis of a full-body scanner. Proceedings of the 23rd USENIX Security Symposium, San Diego. Retrieved March 9, 2015, from https://www.usenix.org/conference/usenixsecurity14/technical-sessions/presentation/mowery

National Research Council (NRC). (1996). *Airline passenger security screening: New technologies and implementation issues* (Publication NMAB-482-1). Washington, DC: National Academy Press.

Ormerod, T. C., & Dando, C. J. (2014). Finding a needle in a haystack: Towards a psychologically informed method for aviation security screening. *Journal of Experimental Psychology: General, 144*(1), 78–84.

Oxburgh, G. E., Myklebust, T., & Grant, T. (2010). The question of question types in police interviews: A review of the literature from a psychological and linguistic perspective. *International Journal of Speech Language and the Law, 17*(1), 45–66.

Parasuraman, R., Molloy, R., Mouloua, M., & Hilburn, B. (1996). Monitoring of automated systems. In R. Parasuraman & M. Mouloua (Eds.), *Automation and human performance: Theory and applications* (pp. 91–115). Mahwah, NJ: Erlbaum.

Pistole, J. S. (2014, April 30). Statement of John S. Pistole, Administrator, Transportation Security Administration, U. S. Department of Homeland Security, before the United States Senate, Committee on Commerce, Science, and Transportation. Retrieved March 9, 2015, from https://www.tsa.gov/sites/default/files/assets/pdf/SenateCommerce_jsp.pdf

Poole, R. W., & Passantino, G. M. (2003, May). *Risk-based airport security policy*. Reason Foundation. Reason Public Policy Institute, Policy Study 308.

Poole, R. W., & Ybarra, S. (2013, July). *Overhauling U. S. airport security screening*. Reason Foundation. Reason Public Policy Institute. Retrieved March 8, 2015, from http://reason.org/news/show/overhauling-us-airport-security-scr

Rasmussen, J. (1986). *Information processing and human-machine interaction: An approach to cognitive engineering*. New York: North-Holland.

Reason, J. (1990). *Human error*. Cambridge, MA: Cambridge University Press.

Schwaninger, A., Hardmeier, D., & Hofer, F. (2005). Aviation security screener's visual abilities & visual knowledge measurement. *IEEE Aerospace and Electronic Systems, 20*(6), 29–35.

Schwaninger, A., & Hofer, F. (2004). Evaluation of CBT for increasing threat detection performance in X-ray screening. In K. Morgan & M. J. Spector (Eds.), *The Internet society 2004, advances in learning, commerce and security* (pp. 147–156) Wessex, UK: WIT Press.

Swezey, R. W., & Salas, E. (1992). Guidelines for use in team-training development. In R. W. Swezey & E. Salas (Eds.), *Teams: Their training and performance* (pp. 219–245). Norwood, NJ: Ablex.

Taylor, P. J., Dando, C. J., Ormerod, T. C., Ball, L. J., Jenkins, M. C., Sandham, A., & Menacere, T. (2013). Detecting insider threats through language change. *Law and Human Behavior, 37*, 267–275. doi:10.1037/lhb0000032

Transportation Security Administration (TSA) (2014a, February). Risk-based security initiatives. Retrieved March 9, 2015, from http://www.tsa.gov/traveler-information/risk-based- security-initiatives

Transportation Security Administration (TSA) (2014b, October). Security technologies: Threat image projection. Retrieved March 7, 2015, from http://www.tsa.gov/about-tsa/security-technologies

Transportation Security Administration (TSA), Office of Intelligence and Analysis (2014c, December). Other agency product of interest (OA-PoI). Retrieved March 9, 2015, from https://prod01-cdn00.cdn.firstlook.org/wp-uploads/sites/1/2015/02/TSA-Bulletin.pdf

Transportation Security Administration (TSA) (2015, February). Layers of U.S. aviation security. Retrieved March 9, 2015, from http://www.tsa.gov/about-tsa/layers-security

Walczyk, J. J., Igou, F. P., Dixon, A. P., & Tcholakian, T. (2013). Advancing lie detection by inducing cognitive load on liars: A review of relevant theories and techniques guided by lessons from polygraph-based approaches. *Frontiers in Psychology, 4*(14), 4–14.

Whitlock, C. (2014). Close encounters on rise as small drones gain in popularity. *Washington Post*, June 23, 2014.

Wiener, E. L. (1988). Cockpit automation. In E. L. Wiener & D. C. Nagel (Eds.), *Human factors in aviation* (pp. 433–461). San Diego, CA: Academic Press.

Wolfe, J. M., & Van Wert, M. (2010). Varying target prevalence reveals two dissociable decision criteria in visual search. *Current Biology, 20*, 121–124.

Wolfe, J. M., Horowitz, T. S., & Kenner, N. M. (2005). Rare items often missed in visual searches. *Nature, 435*, 439–440.

Wolfe, J. M., Horowitz, T. S., Van Wert, M. J., Kenner, N. M., Place, S. S., & Kibbi, N. (2007). Low target prevalence is a stubborn source of errors in visual search tasks. *Journal of Experimental Psychology: General, 136*(4), 623–638. doi:10.1037/0096-3445.136.4.623

Yoo, K. (2009). A study on factors influencing the performance of airport security and on responsibility assignment of security task at international airports. *Journal of Aviation/Aerospace Education & Research, 19*(1), 37–50.

Yoo, K., & Lee, J. (2004). *Airport security: Establishing a clear chain of command*. Aviation Security (AVSEC) 2004 World Conference Proceeding, Vancouver, British Columbia, Canada.

Section III

Research and Development

11 Data Sources and Research Tools for Human Factors

Mark Wiggins, Shayne Loft, and Johanna Westbrook

CONTENTS

Data Collection and Management in Aviation 239
Measures of Central Tendency 242
Distributions of Data 245
The Importance of Continuous Data Acquisition and Analysis 246
Organizational Trend Analyses 248
Data in Different Organizational Contexts 250
Acquiring and Using Data Cautiously 252
The Value of Qualitative Data 253
The Importance of Outliers 254
References 256

To remain competitive in a commercial environment, constantly innovating and searching for efficiencies, sources of data are required that reduce the uncertainty associated with decisions and guide a path toward improvements in performance. The business world is undergoing a revolution driven by the use of data and analytics to guide decision-making. Gone are the days where senior leaders within organizations could decide simply on the basis of intuition to invest in new technologies or a new route structure. In the contemporary industrial environment, regulators, consumers, and investors alike demand evidence-based and, importantly, defensible decisions from organizational leaders.

Defensible decisions, such as those in all walks of life, are dependent upon the availability of meaningful, relevant, and timely sources of data. The difficulty lies in the realization that both the sources and the amounts of data that are now available to aviation managers are almost endless, with aviation data analysts and managers facing a difficult choice in discerning reliable from unreliable data, meaningful from irrelevant data, and valid from invalid data.

Skilled aviation managers, similar to skilled military commanders, are those who have the capacity to quickly and effectively identify the key sources of information to which they should attend and at which time during the business cycle. This chapter explores the types and sources of data and methods of acquisition and analysis that are appropriate for the different conditions that aviation businesses might face now and into the future as they attempt to improve major aspects of their business, from using data to improve efficiency and safety, to improving customer retention.

DATA COLLECTION AND MANAGEMENT IN AVIATION

With 35 million flight departures per year, data are critically important for any planning decision made by aviation management. According to forecasts made by the International Civil Aviation Organization, passenger and freight air traffic will double by 2030 (ICAO, 2013). In the aviation industry, data are available on an almost continuous basis from a number of different sources, including aircraft sensors, air traffic control, passengers, customer search/booking data, and baggage screening. These are data at the micro level and provide the basis for monitoring various components of the ongoing performance of a system. Changes in performance reflect responses to day-to-day or minute-to-minute demands that can be addressed quickly where necessary.

At the macro level, a complex organization might also rely on monthly or annual reports, such as profit and loss statements (Tretheway & Marhvida, 2014). These data are useful insofar they provide a cumulative assessment of the performance of an organization over an extended period and take into account the variability of performance across a period (Sull, 1999).

This notion of using macro data as a barometer of organizational performance is also evident in the safety context where, at its coarsest level, the frequency of incidents and in some cases, accidents over an extended period, is used to assess the performance of an organization in achieving its safety-related goals (Madsen, 2013; Travaglia, Nugus, Greenfield, Westbrook, & Braithwaite, 2011). Similar to profit and loss statements, these data take into account variability over an extended period. This is particularly important and useful in high reliability organizations in which the frequency of incidents and accidents may be relatively low, and/or the incidence is intermittent.

Just as these macro-level data offer useful indicators of performance, they are inevitably outcome-driven and suffer a potential lag between an occurrence and the reporting of that occurrence against a broader dataset (Lindberg, Hansson, & Rollenhagen, 2010). In the absence of other data, the result tends to be reactive and importantly, implemented sometime after an event or series of events have actually occurred. Therefore, the ideal approach is one that involves the collection of data on a cumulative basis, but at a level that is both meaningful and does not draw too heavily on organizational resources.

The importance of cumulative data, rather than a reactive approach to every case that occurs lies in both the efficiency and the effectiveness of the response (Jones, Kirschsteiger, & Bjerke, 1999; Lenne, Salmon, Liu, & Trotter, 2012). If an organization was to respond, in isolation, to every event that occurred, the demands on the organization would increase significantly, to a point where the demands may outstrip the resources available. The advantage of a cumulative assessment over a period of events or time is the opportunity to identify patterns of events, so that interventions impact performance beyond a single occurrence (Edkins, 1998).

The disadvantage associated with sole reliance on cumulative data is the potential delay in initiating a response until such time as a pattern of events emerges. To provide an example, consider a situation in which a single case of a disease with serious consequences is identified in a major population center. For the authorities not to respond immediately would be tantamount to negligence. However, a response that is disproportionate to the threat posed would also be criticized on the basis that valuable resources are being wasted or worse, drawn away from the management of other diseases that then reach epidemic proportions for want of resources. This is all the more important in commercial airline operations, as resources are inevitably constrained, and the costs of interventions can be significant.

This balance between the resources available and potential consequences is an important factor in determining the types of data acquired, the rate at which these data are acquired, and the nature of the response. This process of risk assessment to data acquisition and management is a necessary requirement in both commercial and noncommercial operations but should be approached with a clear understanding of both the severity and the likelihood of an adverse event.

In the case of severe consequences that occur infrequently, the difficulty lies in sourcing data on a sufficiently frequent basis that enables the identification of changes that signal a deterioration in performance. However, this approach presupposes that the precursors to severe events are understood, whether it is a minor conflict in the Middle East that escalates and causes a spike in oil prices, or a low pressure weather system over the Pacific that forms into a typhoon that impacts Hong Kong. In practice, these precursors are often only identified in the case of a system failure, such as an aircraft accident or incident.

An aircraft accident is rarely an isolated event. It is often the outcome of a process of less obvious failures or events that, had they been identified and rectified, might have prevented the accident (Goh, Love, Brown, & Spickett, 2012). These precursors or indicators are precisely the features that data acquisition systems such as Line Operations Safety Audit (LOSA) are intended to identify (Table 11.1).

TABLE 11.1

Summary of the Advantages and Disadvantages of the Various Approaches to Data Acquisition in Organizational Contexts

Approach	Sensitivity Management	Data Cost	Relative	Reliability
Continuous	High	High	High	Low
Cumulative	Moderate	Moderate	Moderate	Moderate
Summary	Low	Low	Low	High

LOSA is an audit process, much like a financial audit. Trained observers located in supernumerary seats on the flight deck record the behavior of pilots in response to the threats, errors, and undesirable states that occur during *normal* aircraft flights (Goodheart & Smith, 2014; Ma et al., 2011). Importantly, data are collected at every stage of the operation so that there is an opportunity to capture issues that might otherwise be overlooked using a more targeted approach. These deidentified data can be collapsed across routes, types of aircraft, or airlines. In addition to specific incidents that are identified as part of this process, comparative analyses are possible both within and across operations (Thomas & Petrilli, 2006).

In addition to more systematic strategies for data collection such as LOSA, it is now the case in most high-technology environments that data are collected automatically from a range of sources. In the context of human performance-related system failure, the masses of data that are acquired, referred to as Big Data, offer an opportunity to identify minor variations in behavior that might subsequently be associated with system failure. However, like systematic strategies for data acquisition, the utility of Big Data lies in the identification of the precursors to events. These precursors are what Walker and Strathie (2016) refer to as *leading indicators*.

In the context of human performance-related system failures, leading indicators are associated with psychological precursors of errors, so that a change in the leading indicator, in effect, becomes a barometer of the risk of a system failure. A practical example of a leading indicator might include changes in an operator's response latency to a series of warnings or alarms. An exceedingly rapid response might be indicative of a bias towards responding to the alarm, perhaps due to the fact that the alarm is always triggered at a particular point in the operational cycle. In the aviation context, the *minimums* warning on final approach to land is an example of an alarm that might be expected to be heard on the flight deck. A response that is too rapid might reveal a lack of conscious consideration of the event, whereas a response that is too slow might reveal a lack of preparation for the landing.

For aviation safety, the usefulness of LOSA and similar data collection systems lies in the capacity to identify *changes* in behavior as potential precursors to system failure. This allows the development and implementation of design, procedural, and/or personnel interventions that are targeted towards a specific issue that poses a threat to operational safety, thereby ensuring the efficient use of limited resources. However, it also benefits organizations in other ways, including the identification of inefficient procedures that increase the risk of delays or otherwise impact adversely the performance of the organization or system (Helmreich, 2000), and the identification of examples of *superior* pilot performance that can provide model behavior for application during training.

In addition to the identification of changes in performance over successful periods of data collection, the outcomes need to be compared against a benchmark. Benchmarks offer an objective standard of performance and can be prescribed by a regulatory agency as a minimum standard, or can be established by summarizing the normal or typical performance of other, similar organizations.

Performance norms are in common use in psychometric assessment and testing, in which it is difficult to prescribe appropriate levels of performance. The value of norms lies in establishing reasonable comparisons or expectations for performance. For example, it would be unreasonable to expect, during the initial stages of training, that a practitioner would demonstrate levels of performance that are comparable to practitioners with much greater levels of experience. In the case of safety-related activities, no incidents or accidents is the ideal. However, in practice, the frequency of errors or incidents is highly variable, and the value of normed data lies in establishing whether a particular series of occurrences is a symptom of an underlying trend or is *normal* or typical (and then managed) given the context within which they occur.

MEASURES OF CENTRAL TENDENCY

Establishing benchmarks begins by establishing measures of central tendency that constitute aggregated representations of a dataset. The most common of these measures of central tendency is the *mean*, which is calculated as the sum of the data points, divided by the number of data points in a dataset. To establish measures of central tendency, it is important first to determine whether the dataset constitutes information from a complete population, such as all of the operators within an organization, or a sample of a larger population.

A sample is used when access to data from the entire population is difficult to obtain. This is not normally the case within organizations, although there are many examples in which large multinational organizations will sample a small group of employees and extrapolate these results to the organization more broadly.

It is this process of extrapolation that means that, inevitably, there is a degree of error. For example, it may be the case that the sample selected was drawn from employees within one particular sector or geographic location that does not necessarily represent the views or capabilities of employees in other locations. Statistically, this possibility can be taken into account by using different statistical analyses.

In the case of a population, the mean is calculated as

$$\mu = \frac{\Sigma(a^1 \cdots a^n)}{n}$$

where:
μ represents the mean
a represents the individual values
n represents the number of values in the dataset

In the case of a sample of a population, the mean is calculated as

$$\overline{X} = \frac{\Sigma(a^1 \cdots a^n)}{n-1}$$

where:
$\overline{X}$ represents the mean
a represents the individual values
n represents the number of values in the dataset

Although the mean is certainly the most common measure of central tendency, it needs to be calculated with caution as it is particularly susceptible to changes when the data are not distributed *normally*. In examining any phenomena for a period of time, behavior will be distributed across a normal or Gaussian distribution (essentially a unimodal, roughly symmetrical, bell-shaped curve) (see Figure 11.1). The nature of this distribution is best described in terms of human performance in which performance occasionally will be high or low but will generally cluster around the *mean*. The occasional forays into exceptionally high performance are reflections of statistical variability

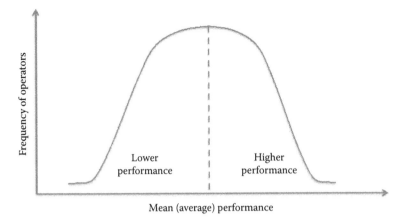

FIGURE 11.1 An example of a normal or Gaussian distribution in which the frequency of operators displaying higher levels of performance broadly matches the frequency of operators displaying lower levels of performance.

that occur for a variety of reasons. Therefore, they are not necessarily a reflection of normal performance. This is why the collection of data from a number of sources is so important. Any single data point may reflect performance that is the exception, rather than the rule.

To illustrate, an airline might be required to calculate the mean frequency of incidents to report to a regulatory authority. Incidents are calculated on a quarterly basis so that, over a 12-month period, the mean number of incidents is calculated as the sum of the total number of incidents, divided by the four quarters. This yields a mean frequency of incidents per quarter.

In a typical year, the number of incidents might be recorded as

Quarter 1–12 Incidents
Quarter 2–9 Incidents
Quarter 3–14 Incidents
Quarter 4–11 Incidents

Substituting,

$$\mu = \frac{\Sigma(12,9,14,11)}{4}$$

$$\mu = \frac{\Sigma(46)}{4}$$

$$\mu = 11.5$$

In this case, the airline experiences a mean 11.5 incidents per quarter.

However, if, during one quarter, the frequency of incidents was to increase markedly, it would influence significantly, the final result. For example, if the number of incidents was distributed as

Quarter 1–12 Incidents
Quarter 2–9 Incidents
Quarter 3–14 Incidents
Quarter 4–27 Incidents

Then, substituting[*],

$$\mu = \frac{\Sigma(12,9,14,27)}{4}$$

$$\mu = \frac{\Sigma(62)}{4}$$

$$\mu = 15.5$$

This result is now much poorer than the previous 11.5. However, the majority of the results in the sample remain identical. The "27" incidents in the case of quarter 4 is referred to as an outlier, and it has affected significantly, the overall assessment of performance. Therefore, the overall result needs to be interpreted with some caution.

The main difficulty in using the mean as a measure of central tendency is the failure to account for outliers that may have skewed a distribution (see Figures 11.2 and 11.3). Although there may be very good reasons why an outlier may have occurred (such as the introduction of new aircraft or a period of poor weather conditions), it does not necessarily represent the pattern of the remaining data points. An alternative approach that better represents the dataset in this case is the median or *middle value*.

To establish the median, the values are ranked in order from least to greatest, and the middle value is taken as the measure of central tendency. In the case of the airline, there are four data points, so the middle value is calculated as the point half-way between the second and the third values (13). Where there is an odd number of items in a dataset, the middle value can be taken as the median.

A third measure of central tendency, although rarely used, is the mode. The mode is the most frequently occurring value in a dataset and is actually useful in establishing whether a ceiling or floor effect was evident. A ceiling effect occurs when a significant proportion of the values are recorded at a maximal value. For example, in an examination, the majority of students might

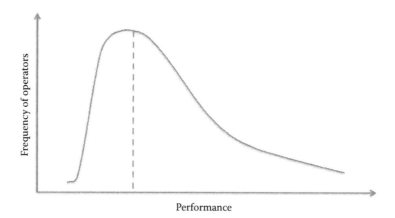

FIGURE 11.2   An example of a positively skewed distribution in which the frequency of operators is distributed across levels of performance. In this case, it is evident that, while the performance of the majority of operators clusters about the mean (signified by the dotted line), a small number of operators display exceptionally high levels of performance.

---

[*] Note that the population mean is used in this case since the data summarize the performance of the organization as a whole.

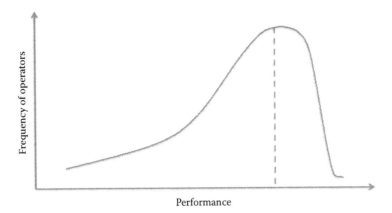

Performance

FIGURE 11.3   An example of a negatively skewed distribution in which the frequency of operators is distributed across levels of performance. In this case, it is evident that, although the performance of the majority of
operators clusters about the mean (signified by the dotted line), a small number of operators display exceptionally low levels of performance.

achieve 100%. Conversely, a floor effect occurs where a significant proportion of data points are
recorded at a minimal value. In either case, it reflects a systematic issue either in the way that the
data were collected or within the examination itself.

Where an organization is safety conscious and is governed by highly practiced policies and procedures, a relatively high frequency of periods of no incidents (floor effect) might be expected. If, at
the same time, the organization is operating as efficiently as is practicable, the frequency of on-time
departures might be maximized (ceiling effect). Although these results are difficult to interpret
from the perspective of the mean or median, they nevertheless represent very useful outcomes in
establishing the effectiveness of policies, procedures, and practices.

DISTRIBUTIONS OF DATA

Although measures of central tendency are useful in summarizing the performance of the majority
of values within a dataset, it is possible that, by focusing on these measures, problems with other
aspects of the performance of an organization may be overlooked. For example, although a mean
might reveal a high level of overall performance in the context of on-time performance, it may also
belie the fact that there remain a number of data points that fall well below the mean and, in fact,
these are likely to represent the key targets for intervention.

Falling below the mean is not necessarily indicative of an outlier. A distribution about the mean
is to be expected in any dataset as values are naturally likely to fall about the mean. It is the breadth
of this distribution that is of greatest interest in practice as it may reflect a lack of consistency within
the cohort. For example, in the previous example, the frequency of incidents ranged from 9 incidents
in Quarter 2 to 27 incidents in Quarter 4.

From an organizational perspective, the first task in determining whether there is an opportunity
for intervention lies in calculating the nature of the distribution. Where the mean is used as a measure of central tendency, the standard deviation is used to represent the distribution. If the median
was used, then the range of scores, from greatest to least, represents the distribution of values.

The standard deviation constitutes the mean squared differences about the mean and is calculated, for a population, as

$$\sigma = \sqrt{\frac{\Sigma\left(X - \bar{X}\right)^2}{N}}$$

where:

 σ is the population standard deviation

 X constitutes the individual values in a dataset

 $\overline{X}$ represents the mean

 N is the number of items in the dataset

Using the following dataset,

 Quarter 1–12 Incidents
 Quarter 2–9 Incidents
 Quarter 3–14 Incidents
 Quarter 4–11 Incidents

the standard deviation is calculated, as

$$\sigma = \sqrt{\frac{13}{4}}$$

$$\sigma = \sqrt{3.25}$$

$$\sigma = 1.80$$

Therefore, in the case of this dataset, the mean deviation about the mean is 1.80. This result can now be compared with data from other periods to determine whether there are any differences in the distribution of scores. A change in the distribution might be indicative of a change in the performance during one or more quarters. This may not have affected the mean, as a loss of performance for one Quarter might be counterbalanced by an improvement in performance during another quarter. However, by examining the distribution, it becomes clear that a change may have occurred that requires further analysis.

Where the data are not normally distributed and the median is selected as the measure of central tendency, the standard deviation is no longer an appropriate measure of variability. In this case, the range needs to be calculated. The range is simply the lowest score subtracted from the highest score. This gives an indication of the breadth of scores.

The standard deviation and variance are only of practical use where the outcomes can be compared against other datasets. For example, if data are collected on an annual basis, it is possible to compare the distributions from one year to another and establish whether any changes have occurred, either in terms of central tendency or in terms of the distribution of values.

From an organizational perspective, differences in the distribution of values, even in the absence of differences in measures of central tendency, constitute an indicator of changes in organizational performance. Specifically, an increase in the standard deviation or range is a reflection of increased variability so that, although there are periods of outstanding performance, there are also periods of poorer performance. As a consequence, examining both measures of central tendency and the distribution can provide the impetus for examinations of the causes of poor performance and the features that enable successful performance. By drawing on the factors that contribute to success, it becomes possible to target areas of poor performance, thereby improving measures of central tendency and reducing the standard deviation.

THE IMPORTANCE OF CONTINUOUS DATA ACQUISITION AND ANALYSIS

For most assessments of organizational performance, the value of measures of central tendency can only be realized once the values are compared with a normed dataset. These are datasets (a) from the same group but at different periods or (b) from different, but comparable groups. It is only by undertaking these types of comparisons that changes or differences within and/or between organizations can be identified and their impact interpreted. However, this comparative process needs to be undertaken with a level of caution to ensure that the assessments are valid.

Clearly, comparative groups need to be considered carefully to ensure that there are no systematic differences that might render the assessment invalid. For example, if the comparative group operates different aircraft, operates to different destinations, or differs substantially in size, then some comparisons, such as the mean number of days lost to workplace injury, might be unreasonable. Even assessments of safety-related performance might lack validity where one organization operates aircraft to and from high-capacity airports where another operates aircraft to uncontrolled airports. In this case, the nature of the hazards will differ and although both organizations will comply with regulatory requirements, the expectations pertaining to the nature and frequency of incidents need to be considered.

Norms overcome the problem of valid comparisons by providing a stratified benchmark so that datasets are comparable. In the case of organizations, this stratification might result in levels of performance that are assessed and are then distributed along one or more dimensions. The number of employees or the number of aircraft operated might constitute two of these dimensions. By matching the characteristics of an organization to the appropriate levels of these dimensions, the analysis becomes comparable.

The main difficulty associated with the establishment of norms is that they require multiple sources of data for each level of a dimension so that means can be calculated that reflect the breadth of a distribution. This process can take time and considerable resources to accomplish, ensuring that the data are collected using the same data collection tools and techniques. LOSA offers an ideal opportunity in this regard, as the process of data collection is standardized, and data are collected from multiple organizations across the world.

Calculating performance against a norm begins by taking into account the relevant dimensions, such as the size of the airline, and then identifying the mean industry performance for airlines of this size. The z score indicates, for a particular organizational value, the number of standard deviations it sits above or below the industry mean. It is calculated as

$$z = \frac{X - \overline{X}}{SD}$$

where:
X represents the organizational value
$\overline{X}$ represents the industry mean
SD represents the standard deviation

Using an organizational value of 11.5 incidents per year, where the mean number of incidents for organizations of a similar size is 13.6, with a standard deviation 2.3, the values can be substituted as

$$z = \frac{11.5 - 13.6}{2.3}$$

$$z = \frac{-2.1}{2.3}$$

$$z = -0.91$$

indicating that the outcome for the organization is almost one full standard deviation below the industry mean. This is a strong result, although it represents performance over a single year.

The question remains as to whether this result is sustained or even improved over time. It requires the collection and analysis of data on a systematic basis by both the organization and the industry more broadly.

An allied issue that needs to be addressed in relation to the collection and analysis of data concerns the rate at which these data are optimally acquired. Inevitably, the implementation of audit systems is costly and needs to be undertaken judiciously, but with sufficient frequency to enable comparative analyses. The value in these types of analyses is both in identifying specific instances and trends that, over time, might constitute a catalyst or precursor to system failure.

ORGANIZATIONAL TREND ANALYSES

Trend analyses are a key component in identifying and responding to conditions where there is a lag between an input and an output. For example, the effects of a restructure at senior levels within an organization may take sometime to translate into changes in organizational performance at the operational level. Therefore, establishing the effects requires the acquisition of data at a range of intervals following the introduction of the change.

Trend analyses are also evident in the context of aviation safety in which the frequency of accidents at a specific point in time is likely to be too few to draw any meaningful conclusions. It is only in the accumulation of data that a pattern becomes evident (O'Hare, Wiggins, Batt, & Morrison, 1994; Wiegmann & Shappell, 2001; Wiggins, Hunter, O'Hare, & Martinussen, 2012). In this case, trend-type historical data are used to predict future outcomes. This pattern might be as simple as a greater frequency of one type of event relative to other events. At a more complex level, it might be the case that a particular event occurs more frequently, but only at a particular point in time. To isolate the issue further, it may become clearer that an event only occurs at a particular time when it is preceded by another factor. Trend analysis has been used to quantify the contribution of the frequency and severity of maintenance errors to aircraft accidents and incidents. In the case of Marais and Robichaud (2012), maintenance-related aircraft accidents were 6.5 times more likely to be fatal than accidents in general and, when fatalities did occur, maintenance accidents resulted in 3.6 times more fatalities on average than aircraft accidents in general.

A relative newcomer to predicting the impact of aviation historical risk factors involves Bayesian Belief Networks (Brooker, 2011). The networks use aviation experts to estimate the probabilities of events. The probabilities are conditional; The chance of something happening given that something else has happened. Probabilities are then combined to model the probabilistic behavior of the system. Bayesian models are well established (for a review see Kruschke, 2015), but part of the challenge is that it can be difficult for aviation experts to estimate rare event conditional probabilities accurately (i.e., Bayesian priors; Kruschke, 2015). Nonetheless, there have been several demonstrations of how Bayesian models can possess a satisfactory range of accuracy in predicting aviation risk (Feng, Sahin, & Karson, 2009; Wang & Gao, 2013). Wang and Gao (2013) used Bayes to quantify the incremental risk to safety caused by flight delays in Chinese airspace, and Feng et al. used Bayes to evaluate the tradeoff between system risk and system cost with different baggage screening systems.

Although Bayes' estimates have been applied successful in a number of contexts, the utility of the approach tends to diminish with increases in complexity or where the specific influence of a variable is unknown. An example of the complexity of these relationships can be drawn from investigation into a series of BEA Comet crashes that occurred in 1954 (Simons, 2013). The Comet was the first commercial jet airliner in service, and much was riding on its success. When, inexplicably, two aircraft broke up over the Mediterranean Sea, it was critical that the cause, or causes, of the failure were established.

The problem for the investigators of the Comet crashes was a problem encountered in subsequent aircraft investigations; the lack of data that had been collected under controlled conditions. Unless specific precursors to the events could be recreated and various combinations compared, it would be unlikely that the specific causes could be identified. In the event, investigators and researchers at the Farnborough Research Laboratories spent many weeks subjecting the mock-up of a Comet Fuselage to different levels of simulated air pressure and importantly, the repeated increase and decrease in pressure that was to constitute the catalyst for fatigue cracks that emerged in and around the passenger windows (Simons, 2013).

The passenger windows in the Comet were a square shape, rather than the oval shape that was used in later, high altitude aircraft. This design, together with the method of manufacture, created the preexisting conditions for fatigue-related stressors. The stress imposed by constant depressurization and repressurization inevitably resulted in the fatigue-related fractures that, in turn, caused explosive decompression and the loss of the aircraft (Withey, 1997). However, it was only due to the methodical collection and analysis of data under a variety of controlled conditions that resulted in the mystery being solved.

The systematic approach to understanding the causes of the so-called Comet crashes is the same approach that researchers use to understand other complex phenomena. It relies on samples collected over time and/or under different conditions. This sampling approach accounts for differences in natural or stochastic variability that influences responses from time to time (Wiggins & Stevens, 1999).

The collection of data across a random sample is the next phase in establishing valid assessments of performance. This is the advantage of tools such as LOSA insofar as data are collected across a broad and random sample of operations and crews. The intention is to establish the level of performance in general, taking into account the performance of crews that are at the extremes (Ma et al., 2011).

Clearly, rare but severe consequences, such as the loss of an aircraft, demand an investment in sources of data that will enable the causes to be identified and addressed. However, even in this context, it is possible to use sample-based data to identify the causes of a failure and/or establish the prevalence of factors that will enable the identification of potential failures in other jurisdictions or amongst other carriers.

A useful example of the acquisition of these data can be drawn from the response of TransAsia Airways following the crash of an ATR-72 following an engine failure immediately after takeoff from Taipei International Airport. The airline undertook to evaluate the performance of its pilots following the accident (Ramzy, 2015). Although the airline probably should have been evaluating its pilots proactively, it does demonstrate the value of a sample-based approach to data acquisition following a major incident or accident.

Although the focus thus far has been directed toward the collection of safety-related data, airlines and aviation-related organizations have access to data well beyond safety and security. Airlines can collect and analyze data from a wide range of sources, including customer search and transactional (demographic data, online search behavior, preferences) data. As commercial concerns, the survival of airlines depends upon a reliable stream of revenue. One of the most important sources of revenue for commercial operations including airlines is daily sales. Sales data are available against daily or weekly costs (Simoudis, 1996). These data translate into the Load Factor that represents the average number of passengers per flight (Jenatabadi & Ismail, 2007). Normally expressed as a percentage or ratio, it allows an organization to quickly and reliably establish the cost effectiveness of a route or scheduled service.

A continuous data stream is a useful means by which changes in, and the performance of, an organization can be quickly and clearly established. The delay between the acquisition of the data and its receipt by analysts, including senior managers, is minimized. This, in turn, reduces the period between the acquisition of information and the implementation of a response.

The ability of organizations to quickly identify and respond to changes in the marketplace is a key precursor to sustained success (Hoogervorst, 2004). This agility is dependent upon the availability of accurate data, in the appropriate form, and at the appropriate time. Importantly, the appropriateness of data is possibly the most vexed in a technical environment where data pertaining to almost any variable are available on an almost continuous basis (Behn & Riley, 1999).

DATA IN DIFFERENT ORGANIZATIONAL CONTEXTS

One of the most significant shifts in air travel in recent years has been the transition from full-service to low-cost carriers. The latter has a lower cost-base, the savings from which can be passed to passengers in the form of lower ticket prices. With less reliance on infrastructure, these organizations are arguably much more agile in the marketplace and can quickly shift routes should there be changes in market demands. However, these carriers are also significantly dependent upon short turnaround times, so that there is little room for delays. Therefore, although the utilization of unproductive resources provides these carriers with a degree of flexibility, it also raises risks since low-cost carriers are especially dependent upon the availability of real-time data that would enable the anticipation of delays or system failures (Reddy & Reddy, 2002).

In the Boeing assembly plant in Seattle, every part that is used in the manufacture of an aircraft is electronically labeled so that, at any one time, it is possible to establish the rates at which particular parts are consumed, when specific parts are required, and the costs of the purchase of parts (Eriksson, 2011; Funk, 1995). Replacement parts are purchased in a just-in-time strategy that ensures their availability without the costs of retaining large quantities of stock. However, this approach is only as effective as are the processes and capabilities of the suppliers of that stock. In the case of Boeing, the materials used are constructed from very high-quality materials and are subjected to rigorous testing. Therefore, data pertaining to the consumption of materials needs to occur in sufficient time to enable manufacturers to undertake the necessary construction and testing, prior to delivery.

Although the consumption of components enables organizations such as Boeing and Airbus to enjoy efficiencies, it also provides a surrogate measure of the performance of different parts of the organization. In effect, the consumption of parts represents a barometer in reaching organizational performance targets which, in the case of companies such as Boeing, Airbus, and Embraer, is the construction of new aircraft. As there is an inevitable delay in the construction of aircraft, the consumption of parts might be employed as a means of identifying gaps and/or delays in the construction process.

Although incident reporting systems provide crucial information for many industries, understanding of the strengths and limitations of incident data is often poor. In the health sector, underreporting is recognized as a serious limitation (Ohrn, Elfstrom, Liedgren, & Rutberg, 2011; Stanhope, Crowley-Murphy, Vincent, O'Connor, & Taylor-Adams, 1999; Westbrook et al., 2015). The incidents reported often fail to represent either the frequency or the distribution of the types of errors that occur. For example, a study comparing medication errors identified during a detailed review of 3,291 patient records with incident reports showed that only 13 of 1,000 errors identified had been reported to the incident system (Westbrook et al., 2015). Evidence in patient records that staff had detected but not reported a serious error occurred at a rate of 219 of 1,000 errors. However, for 78% of the serious medication errors (with the potential to lead to significant patient harm) identified, there was no evidence that staff had in fact detected the errors. Thus, the reliability of the data within incident reporting systems is both heavily reliant upon the reporting culture, and on the vigilance and skills of the staff in error detection.

In some cases, the data from incident management systems can reveal more about an organization than simply the frequency or severity of errors. For example, a relatively high incident rate may actually reflect a positive safety culture and low risk compared to organizations with a relatively lower incident reporting rate. Evidence to support this proposition can be drawn from the medical context in which hospitals with higher incident report rates experience significantly fewer medication errors on audits than sites with lower incident reporting rates (Westbrook et al., 2015). A similar effect is evident in the aviation context where the frequency of incident reports is interpreted as a barometer of the safety health of an airline.

Despite the temptation to draw inferences on the basis of incident data, such analyses, between and across time, or organizations, must be undertaken with caution. The frequency of incident reports pertaining to an event may not necessarily be associated with increased risk, and responding to small numbers of incidents by type over time is likely to reinforce perceptions that small changes in the frequency of reports are significant. Such interpretations may lead decision-makers to assign resources to address incident types that are easy to report, rather than those that pose the greatest threat to safety (Levinson, 2012).

Small numbers of incidents and the absence of denominators exacerbate the risk of misinterpretation of incident data (Shojania, 2008). Many industries experience serious incidents as relatively rare events. A good example is the case of drug maladministration in nuclear medicine (Larcos, Collins, Georgiou, & Westbrook, 2014). Such errors can have catastrophic effects and, as a result, mandatory reporting registries are in place in many countries. However, the incidence of these errors is estimated at only 6 per 100,000 nuclear medicine procedures (Larcos et al., 2014). Therefore, at an organizational level, incidents occur infrequently, and changes in the frequency of these events are very difficult to interpret.

An exciting new development in the study of risk is the work of Didier Sornette and colleagues on what they term *Dragon Kings* (Sornette & Quillon, 2012). These events differ from *Black Swans* (Taleb, 2007). The latter are extreme events that occur infrequently but still conform to the statistical distribution and are caused by the underlying processes of the scenarios under examination. So, in the case of accidents, hazards, or disasters, a Black Swan even though improbable (rare), is still an expectable outcome of the processes that generates more typical events. To give a specific example, earthquakes occur commonly along plate faults. Most are relatively small (even unnoticeable), but a largish earthquake that inflicts damage, although rare, is still to be expected eventually. However, because Black Swans result from the same underlying processes of typical events, they are essentially random and thus, individually essentially unpredictable. We know in the long run, these rare events will occur, but their actual occurrence is essentially random. Dragon Kings, however, are different. They are events that are too large or too impactful to be considered as occurring simply as an uncommon outcome of typical events. Instead, they are the result of run-away processes that cause a new order in the system to emerge.

In the language of complex systems or nonlinear dynamics, Dragon Kings are emergent phenomena that are indicative of a change in regime or the underlying dynamics of the system. These events are the result of a self-organizing dynamical transition which itself is due to run-away processes occurring in the feedback loops and interconnectedness of the system. Unlike Black Swans they are not, therefore, random extreme outcomes of the typical processes. Instead, they are new, emergent features or the result of a phase transition.

The one upside, if this is true, is that, unlike Black Swans that are individually unpredictable (random), Dragon Kings may have telltale cues indicating their emergence and there are available statistical tests to differentiate Dragon Kings from Black Swans (Janczura & Weron, 2012; Sachs, Yoder, Turcotte, Rundle, & Malamud, 2012). In simplistic terms, to detect an oncoming Dragon King, one needs to detect a phase transition in a dynamical system. This statistical approach has been employed in other settings to determine if a phase transitions occurs in individual skill development (Helton, 2011). New ways of doing a task, a new system phase, may be preceded by the cues indicative of phase transitions, namely an increase in performance variability.

As any system transitions from one state to another, a phase transition, there tends to be an increase in overall system variability. Mathematically, this is characterized as a shift in the system to a new attractor or influential point in dynamical space. This may be indicated by techniques that measure an increase in variability in a time series. The new attractor of the system causes system states nontypical of the previous attractor that is still influencing the system until the transition is complete. Tools such as running standard deviations or sample entropy could be employed to detect the increase in system variability prior to the complete phase transition. To accomplish this, however, relevant data need to be collected for a sufficiently long time series to be useful. The key to plausible prediction and thus, intervention, is rich data collection, storage, and ongoing analysis. These techniques are being employed in a wide variety of fields, from natural disasters to medicine, and may be useful to the aviation industry.

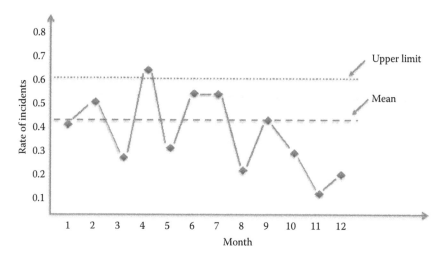

FIGURE 11.4   An example of a process control chart for the rate of incidents within an airline, distributed across a 12-month period and incorporating a designated upper limit that would trigger an intervention.

Statistical process control uses *control charts* to graphically monitor incidents to determine whether a process is in a state of statistical control and may be a useful approach to the monitoring of rare events (Benneyan, Lloyd, & Plsek, 2003; Larcos, Collins, Georgiou, & Westbrook, 2015; Montgomery, 2012). A control chart has a central line that represents the average or target value and an upper and lower line indicating the boundaries for the control limits. This approach recognizes that, for all processes or indicators, some variation is to be expected. The control limits are designed to detect when indicators move beyond natural, random variation and exceed the boundaries as a result of *special cause variation* (Laney, 2002). When a value falls outside the control limits, this signals that action may be required to bring processes back into control. Data patterns that may suggest special cause variation include, for example, one point outside three standard deviations of the mean or a run of at least six data points either increasing or decreasing. There are a range of different types of control charts that can be applied to data of different types. A U-chart is based on a Poisson distribution and presents counts as rates; a C-chart is also based on the Poisson distribution and presents the counts as numbers; and a P-chart is based on a binomial distribution and is used for percentages (Figure 11.4).

ACQUIRING AND USING DATA CAUTIOUSLY

As with all datasets, the information available can be used for a range of different purposes, depending on the nature of the organization and its goals. The difficulty occurs where data are used for a purpose for which they were not designed, or anticipated for, when collected. At the very least, in extrapolating a dataset across an organization, it is important to consider whether the data are representative of the broader population. If the data were sampled on single day, in a single geographic location, and/or amongst a particular group within the organization, it is possible that different responses might have been acquired, had the process of data collection occurred beyond a single day, and across the different geographic locations and assembly lines that might characterize the organization.

To extrapolate from a sample to a population is reasonable, so long as the sample is representative of the broader population to which the outcomes will be applied (Wiggins & Stevens, 1999). This requires some consideration of what are referred to as *extraneous variables*. These are factors that may systematically influence the responses that are elicited. As a case in point, evaluating people on a test of cognitive ability will yield different scores, depending upon the time of day at which the test was administered (Revelle, Humphreys, Simon, & Gilliland, 1980). For the majority of people, testing late in the afternoon after a full day of work will result in mean scores that are lower than the same group tested earlier in the day.

Similar to fatigue, other extraneous variables can impact performance, including age, geographic location, experience, and the type of work undertaken. Therefore, in acquiring data from a representative sample across multiple parts of an organization, it is important to both ensure that the sampling is stratified, and that the impact of the stratification is evaluated prior to conclusions being drawn. Stratification is a process of sampling that involves first, the identification of those extraneous variables that are likely to differentially impact the outcomes. Thereafter, researchers deliberately source data that are representative of these different variables. Finally, comparisons are undertaken to ensure that the proportion of respondents or participants reflects the proportion of participants that comprise the organization. To put it simply, if women comprise 40% of the population of an organization and sex represents an extraneous variable, then females should comprise 40% of the sample.

Where it is not possible to stratify the process of data acquisition, perhaps because the demographic characteristics are unclear or it would undermine the process of data collection, it remains possible to assess the influence of various factors on responses and thereby determine the role of the extraneous variable in contributing to the overall outcome. For example, in assessing the performance of pilots' use of engine-out operations following takeoff, poor performance amongst a sample of pilots might be explained by a lack of recent training experience. Although it may not be possible to stratify the data collection process in this case, by collecting data about pilots' recent training experience, it becomes possible to account for these differences in performance statistically using methods such as Hierarchical Multiple Regression or Analysis of Covariance. These data analytic techniques enable a determination of the level of variability in the outcome variable (e.g., pilot engine-out performance) that is uniquely predicted by certain operator or environmental variables while controlling for other variables (e.g., training).

THE VALUE OF QUALITATIVE DATA

It is often difficult, prior to the collection of quantitative data, to identify explanatory variables and acquire data in a form that will enable the appropriate application of the aforementioned statistical control. Therefore, in addition to quantitative data, it is often useful to collect qualitative data that might explain some of the effects that are identified (Tariq, Georgiou, & Westbrook, 2013; Westbrook, Gosling, & Coiera, 2004). As an example, responses to a questionnaire concerning perceptions of safety within an airline might be influenced by both recent changes in procedures or by changes in management, and this possibility may only become evident following the completion of data collection. Therefore, there needs to be some *a priori* understanding of the context within which data are being collected.

In addition to the features of samples and the need to consider the impact of extraneous variables on the process of data acquisition, it is important to consider the nature of the data being acquired. For example, the acquisition of subjective data, such as attitudes or preferences, can mask behavior that might occur in practice. This is often the case where questions relate to a moral or ethical dilemma (Krumpal, 2013). How people respond to questions regarding safety may belie the fact that, under certain circumstances, safety behaviors do not necessarily reflect the attitudes that are espoused in response to questionnaires. Therefore, the ideal approach is one that draws together subjective responses with objective data. For example, an audit of a maintenance facility might highlight an increased incidence of defects that are undetected before an aircraft is returned to operations. The results of a questionnaire distributed subsequently may provide an explanation insofar as the results reveal a perception of under-resourcing and a lack of support from management.

Although cross-referencing data do not necessarily provide a complete explanation for an effect, at the very least, other forms of data can be used to either confirm or disconfirm an outcome. This is referred to as *triangulation*, and it can be used where there is a limited sample or an outcome is of such importance (such as safety-related issues) that it requires further investigation (O'Connor, Buttrey, O'Dea, & Kennedy, 2011; Tariq, Georgiou, & Westbrook, 2013).

Although there was, for many years in the aviation industry, a propensity toward quantitative data, there has been a shift in recent years toward the collection of more qualitative data both as a means of exploring and explaining the quantitative data and in terms of identifying opportunities for new and potentially disruptive innovations (Wilke, Majumdar, & Ochieng, 2014). Due to the largely subjective nature of both the collection of qualitative data and its interpretation, drawing conclusions exclusively on the basis of qualitative information is fraught with danger, particularly if the data were drawn from a limited sample. However, there are some techniques that can be employed to assist with the interpretation of data and ensure its reliability and validity. The first of these strategies involves a clearly defined process of qualitative data categorization so that different analysts will reach the same or similar conclusions (Roth, 2015). This is important as it provides a degree of confidence to the effect that the interpretations have some basis in reality.

Categorizations can be determined prior to the collection of data or as part of an expectation based on anecdotal observations, previous research, and/or a particular theoretical perspective (Cassell & Symon, 2011). The analysis of air traffic controller self-reported errors conducted by Shorrock (2005, 2007) is a case in point. Shorrock interviewed UK air traffic controllers using an unstructured interview format that allowed controllers to free-recall and explore personal accounts of the kinds of errors they have made in the past, and any factors that have influenced their performance. Shorrock then used an error classification and analysis system called the technique for the retrospective and predictive analysis of cognitive error (TRACEr—Shorrock & Kirwan, 2002) to classify these data into various forms of human memory and perception errors.

However, whether categorizations of qualitative data, such as those documented by Shorrock (2005, 2007), can be proposed depends on the stage of the research at which data are being acquired. For example, in the case of levels of safety climate within a newly established airline, there may be few expectations as to the level of safety climate at a given point in time. Therefore, the process of data acquisition and interpretation is, inevitably, going to be driven by the nature of the data collected. However, as discussed previously, in the absence of a benchmark or expectation, the data do little to inform the organization as to whether or not the outcomes are satisfactory.

At an operational level, establishing a benchmark or expectation represents an anchor point around which future datasets can be compared. In fact, this is precisely the approach that is undertaken in the context of financial reporting, whereby annual profit or income is compared against performance during the preceding year. It is this capacity for a comparative assessment that enables prospective investors to establish the long-term performance of the organization. Similarly, in the context of safety, comparative data are necessary to draw meaning from numerical data (Oster, Strong, & Zorn, 2013).

Although the assessment of data against expectations or a hypothesis offers advantages, it is also possible that data will be interpreted in the context of these expectations and without due regard to the statistical properties of the dataset. For example, drawing inferences based on measures of central tendency, such as the mean (average), can be misleading, particularly if the data are non-normal. The distribution of these collected variables will not match the normal curve perfectly, but for many statistical analyses, referred to as *parametric analyses*, the distribution of the data needs to approximate a normal distribution. Where the data are non-normal, a parametric statistical technique is no longer representative of the dataset and a nonparametric technique needs to be employed or any outliers may need to be transformed or removed from the dataset.

THE IMPORTANCE OF OUTLIERS

By definition, an outlier is a data point that falls three or more standard deviations from the mean. They arise where there are systematic drivers influencing performance that do not impact the performance of other members of the dataset. For example, if pilots' performance during a standard arrival procedure is being assessed, and a small proportion of the sample complete the assessment having just flown the approach, their performance may be systematically greater than pilots who have yet to fly the approach in practice.

In effect, outliers constitute a population distinct from the remainder of the sample. A systematic variation in demographic characteristics or the process of data collection has enabled the acquisition of data from a population distinct from the population under investigation. In some circumstances, it may be appropriate to simply remove extreme outliers from the sample, or to transform the dataset using a mathematical procedure (e.g., a square root transformation), that results in a sample distribution closer to a normal curve. However, it is important to recognize that outliers, although perhaps unrepresentative of a target population, could constitute a window into other populations that may be of interest.

In the context of aviation safety, employees performing particularly poorly or responses that are in marked contrast to the responses of other parts of an organization might be regarded as outliers given the responses from the broader sample. However, they also constitute an area of concern insofar as identifying risks to the organization. In the case of safety climate, outliers might reflect issues that need to be addressed, such as ineffective leadership, the lack of resources necessary to undertake tasks successfully, and/or a lack of appropriate training. In any of these cases, the outlying responses will have revealed an opportunity for intervention.

Outliers may also reveal something about the process of data acquisition and particularly whether the process of data acquisition was consistent across the sample. For example, differences in responses to a survey across different parts of an organization might simply reflect differences in the way in which a survey was distributed. In one case, the survey may have been distributed by a line manager, thereby encouraging a degree of compliance both in completing the survey and providing responses that might be regarded as socially or organizationally acceptable. In another case, the survey may have been distributed by the safety department, and this strategy might result in altogether different responses given that anonymity is assured.

As the identification of dependencies is the goal of most investigations, it is important to establish the extent to which the contribution to a global outcome is influenced by the contributions of different subgroups that comprise a broader cohort. By taking account of behavior at different levels within an organization, it becomes possible to establish the relative contribution of variables. For example, if the population of an airline comprises a series of subgroups, it might be determined that a particular combination of subgroups impacts differently, the overall responses of the cohort. This source of variance is important to establish as it provides a level of explanation that can be useful in determining where, for example, to invest training or other resources.

Referred to as multilevel modeling, or sometimes known as *random coefficient* or *mixed effect* models, this analytical process has been used widely in the biological and social sciences to analyze nested data structures (Hofmann, Griffin, & Gavin, 2000), and in the aviation context, it can be used to identify changes in performance over time whereby an intervention may affect differently a particular group with a cohort. The opportunity in this case is to establish whether interventions have been successful and, importantly, for whom within a cohort they were most successful or unsuccessful. This informs decisions, including how and to whom interventions might be targeted, thereby enabling the efficient and effective utilization of resources.

An example of the practical application of multilevel modeling in aviation can be drawn from an investigation of the workload management of Australian en route controllers (Neal et al., 2014). Managers of air traffic control systems proactively manage controller workload by redesigning airspace and procedures and adjusting rosters to ensure that anticipated task demands can be met. This requires the prediction of controller workload that, in effect, is a prototypical example of a multilevel problem as variability exists across different airspace sectors and between the capacity of different controllers (Loft, Sanderson, Neal, & Mooij, 2007).

Within sectors, air traffic controllers vary in their response to task demands as a consequence of individual differences. There is also a degree of variability over time, as controllers encounter nonroutine events that fall outside the bounds expected under normal conditions. Finally, there are likely to be differences in the way that supervisors and fellow controllers make judgments of workload when observing the performance of their colleagues. Neal et al. (2014) developed a multilevel statistical model capable of predicting the range within which controller's subjective estimates of workload would be likely to fall with a certain probability (prediction interval) given the specific anticipated operational conditions (sector type, daily flight plans, personnel, available automation). Air traffic control management can use this predictive model to analyze historical and projected workflows to dynamically allocate resources to meet changes in demands.

One of the significant criticisms of evidence-based interventions is that the data on which the intervention is based were collected in single instance so that it becomes unclear whether the responses represent systemic issues or whether they are simply a reflection of the state of affairs at a particular point in time. This is important as it may misdirect both the need for an intervention and the scope of any intervention that may be applied.

The solution to the problem of one-off or cross-sectional forms of data collection lies in the acquisition of data over extended or multiple periods. This enables the assessment of intervening variables and ensures that the data collected in one instance are a robust and reliable reflection of the domain.

The assessment of intervening variables can be difficult to establish and requires the application of specialist statistical techniques. In essence, the goal is to determine whether the relationships between two variables are better explained by considering a third or more variables.

REFERENCES

Behn, B. K., & Riley, R. A. (1999). Using nonfinancial information to predict financial performance: The case of the US airline industry. *Journal of Accounting, Auditing & Finance, 14*(1), 29–56.

Benneyan, J. C., Llyod, R. C., & Plsek, P. E. (2003). Statistical control process as a tool for research and healthcare improvement. *Quality & Safety in Health Care, 12*, 458–464.

Brooker, P. (2011). Experts, Bayesian belief networks, rare events, and aviation risk estimates. *Safety Science, 49*, 1142–1155.

Cassell, C., & Symon, G. (2011). Assessing "good" qualitative research in the work psychology field: A narrative analysis. *Journal of Occupational and Organizational Psychology, 84*(4), 633–650.

Edkins, G. D. (1998). The INDICATE safety program: Evaluation of a method to proactively improve airline safety performance. *Safety Science, 30*(3), 275–295.

Eriksson, S. (2011). Globalisation and changes of aircraft manufacturing production/supply-chains—the case of China. *International Journal of Logistics Economics and Globalisation, 3*(1), 70–83.

Feng, Q. M., Shain, H., & Karson, M. J. (2009). Bayesian analysis models for aviation baggage screening. *IIE Transactions, 41*, 995–1006.

Funk, J. L. (1995). Just-in-time manufacturing and logistical complexity: A contingency model. *International Journal of Operations & Production Management, 15*(5), 60–71.

Goh, Y. M., Love, P. E., Brown, H., & Spickett, J. (2012). Organizational accidents: A systemic model of production versus protection. *Journal of Management Studies, 49*(1), 52–76.

Goodheart, B. J., & Smith, M. O. (2014). Measurable outcomes of safety culture in aviation—a meta-analytic review. *International Journal of Aviation, Aeronautics, and Aerospace, 1*(4), 1.

Helmreich, R. L. (2000). On error management: Lessons from aviation. *BMJ: British Medical Journal, 320*(7237), 781.

Helton, W. S. (2011). Animal expertise: Evidence of phase transitions by utilizing running estimates of performance variability. *Ecological Psychology, 23*(2), 59–75.

Hofmann, D. A., Griffin, M. A., & Gavin, M. (2000). The application of hierarchical linear modeling to management research. In K. Klein & S. W. Kozlowski (Eds.), *Multilevel theory, research, and methods in organizations* (pp. 467–511). San Francisco, CA: Jossey-Bass.

Hoogervorst, J. (2004). Enterprise architecture: Enabling integration, agility and change. *International Journal of Cooperative Information Systems, 13*(3), 213–233.

ICAO. 2013. Global Air Transport Outlook to 2030 and Trends to 2040. No. Cir 333, AT/190. Montreal, Canada.

Janczura, J., & Weron, R. (2012). Black swans or dragon-kings? A simple test for deviations from the power law. *The European Physical Journal Special Topics, 205*(1), 79–93.

Jenatabadi, H. S., & Ismail, N. A. (2007). The determination of load factors in the airline industry. *International Review of Business Research Papers, 3*(4), 125–133.

Jones, S., Kirchsteiger, C., & Bjerke, W. (1999). The importance of near miss reporting to further improve safety performance. *Journal of Loss Prevention in the Process Industries, 12*(1), 59–67.

Krumpal, I. (2013). Determinants of social desirability bias in sensitive surveys: A literature review. *Quality & Quantity, 47*(4), 2025–2047.

Kruschke, J. K. (2015). *Doing Bayesian data analysis: A tutorial with R, BUGS and Stan.* Amsterdam, the Netherlands: Elsevier.

Laney, D. B. (2002). Improved control charts for attributes. *Quality Engineering, 14,* 531–537.

Larcos, G., Collins, L. T., Georgiou, A., & Westbrook, J. I. (2014). Maladministrations in nuclear medicine: Revelations from the Australian radiation incident register. *Medical Journal of Australia, 200,* 37–40.

Larcos, G., Collins, L. T., Georgiou, A., & Westbrook, J. I. (2015). Nuclear medicine incident reporting in Australia: Control charts and notification rates inform quality improvement. *Internal Medicine Journal, 45*(6), 609–617.

Lenne, M. G., Salmon, P. M., Liu, C. C., & Trotter, M. (2012). A systems approach to accident causation in mining: An application of the HFACS method. *Accident Analysis & Prevention, 48,* 111–117.

Levinson, D. (2012). *Hospital incident reporting systems do not capture most patient harm.* Office of Inspector General, Department of Health and Human Services.

Lindberg, A. K., Hansson, S. O., & Rollenhagen, C. (2010). Learning from accidents–what more do we need to know?. *Safety Science, 48*(6), 714–721.

Loft, S., Sanderson, P., Neal, A., & Mooij, M. (2007). Modeling and predicting mental workload in en route air traffic control: Critical review and broader implications. *Human Factors, 49,* 376–399.

Ma, J., Pedigo, M., Blackwell, L., Gildea, K., Holcomb, K., Hackworth, C., & Hiles, J. J. (2011). *The line operations safety audit program: Transitioning from flight operations to maintenance and ramp operations* (No. DOT/FAA/AM-11-15). Washington, DC: Federal Aviation Administration.

Madsen, P. M. (2013). Perils and profits a reexamination of the link between profitability and safety in US Aviation. *Journal of Management, 39*(3), 763–791.

Marais, K. B., & Robichaud, M. R (2012). Analysis of trends in aviation maintenance risk: An empirical approach. *Reliability Engineering and System Safety, 106,* 104–118.

Montgomery, D. (2012). *Introduction to statistical quality control.* New York: Wiley.

Neal, A., Hannah, S., Sanderson, S., Bolland, S., Mooij, M, & Murphy, S. (2014). Development and validation of a multilevel model for predicting workload under routine and nonroutine conditions in an air traffic management center. *Human Factors, 56,* 287–305.

O'Connor, P., Buttrey, S. E., O'Dea, A., & Kennedy, Q. (2011). Identifying and addressing the limitations of safety climate surveys. *Journal of safety research, 42*(4), 259–265.

O'Hare, D., Wiggins, M., Batt, R., & Morrison, D. (1994). Cognitive failure analysis for aircraft accident investigation. *Ergonomics, 37*(11), 1855–1869.

Ohrn, A., Elfstrom, J., Liedgren, C., & Rutberg, H. (2011). Reporting of sentinel events in Swedish hospitals: A comparison of severe adverse events reported by patients and providers. *Joint Commission Journal on Quality & Patient Safety, 37*(11), 495–501.

Oster, C. V., Strong, J. S., & Zorn, C. K. (2013). Analyzing aviation safety: Problems, challenges, opportunities. *Research in Transportation Economics, 43*(1), 148–164.

Ramzy, A. (February 12, 2015). Many pilots fail safety test at TransAsia. *New York Times,* p. A8.

Reddy, S. B., & Reddy, R. (2002). Competitive agility and the challenge of legacy information systems. *Industrial Management & Data Systems, 102*(1), 5–16.

Revelle, W., Humphreys, M. S., Simon, L., & Gilliland, K. (1980). The interactive effect of personality, time of day, and caffeine: A test of the arousal model. *Journal of Experimental Psychology: General, 109,* 1–31.

Roth, W. M. (2015). Cultural practices and cognition in debriefing the case of aviation. *Journal of Cognitive Engineering and Decision Making, 9,* 263–278.

Sachs, M. K., Yoder, M. R., Turcotte, D. L., Rundle, J. B., & Malamud, B. D. (2012). Black swans, power laws, and dragon-kings: Earthquakes, volcanic eruptions, landslides, wildfires, floods, and SOC models. *The European Physical Journal-Special Topics, 205*(1), 167–182.

Shojania, K. (2008). The frustrating case of incident-reporting systems. *Quality & Safety in Health Care, 17*(6), 400–402.

Shorrock, S. T. (2005). Errors of memory in air traffic control. *Safety Science, 43,* 571–588.

Shorrock, S. T. (2007). Errors of perception in air traffic control. *Safety Science, 45*, 890–904.

Shorrock, S. T., & Kirwan, B. (2002). Development and application of a human error identification tool for air traffic control. *Applied Ergonomics, 33*, 319–336.

Simons, G. M. (2013). *Comet: The world's first jet airliner.* Barnsley, UK: Pen and Sword.

Simoudis, E. (1996). Reality check for data mining. *IEEE Intelligent Systems, 11*(5), 26–33.

Sornette, D., & Ouillon, G. (2012). Dragon-kings: Mechanisms, statistical methods and empirical evidence. *The European Physical Journal Special Topics, 205*(1), 1–26.

Stanhope, N., Crowley-Murphy, M., Vincent, C., O'Connor, A. M., & Taylor-Adams, S. E. (1999). An evaluation of adverse incident reporting. *Journal of Evaluation in Clinical Practice, 5*(1), 5–12.

Sull, D. (1999). Easyjet's $500 million gamble. *European Management Journal, 17*(1), 20–32.

Taleb, N. N., (2007). *The black swan: The impact of the highly improbable.* London, UK: Penguin.

Tariq, A., Georgiou, A., & Westbrook, J. (2013). Medication errors in residential aged care facilities: A distributed cognition analysis of the information exchange process. *International Journal of Medical Informatics, 82*(5), 299–312.

Thomas, M. J., & Petrilli, R. M. (2006). Crew familiarity: Operational experience, non-technical performance, and error management. *Aviation, Space, and Environmental Medicine, 77*(1), 41–45.

Travaglia, J. F., Nugus, P. I., Greenfield, D., Westbrook, J. I., & Braithwaite, J. (2011). Visualising differences in professionals' perspectives on quality and safety. *BMJ Quality & Safety, 21*, 778–783.

Tretheway, M. W., & Markhvida, K. (2014). The aviation value chain: Economic returns and policy issues. *Journal of Air Transport Management, 41*, 3–16.

Walker, G., & Strathie, A. (2016). Big data and ergonomics methods: A new paradigm for tackling strategic transport safety risks. *Applied Ergonomics, 53*, 298–311.

Wang, H., & Gao, J. (2013). Bayesian network assessment method for civil aviation safety based on flight delays. *Mathematical Problems in Engineering, 2013*, 1–12.

Westbrook, J. I., Gosling, A. S., & Coiera, E. (2004). Do clinicians use online evidence to support patient care? A study of 55,000 clinicians. *Journal of the American Medical Informatics Association, 11*(2), 113–120.

Westbrook, J. I., Li, L., Lehnbom, E., Braithwaite, B. M. J., Burke, R., Conn, C., & Day, R. (2015). What are incident reports telling us? A comparative study at two Australian hospitals of medication errors identified at audit, detected by staff and reported to an incident system. *International Journal of Quality in Health Care, 27*(1), 1–9.

Wiegmann, D. A., & Shappell, S. A. (2001). Human error analysis of commercial aviation accidents: Application of the human factors analysis and classification system (HFACS). *Aviation, Space, and Environmental Medicine, 72*(11), 1006–1016.

Wiggins, M. W., Hunter, D. R., O'Hare, D., & Martinussen, M. (2012). Characteristics of pilots who report deliberate versus inadvertent visual flight into instrument meteorological conditions. *Safety Science, 50*(3), 472–477.

Wiggins, M. W., & Stevens, C. (1999). *Aviation social science: Research methods in practice.* Aldershot, UK: Ashgate Publishing.

Wilke, S., Majumdar, A., & Ochieng, W. Y. (2014). A framework for assessing the quality of aviation safety databases. *Safety Science, 63*, 133–145.

Withey, P. A. (1997). Fatigue failure of the de Havilland comet I. *Engineering Failure Analysis, 4*(2), 147–154.

12 Flight Deck Human Factors

Michelle Yeh, Thomas R. Chidester, and Thomas E. Nesthus

CONTENTS

Introduction...259
Technical Engineering System..261
   Design Philosophy 261
   Automation..262
   Display Location 264
   Alerts..265
   Symbols..268
   Color...271
   Control Devices...272
Personnel Subsystem 276
   Workload 276
   Human Error..278
   Fatigue...280
Organizational/Management Infrastructure 283
   Multicrew Operations...284
   Workforce Interfaces...286
   Key Inputs to Training..286
   Operational Feedback...288
   Responding to Discovered Hazards 289
   Controversies...289
   Summary 291
Environmental Context 291
   Flight, Duty, and Rest Regulations..291
   Accident Prompting Rulemaking Change..292
   Fatigue Risk-Management System 292
   ICAO and Fatigue Management...293
Summary..294
References..294

INTRODUCTION

Human factors research into new and current flight deck technologies and operations enables a data-driven approach to identifying and resolving potential safety issues. Human factors is generally estimated to contribute to approximately 70% of commercial accidents and incidents (e.g., National Research Council, 1998; Shappell, 2006; U.S. Congress, 1988). Human factors is not just one thing but rather a combination of design, training, decision-making, flightcrew interaction, and automation, to name a few.

Advances in technology are intended to improve the safety of operations, but while it may solve some problems, it introduces others. Implementation of new avionics functions or systems must be considered with respect to the flightcrew member to avoid potential vulnerabilities. The purpose of this chapter is to identify and discuss some key human factors considerations in flight deck design and evaluation. This chapter does not intend to provide a comprehensive list of human factors considerations; rather, we focus on those topics that are frequently seen or reported in the approval of flight deck systems or procedures.

The current chapter is organized using a sociotechnical model of the flight deck environment. The sociotechnical model was first proposed in Moray and Huey (1998) and applied to the nuclear regulatory and rail domains (Moray, 2006). We apply the model here to capture flightcrew interaction through a systems perspective. The sociotechnical model consists of four subsystems, shown in Figure 12.1: the technical/engineering system, the personnel subsystem, the organizational/ management infrastructure, and the environmental context. The four elements in the sociotechnical framework interact such that a change in one layer will exert influence on the other layers. For each layer, we have identified a corresponding aspect of flight deck human factors. By considering each layer and the interaction among these layers, flight deck human factors can be viewed as an integrated systems approach intended to ensure safe operations.

At the center of the model is the *technical/engineering system*. This layer defines the physical system—the flight deck, and the boundary between this layer and the next, the personnel subsystem, represents the human–machine interface. As defined here, this layer consists of the flight deck avionics systems that are used by the pilot/flightcrew and the design and evaluation of those systems. Much of this chapter is devoted to human factors aspects related to the technical/engineering system. In the model set forth here, this layer includes aspects of user interface design, for example, a *design philosophy* to guide system development and in particular, the introduction and use of automation on the flight deck. This layer also addresses the *location* of the avionics display. Finally, we address some recurring flight deck human factors issues in the design and use of *alerts, symbols, color,* and *labels and controls.*

Factors that influence how the pilot processes the information are represented by the second layer, the *personnel subsystem.* Pilot behavior on the flight deck is influenced by factors other than the design of displays and controls. This layer addresses factors that affect the cognitive ability of the pilot to respond appropriately, such as *fatigue* and *workload.* We also consider the role of *human error* and how to improve error prevention, detection, and recovery.

Flightcrew behavior will also be influenced by the *organizational/management infrastructure,* which define pilot performance standards. This layer addresses teamwork and group behavior, for example, pilot's interaction with other flightcrew members as well as interaction with dispatchers, mechanics, flight attendants, and airport personnel. This chapter focuses specifically on pilot training and the different approaches and perspectives for training.

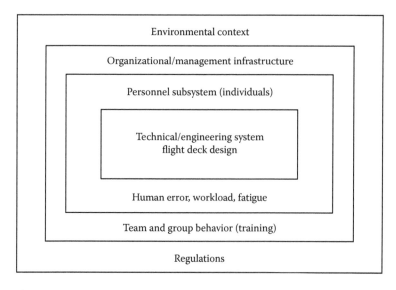

FIGURE 12.1 Sociotechnical model of the flight deck environment.

All these layers function within a political, cultural, and social environment, as highlighted by the outermost layer of the model, the *environmental context*. Regulatory oversight and public support or opposition to specific policies and actions also influences safety. Regulatory influences are addressed in more detail in Chapter 9, but we provide one example of how the layers interact with respect to pilot fatigue to influence flight, duty, and rest regulations.

Note that the sociotechnical framework applied here is only one way to describe flightcrew behavior and interactions. In addition, the sociotechnical system can be described in various ways and at many levels. Several models have been proposed to describe human capabilities and limitations, for example, SHEL (Hawkins, 1987) and the "Swiss Cheese model" (Reason, 1990), and can serve as valuable frameworks in other contexts.

TECHNICAL ENGINEERING SYSTEM

This *technical/engineering system* addresses considerations in the design of the avionics interface and the interaction between the flightcrew and the avionics. This section will begin with a discussion of the importance of a *design philosophy*. The technical/engineering system also includes technology considerations with focus here on lessons learned in the implementation and use of *automation*. We also consider how the *display location* influences the effectiveness of and attention given to an avionics display. Finally, we identify some recurring human factors issues related to flight deck design.

DESIGN PHILOSOPHY

The process for designing a flight deck is complex and varies from one manufacturer to another. Over the years, the design of flight decks has shifted from the simple replacement of units (with newer displays and controls) to redesign with consideration of human perception and performance. A design philosophy outlines the assumptions for the flight deck and defines guiding principles that can improve consistency and compatibility across flight deck systems. A flight deck design philosophy describes how information should be presented, how color is used and applied on the displays, how information is managed, how the pilot/flightcrew will interact with the displays, and how failures are accommodated or mitigated. A flight deck design philosophy may also describe assumptions about roles and responsibilities, for example, the allocation of functions between the flightcrew and the automation.

Establishing a design philosophy for the flight deck provides an overarching "style guide", which can change the focus from technological implementation only to considerations of user interface and flightcrew interaction. This is especially important when systems made by different avionics vendors with potentially different underlying design philosophies are integrated onto one flight deck. Federal Aviation Administration (FAA) Advisory Circular (AC) 25-11B, *Electronic Flight Displays*, recommends that the design philosophy addresses five aspects of the user interface (FAA, 2014):

1. Information presentation, for example, "quiet dark" flight deck.
2. Color of electronic displays: What is the meaning and intended interpretation of the different colors?
3. Information management, for example, when should the pilot take action to retrieve information, should the information be brought up automatically?
4. Interactivity, for example, when is feedback provided?
5. Redundancy management: How are failures accommodated (e.g., single and multiple display failures, power supply failures, etc.)?

Each airplane manufacturer specifies a design philosophy, and the design philosophy may differ from one to another. Airbus and Boeing are known for having different design philosophies, particularly when it comes to the implementation of automation, although both philosophies recognize that the pilot is ultimately responsible for safe flight.

Airbus has described its automation philosophy as follows (FAA, 2005a):

- Automation must not reduce overall aircraft reliability; it should enhance aircraft and systems safety, efficiency, and economy.
- Automation must not lead the aircraft out of the safe flight envelope, and it should maintain the aircraft within the normal flight envelope.
- Automation should allow the operator to use the safe flight envelope to its full extent, should this be necessary due to extraordinary circumstances.
- Within the normal flight envelope, the automation must not work against operator inputs, except when absolutely necessary for safety.

Boeing's automation philosophy addresses the following (FAA, 2005a):

- The pilot is the final authority for operation of the plane.
- Both crewmembers are ultimately responsible for the safe conduct of the flight.
- Flight crew tasks, in order of priority, are safety, passenger comfort, and efficiency.
- Design for crew operations based on pilot's past training and operational experience.
- Design systems to be error tolerant.
- The hierarchy of design alternatives is simplicity, redundancy, and automation.
- Apply automation as a tool to aid, not replace the pilot.
- Address fundamental human strengths, limitations, and individual differences—for both normal and non-normal operations.
- Use new technologies and functional capabilities only when they result in clear and distinct operational or efficiency advantages and there is no adverse effect to the human–machine interface.

Based on their automation design philosophy, Airbus' approach has tended toward greater flight management system (FMS) authority and enforcement of "hard" limits, in which the aircraft does not exceed the flight envelope regardless of the pilot's control inputs. On the other hand, Boeing's automation philosophy has been to implement automation as a support tool rather than replacement for the flightcrew. Other manufacturers may have different philosophies from Boeing and Airbus. Regardless, the design philosophy should be documented, so the pilot can understand the design philosophy for the aircraft being flown and the level of authority being granted.

AUTOMATION

The job of the pilot has changed as automation in the flight deck has increased. The term *automation* is used generally to describe different types of automated systems, including control automation, information automation, and management automation. *Control automation* is automation that is intended to aid or replace a pilot in controlling the aircraft. *Information automation* describes the integration, filtering, and transformation of data and includes avionics displays for communication, navigation, and surveillance. Finally, *management automation* refers to automation that helps pilot manage the mission strategically versus tactically (Billings, 1997). In this chapter, we use the term *automation* generally when it applies to all three types of automation.

Control, information, and management automation provide alternatives for accomplishing tasks, but at the same time, these forms of automation changes the nature of the tasks to be performed and methods of operation. Automation is often introduced to reduce workload, but it may actually have the opposite effect as it simply shifts flightcrew tasks from physical actions to cognitive ones by creating new tasks related to controlling or monitoring automated functions. That is, the pilot takes on a supervisory role with respect to the automation and is expected to manage the systems' operation (FAA, 2005a; Wickens, Hollands, Banbury, & Parasuraman, 2013; Wood, 2004).

As noted in the previous section, the pilot should understand the rules and design philosophy governing the automation's behavior for the aircraft being flown. However, the internal processes may only be partially revealed to the pilot. Previous research has identified vulnerabilities related to how flightcrews manage the automation, specifically whether pilots understand and can anticipate the automation's behavior as well as when it is appropriate to use the automation and the appropriate level of automation in unusual/non-normal situations (FAA, 2013b).

Vulnerabilities may result from a miscalibration of trust in the automation. In some cases, the pilot may think the automation is more reliable than it is and *overtrust* the automation. Overtrust tends to lead to *overreliance*, which can cause the pilot to accept the advice of the automation without confirmation. Overreliance on flight deck technology and failure to understand the changes in the levels of automation is one of the biggest potential threats to airliner safety. The PARC (Performance-based operations Aviation Rulemaking Committee)/CAST (Commercial Aviation Safety Team) Flight Deck Automation Working Group conducted an analysis to identify potential vulnerabilities in the operational use of flight deck systems for flight-path management (FAA, 2013b). As part of that analysis, the Working Group reviewed 734 relevant incident reports in the aviation safety reporting system (ASRS) incident database from 2002 to 2007, 46 accidents and major incidents, as well as Line Operations Safety Audit (LOSA) aggregated data for over 9,000 flights from 2001 to 2009. The Working Group concluded that pilots overrelied on automation and delegated authority to those systems, sometimes deviating from the desired flight path. In approximately 25% of the accidents, the Working Group analyzed, participants were overconfident in the automation and were sometimes hesitant to intervene. Furthermore, in 50% of the accidents identified by the Working Group, pilots were out of the control loop and were not prepared to assume control of the aircraft. The PARC/CAST Working Group noted that the perceived high system reliability resulted in insufficient cross-check of the data, a failure to recognize when the autopilot or autothrottle had disengaged, or a failure to maintain speed, heading, or altitude.

On the other hand, the pilot may think that the system is not reliable and *undertrust* it, failing to use the system when it is appropriate to do so. Undertrust can lead to underreliance on automation such that the pilot turns off the automation or ignores it when it may help. This may happen if the automation produces too many false alarms and becomes a distraction (Billings, 1997; FAA, 2010; Parasuraman & Riley, 1997). This was particularly a problem for early traffic collision and avoidance systems (TCAS) and ground proximity warning systems (GPWS).

Another potential vulnerability with respect to flightcrew awareness of automation behavior has been ensuring that the pilot understand what mode the automated system is in (FAA, 2013b). A loss of mode awareness can lead to a mode error, that is, an incorrect action for the current system state that would have been appropriate had the automation been in the assumed state. The PARC/CAST Flight Deck Automation Working Group noted autoflight mode selection, awareness, and understanding as a continued vulnerability, and that pilots' lack of knowledge about the automation led them to misuse automation in certain circumstances (FAA, 2013b). For example, the flightcrew of Pinnacle Airlines Flight 3701 that crashed on October 14, 2004, outside Jefferson City, Missouri, inappropriately used the vertical speed mode under autopilot control while climbing from 37,000 to 41,000 feet rather than the autopilot airspeed mode that could have prevented the loss of airspeed. Consequently, the aircraft was in a low-energy state when it reached 41,000 feet. The National Transportation Safety Board (NTSB) concluded that the use of vertical speed during the climb was a misuse of automation and this was compounded by the fact that the pilots did not understand how airspeed affected aircraft performance (NTSB, 2007).

Loss of mode awareness may result from an incomplete or inaccurate mental model of the automation behavior, inadequate feedback from the automated system, or complexity in the design of the automated system. In their analysis, the Working Group identified 27% of the accidents related to *mode selection errors*. The analysis also attributed 12% of ASRS reports and 50% of major accidents and incidents to *pilots were out of the loop*, indicating that pilots were not aware of the airplane state or situation (FAA, 2013b). Prior to the Working Group's analysis, there were also numerous reports of autopilot systems that changed modes without sufficient indication to the flightcrew, particularly in high workload situations; for example, an annunciator light alone that indicated the current mode was not always salient (FAA, 1996, 2006a).

The interaction between flightcrews and aircraft has changed as more functions are delegated to or shared with automated systems. The vulnerabilities may be mitigated though system design. For example, providing feedback on automation states and behavior may help the pilot's mode awareness. In addition, making the automation behavior more "transparent" or providing reliability information could help the pilot better calibrate trust in the system (Mumaw, Sarter, & Wickens, 2001; Wickens, Hollands, Banbury, & Parasuraman, 2013). The issues discussed here are common across the different types of automation (control, information, or management), but the importance of the issue and the appropriate mitigation may differ from one type of automation to another.

DISPLAY LOCATION

The location of the display can significantly affect the attention paid to the information on the display as well as the readability of the display. As new avionics and functions are introduced, limited space on the flight deck had led to compromises when integrating and installing new systems. In particular, the use of portable electronic technologies has broadened the flight deck visual field, sometimes increasing the pilot's normal viewing area. The FAA requires that "each flight, navigation, and powerplant instrument for use by any pilot must be plainly visible to him from his station with the minimum practicable deviation from his normal position and line of vision when he is looking forward along the flight path" (14 CFR 25.1321(a)). Furthermore, displays should be placed in such a way that they can be monitored by minimal head and eye movement. The field-of-view is measured by establishing the design eye position, which is the point at which the midpoint of the pilot's eyes would be located when properly seated at the normal position (FAA, 2011a, 2014c). The visual field for each eye is approximately 135 degrees vertically and 160 degrees horizontally (FAA, 2002c). However, it is only in the fovea, an area approximately 22 degrees in visual angle in the center of the visual field, where fine detail can be resolved (Wickens, Lee, Liu, & Gordon-Becker, 2004).

The literature defines two specific regions in the visual field: the *primary field-of-view* and the *secondary field-of-view*. The primary field-of-view (sometimes referred to as the optimum field-of-view) is generally defined as a region with a 15 degrees radius extending from the normal line of sight, established 15 degrees below a line extending horizontally from the eye (Cardosi & Huntley, 1993; Department of Defense, 2012; FAA, 2013a, 2014c). This region is shown in Figure 12.2. The primary field-of-view is the area of the visual field with the highest acuity. From the center of the eye, the density of receptor cells decreases toward the periphery.

FAA regulatory and guidance material requires primary flight instruments and alerts, specifically warnings and cautions that require flightcrew awareness and response, to be presented in the primary field-of-view (e.g., see FAA, 2010, 2011b, 2014c). In older Part 23 aircraft, powerplant information was sometimes located on the right side of the flight deck, and these displays created not only head and eye movement but were also difficult to read. The FAA guidance is supported by

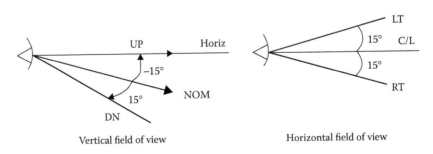

Vertical field of view   Horizontal field of view

FIGURE 12.2   Primary field-of-view.

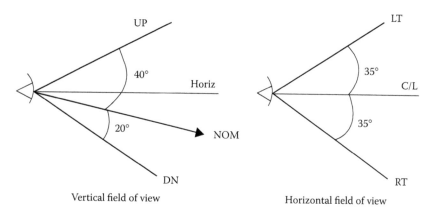

FIGURE 12.3   Secondary field-of-view.

the research literature; in general, the closer information is to the primary field-of-view, the more likely it is that the information will be scanned, detected, or noticed compared with information that is further away (Cardosi & Huntley, 1993; Wickens et al., 2016). In fact, a meta-analysis of research examining detection rate as a function of alert location indicated that the miss rate was unchanged from 0 to 15 eccentricity (i.e., the degrees of visual angle away from a central point), but increased as the alerting stimulus was presented farther outside the primary field-of-view (Wickens et al., 2016).

The secondary field-of-view (also referred to as the maximum field-of-view) is a region around the primary field-of-view that extends vertically +40 degrees up and −20 degrees down and horizontally +35 degrees, as shown in Figure 12.3. There is less guidance as to what information should be in the secondary field-of-view, compared with the primary field-of-view.

The location of avionics displays on the flight deck must be balanced so that the information is visually accessible, as appropriate. To determine how visually accessible information is, one can measure the angle with which the avionics display is offset from the pilot's centerline of vision (FAA, 2002d). It is important to consider that detectability of information may be influenced by the expectancy of that information, salience, and workload. That is, recommendations for presenting alerts in the primary field-of-view support detection of a salient, "expected" event under modest workload, but if the event is surprising, if the alert is not salient, or if workload is high, alerting information may need to be presented closer to the center of the visual field (Wickens et al., 2016). We discuss the presentation of alerts further in the next section.

ALERTS

The number of alerts and annunciations on the flight deck has proliferated as more complex avionics are introduced. In the past, annunciator panels contained all the systems' alert information in one place, and discrete lights would illuminate to indicate non-normal conditions. The pilot could see immediately which system experienced a problem. As avionics systems have become more integrated, the alerting function has been incorporated within the display, creating operational complexity (FAA, 2002a; Veitengruber, Boucek, & Smith, 1977).

The term *alert* is used in different ways in literature and in regulatory and guidance material. It can be used generally to refer to a wide range of annunciations, some of which may be *normal* conditions, or it can also be used more specifically and refer only to indications of more serious or *non-normal* events that require some type of pilot action or awareness. This chapter uses the term *alerts* in the most generic sense; it is the activation of any visual or aural indication, annunciation, or alarm that is intended to make the pilot aware of an event that requires his/her awareness and possibly provide advice as to potential actions to take.

The purpose of an alert is to draw the pilot's attention to a specific condition and report the nature of the condition (FAA, 2010). An alert may take many forms (e.g., switches, lights, flags, prompts, or messages), be presented in several modalities (visual, auditory, and haptic), and vary in their criticality. The effectiveness of an alert will depend on the design of the complete alerting function, including the condition(s) required to trigger the alert, the urgency and priority assigned to the alert, the presentation of the alert, and whether the presentation is consistent with the urgency. Critical alerts should be easily distinguishable from noncritical alerts; alerts should not be generated for conditions that do not require pilot's awareness (FAA, 2010; McAnulty, 1995). An alerting philosophy can help integrate the various alerts across flight deck systems to reduce the number of alerts and annunciations presented and ensure that the pilot is not given contradictory information. Alerts should be prioritized so that the most urgent alert is presented first (Boucek, Erickson, Berson, Hanson, & Leffler, 1980; FAA, 2010).

The FAA classifies alerts in three categories—warnings, cautions, and advisories (see 14 CFR 25.1322):

- Warning: for conditions that require immediate flightcrew awareness and immediate flightcrew response
- Caution: for conditions that require immediate flightcrew awareness and subsequent flightcrew response
- Advisory: for conditions that require flightcrew awareness and may require subsequent flightcrew response

Alerts must be prioritized in two ways: first by grouping into categories of warning, caution, or advisory, and second by evaluating the importance of the alert within the category. There is little empirical work on how alerts can be prioritized by flight phase. Consideration during prioritization should be given to the urgency of flightcrew awareness, urgency of flightcrew response, the speed of the response required, and the potential consequences of failing to detect the alerting condition (Cardosi & Huntley, 1993; Cardosi & Murphy, 1995; FAA, 2010; Palmer et al., 1995). Failure to prioritize or standardize the presentation of alerts on the flight deck can result in confusion and recognition errors.

Furthermore, warnings must be presented in red and cautions in amber/yellow (14 CFR 25.1322 (e)(1)), based on results of previous research that showed red lights were typically detected as fast or faster than other lights (Boucek et al., 1980). Use of the colors red and amber/yellow (or colors confusable with red or amber/yellow) for means other than alerting has been hypothesized to desensitize pilots to the urgency of the alerts and diminish the effectiveness of those colors (Boucek, Veitengruber, & Smith, 1977; Gabree, Chase, & Cardosi, 2014; Veitengruber, Boucek, & Smith, 1977; Widdel & Post, 1992).

It may not always be clear when a situation should be classified as a warning or a caution, however. The distinction between warnings and cautions is subtle, although the consequences are quite different. In addition, the increase in the number of aircraft systems has also caused an increase in the number of warnings, cautions, and advisories, making it more difficult for pilots to identify which system generated the alert. Careful consideration should be paid to which situations are warnings and which are cautions and how to make it apparent as to which event(s) generated the warning or caution (FAA, 2002a).

The detectability of an alert depends on where, when, and how it is presented. Evaluation may be needed to ensure that alerts are detected in a manner timely to the information provided. Warning and caution alerts must be indicated through two different senses using a combination of visual, aural, or tactile indication (14 CFR 25.1322(c)(2)). The FAA requires visual indications of warnings and cautions to be placed in the pilot's primary field of view to maximize the likelihood they will be detected (14 CFR 25.1322). Boucek et al. (1980) further recommend that high priority signals be placed within 15 degrees of the pilot's centerline of vision, and lower priority signals be presented within 30 degrees. Consequently, space limitations on the flight deck may result in other alerts placed in less desirable locations where they may not be easily noticed. Several visual coding methods can be used in combination with location to attract attention, including blinking or flashing, reverse video, size coding, and color. The effectiveness of these coding methods will vary depending on the context. For example, blinking or flashing lights are generally detected faster than steady lights, but the detection time for a blinking or flashing light increased if the background contained other blinking or flashing lights (Boucek et al., 1980). In general, each of these coding methods must be applied carefully, however, as these coding methods are by their nature intended to be distracting, and overuse of a coding method or combining too many methods can minimize the effectiveness of any one method. In particular, blinking or flashing must be applied carefully as indiscriminate use can reduce legibility and lead to visual fatigue. Blinking or flashing is recommended only for the most urgent warnings (Cardosi & Murphy, 1995; GAMA, 2000; Garner & Assenmacher, 1997; McAnulty, 1995).

Aural alerts have the advantage over visual alerts of being omnidirectional; that is, the signal can be perceived regardless of head or eye orientation (Wickens, Lee, Liu, & Gordon-Becker, 2004). However, because aural alerts are transient, they are more effective in drawing attention to information on a display rather than being the sole source of information (FAA, 2010). The volume of aural alerts is an important design consideration; the volume should be able to be heard above the ambient noise but not be so intense that it is above the danger level for hearing or that it will distract or cause discomfort (Mitman, Neumeier, Reynolds, & Rehmann, 1994; Wickens, Lee, Liu, & Gordon-Becker, 2004). A review of aural alert-related incidents and feedback from pilots regarding aural alert presentation highlighted the following issues (Cardosi & Murphy, 1995; McAnulty, 1995, Mitman et al., 1994; Patterson, 1982; Peryer et al., 2005):

- Aural alerts were too loud and interrupted flightcrew's performance of an ongoing task, masked other important aural information, or startled the flightcrew.
- The onset and offset of aural alerts were too abrupt, with the potential to startle the flightcrew and disrupt ongoing tasks.
- The temporal patterns used for warnings were not distinctive and were confusable. Similarity will make it more difficult for the pilot to identify which system generated the alert and increase workload in diagnosing the alert.
- The length of the aural alert was too long, unnecessarily adding to the noise level on the flight deck.
- The spectral characteristics (tones) of the alerts were such that lower priority warnings were perceived as being more urgent than higher priority alerts.

Recommendations have been provided to address these concerns in the design of aural alerts. First, the volume of an aural alert should range from 20 to 30 dB greater than the ambient noise level (FAA, 2010; McAnulty, 1995; Wickens, Lee, Liu, & Gordon-Becker, 2004). Second, the rise in tone, which characterizes a non-normal event, should be approximately 20–30 milliseconds for the portion of the tone in which the sound level is above the threshold noise level (Patterson, 1982). Third, aural alerts may be designed to be unique from other sounds from the flight deck by varying one or more of four dimensions: the pitch or frequency, the envelope (e.g., is the tone rising or constant), rhythm, and location (Cardosi & Huntley, 1993). Recommendations for how long a tone should be are less clear; the detection of a tone may take up to 50 milliseconds, and presenting a signal beyond 300 milliseconds does not provide much benefit (Patterson, 1982). The minimum duration should be 50 milliseconds, but beyond that the signal duration may vary depending on the urgency level and the type of response required (FAA, 2010; McAnulty, 1995).

Voice alerts may be used to indicate conditions demanding immediate flightcrew awareness, because the alerts provide the information directly. However, voice messages have some limitations. First, voice alarms are likely to be more confusable with other voice communications (e.g., flightcrew communications, air traffic control [ATC]) than other aural alerts. The artificiality of computer synthesized speech may help one distinguish the message from other voice communications but may also make the message difficult to understand. Second, too many different voices can be a nuisance and a distraction, similar to having too many aural alerts. Finally, depending on the length of the message, understanding a voice message may take more time than reading a visual message (Cardosi & Murphy, 1995; McAnulty, 1995; Wickens, Lee, Liu, & Gordon-Becker, 2004).

As alerts may be distracting, it may be helpful to sometimes inhibit alerts that are inappropriate or unnecessary for a particular phase of operations to minimize distractions to the flightcrew from what may be perceived as a nuisance alert. For example, warnings, annunciations, and messages that are not critical to the safety of instrument approaches or missed approaches may be suppressed during those phases (RTCA, 2006). In fact, surveys of pilots have indicated that it is a potential for the presentation of too many noncritical alerts during critical phases of flights, when the alerts would be perceived to be a distraction, and agree with the need to inhibit noncritical alerts during these critical phases of flight (Boucek et al., 1980). Whether to allow the flightcrew to inhibit, cancel, or defer alerts must be considered carefully, however, because this action essentially defeats the presentation of the alert in the first place (Boucek et al., 1980). The ability to inhibit alerts is used on all air transport aircraft to minimize nuisance alerts. Alerts should not suppress or inhibit other displays or alerts requiring immediate flightcrew attention (FAA, 2010).

Finally, the integrity and reliability of the alerting system should be evaluated, as perceived trust and credibility will decrease as the number of false alarms increase. Alerts are one form of automation, and consequently, the system will sometimes make mistakes and signal that there may be a problem when one does not exist. A high number of false alerts or alerts that provide inaccurate information may increase workload and slow response time in the case of a real alert. As noted in the Automation section earlier, pilots have suppressed or ignored alerts because the number of false alarms was too high (see also Parasuraman & Riley, 1997; Wickens & Dixon, 2005). Thus, care should be taken when setting the alerting threshold to prevent the likelihood of false or nuisance alerts. In addition, training to the user to help explain the tradeoff between misses and false alarms may help pilots understand the inevitability of false alarms rather than view them as a system failure (Wickens, Hollands, Banbury, & Parasuraman, 2013).

SYMBOLS

The design of symbols is complex due to the wide range of display technology and functionality on which they may be shown. A consistent symbol design across manufacturers and chart providers will facilitate recognition of symbols. Historically, however, the symbols used by manufacturers and chart providers differ slightly. Although there is no standard set of symbology for navigation information, there are common properties in the symbols used by manufacturers and chart providers (e.g., in terms of shape or fill). Symbols currently in use by various avionics and chart manufacturers for navigation aids and airports as well as line and linear patterns are documented in a Volpe Center report, *Survey of Symbology for Aeronautical Charts and Electronic Displays: Navigation Aids, Airports, Lines, and Linear Patterns* (DOT/FAA/AR-07/66; DOT-VNTSC-FAA-08-01). Industry recommendations and guidelines for symbology are provided in the following:

1. SAE ARP4102/7, *Electronic Displays*, Appendices A through C (for primary flight, navigation, and powerplant displays)
2. SAE ARP5289A, *Electronic Aeronautical Symbols*, (for depiction of navigation symbology)
3. SAE ARP5288, *Transport Category Airplane Head Up Display Systems*, (for Head Up Display symbology)

To promote consistency, displays should use symbols similar to those shown on published charts and sectionals or with commonly accepted aviation practices, when possible. New symbols, a new design, or a new symbol for a function historically associated with another symbol, should be tested for flightcrew comprehension, retention, and ability to distinguish from other symbols. In particular, considerations for new symbol design should include the *legibility*, the *distinctiveness*, and *interpretability* of that design.

Legibility: The legibility of information on an electronic display is influenced by several factors including the viewing angle, pixel density, and contrast ratio (Nelson et al., 2014; Yeh & Chandra, 2005). Some symbols may have fine details that are difficult to see when viewed off-angle or under degraded display conditions. The *viewing angle* refers to the angle at which a symbol subtends the eye, measured in degrees of arc, and captures the relationship between symbol size and viewing distance. That is, a small symbol viewed at a close distance could have the same visual angle as a larger symbol viewed from farther away. The *pixel density* describes the number of pixels per inch of a display. Five pixels are needed to draw a letter on a display, but six pixels are needed for the letter to be "fully recognizable" (Nelson et al., 2014). Finally, *contrast ratio* describes the difference in luminance between the symbol and the background. Contrast ratio varies depending on the display hardware, the information being shown, and the amount of light on the flight deck.

Nelson et al. (2014) incorporated these factors into two tools to objectively evaluate the legibility of electronic charts on portable electronic devices. One addresses legibility with respect to visual angle and specifies whether the electronic chart is legible to a person with 20/40 vision by measuring the smallest text on the chart and considering it with respect to the viewing distance. The other tool examines whether the pixel density is sufficient to display the information clearly.

Distinctiveness: A symbol is distinctive if it can be easily discriminated from other symbols. Symbols should have a basic shape or characteristic that can be recognized with and without context. If a symbol is identifiable *only* with context, then the pilot may be relying on context cues, and the meaning may not be intuitive. Distinctiveness should be considered within and across symbol sets to ensure *consistency* in the use of a symbol, regardless of the manufacturer or chart provider, and to prevent *confusability* of a symbol with other symbols (Yeh & Chandra, 2004). Pilots must be able to identify and understand information conveyed by symbols for flight planning, position awareness, and navigation.

There have been instances in which a symbol shape used by one manufacturer is confusable with a symbol shape used by another. The Figure 12.4 provides an example of using the same symbol shape for two different meanings.

In 1999, the FAA identified the potential for confusion because the U.S. representation for a fly-by waypoint looks similar to the International Civil Aviation Organization (ICAO) representation for a fly-over waypoint. The meanings are quite different, however. A *fly-by* waypoint allows the pilot to anticipate a turn to avoid overshooting the next flight segment. A *fly-over* waypoint is used when the aircraft must fly over the point prior to initiating a turn. If these symbols were misinterpreted, the resulting flight path deviation could have safety consequences (Yeh & Chandra, 2008).

The Volpe Center conducted an exploratory paper-based study to evaluate the saliency of the U.S. and ICAO fly-over and fly-by waypoint symbols, shown in Figure 12.4. Pilots were asked to find fly-over and fly-by symbols in cluttered paper charts; saliency was measured as the accuracy with which the symbols were detected. The results showed that the ICAO *fly-by* waypoint symbol was detected *more* accurately than the USA symbol for a *fly-by* waypoint, but the ICAO *fly-over* waypoint symbol was detected *less* accurately than the USA *fly-over* symbol. However, the lower

	USA symbols	Previous ICAO symbols
Fly-by waypoint		
Fly-over waypoint		

FIGURE 12.4 Fly-by and fly-over symbols.

accuracy for finding the ICAO fly-over waypoint symbol was due primarily to detection performance when that symbol was drawn at the runway and the size of the symbol was reduced so that the runway could remain visible. In 2000, this discrepancy and potential safety risk was resolved when ICAO added a circle to their recommended fly-over waypoint symbol (Yeh & Chandra, 2004).

Yeh and Chandra (2005) examined the role of representativeness in symbol design by identifying key defining features for eight navigation aid symbols (DME, fix, NDB, TACAN, VOR, VORDME, VORTAC, and waypoint). Pilots were presented with symbol shapes currently in use by aviation display manufacturers and chart providers and asked if they considered whether a symbol shape was representative of a specific symbol type. Symbol recognition was evaluated on the basis of the frequency with which a test symbol shape was considered to be representative of a symbol type. The results of the study are shown in Table 12.1. Pilots identified a representative symbol shape for seven of the eight symbols; no representative shape was identified for the DME, likely because DMEs are typically drawn in conjunction with another symbol (e.g., the VORDME). Closer examination of the data showed that symbol shape was the key factor for classifying symbols, despite variations in size, color, and orientation.

Interpretability: Symbols may be designed so that one symbol contains multiple features that convey information. Using the example in Figure 12.4, a single symbol shape (the four-pointed star) is used to convey what the symbol represents (a waypoint), but the fill of the symbol also conveys information (i.e., whether it is fly-by or fly-over). Variations in fill, color, and border (e.g., a circle surrounding the symbol) have been proposed to convey information. (Note, however, that color alone should not be used to convey meaning. This is an issue that will be discussed further in the next section.) The amount of instruction that pilots are given about how the rules are applied may influence how easily they learn the rules. In addition, key features of the symbol (e.g., fill or color) should be applied clearly and not be confusable with other features. This is a particular issue for Cockpit Display of Traffic Information (CDTI) displays, which may use the color brown/tan to represent aircraft on the ground, and in some cases (e.g., lighting conditions, viewing angle, etc.), the brown/tan color is confused with yellow/amber that represents a warning or caution (Gabree, Chase, & Cardosi, 2014).

One consideration is ensuring symbology on new avionics systems is consistent with symbology on older, more familiar systems. For example, symbology for new CDTI may or may not be based on

TABLE 12.1

Representative Navigation Aid Symbol Shapes

Symbol Type	Representative Shape
DME	None identified
Fix	◭
NDB	◉
TACAN	⬡
VOR	⬡
VORDME	▣
VORTAC	⬡
Waypoint	✧

traffic symbology used for TCAS, an air-to-air surveillance system, but because of previous pilot experience with TCAS, it is desirable to consider aspects of how TCAS symbology is designed and coded in the design of CDTI symbols (Zuschlag, Chandra, & Grayhem, 2013). Thus, when expanding the definitions of current symbols, integrating symbols, or remapping symbol shapes, the design of the symbol needs to be evaluated to determine whether the meaning conveyed by the combined features is clear.

COLOR

Display technology allows the presentation of a myriad of colors with good fidelity, allowing manufacturers to use a large number of different colors on their systems. Color is often applied with the intention of differentiating among the information elements displayed, and it can be an effective method for coding and enhancing the understanding of information, when applied appropriately. However, excessive or inappropriate use of color can decrease the effectiveness of a display, add more complexity, increase visual search time, and increase the potential for misinterpretation.

A common issue that has been observed in avionics submitted for approval is the inappropriate use of color, particularly in the use of red for elements that may *not* require an immediate response, or the use of amber/yellow for elements that may *not* need corrective action (FAA, 2002a). The consistent use of color within an application and across all flight deck displays is encouraged. The FAA reserves the use of red and amber/yellow for warnings and cautions, specifically (see 14 CFR 25.1322; FAA, 2010).

There are also special considerations with the use of the color blue. The human eye has the least number of blue-light sensitive cells, and those cells are located in the periphery of the eye rather than the fovea, so it takes more time to bring that color into focus. Consequently, the color blue should not be used for small, detailed symbols. In addition, blue is at the far end of the visual spectrum, so the eye may have difficulty focusing on blue at the same time as other colors. In particular, red and blue should not be used together because the two colors are at opposite ends of the visual spectrum; so when one color is in focus, the other will be slightly out of focus (Gabree, Chase, & Cardosi, 2014).

Another consideration related to color is the perception of color, which varies considerably from one individual to another. The *2010 Aeromedical Certification Statistical Handbook* (Table 22) estimates that there were 4,438 active airmen (male and female) with color vision deficiencies as of December 31, 2010, with 1,553 of these holding firstclass medicals, 991 holding secondclass medicals, and 1,894 holding thirdclass medicals (Skaggs, Norris, & Johnson, 2012). In addition, aging of the eye causes the lens to yellow and reduces the ability to distinguish between colors (Salvi, Akhtar, & Currie, 2006). To accommodate the proportion of pilots that are colorblind or color deficient, color should be applied redundantly, for example, with shape, location, or fill. Use of two or more coding techniques will also improve recognition, identification, and interpretation.

In general, no more than six colors should be used if information is color coded to avoid errors in judgment. Research shows that although we can discriminate among colors when they are placed side by side, it is harder to identify a specific color alone. In other words, the human observer may not be able to accurately judge the level of a color with precision if there are more than six levels (Wickens, Hollands, Banbury, & Parasuraman, 2013; FAA, 2014c). In addition, colors that are assigned meaning should be identifiable when presented alone and with respect to all backgrounds and in all viewing conditions (Gabree, Chase, & Cardosi, 2014). Color coding is most effective if the intended meaning of the coding is immediately understood, for example, when color is a natural representation of the information or conforms to pilot stereotypes.

Colors should be selected so that they are easily discriminable to reduce the potential for misinterpretation. Color is defined by hue, saturation, and brightness. The term hue is generally synonymous with color; it describes the degree of "red"-ness or "blue"-ness. Saturation is the purity of the color, and brightness is how light or dark a color is. The Commissions Internationale de L'Eclairage developed a model defining color by its u', v' coordinates, as shown by the Commissions Internationale

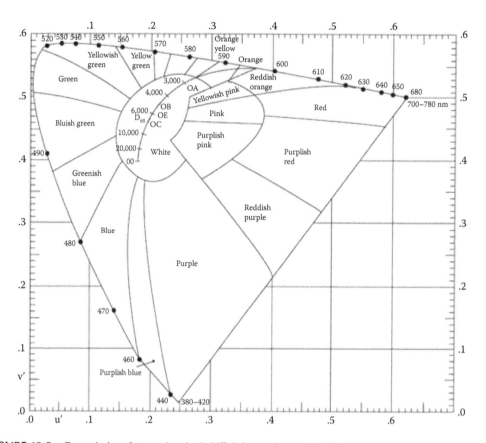

FIGURE 12.5 Commissions Internationale de L'Eclairage chromaticity diagram.

de L'Eclairage chromaticity diagram in Figure 12.5. The distance between two colors in the figure is a direct reflection of the perceptual difference between those two colors. Colors that are widely spaced apart in the figure will appear more different from one another.

The topics so far have focused on the information display. We now turn to control devices and the methods for interacting with the information.

CONTROL DEVICES

Control devices are the primary means for inputting information. The term *control* in this chapter refers to input devices rather than flight controls, such as a yoke or rudder pedals. A modern flight deck may have more than 200 control devices, such as buttons, knobs, keyboards, switches, cursor-control devices (e.g., mouse, touchpad, trackball, and joystick), or touch screens. Each control device has unique characteristics that will influence its effectiveness for an application or avionics system. Previous experience and expectations affect usability of controls, so control design that is consistent within a system and across the flight deck can reduce the chance of confusion. A design philosophy for controls will help promote this consistency.

The usability of a control is influenced by its function, operation, and arrangement with respect to the operating environment. Each of these factors will be considered in turn. An application of these principles to touch screens is also included to address some of its unique characteristics.

Function: The function of a control device should be identified quickly and accurately, as appropriate for the task. Control labels are the most common means for identifying and describing control functions so that the pilot can easily identify what each control does. The labels should be readable and legible from the pilot's normal operating position and in all lighting conditions (see FAA, 2014c). As the number of separate controls has increased, creative approaches have been used to communicate a control's function. In many cases, the length of the labels introduced new abbreviations and acronyms, and some times, the labels did not sufficiently describe the function performed.

One solution to the labeling problem is to use icons, but most functions do not have a universally accepted icon (FAA, 2011a). Another solution was to reduce the number of controls by using multifunction controls, in which one device controls several functions or systems. However, a disadvantage for multifunction controls is that it may not be obvious what is being controlled, increasing the potential that the pilot may inadvertently activate the wrong function or provide input to the wrong system. Thus, proper labeling is quite important to indicate the active function (FAA, 2011a). Multifunction controls with hidden functions should be avoided as they increase the potential for error and workload (European Aviation Safety Agency, 2007).

Operation: The interaction between a control device and the system or display being controlled should be apparent and understandable. That is, the movement required to operate a control and the resulting movement or action on the system or display should be obvious. The conventional relationships between a control function and the expected direction of movement are provided in Table 12.1. The control should be installed in an orientation such that the direction of the control movement is consistent with what is being controlled.

Control operation must also consider response gain, that is, the sensitivity with which the movement of a control or the force applied to a control is transformed to produce an output. A control that has a high gain requires only a small movement of the control to produce a large change in the display output, which allows more rapid inputs but can lead to overshooting the target. Low-gain controls require more input or force than high-gain controls but provide for precision. The response time and accuracy with which the task must be performed need to be traded off. Some controls provide variable gain to support high-gain task when the control is moved quickly and low-gain tasks when the control is moved slowly (Table 12.2).

Controls should also provide feedback about the results of the actions. Feedback provides information to the pilot about whether a control has been activated and confirms whether the input has been accepted or not. Response time for feedback is important in terms of system acceptability. There is an expectation that system response time will be faster for tasks that are considered to

TABLE 12.2

Conventional Relationships between Control Function and Direction of Movement

Function	Direction of Movement
Increase	Up, right, forward, clockwise, push
Decrease	Down, left, rearward, counterclockwise, pull
On	Up, right, forward, pull, depress, rotate clockwise
Off	Down, left, rearward, push, release, rotate counterclockwise
Right	Right, clockwise
Left	Left, counterclockwise
Up	Up, forward
Down	Down, rearward
Retract	Rearward, pull, counterclockwise, up
Extend	Forward, push, clockwise, down

Source: Federal Aviation Administration (FAA), *Controls for flight deck systems* (Advisory Circular 20–175), Washington, DC, 2011a, Retrieved from: http://www.faa.gov/documentlibrary/media/advisory_circular/AC%2020–175.pdf.

be simple than those that require more complex calculations. However, long or variable response times can lead to a negative perception for usability (Nielsen, 1994). The response time and appropriateness of feedback varies depending on the task being performed and the specific information required for successful operation (FAA, 2011a).

Arrangement: Limited space on the flight deck may lead to controls that are placed in less-than-ideal locations. Controls should be arranged as a function of the tasks that need to be performed and the sequence with which those tasks are performed. 14 CFR 25.777 requires that "each cockpit control must be located to provide convenient operation and to prevent confusion and inadvertent operation." To comply with this rule, the position of the control must fall within the reach envelope of the intended pilot population. Anthropometric data may provide a means for accounting for differences in human size, including reach, hand, and finger size. Dedicated controls should be as close to the display being controlled as possible. Controls should be placed so that the function being controlled and related elements (e.g., indications and labels) can be seen when manipulating the controls. Controls placed below the display or to one side could minimize visual obstruction when operating the control.

Inadvertent activation of controls should be prevented to the extent possible, although controls will be operated inadvertently at some point. For example, the pilot may accidentally bump a control or activate one control while activating another. Common means for preventing inadvertent activation include spacing buttons of an ¼ inch (6.25 millimeter) apart, providing physical protection for hard control devices, and logical protection for software-based controls (FAA, 2011b). The designer must consider the tradeoff between preventing inadvertent activation and the control's operation as methods to prevent inadvertent activation may make controls more difficult to operate (Cardosi and Murphy, 1995; FAA, 2002a, 2002d).

Operating Environment: The usability of control devices must be considered with respect to the operating environment. Some considerations are (FAA, 2011a) as follows:

- The target population or end user, such as body size, previous experience with a specific control, and training
- Use in lighting conditions from bright sunlight to darkness
- Use of gloves while operating the control, which influence factors such as button size and spacing, and use of capacitive touch screens
- Effect of turbulence or other vibrations
- Potential interruptions while performing a task
- Objects that could interfere with control movement, such as interference from pilot clothing or flight deck equipment
- Pilot incapacitation for aircraft designed for multicrew operation, specifically whether controls are viewable, reachable, and operable by both flightcrew members
- Use of nondominant hand and
- Excessive noise

It will not be possible to consider *all* environmental and use cases, but a representative set should be considered that addresses the environment in which the controls are expected to be used, including normal and non-normal conditions.

Touch Screens: As technology has evolved, traditional controls (such as knobs, push buttons, and switches) are being replaced with touch screens. A touch screen provides a direct relationship between input and output and is helpful when space is limited. The use of touch screens for installed systems on the flight deck emerged with the advent of touch screen portable electronic devices, such as the Apple iPad and the Microsoft Surface, which host applications such as electronic charts, electronic documents, and airport moving maps.

There are several touch screen technologies that vary in how they sense and respond to touch: resistive, capacitive, and infrared. A touch screen device may be developed from one or a combination of these technologies. *Resistive* touch screens are the most common of the three. It is composed of several layers of electrically conductive material. A "touch" applies pressure so that the layers come into contact with each other, and this action completes a circuit and is processed as input. The pressure required to produce a response varies from one resistive touch screen to another. A *capacitive* touch screen is coated with conductive material. A *touch*, usually skin contact, creates a change in capacitance, which is processed as input. Consequently, use of gloves or use of a stylus that is not specifically designed for the touch screen will not create a measurable change in capacitance. Finally, an *infrared* touch screen has infrared beams across the surface of the touch screen, and a *touch* is registered when the beams are disrupted (e.g., by a finger). In this case, touching the screen is *not* necessary, because the beams of light are slightly above the screen.

Touch screens are not appropriate for all tasks and operating environments, however. They have unique considerations from traditional controls related to readability in various lighting conditions, surface contamination (e.g., smudges and scratches), inadvertent touch, and feedback. Specifically, fingerprint smudges may create glare in some lighting conditions, reducing display readability. In addition, touch inaccuracy may be higher relative to traditional controls, thereby increasing the potential for error. With respect to feedback, touch screens generally do not provide tactile feedback, so the pilot may not know whether his/her input was accepted without looking directly at the display (Dodd et al., 2014).

It is important to consider factors such as how quickly a response is needed, the force needed, feedback, and inadvertent activation. Dodd et al. (2014) conducted a series of studies to examine the effects of touch screen location (overhead, to the left of the pilot—"outboard", forward panel, center panel), touch screen size (8″ or 15″), touch target size (0.25″ or 0.50″), and touch screen technology (projected-capacitive or resistive) on pilot usability with and without moderate turbulence. Pilots flew a simulated path while performing data entry, panning, and menu navigation tasks on a touch screen display. So that the study simulated a traditional flight, the primary task required maintaining a straight and level path, and interacting with the touch screen display was secondary.

Not surprisingly, the time needed for data entry and the number of data entry errors increased in turbulence than without turbulence. However, touch target size moderated these effects, such that pilots using large touch targets in moderate turbulence made fewer errors than pilots using small touch targets in no turbulence. In addition, pilots completed the tasks quickly and more accurately with the resistive touch screen than the projected-capacitive touch screen.

With respect to display location, touch displays on the center panel and overhead locations led to higher subjective ratings of fatigue relative to the forward panel. There was no difference with the outboard location relative to the other locations. More importantly, observation of the pilots during the study showed the need for hand rests; Figure 12.6 shows photos taken during the study of how pilots stabilized their hands based on display size and location. As the figure shows, pilots tended to use the edges of the touch screen to stabilize their hand, regardless of display location. The data collected showed that this stabilization improved input accuracy and minimized fatigue.

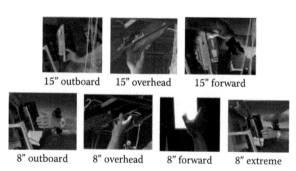

15" outboard   15" overhead   15" forward

8" outboard   8" overhead   8" forward   8" extreme

FIGURE 12.6 Method for hand stabilization (Courtesy of Dodd).

The discussion so far has focused on flight deck design and the information presentation—that is, the technical system itself. Flight deck design cannot be considered in isolation, however, and any design must consider the knowledge and skills of the pilots who will be operating the aircraft. Any assumptions about what the pilot is expected to do should be explicitly identified, and individual characteristics inherent to the end users must be considered. This is a topic that is addressed by the second layer of the sociotechnical model, the personnel subsystem.

PERSONNEL SUBSYSTEM

The second layer of the sociotechnical system addresses the people—*personnel system*. Although there are several types of people involved in flight deck interactions (the air traffic controller, maintenance personnel, dispatch, etc.), we focus here on the pilot; the maintainer was addressed in Chapter 5, and the air traffic controller is addressed in the next chapter. Specifically, we address three factors that influence the pilot's ability to apply knowledge, skills, and abilities to respond and perform a given task: workload, potential for error, and fatigue.

WORKLOAD

Workload describes the mental and physical demand imposed by pilot duties, work, and the number of tasks to be accomplished given the time constraints within which those duties, work, and tasks must be completed. Workload may be used as an indication of how busy the pilot is as well as the complexity of the task as it reflects the mental and physical resources needed by the task relative to the amount of resources available. Each task performed imposes some amount of workload, but the level of workload experienced will vary depending on the pilot's experience with the task, training, and skill. In addition, the workload experienced for one task may vary at different times depending on what other tasks are being performed concurrently. Each system on the flight deck should be evaluated in isolation as well as in combination with other systems to understand its impact on pilot workload in both normal and non-normal situations.

14 CFR 25.1523 addresses flightcrew workload, requiring that:

The minimum flight crew must be established so that it is sufficient for safe operation, considering

(a) The workload on individual crewmembers;
(b) The accessibility and ease of operation of necessary controls by the appropriate crewmember; and
(c) The kind of operation authorized under § 25.1525.

Demonstrating compliance to 14 CFR 25.1323 requires evaluating the predicted overall flightcrew workload as well as measuring the workload for each individual crewmember. There are several ways to measure workload using objective measures, subjective measures, or a combination of the two. Objective measures include collecting data on primary task performance, secondary task performance, or physiological data. The *primary task* is typically flight performance, that is, how well is the pilot maintaining the predefined flight path. Measures such as deviations from the flight path or control activity (the amount of displacement on the flight controls) are indicators of workload. A *secondary task* may also be imposed; a secondary task is one that is not the focus of the study but rather is introduced to measure the "spare" mental or physical resources. Decrements in task performance on the secondary task (e.g., response time and response accuracy) are assumed to be indications of higher workload, because as workload increases, pilots will focus on the highest priority task (the primary one). Physiological measures, such as changes in heart rate, heart-rate variability, or evoked brain potential, have also been proposed; increases in workload would be reflected to a higher heart rate, more variability, or increased brain activity.

Workload is also assessed using subjective measures, in which participants provide an estimate of their workload on different tasks. Common scales for subjective workload measures are the Bedford workload scale (Roscoe & Ellis, 1990), NASA-TLX (National Aeronautics and Space Administration-task load index) (Hart & Staveland, 1988) and the subjective workload assessment technique (SWAT; Reid & Nygren, 1988). These scales differ in their dimensionality. The Bedford workload scale is unidimensional and provides "one number" that describes the workload experienced based on a series of questions: Was it possible to complete the task? Was workload tolerable for the task? Was workload satisfactory? On the other hand, the NASA-TLX and SWAT are multidimensional, assuming that several dimensions contribute to workload. The NASA-TLX measures mental demand, physical demand, temporal demand, performance, effort, and frustration. The SWAT gathers feedback on the time demand imposed by the task, the mental demand, and the stress imposed.

The problem with subjective workload measures, however, is the potential for response bias. Every pilot is different, and the individual perception of workload can vary considerably. In addition, subjective workload measures may not always agree with performance measures (Wickens, 1993). Therefore, we generally recommend that subjective data be used in conjunction with objective data to increase the reliability of the results.

The issue of workload is often mentioned with respect to the introduction and use of automation; it is generally believed that one way to reduce flightcrew workload is to introduce automation. Although there are in fact periods of time in which automation does decrease workload, there are also periods when task demand is high in which use of the automation adds *more* complexity and workload to the tasks being performed (Parasuraman & Riley, 1997; Wiener, 1988). In particular, programing of the automation has shifted workload from takeoff and approach and landing to the time periods prior to these phases of flight, but this is also the time period when ATC may notify the pilot of a route or approach change. Consequently, the flightcrew may need to reprogram the automation at a time when they also need to be managing the aircraft and lead to aircraft mishandling errors (FAA, 2009).

Furthermore, the PARC/CAST Working Group noted that workload management is an important consideration for flight-path management. Although workload and distractions are discussed in flightcrew training courses, the Working Group was concerned that flightcrews are not adequately prepared to manage tasks on the flight deck when workload is high. In particular, task prioritization for aviating, navigating, and communicating is harder to operationalize when there are many tasks to be performed in each category, and the tasks overlap, or are queued (FAA, 2013b). For example, on July 6, 2013, Asiana Airlines Flight 214 struck a seawall at San Francisco International Airport when the flightcrew mismanaged the aircraft's vertical profile during the approach, relied on automated flight controls without a full understanding of the system, and did not notice their speed was too slow. Although there were many contributing factors to the accident (including pilot fatigue), NTSB investigators found a period of increased flightcrew workload due to the complexity of the flight control computers, so none of the pilots noticed that the automatic airspeed control had deactivated (NTSB, 2014b).

This section has focused on those cases in which workload was too high, but workload that is too low—*underload*—must also be considered. Underload refers to long periods of relative inactivity, for example, as on transoceanic flights where the pilot does not have much to do apart from monitoring displays. The focus on the flight deck has generally been on reducing workload, but maintaining vigilance in low workload situations may be fatiguing as well (Hancock & Warm, 1989). Low workload is generally more difficult to detect than high workload, but the consequences are similar to those observed in high-workload conditions such that pilots are less efficient and do not perform the task at a commensurate level of effort (Hancock & Verwey, 1997). As the level of cognitive activity dedicated to a task is influenced by the task demands, as demands change, so too does the effort invested in a task to maintain "level" performance. Complacency and boredom reduce vigilance and may hinder pilots from effectively responding to surprises and non-normal events on the flight deck. In fact, Young and Stanton (2002) have proposed that attentional resources are reduced in low-workload situations and may not return to the maximum attentional capacity that would be normally available, even when task demand increases. In addition, pilots who are sleep deprived do not perform as well as pilots who are well rested under low-workload conditions (Wickens, 1993).

There are fewer studies examining the effects of low workload, and measuring low workload is not as clear as measuring high workload. Research is needed to evaluate the effect of designs for susceptibility to underload situations. In particular, the Working Group reported hearing concerns from pilots about low workload and attempts to engage the flightcrew, particularly when pilots are further out of the control loop.

HUMAN ERROR

On October 31, 2014, Virgin Galactic's SpaceShipTwo crashed in the Mojave Desert due to the premature unlocking of the spaceship's feather system. The NTSB attributed the crash to a failure to consider and protect against human error. Specifically, the NTSB noted that the design of the system had created a single-point human error, and that the hazard analysis did not account for the possibility that the feather could be unlocked prematurely, consequently extending in conditions that could lead to a catastrophic failure of the vehicle. In addition, the manuals and procedures did not provide information about the consequences about unlocking the feather early (NTSB, 2015).

Human performance is often the largest contributor to system variability, and human error is the most common accident theme, occurring in one form or another in nearly all investigated accidents (Abbott, 2001; FAA, 2015a; U.S. Congress, 1988). Accidents generally result not from one error but rather from a combination of factors, only some of which are human errors. The potential for error is influenced by system design, previous experience, and training. Preventing *all* human error is not possible. The most qualified and well-trained flightcrews will commit errors even if the flight deck is well designed. Fortunately, most errors are detected and mitigated and will not result in a safety event (FAA, 2013a). Rather, an understanding of the issues created by the intersections between and systems and end users is needed and is critical to preventing the risk of errors mitigating the ones that do occur.

Human error is an action or omission of an action that was not intended, expected, or desired. Human error is often classified by examining behavior in terms of skills, rules, and knowledge (Rasmussen, 1983; Reason, 1990). These three categories generally refer to the amount of conscious control required to perform a task. Skill-based tasks are "automatic", highly practiced, and generally not under conscious monitoring once the intention to act has been formed. Rule-based tasks are characterized by the application of a rule or procedures (e.g., instructions, checklists) used to determine a course of action. Knowledge-based tasks are generally novel and unexpected and require conscious thought (Wickens, 1993).

Errors in skill-based tasks are generally errors of execution; that is, a plan of action has been identified and the intention is correct, but the plan is executed incorrectly. There are two types of errors: slips and lapses. A slip is a correct action that is carried out inappropriately, such as activating the wrong control. Slips may result because there is a slight deviation from an expected sequence of behavior, because the intended action is similar to the conditions of a more frequent action, or because the action is so "automatic" that attention was directed elsewhere. A lapse is a memory failure that results in an action that is not taken, for example, a missed step in a checklist.

A mistake is the result of an incorrect intention that is executed. A mistake may be rule based, in which the pilot applies a "rule" to the wrong situation or applies the wrong *rule* (reference). Mistakes may also be knowledge based, in which the diagnosis of the situation is incorrect so the plan (execution) is inappropriate. Planning errors may be due to decision-making biases, in which pilots on only a subset of available information that is consistent with the original diagnosis (a confirmation bias) or gather data from what is available rather than what is most reliable and relevant (availability). Mistakes may have more serious consequences than slips and lapses because the operator committing the mistake believes that s/he is taking a correct action, regardless of other signs that may tell otherwise. Note that a mistake is different from a *violation*, which is an action that is deliberate and known to violate rules or procedures (Reason, 1990).

Reason's framework is only one approach to looking at human error. There are several other models for predicting human error that can be applied early in the design process. THERP (Technique for Human Error Rate Prediction) is a human reliability assessment technique that assumes human error is a result of omission, commission, selection, sequence, timing, and quantity and requires examination of "performance shaping factors" that influence the probability of human error (Swain & Guttman, 1983). Systematic Human Error Reduction and Prediction Approach is another predictive method that attempts to link methods to reduce error to the underlying cause of the error. Systematic human error reduction and prediction approach combines hierarchical task analyses methods with error taxonomies (e.g., the skill, rules, and knowledge framework defined by Rasmussen) to identify credible errors (Harris, Stanton, Marshall, Demagalski, & Salmon, 2005). The purpose of this chapter is not to provide a comprehensive discussion of the many models for classifying human error but to generally recognize their contributions for managing and mitigating errors.

Error Management refers to how the system design facilitates error prevention, detection, and recovery. To effectively manage errors, the system designer must understand what types of human error have occurred on the basis of previous accidents and incidents, and how to compensate. *Error Prevention* is intended to avoiding errors and is generally accomplished through system design, for example, by switch guards, confirmation actions, or interlocks, or by reducing the occurrence of cognitive factors that contribute to error (e.g., distraction, workload). *Error Detection* is intended to ensure that errors can be detected quickly, enabling recovery. Feedback is particularly important for error detection; immediate feedback can provide information to the flightcrew about the correctness of their actions, for example, with alerts to a specific error or system condition, annunciations of aircraft state information during normal operations, or indications of external hazards. Finally, ERROR RECOVERY addresses how easy it is to recover or return to a safe state after the error (Human Factors Harmonization Working Group Final Report, 2004). Procedures and checklists that remind the pilot of the sequence of tasks that need to be completed can support error recovery.

In a complex system like the flight deck, there are multiple defenses to mitigate known errors. *Error Mitigations* may be in the form of changes in equipment design or establishing new operations and procedures. Mitigations have also been proposed in the form of "error tolerant" and "error resistant" systems. "Error tolerant" systems help to mitigate the errors that are committed. This approach to human error assumes that errors are unpredictable and inevitable. It gives the pilot flexibility in maintaining safe and level flight, and although that flexibility sometimes may lead to a wrong action, in general, that flexibility is advantageous, and the system should respond in a way that tolerates and forgives the errors that do occur. The automation monitors the system and informs the pilot of errors, if they occur. As an example, the Boeing 757/767 aircraft shows its error tolerance by presenting an alert when a pilot enters a destination into the flight management computer (FMC) that is not compatible with the fuel load (Wickens, 1993; Weiner & Nagel, 1988).

On the other hand, "error resistant" systems attempts go one step beyond error tolerant system and may control and correct for pilot inputs that are considered to be erroneous. For example, automation on the Airbus A-320 prevents the pilot from exceeding the aircraft's operating envelope, regardless of the control stick input. However, by seizing control, these systems become potential sources of error (U.S. Congress, 1988).

Finally, training is often proposed as a mitigator for human error, but training can only reduce errors, not eliminate them. Training influences the knowledge, skills, and abilities of pilots, but the method by which this occurs is influenced by the airlines and evaluated by the FAA or appropriate regulatory authority. This topic is addressed by the organizational/management infrastructure.

FATIGUE

Fatigue is a topic of interest for human factors professionals directing or advising any workforce regarding hours of operation and work schedules, and particularly for industries that must cover 24/7 operations. Fatigue effects on performance and safety have been recognized and are well documented (Bonnet, 2000; Carskadon & Dement, 1987; Dinges & Kribbs, 1991; Dinges, 1992; Horne & Reyner, 1985; Naitoh, 1975). We find that the length of one's work period plays a role in increasing fatigue, as does the timing of the shift as it occurs within each 24-hour period. Personnel who are required to work in a 24/7 operations setting are particularly vulnerable to working during hours of the day when they would ordinarily be sleeping. Most people know how difficult it is to remain alert between the hours of midnight and 6 a.m. We see that when changing or rotating from one shift schedule start time to another, one might experience what is called shift lag that is not unlike the experience of flightcrew who have traversed multiple time zones and experience the desynchronous effects of jet lag (Caldwell, 1997; Commercial Transportation Operator Fatigue Management Reference, 2003). Over the last decade or so, we have seen significant advancements in the science of fatigue including sleep, circadian rhythms, and chronobiology (Kecklund, Di Milia, Axelsson, Lowden, & Akerstedt, 2012). We have also seen that the integration of this knowledge into the operational aviation environment has remained difficult and challenging.

Definitions: There are about as many definitions of fatigue as the number of individuals questioned. You may have developed your own personal definition from experiences with your work and life activities. Even the research literature espouses definitions with slight variations. With regard to aviation, the FAA defines fatigue as a "physiological state of reduced mental or physical performance capability resulting from a lack of sleep or increased physical activity that can reduce a flightcrew member's alertness and ability to safely operate an aircraft or perform safety-related duties" (FAA, 2012). The ICAO expands the definition a bit by stating that fatigue is also the result of "sleep loss or extended wakefulness, circadian phase, or workload (mental and/or physical activity) that can impair a crew member's alertness and ability to safely operate an aircraft or perform safety related duties" (ICAO, 2012).

So, respectively, it is clear that fatigue is not based on a single dimension, but rather the result of a combination of several factors related to an individual's physiological sleep needs, internal biological rhythms, and the kind of work in which one is actively involved. Significantly, even though fatigue is relatively complex and difficult to define, the operational causes and consequences are surprisingly very consistent across different aviation operations and personnel (Battelle Report, 1998; Miller, 2006).

Research results illustrate how operator performance is negatively affected by factors associated with fatigue, such as lengthy duty times, times of the day that the work period is scheduled (e.g., working at night), acute sleep loss from getting less than a normal amount of sleep in a single night, and an accumulated sleep debt that occurs over several days/nights of restricted sleep (Eriksen and Akerstedt, 2006; Powell, Spencer, Holland, Broadbent, & Petrie, 2007; Samel, Wegmann, & Vejvoda, 2005). In addition, we find that time on task and workload issues contribute to fatigue, though presently it is not clear how to quantify these effects. For the crewmember, Caldwell et al. (2009) state, "… both long- and short-haul pilots commonly associate their fatigue with night flights, jet lag, early wakeups, time pressure, multiple flight legs, and consecutive duty periods without sufficient recovery breaks." Goode (2003) and Powell, Spencer, Holland, and Petrie (2008) discuss elevated risks associated with lengthy flight and duty time, and the need for limiting these flight duty periods (FDP) and flight times. So, it is clear that within the aviation industry, running 24/7 combined operations contributes to fatigue at the onset, and that science can certainly share the responsibility with industry to develop mitigation and management strategies to minimize the effects of fatigue to improve flightcrew alertness and performance, and maintain safety.

The following paragraphs provide brief descriptions of a few key elements in the study and understanding of fatigue:

Sleep: Humans are normally active during the daylight hours (diurnal) and sleep during the night. The optimal length of our sleep period is generally around eight hours, though some people appear to need less or more sleep than others. We typically cycle through different stages of sleep during the sleeping period. Generally, we fall asleep during stages 1 and 2 and cycle into progressively deeper sleep stages 3 and 4 before returning to lighter sleep and Rapid Eye Movement (REM) sleep in around 90–120 minutes. This cycle repeats itself four to five times throughout the night. We experience more non-REM sleep stages (i.e., stages 1–4) during the first half of the night with more REM sleep occurring in the latter half of the sleep period. As we age (40–45 years +), we experience less deep or slow-wave sleep and tend to spend more time in the lighter sleep stages, from which we are more easily awakened (Carskadon & Dement, 2011).

During non-REM or deep, slow-wave sleep, biological reconstruction and physical maintenance occurs through nocturnal increases in protein synthesis and cell division (Van Cauter & Tasali, 2011). Moreover, during non-REM and REM sleep, the restoration of higher cognitive elements, emotional balance, and mood occurs along with what we believe is a process for the consolidation of declarative and spatial memories (non-REM) and nondeclarative memories (REM) (Peigneau & Smith, 2011). We need to cycle through both types of sleep staging (i.e., REM and non-REM) to feel fully rested. Studies show that we respond best to a single, consecutive period of sleep, though some work has suggested that if part of a split-sleep schedule is *anchored* to our normal nighttime and the remaining sleep occurs at another time and adds up to a total of 8 hours within that 24-hour period, our performance is not significantly degraded. The results of the Mollicone et al. (2008) study suggest that split sleep schedules are feasible and can be used to enhance the flexibility of sleep/work schedules for operations that might involve restricted nighttime sleep due to work period and task scheduling. But this remains somewhat controversial and not without what appears to be sleep restriction costs over several days and nights.

Circadian rhythms: Circadian rhythms refer to the changes in physiology, alertness, and performance that are regulated by our internal (biological) pacemaker or body clock (i.e., suprachiasmatic nucleus of the hypothalamus). The neuronal and hormonal activities that the suprachiasmatic nucleus generates regulate many different body functions in a 24-hour cycle. Fatigue-related performance effects are also found to fluctuate across the daily cycle and include changes in subjective alertness and sleepiness, performance of mental and physical work, and our sleep and wake cycle (Akerstedt, 1995). Briefly mentioned before, the risk of significant impairment in performance has been shown to be greatest for individuals working at night when normally sleeping. Elevated risk also occurs during the morning hours prior to an early show time (Battelle Report, 1998; Dinges, 1992). Going to bed earlier in preparation for early morning work is difficult because you are trying to begin sleep during the "evening wake maintenance zone," identified in the few hours before your usual bedtime and associated with high alertness and an elevated core body temperature (Gander, Myhre, Graeber, Andersen, & Lauber, 1989). Moreover, greater risks of impaired performance are associated with the repeated effects of consecutive night time schedules in which an accumulation of sleep debt typically occurs because we do not fully adapt to sleeping during the daytime and working at night (Naitoh, 1975).

Desynchrony: So, our circadian rhythms are normally entrained by external time cues (i.e., Zeitgebers) and specifically by the light and darkness cycle. Jet lag among flightcrew is commonly described as the feelings of malaise and fatigue that accompany a time zone change. This primarily occurs during that period of resynchronization of your circadian rhythms to the *new* external time cues in the new location (i.e., again, to light and darkness). For other aviation workers like those in ATC or technical operations, shift lag is described as the feelings of malaise and fatigue that accompany changes in shift schedules, for example, from daytime work to nighttime work, and vice versa. Shift lag occurs during the period of *forced* resynchronization of circadian rhythms to the new external time cues due to the work schedule. Compared with jet lag, however, the attempt to resynchronize to a nighttime work and a daytime sleep schedule occurs more slowly and is much less successful because the main time cues (i.e., daylight and darkness) tend to inhibit the resynchronization process (Miller, 2006).

Aviation relevance: Alertness and performance degradation occur when individuals are scheduled for work during times that encroach on their normal sleep times, either by extending and completing a flight duty period beyond midnight or by beginning a flight duty period very early in the morning (Gander, van den Berg, Mulrine, Signal, & Mangie, 2013). These times when the body clock is expecting sleep, for example, when flight duty periods and flight time encroach on the *back side of the clock*, we experience significantly degraded performance during that period of maximum sleepiness, identified as our window of circadian low (between 0200–0600) and during the resynchronization period following time zone changes (Gander, Graeber, Foushee, Lauber, & Connell, 1994).

Degraded performance can also occur with early morning *show times*, for example, before 6 a.m., because this schedule essentially restricts your sleep time by trying to sleep earlier than normal during the wake maintenance zone. Lastly, if flightcrew are required to sleep during the daytime when they are normally awake, sleep is found to be qualitatively less restful, and quantitatively less likely to occur consecutively for up to eight hours. Sleep would be difficult during the morning hours when our alertness is normally rising or outside of the secondary window of circadian low (i.e., 1400–1700). Repeating any of these restricted sleep scenarios over the few days and/or nights of a trip sequence would certainly produce a cumulative effect that would trigger the need for *recovery sleep* before another trip sequence could begin (Battelle Report, 1998; Bonnet, 2000; Hursh & Van Dongen, 2010; Dinges, 1989).

Ultralong Range (ULR) Flight/Safety implications: Flight crewmembers are particularly vulnerable to schedules that place their work and wakefulness at odds with their circadian rhythm. Previous 14 CFR Part 121 flight crewmember duty and rest regulations had not originally considered much of what is now known about our internal body clock and scheduled flight operations that occur during hours of reduced alertness and performance.

With modern technology and manufacturing developments, commercial aircraft became capable of flying longer segments without the requirement to refuel. Hence, long-haul and ULR flight operations became possible. Fatigue and human alertness limitations then became more imposing to the FAA and the international aviation community. In 1993, Air New Zealand was interested in a progressive and innovative data-driven approach for developing crew scheduling, particularly for long-haul flight operations. Most every route they flew was a long-haul trip. They teamed with British, New Zealand, and NASA scientists to develop this new approach. They also formed an internal multidisciplinary team to implement the approach and established an external oversight panel. These activities contributed to the aviation industry to *think outside of the box*, that is, beyond the prescriptive limitations of regulations, to manage fatigue.

In 1998, an International ULR Crew Alertness Steering Committee was formed at around the same time that Singapore Airlines requested approval from the Civil Aviation Authority of Singapore to fly ULR flights from Singapore to Los Angeles. This city pair route required flight times of 16 hours and flight duty periods of 18–22 hours (Flight Safety Digest, 2005). This had never before been done. Three ULR Crew Alertness Workshops were held from 2001 to 2003 with the objectives to identify common fatigue mitigation approaches, develop a technical basis for operational and regulatory guidance to maintain crewmember alertness, and seek global multistakeholder consensus. The workshops attracted as many as 90 participants, including many prominent scientists and flight safety personnel from 14 countries. Eventually, Singapore Airlines was granted a *safety of flight operation* status based on the carefully directed efforts of the International Crew Alertness Steering Committee, and they began ULR flights to both Los Angeles and New York City in 2004. The important operational experience, data collection and analysis, and scientific knowledge from the steering committee's work, and the work of Australian scientists, established a basis for a new operational model and an approach to monitor and mitigate flightcrew fatigue. The procedure is now known as fatigue risk-management systems (FRMS). This nonprescriptive approach has become an exemplary basis for the U.S. FAA's authorization of the ULR Operation Specification

A-332, first issued to Delta Air Lines in the development of its New York City to Mumbai city pair route in 2006. We discuss FRMS in more detail later in this chapter in the context of flight, duty, and rest regulations.

Summary: The definition of fatigue is found to be varied and complex for many reasons, and it seems that the more that is understood about fatigue, the more difficult it becomes to manage. The operational definition used by the FAA touches on important aspects of fatigue to include that it is a "physiological state of reduced mental or physical performance capability resulting from a lack of sleep or increased physical activity that can reduce a flightcrew member's alertness and ability to safely operate an aircraft or perform safety-related duties." Expanding this definition a bit, ICAO includes the influence and impact of extended wakefulness, circadian rhythms, and workload. It is also important to recognize that there are both work and nonwork related components and many other factors that interact with each individual's tolerance to fatigue.

The discussion with respect to the personnel subsystem has focused on individual behavior, but operating an aircraft is not a task conducted in isolation but rather as part of a system that requires coordination with other pilots, airline operators, dispatchers, and air traffic controllers, to name a few. This coordination is addressed in the third layer of the sociotechnical model, the organizational/management infrastructure.

ORGANIZATIONAL/MANAGEMENT INFRASTRUCTURE

The organization/management infrastructure layer of the sociotechnical model acknowledges the role of teamwork and group behavior. In this chapter, we focus specifically on pilot training and the development of pilot performance standards for air transport operations. In addition, we examine the means for understanding human errors when they occur and identifying areas where additional training may be needed.

Requirements for pilot training appear in 14 Code of Federal Regulations (CFR) Part 61 and are administered by the FAA. They differ by categories of operations, resulting in different certificates (airline transport, commercial, private, recreational, sport, and student pilots), ratings (airplane, rotocraft, glider, lighter than air, etc.), and classes (single-engine, multiengine, land, and sea), some with training supplements in their own parts of the CFR, including 121 and 135 for commercial airlines. Training regulations are written to ensure pilots acquire individual technical skills necessary to their intended certificate, rating, and class. At a fundamental level—definition of certificates, ratings, authorizations, endorsements, and so on—these requirements have been fairly stable since the dawn of the jet age and establishment of the Federal Aviation Agency in 1958. But the science and practice of human factors engineering and psychology have impacted training in a number of key ways:

- Training has evolved from a solely individual skill and maneuver focus to incorporate skills necessary to perform in a multicrew cockpit within a realistic operational environment.
- This has further expanded to consider the interfaces with other front-line employee groups within airline operations, including dispatchers, mechanics, flight attendants, and airport personnel.
- Hazard and task analyses and instructional system design (ISD) have influenced the design of operating procedures and the focus of training, especially with increasing levels of cockpit automation and recognition of unique skills for pilot flying and monitoring roles.
- Systems of feedback on operational performance have become a driver of training emphasis.
- Specific issues identified through feedback systems and accidents have required significant additional training modules and programs.
- Controversies remain and evolution of training is ongoing.

Multicrew Operations

Although solo private pilots may transport friends and family and solo commercial pilots may move small numbers of passengers and quantities of cargo for hire, most passenger transport is accomplished in multipilot aircraft—at a minimum, a pilot in command or Captain, and a copilot or first officer. The crew may be larger on older aircraft requiring a flight engineer or for longer duration, typically international flights requiring relief crewmembers. Foushee (1984) argued that air transport crews share capabilities and are susceptible to errors typical of small work groups, meaning much of social psychological research should be considered to apply. A series of accidents in the 1970s and 1980s resulted in classroom and simulator training focusing on crew resource management (Cooper, White, & Lauber, 1980). Prior to this point, training and performance evaluation were very much individually focused and maneuver based. The practical test standards for air transport pilots (FAA, 2008) require that an applicant for a certificate be able to perform each maneuver deemed necessary to safely operate an aircraft and focuses minimally on the coordination with other pilots or the context in which non-normal maneuvers might be required. So for example, a pilot will demonstrate the ability to perform a nonprecision approach to a runway by these criteria:

1. Exhibits adequate knowledge of nonprecision approach procedures representative of those the applicant is likely to use
2. Accomplishes the nonprecision instrument approaches selected by the examiner
3. Establishes two-way communications with ATC as appropriate to the phase of flight or approach segment and uses proper communications phraseology and techniques
4. Complies with all clearances issued by ATC
5. Advises ATC or the examiner any time the applicant is unable to comply with a clearance
6. Establishes the appropriate airplane configuration and airspeed, and completes all applicable checklist items or coordinates with crew to ensure completion of checklist items in a timely manner and as recommended by the manufacturer
7. Maintains, prior to beginning the final approach segment, the desired altitude ± 100 feet, the desired airspeed ± 10 knots, the desired heading ± 5; and accurately tracks radials, courses, and bearings
8. Selects, tunes, identifies, and monitors the operational status of ground and airplane navigation equipment used for the approach
9. Applies the necessary adjustments to the published minimum descent altitude (MDA) and visibility criteria for the airplane approach category when required, such as
   a. Notices to airmen (NOTAM), including Flight Data Center Procedural NOTAMs
   b. Inoperative airplane and ground navigation equipment
   c. Inoperative visual aids associated with the landing environment
   d. National Weather Service reporting factors and criteria
10. Establishes a rate of descent that will ensure arrival at the MDA (at, or prior to reaching, the visual descent point , if published) with the airplane in a position from which a descent from MDA to a landing on the intended runway can be made at a normal rate using normal maneuvering
11. Allows, while on the final approach segment, not more than quarter-scale deflection of the course deviation indicator or (five inches the case of the Radio Magnetic Indicator [RMI] or bearing pointer and maintains airspeed within) 5 knots of that desired
12. Maintains the MDA, when reached, within $-0, +50$ feet to the missed approach point
13. Executes the missed approach at the missed approach point if the required visual references for the intended runway are not unmistakably visible and identifiable at the missed approach point
14. Executes a normal landing from a straight-in or circling approach when instructed by the examiner (pp. 58–59)

Performance standards for emergencies precipitated by other than, say, an engine failure at a critical altitude are more vaguely specified. At best, pilots are presented with a sampling of possible failures over the course of their career, most likely within the vicinity of an airport at relatively low altitude. These standards are now supplemented by training requirements and advisory circulars on crew resource management (FAA, 2004a) and line operational simulation (FAA, 2015b). This guidance emphasizes how to train and test skills necessary to perform as a crewmember and within the context of flight.

The 2009 Air France accident over the southern Atlantic Ocean illustrates some of the challenges (Bureau d'Enquêtes et d'Analyses, 2012). After an uneventful takeoff and ascent to a cruise altitude of 35,000 feet, the captain turned over the controls to his copilot and relief officer and left the cockpit for planned crew rest. Shortly thereafter, the aircraft entered some turbulence associated with a strong weather system common to the area during summer months, and airspeed indications became inconsistent due to ice accumulation in the pitot probes (used to measure airspeed), not perceived by the crew. On account of the airspeed inconsistency, the following occurred: the autopilot and autothrust control systems disengaged, several messages alerting failure of systems that use airspeed data were displayed to the crew, and the flight controls exited *normal law* and entered *alternate law* (discussed next). The copilot-flying made several pitch and roll changes. Roll inputs were most likely intended to maintain wings level; initial pitch inputs were most likely intended to correct a slight pitch down and decrease in indicated altitude when airspeed indications were lost and the autopilot disconnected. The copilot-flying also increased engine power to takeoff/go-around thrust. Aircraft pitch attitude slowly increased beyond 10 degrees nose-up, and airspeed decreased from 275 to 60 knots. Persistence of the nose-up input could not be explained by the investigators, though they presented multiple hypotheses. The aircraft climbed until it entered an aerodynamic stall, and then descended rapidly until impact with the ocean four minutes later. Although aerodynamic stall is precluded on modern Airbus aircraft operating under *normal law*, in which the flight controls mitigate pitch commands that exceed critical angle of attack, this function is not possible in *alternate law*. The crew apparently never gained a full understanding of the stall and did not lower the nose of the aircraft sufficiently to recover lift. Potentially contributing to the crew's confusion, stall warnings are inhibited when angle of attack is so extreme as to be considered an invalid indication. In this case, this produced a paradoxical result: When pitch was reduced sufficiently to produce an angle of attack considered valid, stall warnings resumed, giving the appearance that an appropriate corrective action was making the situation worse. The investigation examined information displayed and warnings announced to the crew, functionality of the aircraft in these conditions, how crewmembers were trained, industry expectation of crew actions in that context, and the actions of crews that had previously encountered loss of airspeed information. The investigation said of crew actions during the accident sequence:

> The occurrence of the failure in the context of flight in cruise completely surprised the pilots of flight AF 447. The apparent difficulties with aeroplane handling at high altitude in turbulence led to excessive handling inputs in roll and a sharp nose-up input by the PF (copilot flying). The destabilization that resulted from the climbing flight path and the evolution in the pitch attitude and vertical speed added to the erroneous airspeed indications and ECAM (Electronic Centralized Aircraft Monitoring) messages, which did not help with the diagnosis. The crew, progressively becoming de-structured, likely never understood that it was faced with a *simple* loss of three sources of airspeed information. (p. 199)

To prevent such an accident, one must focus on ensuring each pilot understands the aircraft systems and how such a failure might be diagnosed, the procedures for recovering from both the failure and a resulting extreme aircraft attitude or upset, and aircraft performance at high altitudes. Portions of each would be present in ground school and simulator-based training and are reflected in approach to stall, stall recovery, and unusual attitude recovery, as required by the practical test standards. However, none of this ensures training in the specific or generic events in the context of this situation—stall recovery is typically trained at low altitudes because aircraft operate at low speed when departing or arriving at an airport, making departure and approach stalls more likely than high-altitude stalls. But perhaps, more importantly from a human factors perspective, the availability of more than one crewmember should make detection, diagnosis, and recovery easier. A threat to safe flight or an error by one crewmember should be detected and corrected by the other(s) (Helmreich, 1998). This did not occur, even when the Captain returned to the cockpit.

Methods for ensuring each crewmember's functions not just as a technically competent individual, but as a contributor to a work team, complementing each other's capabilities and limitations are key human factors contributions to flight deck safety and are now reflected in FAA ACs. Key knowledge, skill, and abilities described in AC are communication processes and decision behavior, team building and maintenance, and workload management and situation awareness. Their antecedent research bases are documented in Wiener, Kanki, and Helmreich (1993), and Orlady and Orlady (1999). Continued evolution of these concepts in research is illustrated by Dismukes, Berman, and Loukopoulos (2007). Research and industry collaboration and guidance culminated in advanced qualification training (SFAR 58; GPO, 2003), wherein aircraft and crew position specific task analyses by phase of flight lead to definition of training proficiency objectives that must be demonstrated in maneuver validation and line-oriented evaluation.

WORKFORCE INTERFACES

Any human interface with the cockpit crew introduces both capabilities and limitations and can become a source of error if not managed effectively. Consider the role of the flight dispatcher. Were air transport pilots to self-dispatch, they would require an hour or more before each flight to consider weather conditions, choose an optimal route, determine the passenger, baggage, and freight load for the flight, determine how much fuel to carry, and identify any limitations of aircraft or ground equipment impacting their flight plan. Instead, 14 CFR Part 65 provides for aircraft dispatchers who exercise responsibility with the pilot in command in the operational control of the flight—planning, monitoring, advising of changing circumstances, and terminating the flight at destination or an alternate airport. The dispatcher provides information to the pilots and monitors their progress. This represents a relief of workload, a profoundly valuable source of advisory information, a coordination challenge, *and* a potential source of error in information exchange. Training has expanded to accommodate this and other interfaces—flight attendant, mechanic, ground crew, and so on. This is reflected in FAA ACs on dispatch (FAA, 2005b) and maintenance (FAA, 2000) resource management.

KEY INPUTS TO TRAINING

If pilot qualifications are defined in terms of maneuvers, the training design task for introducing a new aircraft type may be defined as tailoring the footprint of ground instruction and simulator or aircraft training developed for previous generations of aircraft. This has often been successful. Airlines have developed and gained approval for training footprints describing a certain number of days of ground school and simulator sessions and then generalized this to new fleets. A notable exception was the introduction of FMC aircraft, which integrate area-navigation functions with aircraft performance management, allowing computer entries to guide or control the aircraft in four dimensions and integrate warning and alerting systems for mechanical problems. Models include Boeing's 757, 767, 777, 787, and recent generations of the 737; Airbus's A-300-600, 310, 320, 330, and 340; Fokker's F-100; and McDonnell-Douglas's MD-88 and MD-11. They incorporate increasing levels of flight guidance, flight control, and system automation, requiring accommodation in training. Chidester (1999) argued that these aircraft produced a two-decade accommodation in policy, procedure, and training. Two FAA human factors Working Groups (FAA, 1996, 2013b) documented a range of issues encountered in the operation of FMC aircraft.

But the broader lessons for training were found in task analysis and ISD, typically implemented under the advanced qualification program. Training must become an end result of a comprehensive application of human factors to the aircraft, its procedures, and its intended interface with the airspace system. Hazard analyses, task analyses, and procedures for normal and abnormal operations are antecedents of training design. If reconfiguring aircraft controls (flaps, slats, gear, etc.) or systems (pressurization, hydraulics, etc.), or coordinating with ATC is required during flight, what are the designed means of configuration or coordination? What are the impacts of not doing so in a timely or correct fashion? What safeguards may be put into place to prevent or correct error? Where safeguards cannot be engineered, procedures and training are required.

Pilot flying and monitoring roles have become a special emphasis of procedures and training, perhaps in part due to the effect of increasing automation. Having more than one pilot in the cockpit accommodates workload, requires some division of labor, and enables detection and correction of error. Typically, one pilot serves as a flying pilot (PF), the other as a monitoring pilot (PM). This is independent of the pilot-in-command concept; the Captain and First Officer may serve in either PF or PM role, but the Captain always retains pilot-in-command authority and responsibility. The PF controls the flight path of the aircraft by direct inputs to the control wheel, yoke, and pedals by entering targets into a flight guidance panel, or by making changes to the control-display unit of the FMC, the latter two coupled to the autopilot. The PM crosschecks data entry, monitors, and calls out key events or deviations from target values, manages configuration changes and checklists, and communicates with ATC. Monitoring skills are complex, challenging to apply, and historically underemphasized to the point of accident involvement (Sumwalt, Thomas, & Dismukes, 2003). When flying an approach to landing in instrument conditions, for example, the PF must shift gaze and attention between the instrument panel and the windscreen. Looking out, the PF searches for the runway environment. Looking at the instrument panel, the PF must ensure the flight path and speed remain appropriate until a safe transition to visual maneuvering to landing can be accomplished. What should the PM look at? Should they watch to ensure the runway environment is acquired, that approach constraints are honored and called out, that the aircraft is appropriately configured, that checklists are complete before landing, that landing clearance has been received, and that weather conditions remain as expected? The accident literature abounds with failures in each of these tasks. The Flight Safety Foundation (2014) offered a comprehensive set of recommendations to improve pilot monitoring, including, among others:

- Brief flight path–related plans and announce any deviations from the prebriefed plans
- Manage workload to prioritize flight path monitoring
- Be particularly attentive to the flight guidance automation
- Clearly define the monitoring role of each pilot, recognizing this may change when autoflight systems are engaged
- Emphasize predictable areas during flight where risk of flight path deviation increases, heightening the importance of task and workload management
- Implement policies and practices that protect flight-path management from distractions and interruptions
- Train pilots about why they are vulnerable to errors and monitoring lapses

Perhaps, most importantly, an airline must provide guidance and training that addresses the focus of the flying and PM by phase of flight. Otherwise, a simple deviation from a constraint can cascade to an accident. Consider these excerpts from the NTSB (2014a) accident sequence narrative of a United Parcel Service A-310 at Birmingham, Alabama:

> On August 14, 2013, about 0447 central daylight time (CDT), UPS flight 1354, an Airbus A300–600, N155UP, crashed short of runway 18 during a localizer nonprecision approach to runway 18 at Birmingham-Shuttlesworth International Airport (BHM), Birmingham, Alabama.

The captain was the pilot flying, and the first officer was the pilot monitoring. Before descent, while on the direct-to-KBHM leg of the flight, the captain briefed the localizer runway 18 nonprecision profile approach, and the first officer entered the approach into the airplane's flight management computer (FMC).

However, although the flight plan for the approach had already been entered in the FMC, the captain did not request and the first officer did not verify that the flight plan reflected only the approach fixes; therefore, the direct-to-KBHM leg that had been set up during the flight from Louisville remained in the FMC. This caused a flight plan discontinuity message to remain in the FMC, which rendered the glideslope generated for the profile approach meaningless.

Had the FMC been properly sequenced and the profile approach selected, the autopilot would have engaged the profile approach and the airplane would have begun a descent on the glidepath to the runway. However, this did not occur.

When the autopilot did not engage in profile mode, the captain changed the autopilot mode to the vertical speed mode, yet he did not brief the first officer of the autopilot mode change.

About seven seconds after the first officer completed the Before Landing checklist, the first officer noted that the captain had switched the autopilot to vertical speed mode; shortly thereafter, the captain increased the vertical descent rate to 1,500 feet per minute. The first officer made the required 1,000-feet above-airport-elevation callout, and the captain noted that the decision altitude was 1,200 feet msl. but maintained the 1,500 feet per minute descent rate. Once the airplane descended below 1,000 feet at a descent rate greater than 1,000 feet per minute, the approach would have violated the stabilized approach criteria defined in the UPS flight operations manual and would have required a go-around. As the airplane descended to the minimum descent altitude, the first officer did not make the required callouts regarding approaching and reaching the minimum descent altitude, and the captain did not arrest the descent at the minimum descent altitude.

The airplane continued to descend, and at 1,000 feet msl. (about 250 feet above ground level), an enhanced ground proximity warning system (EGPWS) *sink rate* caution alert was triggered. The captain began to adjust the vertical speed in accordance with UPS's trained procedure, and he reported the runway in sight about 3.5 seconds after the *sink rate* caution alert. The airplane continued to descend at a rate of about 1,000 feet per minute. The first officer then confirmed that she also had the runway in sight. About two seconds after reporting the runway in sight, the captain further reduced the commanded vertical speed, but the airplane was still descending rapidly on a trajectory that was about one nautical mile short of the runway. Neither pilot appeared to be aware of the airplane's altitude after the first officer's 1,000-feet callout. The cockpit voice recorder then recorded the sound of the airplane contacting trees followed by an EGPWS *too low terrain* caution alert (pp. 10–12).

The sequence involves multiple errors by both the PF and PM, but the monitoring function—to call out the MDA and deviations from the stabilized approach requirements—was intended as a final layer of safety protection by the crew, backed up by the GPWS. Error, failure to catch and correct error, and last minute terrain warnings are typical of controlled flight into terrain accidents; emphasizing and directing monitoring is one key solution (Flight Safety Foundation, 2014). In a sense, this might be construed as merely a more substantial definition of crew resource management. More broadly, developing procedures for function within operating constraints and emphasizing required coordination inside and outside the cockpit have been emphasized by Degani and Wiener (1993) and Barshi, Dismukes, and Loukopoulos (2012).

OPERATIONAL FEEDBACK

In the 1990s, airlines began routine collection and analyses of safety reports and recorded flight data. Aviation Safety Action Programs (ASAP; FAA, 2002b) allow certificated airmen to report any safety concern they encounter, even if they inadvertently caused the concern. Pilots, dispatchers, mechanics, and others report events, deviations, or observations during line operations, and these are investigated collaboratively by airline, FAA, and employee representatives. Flight Operational Quality Assurance programs (FOQA; FAA, 2004b) routinely collect and analyze digital flight data downloaded from aircraft after flight. They enable airlines and the FAA to identify locations and contributing factors to exceedances of the desired flight envelope, such as unstable approaches. Line operations safety audits (LOSA; FAA, 2006b) provide a cadre of observers trained to evaluation standards, who review and provide feedback on flight deck, maintenance, and ground operations.

All of these programs attempt to discover developing issues in an airline's operations that may be precursors of incidents or accidents. Antecedents of these programs can be found in the ASRS managed by NASA on behalf of the FAA since 1975, and NTSB use of flight data in accident investigation since the 1960s. ASAP, FOQA, and LOSA have both become a driver of training issues and provided a compelling method of presentation for training. A good example is the emphasis on stabilized approaches. Most airlines require that before descending below 500 feet above the runway in visual, or 1,000 feet in instrument meteorological conditions, the aircraft be on the correct flight path, at target airspeed, configured for landing, descending at target rate, and with engines powered to target thrust value (Flight Safety Foundation, 2000). Many approach and landing accidents have been preceded by an unstable approach. FOQA programs have identified airports and runways where unstable approaches are more likely, and ASAP reports have often explained why they occurred. Airlines have dedicated a great deal of training emphasis as a result of feedback, and recreations revealing precipitating factors, progress of the approach, and outcomes in difficult landings, or execution of missed approaches have come from FOQA data and ASAP reports.

RESPONDING TO DISCOVERED HAZARDS

A broader phenomenon of training as a key method of responding to discovered hazards can be observed in accident investigation and recommendations by the NTSB. When Delta Air Lines Flight 191 crashed during a thunderstorm at Dallas-Ft. Worth airport in 1985, the airline industry gained an understanding of microburst and wind shear hazards (NTSB, 1986). The legacy is escape maneuvers trained in initial and recurrent aircraft training incorporated into the practical test standards for air transport pilots. When Air Florida flight 90 crashed into the Potomac river shortly after takeoff from Washington National Airport in 1982 (NTSB, 1982) and USAir Flight 405 crashed on takeoff from New York LaGuardia airport, the airline industry gained an understanding of the impact of ice accumulation on wings and engine nacelles during ground operations (NTSB, 1993).The legacy is strong deicing procedures, holdover times allowed between deicing and takeoff, and emphasis in airline classroom and simulator training. A similar pattern has followed loss of control accidents, such as Colgan Air flight 3407 (NTSB, 2010), which resulted in a statutory requirement to implement a formal rule requiring stall and upset prevention and recovery training (Federal Register, 2014a).

Responsive training interventions will continue, possibly accelerate, through collaborative review of shared safety data using the aviation safety information analysis and sharing (see http://www.asias.faa.gov/pls/apex/f?p=100:1:) program, led by the CAST (see http://www.cast-safety.org/). Their methodology focuses on risks identified through incident and report review and will likely impact training requirements and emphasis. Interventions have already been accomplished as a result of GPWS and TCAS events observed in data.

CONTROVERSIES

Training and human factors contributions to training are not without unresolved controversy. Here are three. First, consensus defining optimal approaches to training for aircraft with advanced automated systems appears to have been destabilized by continuing advances. Chidester (1999) describes airline consensus around uses of levels of automation, as documented by an Air Transport Association Human Factors Subcommittee:

> When immediate, decisive, and correct control of aircraft path is required, the lowest level of automation – hand flying without flight director guidance – may be necessary. Such instances would include escape or avoidance maneuvers (excepting aircraft with flight director wind shear guidance)

and recovery from upset or unusual attitudes. With the exception of visual approaches and deliberate decisions to maintain flying proficiency, this is essentially a non-normal operation for flight guidance or FMS-generation aircraft. Is should be considered a transitory mode used when the pilot perceives the aircraft is not responding to urgent aircraft demands. The pilot can establish a higher level of automation as soon as conditions permit.

When used with flight director guidance, hand flying is the primary takeoff and departure mode. It is also the primary mode for landings, except autolands.

Where short-range tactical planning is needed (i.e., radar vectors for separation or course intercept, short-range speed or climb rate control, etc.), Mode Control or Flight Guidance inputs may be most effective. This level should be used predominantly in the terminal environment when responding to clearance changes and restrictions, including in-close approach/runway changes.

Autoflight coupled to the FMS/GPS is the primary mode for nonterminal operations and should be established as soon as *resume own navigation* or similar clearance is received. This level exploits programming accomplished preflight. Where the longer-range strategic plan is changed (i.e., initial approach and runway assignment, direct clearances, etc.), Flight Management inputs remain appropriate. However, when significant modifications to route are issued by ATC, the pilot should revert, at least temporarily, to lower levels of automation (p. 180).

The Flight Deck Automation Working Group (FAA, 2013b) suggested that the issues have expanded, making this consensus less clear. While recognizing the extent to which pilots mitigate operational risk, the Working Group reported concerns including maintenance of manual flying skills, managing malfunctions, use of automated systems, communication within the cockpit and with ATC, usability of standard procedures, data entry and verification, operating policies, workload management, flight path knowledge and skills, training time and content, and equipment design and standardization. Pilots must be provided opportunities to practice manual and flight path monitoring skills. Mode awareness is challenging, even when choices among levels of flight path automation are deliberate. Much automation progress has involved information rather than flight control; little guidance is available for optimizing use of electronic flight bags, map displays, and electronic checklists. This requires a new round of hazard and task analysis and development of new guidance. The Working Group made 18 recommendations involving design and certification, procedures and training, interfaces with air traffic services, and regulatory oversight. The issues are broader than training, and existing training accommodations are not comprehensive.

Second, it is not clear when and how much maneuver-based versus line-oriented training is needed. Maneuver-based training is efficient. By having a pilot fly in the vicinity of the airport, conducting many takeoffs, approaches, and landings, an instructor can accomplish a great deal of training, and an evaluator can observe many events in a typically four-hour block of simulation. From the pilot's perspective, this can be like batting practice—having many problems thrown at them in succession and honing their responses. But this form of training offers little context. Every real flight presents some unique challenges and each requires only a single takeoff and landing. Line oriented simulation allows practice and evaluation in context. Industry consensus reflected under advance qualification programs appears to embrace both, culminating in a maneuvers validation and a line-oriented evaluation in both initial/qualification and recurrent/continuing qualification training. But examination across airlines would reveal different numbers of maneuver and line-oriented events preceding validation, suggesting an optimal balance has yet to be found.

Third, some industry challenges butt up against the limits of simulation, especially training for recovery from extreme upsets or unusual attitudes. The performance characteristics of air transport aircraft have been measured and modeled for known flight envelopes defined in pitch, roll, yaw, acceleration, and speed. When an aircraft is upset beyond those limits, its performance is not quantified and modeled in a way that can be incorporated into simulation. So, the simulator must extrapolate from nearby known values and might perform differently than the aircraft, leading the pilot to dysfunctional recovery strategies. Further, forces exerted on the aircraft and its occupants in extreme upset cannot be recreated in simulators. As a result, the pilot may be surprised, startled, or misled in interpreting forces encountered in an actual event. Industry collaboration resulting in the Airplane Upset Recovery Training Aid (Upset Recovery Industry Team, 2008) offers a set of maneuvers that can be practiced within current simulation envelopes. FAA (2013b) will require extended envelope training, including upset recovery and recovery from full stalls and stick pusher activation. This necessitates a revision to requirements for flight simulation training devices; a notice of proposed rulemaking (NPRM) for 14 CFR Part 60 has been published in the Federal Register (2014b).

SUMMARY

Requirements for flight training are both well established and evolving in response to systems of feedback. Human factors science and practice has improved training by expanding its focus from individual and maneuvers-based events to line-oriented training and evaluation, strengthening the ability to perform as a crewmember in context. Hazard and task analyses and ISD have influenced the design of operating procedures and the focus of training, especially with increasing levels of cockpit automation and recognition of unique skills for pilot flying and monitoring roles. FOQA, ASAP, and LOSA have become a driver of training emphasis. Specific issues identified through feedback systems and accidents have required significant additional training modules and programs. Controversies remain, but human factors researchers and practitioners are closely involved in developing solutions.

ENVIRONMENTAL CONTEXT

The outermost layer of the sociotechnical model is the environmental layer that encompasses the regulatory, social, and cultural influences in aviation safety. Chapter 9 describes regulatory aspects in detail. In this chapter, we take one example—pilot fatigue—and explore the interaction of these factors in the development of 14 CFR 117, *Flight and Duty Limitations and Rest Requirements: Flightcrew Members.*

FLIGHT, DUTY, AND REST REGULATIONS

Experience and research results gathered over the years have established that pilot fatigue interacts with essentially all aspects of flight operations, and as mentioned earlier, its impact depends on the scheduled flight time, how long the pilot has been awake, how much sleep was attained in a 24-hour period, and if sleep had been reduced multiple times over a trip sequence before recovery rest opportunities were received (Dinges, Graeber, Rosekind, Samel, & Wegmann, 1996; Mallis, Banks, & Dinges, 2010). But the U.S. regulations for flight, duty, and rest remained complex and unchanged until an effort had been initiated in the early 1990s.

The FAA considered revisions to Title 14 CFR part 121 flight, duty, and rest regulations in June 1992. They chartered an Aviation Rulemaking Advisory Committee to determine a path for changing the regulations to improve the mitigation of fatigue and maintain alertness and flight safety. Convened over a rather long period, the Aviation Rulemaking Advisory Committee members could not agree on recommendations or the path forward. Nevertheless, the FAA issued a NPRM in 1995 and received more than 2,000 comments from industry associations, pilot union groups, and the flying public. Although the commenters, including the NTSB, NASA, Air Line Pilots Association, and Allied Pilots Association, agreed that the proposal would enhance safety, many industry associations opposed the 1995 NPRM, stating the FAA lacked safety data to justify the rulemaking and that compliance would impose significant costs to the industry. The FAA never finalized the 1995 rulemaking and eventually withdrew it because it was outdated and raised many significant issues that needed to be reconsidered before proceeding with a final rule.

ACCIDENT PROMPTING RULEMAKING CHANGE

A national debate was triggered on the effects of fatigue and flight safety in 2009 following the investigation of the crash of Colgan Air Flight 3407. At that time, the NTSB had listed fatigue as one of the top 10 safety issues since the early 1990s. The NTSB database contained 23 accidents (with a total of 349 fatalities) since 1991, as having causal or contributing fatigue-related factors.

The accident statistics combined with the knowledge that sleep, circadian rhythm, and fatigue sciences were never considered in the original flight, duty, and rest regulations, prompted the FAA to charter an aviation rulemaking committee on June 24, 2009. This committee, comprised of members from industry and labor, met with the FAA and science representatives weekly between July and August. The aviation rulemaking committee provided its recommendations to the FAA on September 9, 2009. Transcribing the recommendations into rulemaking language took nearly a year, and as a result, the FAA published an NPRM on September 14, 2010, in the Federal Register. Even with a brief period for public review and comment, the FAA received more than 8,000 comments in response to the NPRM. During resolution of all those comments, the FAA made a number of functional changes to the regulatory provisions proposed in the NPRM. The final rule was published in the Federal Register January 4, 2012, and Title 14 CFR Parts 117, 119, and 121 flightcrew member duty and rest requirements was fully implemented two years later on January 4, 2014. Part 117 now governs flight and duty limitations and rest requirements for all certificate holders and flightcrew members conducting passenger operations under Part 121.

It should be mentioned that during the rulemaking timeframe, the President signed into Public Law111-216, the Airline Safety and Federal Aviation Administration Extension Act of 2010. This mandate was published in FAA Order 8900.1 CHG 301, Section 1, review and acceptance of fatigue risk-management plans (FRMP). In Section 212(b), the mandate required each air carrier conducting operations under 14 CFR part 121 to develop, implement, and maintain an FRMP, which, effectively, is an air carrier's fatigue management plan for reducing the potential effects of flightcrew member's fatigue and improving flightcrew member alertness on a day-to-day basis within the structure of the flight, duty, and rest regulation.

Several ACs were also developed during this timeframe and released as supporting guidance and information relevant to fatigue mitigation and training, including these documents:

- AC 120–100, Basics of Aviation Fatigue
- AC 117–1, Flightcrew Member Rest Facilities
- AC 117–2, Fatigue Education and Awareness Training Program
- AC 117–3, Fitness for Duty
- Clarification of the Flight, Duty, and Rest Requirements of Part 117 (Docket No. FAA-2012-0358)

FATIGUE RISK-MANAGEMENT SYSTEM

In addition to the new rule and the mandate for FRMP, Section 117.7, FRMS of the rule contains an optional provision that permits air carriers the flexibility for conducting operations that might exceed the rule's limitations. An FRMS can be proposed to the FAA with enhanced fatigue management approaches that provides an *alternative method of compliance* (AMOC) to the rule. Guidance for the authorization process is found in the FAA AC 120-103A—FRMSs for aviation safety. This AC provides certificate holders with detailed guidance and step-by-step procedures required for the development of an FRMS proposal and the approval process. The FAA's five-step FRMS authorization process requires a systematic and progressive approach, including scientific review and evaluation of the proposed alternative methods.

So, an FRMS employs multilayered defensive strategies to manage fatigue-related risks regardless of their source. For the FAA, the FRMS is an AMOC based on specific numerical standards, for example, the maximum allowable time of the flight duty period and/or the amount of rest required before and after specific types of operations. The AMOC must target the specific limitation(s) from which the carrier seeks relief and justify its request for exceeding those limits. In addition, the AMOC must provide a demonstration of an equivalent level of safety. This is accomplished through a required data collection phase of the authorization process that is specific to the proposed AMOC. Each flight exemption must be examined prior to data collection, and the resultant data analyses are *validated* by the FAA before the air carrier is granted an FRMS authorization in the form of Operation Specification A318.

In essence, the air carriers seeking an FRMS authorization must collect data to verify that the sleep and performance results are equivalent or better than found for the targeted limitation(s). Essentially, the carrier must demonstrate that its proposed AMOC with enhanced fatigue mitigations produces the same outcomes as a safety standard operation, which is defined as a flight operation having comparable features, such as departure/arrival times, direction of flight, (approximate) flight time, and operating within the limitations of Part 117.

In addition to reviewing all of the FRMS documentation submitted by the carrier, the FAA must review the empirical sleep and performance data and its analysis to validate the AMOC. Throughout the latter stages of the FRMS process, the carrier is responsible for submitting those data to the FAA at various intervals, which range from monthly, during the initial data collection and validation steps; quarterly, once an authorization is issued; then semiannually; and then yearly. The data collection requirements continue until the carrier no longer requires the FRMS authorization or the rule changes.

ICAO AND FATIGUE MANAGEMENT

In parallel with the United States developments of improved fatigue management, the international aviation community has been addressing similar issues and making its own regulatory revisions and recommendations for Member States to incorporate nonprescriptive approaches, such as FRMS, to manage flightcrew fatigue. In 2011, ICAO released two guidance documents titled "FRMS Implementation Guide for Operators" and the "FRMS Manual for Regulators" following a major international task force effort chaired by Dr. Curt Graeber.

Very similar to the FAA approach, the resultant ICAO standards and recommended practices relating to the management of fatigue experienced by flight and cabin crew provide a high-level regulatory framework for both prescriptive flight and duty limitations and FRMS as methods for managing fatigue risk. Both the United States and the International methods share two important basic features:

1. They are required to take into consideration the dynamics of transient and cumulative sleep loss and recovery, the circadian biological clock, and the impact of workload on fatigue along with operational requirements.
2. As fatigue is affected by all waking activities and not only work demands, regulations for both are necessarily predicated on the need for shared responsibility between the operator and individual crewmembers for its management. So, whether complying with prescriptive flight and duty limitations or using an FRMS, operators are responsible for providing schedules that allow crewmembers to perform at adequate levels of alertness, and crewmembers are responsible for using that time to start work well-rested (IATA, ICAO, IFALPA, 2011).

After a few years of experience by some member states to develop and implement functional FRMS programs, it became clear that countries varied significantly in their approaches, with some facing greater challenges than others. ICAO is now building on its experiences and is developing new detailed guidance documents that provide more instructions on the *how to* develop and implement an FRMS to complement its recommendations of fatigue-smart regulations. These taskforce efforts were led by ICAO's Dr. Michelle Millar and released in April 2016.

SUMMARY

A great deal of progress has been made in recent history toward the recognition and establishment of sound, science-based fatigue risk-management practices. First, the U.S. FAA implemented the FRMP mandate for air carriers to manage fatigue on a day-by-day basis. Second, the FAA fully implemented a new flight time, duty, and rest regulation with its internal interacting components of fatigue mitigation. And third, the *optional* FRMS procedures are available for air carriers to propose enhanced fatigue mitigation strategies with required data collection for those special instances in which flight operations exceed the limitations of the rule.

The FAA finds that a comprehensive program for the management of fatigue risk in aviation requires an integrated, multifaceted, and shared approach. Based on the successes of many organizations within the aviation industry, it is clear that with genuine effort, the risks associated with fatigue can be managed in ways that can benefit all. But it does require a commitment from all parties, including the regulators, the air carriers, and the flight crewmembers. First and foremost, science and operationally based regulations and flight scheduling practices go a long way toward managing fatigue. Moreover, contributing to success is the air carrier's leadership and its demonstration of commitment to the management of fatigue by providing proper resources including a disciplined, risk-based approach (e.g., FRMP) that is carried out with appropriate data reporting and communications with all relevant stakeholders. And finally, success comes from the professionalism of the flightcrew, its operational understanding of fatigue, and the importance of exercising personal fatigue risk-management principles and good sleep hygiene. All combined, the aviation industry now has the essential tools to improve the management of fatigue and to maintain safety within the National Airspace System and beyond its borders.

REFERENCES

Abbott, K. H. (2001). Human factors engineering and flight deck design. In C.R. Spitzer (Ed.), *Digital avionics handbook* (pp. 9-1–9-15). Boca Raton, FL: CRC Press.

Akerstedt, T. (1995). Work hours, sleepiness and underlying mechanism. *Journal of Sleep Research, 4*(S2), 15–22.

Barshi, I., Dismukes, R. K., & Loudopoulos, L. D. (2012). *The multitasking myth: Handling complexity in real-world operations.* Aldershot, UK: Ashgate Publishing.

Battelle Report (1998). *An overview of the scientific literature concerning fatigue, sleep, and the circadian cycle.* Battelle Memorial Institute. Prepared for the Office of the Chief Scientific and Technical Advisor for Human Factors, Federal Aviation Administration. Retrieved from: http://cf.alpa.org/internet/projects/ftdt/backgr/batelle.htm.

Billings, C. E. (1997). *Aviation automation: The search for a human-centered approach.* Mahwah, NJ: Lawrence Erlbaum Associates.

Bonnet, M. H. (2000). Sleep deprivation. In M. H. Kryger, T. Roth, & W. C. Dement (Eds.), *Principles and practice of sleep medicine* (3rd ed., pp. 53–71). Philadelphia, PA: Saunders.

Boucek, G. P., Erickson, J. B., Berson, B. L, Hanson, D. C., & Leffler, M. F. (1980). *Aircraft alerting systems standardization study phase 1 final report* (FAA-RD-80-68). Washington, DC: Federal Aviation Administration.

Bureau d'Enquêtes et d'Analyses. (2012). *Final report on the accident to the air bus A330-203 Flight AF 447 on 1st June 2009.* Retrieved from http://www.bea.aero/docspa/2009/f-cp090601.en/pdf/f-cp090601.en.pdf.

Caldwell, J. A. (1997). Fatigue in the aviation environment: an overview of the causes and effects as well as recommended countermeasures. *Aviation, Space, and Environmental Medicine, 68*, 932–938.

Caldwell, J. A., Mallis, M. M., Caldwell, J. L., Paul, M. A., Miller, J. C., & Neri, D. F. (2009). Aerospace medical association aerospace fatigue countermeasures subcommittee of the human factors committee. Fatigue countermeasures in aviation. *Aviation, Space, and Environmental Medicine, 80*, 29–59.

Cardosi, K., & Huntley, M. S. (1993). *Human Factors for Flight Deck Certification Personnel* (DOT-VNTSC-FAA-93-4; DOT-FAA-RD-93-5). Cambridge, MA: US DOT Volpe National Transportation Systems Center. Available from: http://ntl.bts.gov/lib/33000/33500/33543/33543.pdf.

Cardosi, K. M., & Murphy, E. D. (1995). *Human Factors in the Design and Evaluation of Air Traffic Control Systems* (DOT-VNTSC-FAA-95-3; DOT/FAA/RD-95-3). Washington, DC: US DOT Office of Aviation Research. Available from: http://ntl.bts.gov/lib/33000/33600/33633/33633.pdf.

Carskadon, M. A., & Dement, W. C. (1987). Daytime sleepiness: Quantification of a behavioral state. *Neuroscience & Biobehavioral Reviews, 11,* 307–317.

Carskadon, M. A., & Dement, W. C. (2011). Normal human sleep: An overview. In M. H. Kryger, T. Roth, & W. C. Dement (Eds.), *Principles and practice of sleep medicine* (5th ed., pp. 16–28). Canada: Elsevier.

Chidester, T. R. (1999). Introducing FMS aircraft into airline operations. In S. Dekker & E. Hollnagel (Eds.), *Coping with computers in the cockpit* (pp. 153–194). Aldershot, UK: Ashgate Publishing.

Commercial Transportation Operator Fatigue Management Reference. (2003). Prepared for: U.S. Department of Transportation, Research and Special Programs Administration. K. Thirumalai and C. Comperatore. Prepared by: M. McCallum, T. Sanquist, M. Mitler, and G. Krueger. Work Performed Under Other Transaction Agreement No. DTRS56-01-T-003.

Cooper, G. E., White, M. D., & Lauber, J. K. (Eds.) (1980). Resource management on the flightdeck. *Proceedings of a NASA/Industry Workshop*, (National Aeronautics and Space Administration CP-2120). Moffett Field, CA.

Degani, A., & Wiener, E. L. (1993). Cockpit checklists: Concepts, design, and use. *Human Factors, 35,* 345–359.

Department of Defense. (2012). *Design criteria standard: Human engineering* (MIL-STD-1472G). Washington, DC: Department of Defense.

Dinges, D. F. (1989). The nature of sleepiness: Causes, contexts and consequences. In A. Stunkard & A. Baum (Eds.), *Perspectives in behavioral medicine: Eating, sleeping and sex* (pp. 147–179). Hillsdale, NJ: Lawrence Erlbaum.

Dinges, D. F. (1992). Probing the limits of functional capability: The effects of sleep loss on short duration tasks. In R. J. Broughton & R. D. Ogilvie (Eds.), *Sleep, arousal, and performance* (pp. 177–188). Boston, MA: Birkhauser.

Dinges, D. F., Graeber, R. C., Rosekind, M. R., Samel, A., & Wegmann, H. M. (1996). Principles and guidelines for duty and rest scheduling in commercial aviation; NASA Technical Memorandum 110404; 1996.

Dinges, D. F., & Kribbs, N. B. (1991). Performing while sleepy: Effects of experimentally-induced sleepiness. In T. Monk (Ed.), *Sleep, sleepiness, and performance* (pp. 98–128). Chichester, UK: John Wiley & Sons.

Dismukes, R. K., Berman, B. A., & Loukopoulos, L. D. (2007). *The limits of expertise: Rethinking pilot error and the causes of airline accidents.* Aldershot, UK: Ashgate Publishing.

Dodd, S., Lancaster, J., Nelson, E., Rogers, B., DeMers, B., Groethe, S., Miranda, A., & Almquist, J. (2014). *Human Factors Issues with NextGen Flight Deck System Controls.* Minneapolis, MN: Honeywell International.

Eriksen, C. A., & Akerstedt, T. (2006). Aircrew fatigue in trans-Atlantic morning and evening flights. *Chronobiology International, 23*(4), 843–858.

European Aviation Safety Agency. (2007). *Certification specifications for large aeroplanes CS-25,* Amendment 4. Cologne, Germany: EASA.

Federal Aviation Administration (FAA). (1996). *The interface between flightcrews and modern flight deck systems.* Washington, DC. Retrieved from: http://www.tc.faa.gov/its/worldpac/techrpt/hffaces.pdf.

Federal Aviation Administration (FAA). (2000). *Maintenance resource management training* (Advisory Circular 120–72). Washington, DC. Retrieved from: https://www.faa.gov/regulations_policies/advisory_circulars/index.cfm/go/document.information/documentID/23217.

Federal Aviation Administration (FAA). (2002a). *Addressing human factors/pilot interface issues of complex, integrated avionics as part of the technical standard order (TSO) process* (N 8110.98). Washington, DC: Federal Aviation Administration.

Federal Aviation Administration (FAA). (2002b). *Aviation safety action program* (Advisory Circular 120-66B). Washington, DC. Retrieved from: https://www.faa.gov/regulations_policies/advisory_circulars/index.cfm/go/document.information/documentID/23207.

Federal Aviation Administration (FAA). (2002c). *Pilot vision.* Oklahoma City, OK: FAA Civil Aerospace Medical Institute.

Federal Aviation Administration (FAA). (2002d). *Public statement number PSACE100- 2001–004 on guidance for reviewing certification plans to address human factors for certification of part 23 small airplanes.* Washington, DC: Federal Aviation Administration.

Federal Aviation Administration (FAA). (2004a). *Crew resource management training* (Advisory Circular 120-51E). Washington, DC. Retrieved from: http://www.faa.gov/documentLibrary/media/Advisory_Circular/AC120-51e.pdf.

Federal Aviation Administration (FAA). (2004b). *Flight operational quality assurance* (Advisory Circular 120-82). Washington, DC. Retrieved from: https://www.faa.gov/regulations_policies/advisory_circulars/index.cfm/go/document.information/documentID/23227.

Federal Aviation Administration (FAA). (2005a). Cockpit automation issues. In *Aviation Human Factors* (Chapter 8.4). Washington, DC: FAA.

Federal Aviation Administration (FAA). (2005b). *Dispatch resource management training* (Advisory Circular 121-31A). Washington, DC. Retrieved from: https://www.faa.gov/regulations_policies/advisory_circulars/index.cfm/go/document.information/documentID/22513.

Federal Aviation Administration (FAA). (2006a). *Approval of flight guidance systems* (Advisory Circular 25.1329-1B). Washington, DC. Retrieved from: http://www.faa.gov/documentLibrary/media/Advisory_Circular/AC%2025.1329-1B.pdf.

Federal Aviation Administration (FAA). (2006b). *Line operations safety audits* (Advisory Circular 120-90). Washington, DC. Retrieved from: https://www.faa.gov/regulations_policies/advisory_circulars/index.cfm/go/document.information/documentID/22478.

Federal Aviation Administration (FAA). (2008). *Airline transport pilot and aircraft type rating practical test standards for airplane* (FAA-S-8081-5F). Washington, DC. Retrieved from: http://www.faa.gov/training_testing/testing/test_standards/media/atp_pts.pdf.

Federal Aviation Administration (FAA). (2009). *Risk management handbook* (FAA-H-8083-2). Federal Aviation Administration: Washington, DC. Retrieved from: https://www.faa.gov/regulations_policies/handbooks_manuals/aviation/risk_management_handbook/.

Federal Aviation Administration (FAA). (2010). *Flightcrew alerting* (Advisory Circular 25.1322–1). Washington, DC. Retrieved from: http://www.faa.gov/documentLibrary/media/Advisory_Circular/AC%2025.1322-1.pdf.

Federal Aviation Administration (FAA). (2011a). *Controls for flight deck systems* (Advisory Circular 20-175). Washington, DC. Retrieved from: http://www.faa.gov/documentlibrary/media/advisory_circular/AC%2020-175.pdf.

Federal Aviation Administration (FAA). (2011b). *Installation of electronic display in part 23 airplanes* (Advisory Circular 23.1311-1C). Washington, DC. Retrieved from: http://www.faa.gov/documentLibrary/media/Advisory_Circular/AC%2023.1311-1C.pdf.

Federal Aviation Administration (FAA). (2012). 14 CFR, Parts 117, 119, and 121: Flightcrew Member Duty and Rest Requirements. Docket No. FAA-2009-1093; Amendment Nos. 117-1, 119-16, 121-357; RIN 2120-AJ58.

Federal Aviation Administration (FAA). (2013a). *Installed systems and equipment for use by the flightcrew* (Advisory Circular 25.1302-1). Washington, DC. Retrieved from: http://www.faa.gov/documentLibrary/media/Advisory_Circular/AC_25.1302-1.pdf.

Federal Aviation Administration (FAA). (2013b). *Operational use of flight path management systems*. Washington, DC. Retrieved from: http://www.faa.gov/about/office_org/headquarters_offices/avs/offices/afs/afs400/parc/parc_reco/media/2013/130908_PARC_FltDAWG_Final_Report_Recommendations.pdf.

Federal Aviation Administration (FAA). (2014c). *Electronic flight displays* (Advisory Circular 25-11B). Washington, DC. Retrieved from: http://www.faa.gov/documentLibrary/media/Advisory_Circular/AC_25-11B.pdf.

Federal Aviation Administration (FAA). (2015a). *Flightcrew member line operational simulations: Line-oriented flight training, special purpose operational training, line operational evaluation* (Advisory Circular 120-35D). Washington, DC. Retrieved from: http://www.faa.gov/documentLibrary/media/Advisory_Circular/AC_120-35D.pdf.

Federal Aviation Administration (FAA). (2015b). *Lessons learned from transport airplane accidents*. Retrieved from: http://lessonslearned.faa.gov/.

Federal Register. (2014a). *Final rule; technical amendment by the federal aviation administration: qualification, service, and use of crewmembers and aircraft dispatchers* (14 CFR Subchapter I, Part 121). Washington, DC. Retrieved from: https://www.federalregister.gov/articles/2014/03/07/2014-04902/qualification-service-and-use-of-crewmembers-and-aircraft-dispatchers.

Federal Register. (2014b). *Notice of proposed rulemaking: Flight simulation training device qualification standards for extended envelope and adverse weather event training tasks* (14 CFR Part 60). Washington, DC. Retrieved from: https://www.federalregister.gov/articles/2014/07/10/2014-15432/flight-simulation-training-device-qualification-standards-for-extended-envelope-and-adverse-weather.

Flight Safety Foundation. (2000). *Approach and landing accident reduction tool kit*. Alexandria, VA. Retrieved from: http://flightsafety.org/files/alar_bn7-1stablizedappr.pdf.

Flight Safety Foundation. (2014). *A practical guide for improving flight path monitoring*. Alexandria, VA. Retrieved from: http://flightsafety.org/files/flightpath/EPMG.pdf.

Foushee, H. C. (1984). Dyads and triads at 35,000 feet. *American Psychologist, 39*, 885–893.

Gabree, S., Chase, S., & Cardosi, K. (2014). *Use of color on airport moving maps and cockpit displays of traffic information* (CDTIs) (DOT-VNTSC-FAA-14-13 & DOT/FAA/TC-14/18). Cambridge, MA: US DOT Volpe Center.

Gander, P., van den Berg, M., Mulrine, H., Signal, L., & Mangie, J. (2013). Circadian adaptation of airline pilots during extended duration operations between the USA and Asia. *Chronobiology International, 30*(8), 963–972.

Gander, P. H., Graeber, R. C., Foushee, H. C., Lauber, J. K., & Connell, L. J. (1994). *Crew factors in flight operations II: Psycho physiological responses to short haul air transport operations*. NASA Technical Memorandum 108856.

Gander, P. H., Myhre, G., Graeber, R. C., Andersen, H. T., & Lauber, J. K. (1989). Adjustment of sleep and the circadian temperature rhythm after flights across nine time zones. *Aviation, Space, and Environmental Medicine, 60*, 733–743.

Garner. K. T., & Assenmacher, T. J. (1997). *Situational Awareness Guidelines*. Naval Air Systems Command: Washington, DC.

General Aviation Manufacturers Association (GAMA). (2000). *Recommended Practices & Guidelines for Part 23 Cockpit/Flight Deck Design* (GAMA Publication No. 10). Washington, DC: General Aviation Manufacturers Association.

Goode, J. H. (2003). Are pilots at risk of accidents due to fatigue? *Journal of Safety Research, 34*, 309–313.

Government Publishing Office (GPO). (2003). Special federal aviation regulation 58: Advanced qualification program. *United States Code of Federal Regulations*, 14 CFR Subchapter I Part 121. Retrieved from: http://www.gpo.gov/fdsys/pkg/CFR-2004-title14-vol2/pdf/CFR-2004-title14-vol2-part121-appFederal-id903.pdf.

Hancock, P. A., & Verwey, W. B. (1997). Fatigue, workload and adaptive driver systems. *Accident Analysis and Prevention, 29*(4), 495–506.

Hancock, P. A., & Warm, J. S. (1989). A dynamic model of stress and sustained attention. *Human Factors, 31*, 519–537.

Harris, D., Stanton, N. A., Marshall, A. Y. M. S., Demagalski, J., & Salmon, P. (2005). Using SHERPA to predict design-induced error on the flight deck. *Aerospace Science and Technology, 9*(6), 525–532.

Hart, S. G., & Staveland, L. E. (1988). Development of NASA-TLX (task load index): Results of empirical and theoretical research. In P. A. Hancock & N. Meshkati (Eds.), *Human mental workload*. Amsterdam, the Netherlands: North Holland Press.

Hawkins, F. (1987). *Human factors in flight*, 2nd ed. Aldershot, UK: Avebury Aviation.

Helmreich, R. L. (1998). Error management as organisational strategy. *Proceedings of the IATA Human Factors Seminar* (pp. 1–7). Bangkok, Thailand: International Air Transport Association.

Horne, J. A., & Reyner, L. A. (1985). Sleep related vehicle accidents. *British Medical Journal, 310*, 565–567.

Human Factors Harmonization Working Group. (2004). Flight Crew Error / Flight Crew Performance Considerations in the Flight Deck Certification Process.

Hursh, S. R., & Van Dongen, H. P. A. (2010). Fatigue and performance modeling. In M. H. Kryger, T. Roth, & W. C. Dement (Ed.), *Principles and practice of sleep medicine*, 5th ed (pp. 745–752). Philadelphia, PA: Saunders.

International Air Transport Association (IATA), International Civil Aviation Organization (ICAO), International Federation of Air Line Pilots' Associations (IFALPA). (2011). *Fatigue risk management systems: Implementation guide for operators*. 1st ed. Montreal: ICAO.

International Civil Aviation Organization (ICAO). (2012). *Fatigue risk management systems—Manual for regulators*, 1st ed (Doc. 9966). Montreal: International Civil Aviation Organization (ICAO).

Kecklund, G., Di Milia, L., Axelsson, J., Lowden, A., & Akerstedt, T. (2012). 20th international symposium on shiftwork and working time: Biological mechanisms, recovery, and risk management in the 24-h society. *Chronobiology International, 29* (5), 531–536.

Mallis, M. M., Banks, S., & Dinges, D. F. (2010). Aircrew fatigue, sleep need and circadian rhythmicity. In E. Salas, T. Allard, & D. Maurino (Eds.), *Human Factors in Aviation,* 2nd ed. Burlington, MA: Elsevier.

McAnulty, D. M. (1995). *Guidelines for the Design of GPS and LORAN Receiver Controls and Displays* (DOT-VNTSC-FAA-95-7; DOT/FAA/RD-95-1). Cambridge, MA: US DOT Volpe National Transportation Systems Center. Available from: http://ntl.bts.gov/lib/33000/33500/33504/33504.pdf.

Miller, J. C. (2006). *Fundamentals of shiftwork scheduling.* Human Effectiveness Directorate Biosciences and Protection Division, Biobehavioral Performance Branch Brooks City-Base, TX. AFRL-HE-BR-TR-2006-0011.

Mitman, R., Neumeier, M., Reynolds, M., & Rehmann, A. (1994). *Flight Deck Crew Alerting: Problems and Concerns as Reported in the Aviation Safety Reporting System* (SAE technical report no. 942096).

Mollicone, D. J., Van Dongen, H., Rogers, N. L., & Dinges, D. F. (2008). Response surface mapping of neurobehavioral performance: Testing the feasibility of split sleep schedules for space operations. *Acta Astronaut, 63*(7–10), 833–840.

Moray, N. (2006). Culturing safety for railroads. *Transportation Research Circular E-C085, Railroad Operational Safety: Status and Research Needs.* Washington, DC: Transportation Research Board.

Moray, N. P., & Huey, M. (1988). *Human factors research and nuclear safety.* Washington, DC: National Academy Press.

Mumaw, R. J., Sarter, N. B., & Wickens, C. D. (2001). Analysis of pilots' monitoring and performance on an automated flight deck. *Proceedings of the 11th International Symposium on Aviation Psychology.* Columbus, OH: The Ohio State University.

Naitoh, P. (1975). Sleep deprivation in humans. In P. H. Venables & M. J. Christie (Eds.), *Research in psychophysiology.* London, UK: John Wiley & Sons.

National Research Council. (1998). *Improving the continued airworthiness of civil aircraft: A strategy for the FAA's aircraft certification service.* Washington, DC: National Academy Press. Retrieved from: http://www.nap.edu/catalog/6265/improving-the-continued-airworthiness-of-civil-aircraft-a-strategy-for.

National Transportation Safety Board (NTSB). (1982). *Air Florida, Inc., Boeing 737-222, N62AF* (Accident Report NTSB-AAR-82-8). Washington, DC. Retrieved from: http://libraryonline.erau.edu/online-full-text/ntsb/aircraft-accident-reports/AAR82-08.pdf.

National Transportation Safety Board (NTSB). (1986). *Delta Air Lines, Inc., Lockheed L-1011-385-1, N726DA* (Accident Report NTSB/AAR-86/05). Washington, DC. Retrieved from: http://libraryonline.erau.edu/online-full-text/ntsb/aircraft-accident-reports/AAR86-05.pdf.

National Transportation Safety Board (NTSB). (1993). *Takeoff stall in icing conditions* (Accident Report NTSB/AAR-93/02 PB93-910402). Washington, DC. Retrieved from: http://libraryonline.erau.edu/online-full-text/ntsb/aircraft-accident-reports/AAR93-02.pdf#page=85.

National Transportation Safety Board (NTSB). (2007). *Crash of Pinnacle Airlines Flight 3701, Bombardier CL-600-2B19, N8396A, Jefferson City, Missouri, October 14, 2004* (Accident Report NTSB/AAR-07/01. PB2007-910402). Washington, DC: National Transportation Safety Board.

National Transportation Safety Board (NTSB). (2010). *Loss of control on approach* (Accident Report NTSB/AAR-10/01 PB2010-910401). Washington, DC. Retrieved from: http://www.ntsb.gov/investigations/AccidentReports/Reports/AAR1001.pdf.

National Transportation Safety Board (NTSB). (2014a). *Crash during a nighttime nonprecision instrument approach to landing* (Accident Report NTSB/AAR-14/02, PB2014-107898). Washington, DC. Retrieved from: http://www.ntsb.gov/investigations/AccidentReports/Reports/AAR1402.pdf.

National Transportation Safety Board (NTSB). (2014b). *Descent below visual glidepath and impact with seawall* (Accident Report NTSB/AAR-14/01, PB2014-105984). Washington, DC. Retrieved from: http://www.ntsb.gov/investigations/AccidentReports/Reports/AAR1401.pdf.

National Transportation Safety Board (NTSB). (2015). In-flight breakup during test flight scaled composites spaceshiptwo, N339SS near Koehn Dry Lake, California. Washington, DC. Retrieved from: http://www.ntsb.gov/news/events/Documents/2015_spaceship2_BMG_abstract.pdf.

Nelson, E., DeMers, B., Dodd, S., Grothe, S., Lancaster, J., Miranda, A., & Rogers, B. (2014). *Chart legibility on portable electronic devices.* Retrieved from http://fsims.faa.gov/wdocs/other/chart%20legibility%20on%20portable%20electronic%20devices.pdf.

Nielsen, J. (1994). *Usability engineering.* San Francisco, CA: Morgan Kaufmann.

Orlady, H. W., & Orlady, L. M. (1999). *Human factors in multi-crew flight operations.* Aldershot, UK: Ashgate Publishing.

Palmer, M. T., Rogers, W. H., Press, H. N., Latorella, K. A., & Abbott, T. S. (1995). *PA crew-centered flight deck design philosophy for high-speed civil transport (HSCT) aircraft,* (NASA Technical Memorandum 109171). Hampton, VA: NASA Langley Research Center.

Parasuraman, R., & Riley, V. A. (1997). Humans and automation: Use, misuse, disuse, abuse. *Human Factors, 39,* 230–253.

Patterson, R. (1982). *Guidelines for auditory warning systems on civil aircraft* (CAA Paper 82017). London, UK: Civil Aviation Authority.

Peigneau, P., & Smith, C. (2011). Memory processing in relation to sleep. In M. H. Kryger, T. Roth, & W. C. Dement (Eds.), *Principles and practice of sleep medicine* (5th ed., pp. 335–347). Canada: Elsevier.

Peryer, G., Noyes, J., Pleydell-Pearce, K., & Lieven, N. (2005). Auditory alert characteristics: A survey of pilot views. *The International Journal of Aviation Psychology, 15*(3), 233–250.

Powell, D. M. C., Spencer, M. B., Holland, D., Broadbent, E., & Petrie, K. J. (2007). Pilot fatigue in short-haul operations: Effects of number of sectors, duty length, and time of day. *Aviation Space Environmental Medicine, 78,* 698–701.

Powell, D. M. C., Spencer, M. B., Holland, D., & Petrie, K. J. (2008). Fatigue in two-pilot operations: Implications for flight and duty time limitations. *Aviation Space Environmental Medicine, 79*(11), 1047–1050.

Radio Technical Commission for Aeronautics (RTCA). (2006). *Minimum operational performance standards for global positioning system/wide area augmentation system airborne equipment* (RTCA DO-229D). Washington, DC: RTCA.

Rasmussen, J. (1983). Skills, rules, and knowledge; Signals, signs, and symbols and other distinctions in human performance models. *IEEE Transactions on Systems, Man, and Cybernetics, 13*(3), 257–266.

Reason, J. (1990). *Human error.* Cambridge University Press: Cambridge.

Reid, G. B., & Nygren, T. E. (1988). The subjective workload assessment technique: A scaling procedure for measuring mental workload. In P. A. Hancock & N. Meshkati (Eds.), *Human mental workload.* (pp. 185–218). Amsterdam, the Netherlands: North Holland Press.

Roscoe, A. H., & Ellis, G. A. (1990). *A subjective rating scale for pilot workload in flight: A decade of practical use* (Technical Report TR 90019). Farnborough, Hampshire: Royal Aerospace Establishment.

Salvi, S. M., Akhtar, S., & Currie, Z. (2006). Ageing changes in the eye. *Postgraduate Medical Journal, 82*(971), 581–587.

Samel, A., Wegmann, H. M., & Vejvoda, M. (2005). Aircrew fatigue in long-haul operations. *Accident Analysis & Prevention, 29*(4), 439–452.

Shappell, S. (2006). *Human error and commercial aviation accidents: A comprehensive, fine-grained analysis using HFACS* (DOT/FAA/AM-06/18). Washington, DC: Federal Aviation Administration.

Skaggs, V., Norris, A., & Johnson, R. (2012). *2010 Aeromedical certification statistical handbook* (DOT/FAA/AM-12/3). Oklahoma City, OK: FAA Civil Aerospace Medical Institute.

Sumwalt, R. L. III., Thomas, R. J., & Dismukes, R. K. (2003). The new last line of defense against aviation accidents. *Aviation Week and Space Technology, 159,* 66.

Swain, A. D., & Guttmann, H. E. (1983). *Handbook of human reliability analysis with emphasis on nuclear power plant applications.* NUREG/CR-1278. Washington, DC.

ULR Crew Alertness Steering Committee and FSF Editorial Staff (2005). Lessons from the dawn of ultra-long-range flights. *Flight Safety Digest, 24,* 8–9, (August–September 2005).

Upset Recovery Industry Team. (2008). *Airplane upset recovery training aid, revision 2.* Federal Aviation Administration & Airline Industry Sponsors, Washington, DC. Retrieved from: https://www.faa.gov/other_visit/aviation_industry/airline_operators/training/media/AP_UpsetRecovery_Book.pdf.

U.S. Congress, Office of Technology Assessment. (1988) *Safe skies for tomorrow: Aviation safety in a competitive environment,* OTA-SET-381. Washington, DC: U.S. Government Printing Office.

Van Cauter, E., & Tasali, E. (2011). Endocrine physiology in relation to sleep and sleep disturbances. In M. H. Kryger, T. Roth, & W. C. Dement (Eds.), *Principles and practice of sleep medicine* (5th ed., pp. 291–311). Ontario, Canada: Elsevier.

Veitengruber, J. E., Boucek, G. P. Jr., & Smith, W. D. (1977). *Aircraft Alerting Systems Criteria Study Volume 1: Collation and Analysis of Aircraft Alerting System Data.* (FAA-RD-76-222I). Washington, DC: US DOT Federal Aviation Administration.

Weiner, E. L., & Nagel, D. (Eds.). (1988). *Human factors in aviation.* San Diego, CA: Academic Press.

Wickens, C. D. (1993). Timesharing, workload, and human error. In K. Cardosi & M. S. Huntley (Eds.), *Human factors for flight deck certification personnel* (DOT/FAA?RD-93/5; DOT-VNTSC-FAA-93-4). Cambridge, MA: US DOT Volpe Center.

Wickens, C. D., & Dixon, S. R. (2005). *Is there a magic number 7 (to the minus 1)? The benefits of imperfect diagnostic automation: A synthesis of the literature* (Tech. Rep. AHFD-05-01/MAAD-05-1). Savoy, IL: University of Illinois, Aviation Human Factors Division.

Wickens, C. D., Hollands, J. G., Banbury, S., & Parasuraman, R. (2013). *Engineering psychology and human performance.* New York: Taylor and Francis.

Wickens, C. D., Lee, J., Liu, Y. D., & Gordon-Becker, S. (2004). *Introduction to human factors engineering.* Upper Saddle River, NJ: Pearson.

Wickens, C. D., Sebok, A., Walters, B., McDermott, P., Sarter, N., Wan, Y., & Fennell, K. (2016). *Alert Design Considerations*. Boulder, CO: Alion Science and Technology.

Widdel, H., & Post, D. L. (1992). *Color in electronic displays.* New York: Plenum Press.

Wiener, E. (1988). Cockpit automation. In E. L. Wiener & D. C. Nagel (Eds.), *Human factors in aviation.* (pp. 433–461). New York: Academic.

Wiener, E. L., Kanki, B. G., & Helmreich, R. L. (Eds.). (1993). *Cockpit resource management.* San Diego, CA: Academic Press.

Wood, S. (2004). *Flight crew reliance on automation* (CAA Paper 2004/10). West Sussex, UK: Civil Aviation Authority.

Yeh, M., & Chandra, D. C. (2004). Issues in symbol design for electronic displays of navigation information. *Proceedings of HCI-Aero 2004* (September 29–October 1). Toulouse, France.

Yeh, M., & Chandra, D. C. (2005). *Designing and evaluating symbols for electronic displays of navigation information: Symbol stereotypes and symbol-feature rules* (DOT/FAA/AR-05/48; DOT-VNTSC-FAA-05-16). Cambridge, MA: US DOT Volpe National Transportation Systems Center. Retrieved from: http://ntl.bts.gov/lib/47000/47500/47566/dot-vntsc-faa-05-16.pdf.

Yeh, M., & Chandra, D. C. (2008). *Survey of symbology for aeronautical charts and electronic displays: Navigation aids, airports, lines, and linear patterns* (DOT/FAA/AR-07/66; DOT-VNTSC-FAA-08-01). Cambridge, MA: US DOT Volpe National Transportation Systems Center. Retrieved from: http://ntl.bts.gov/lib/34000/34900/34909/DOT-VNTSC-FAA-08-01.pdf.

Young, M. S., & Stanton, N. A. (2002). Malleable attentional resources theory: A new explanation for the effects of mental underload on performance. *Human Factors, 44*(3), 365–375.

Zuschlag, M., Chandra, D., & Grayhem, R. (2013). *The usefulness of the proximate status indication as represented by symbol fill on cockpit displays of traffic information* (DOT-VNTSC-FAA-13-03; DOT/FAA/TC-13/24). Cambridge, MA: US DOT Volpe National Transportation Systems Center. Retrieved from: http://ntl.bts.gov/lib/47000/47600/47675/DOT-VNTSC-FAA-13-03.pdf.

13 Situation Awareness in Air Traffic Control

Kim-Phuong L. Vu and Dan L. Chiappe

CONTENTS

NextGen and Changes to the Air Traffic Management System 302
Air Traffic Controller Situation Awareness..303
    Situation Awareness in Air Traffic Controller Activities..303
    Situation Awareness Theory and Measurement in Air Traffic Control Contexts 304
        Situation Awareness Global Assessment Technique 307
        Situation Present Assessment Method..307
    Principles for Designing to Support SA...308
    Summary 310
Research on Impact of Potential NextGen Tools and Concepts of Operations............................310
    Effects of NextGen Concepts on SA of Experienced Air Traffic Controllers.........................311
        NextGen Scenario and Concepts of Operation..311
        SA Probe Questions 312
    Impact of Concept on Air Traffic Controller Workload and Situation Awareness 312
    Summary 313
Training Air Traffic Controllers with Potential NextGen Tools..313
    Training Approach...313
    Training Manipulations and Findings 314
    Summary 316
References..316

Air traffic control (ATC) is a highly complex activity that requires air traffic controllers (ATCos) to interact with various technical and social elements of a system (e.g., radar screens, Ultra High Frequency/Very High Frequency (UHF/VHF) radios, automation tools, pilots, other ATCos, and air traffic management [ATM] personnel). This challenges their information processing abilities as they seek to acquire and maintain their situation awareness (SA) regarding the ongoing traffic situation (Durso & Alexander, 2010; Endsley & Smolensky, 1998; O'Brien & O'Hare, 2007; Prevot, Homola, Martin, Mercer, & Cabrall, 2012). For example, to ensure that aircraft (AC) are not in danger of losing separation, ATCos have to scan their 2D radar screens to assess the current and future 3D locations of various AC. In projecting these locations, they have to take into account multiple factors such as wind speed, weather, and the performance characteristics of the AC. Despite being a challenging task, aviation remains one of the safest areas in transportation (see accidents rates from the National Transportation Safety Board; http://www.ntsb.gov/). This is due to the fact that human factors has played an important part in shaping the roles, tasks, training, and technology used by ATCos (Roske-Hofstrand & Murphy, 1998). This continues to be the case under the airspace modernization plan known as the Next Generation Airspace System, the goal of which is to increase the capacity of the airspace without compromising its safety (JPDO, 2011). We begin the chapter by outlining some of the changes that are envisioned as part of the NextGen ATM system. Then, we discuss some of the key areas of human factors research that have been carried out, focusing on SA and best practices for training student ATCos in the use of NextGen technologies.

NEXTGEN AND CHANGES TO THE AIR TRAFFIC MANAGEMENT SYSTEM

The main functions of an ATCo include maintaining the safety of flight by keeping AC separate from each other, making traffic flow efficiently, and providing traffic and weather information to pilots (Nolan, 2010). There are three general types of ATCos, operating at different altitudes. First, there are Tower Controllers, who manage traffic on the ground and until AC are a few thousand feet from the ground. Second, there are Terminal Radar Approach Control ATCos, who handle AC from 30 to 50 miles around an airport, and up to 10,000 feet. Third, there are *en route* ATCos who control AC during the high-altitude phases of flight (Durso & Manning, 2008). ATC services are provided only in specified airspaces (i.e., in controlled and positively-controlled airspace). In the United States, the National Airspace System (NAS) is divided into a series of 3D sectors. Each sector is managed by one or more ATCos, and the responsibility for providing services to an AC is passed from one sector controller to the next. Sectors with little traffic often do not need the support of a highly technical ATM system, but traffic in major cities and commercial hubs requires complex coordination and control of AC. Due to the workload placed on controllers in these high-density sectors and the projected increase in traffic demand brought about from passenger travel and e-commerce transport at the start of the millennium, the number of AC estimated to be operating in the U.S. NAS in 2025 is likely to exceed the capacity of the current ATM system. As a result, the Joint Planning and Development Office was formed in 2003 to develop the Next Generation Air Transportation System (NextGen; JPDO, 2011).

NextGen reflects a transformation of the current, radar-based ATM system to a state-of-the-art, satellite-based system for improving the efficiency, security, maintenance, and safety of the airspace. NextGen innovations are already being introduced into the system and will continue to be incrementally implemented over the next several decades. The innovations include changes to existing airspace structures and increased reliance on automation (JPDO, 2011; Kopardekar & Quon, 2011). As automation becomes more complex and capable of higher information-processing tasks, however, it does not replace humans but rather changes the nature of the work they carry out (Parasuraman, Sheridan, & Wickens, 2008). A human operator is typically required to interact with automation as a supervisor and collaborator/team member. Successful implementation of automation for achieving NextGen goals demands that operators use the tools appropriately, taking into account characteristics of the automation and of human operators themselves (Lee & Moray, 1994; McCarley & Wickens, 2005).

Effective operations in complex environments such as NextGen thus require that ATCos rely heavily on automated components of the system. *Automation* refers to when functions that would otherwise be carried out by a human are allocated to a machine (Parasuraman, et al., 2008; Sheridan, 1992). Functions that are candidates for automation span the full range of human information-processing: Information acquisition and analysis, decision-making, and action implementation (Parasuraman, Sheridan, & Wickens, 2000). Automated tools are expected to reduce ATCo workload, a major bottleneck limiting the capacity of current day operations (Kopardekar & Quon, 2011). Successful implementation of these automation solutions for achieving NextGen goals, however, demands that operators use these tools appropriately. In particular, for optimal use, the operator must be aware of, and understand, several characteristics of an automation tool such as its current reliability, false alarm rate, and miss rate (Lee & See, 2004; Manzey, Reichenbach, & Onnasch, 2012; Metzger & Parasuraman, 2005; Parasuraman & Riley, 1997). Less than optimal usage occurs when the operator is not calibrated with respect to the reliability of the automation tool. Overreliance on automation introduces problems such as complacency, loss of SA, and the out-of-the-loop syndrome (e.g., Onnasch, Wickens, Li & Manzey, 2014; Parasuraman et al., 2008; Wickens, 1999). Underreliance prevents automated systems from achieving the purpose for which they were designed. Therefore, to be effective in achieving the goals of NextGen, it is critical that proper human–automation teaming mechanisms and training strategies are in place to ensure that the increased traffic demands being made on the NAS are achieved in a safe and efficient manner.

Several strategies for automating ATC tasks have been proposed to assist controllers in NextGen environments (e.g., Prevot et al., 2012; Willems & Heiney, 2002). For example, a critical automation tool for ATC operations is the conflict detection and resolution (CD&R) system that will partly relieve ATCos of the separation assurance task. The CD&R system, however, can be configured so as to feature different levels of automation, depending on how much autonomy and responsibility is allocated to the machine and how much cognitive and physical activity is required of the human operator (e.g., Endsley & Jones, 2012; Parasuraman et al., 2008). For instance, the tool could be formatted to only alert the ATCo to potential conflicts but leave the ATCo responsible for resolving them. Another possible implementation of the tool is that it could alert the ATCo and suggest conflict resolutions that the controller could accept (management by consent) or reject (management by exception). Furthermore, a fully automated CD&R tool could alert the ATCo of conflicts, but the system would resolve them by sending commands to the pilot without any human intervention. The latter concept would produce the greatest reductions in workload, depending on the extent to which the ATCo trusts and uses the CD&R system.

It is important, however, to examine broader consequences of adopting different levels of automation. In their meta-analysis, Onnasch et al. (2014) found that higher degrees of automation tended to lead to improved performance and lower workload when the automation functioned properly, but it also compromised SA. Moreover, when automation failed, the higher the degree of automation, the greater the negative impact on performance, especially when automation supported action selection. Designers therefore need to consider not only how a tool will improve workload and performance when it is functioning normally but also how performance will be affected when automation fails. As automation-related problems can arise under many conditions, it is important for designers to take into account the entire human–automation system, including level of automation, operator training requirements, and human–automation interfaces (Lee & Seppelt, 2009; Madhavan & Wiegmann, 2007; Vu & Chiappe, 2015).

This type of system approach therefore calls for a user-centered design philosophy (Salvendy, 2012; Maguire, 2001). That is, rather than expecting users to conform to the characteristics of the system, design is based on the needs, wants, and limitations of the end user, in this case, the ATCos. These factors are taken into account during each stage of the system design and development lifecycle, not just at the end. In this way, product development and lifecycle costs are minimized, whereas operator performance is optimized (Vu & Chiappe, 2015). For example, with respect to ATCo displays, instead of presenting information in a way that is organized around the sensors and tools that generate it, information is displayed with an understanding of the goals and tasks of the ATCo. It is also important to keep in mind that ATCos may have multiple operational goals to satisfy at any given time, and goals may quickly change as a task unfolds. A user-centered design approach is one that is sensitive to the dynamics of user goals and involves ensuring that displays organize information around the multiple goals of the user (Endsley & Jones, 2012).

In what follows, we will describe two different areas of human factors concern in ATC operations. We begin with an overview of SA and its measurement and how the concept has been applied to evaluation of potential NextGen concepts and technologies for ATC, and we end with a review of a series of training studies we have conducted with students studying to be ATCos, and how their training can be affected by the introduction of NextGen Tools and Concepts. We acknowledge that human factors researchers have contributed to other areas of ATC operations, but we limit our focus to these two topics in the present chapter.

AIR TRAFFIC CONTROLLER SITUATION AWARENESS

SITUATION AWARENESS IN AIR TRAFFIC CONTROLLER ACTIVITIES

A critical human factors concern in ATC is designing displays, tools, concepts of operation, and training regimens that enhance and do not undermine SA (Endsley & Smolensky, 1998). SA is an operator's dynamic understanding of what's going on as he or she is managing a complex system

(Chiappe, Strybel, & Vu, 2012, 2015; Durso, Rawson, & Girotto, 2007; Endsley, 1995a, 2015). *En route* ATCos, for example, need to possess an ongoing awareness of traffic, which they refer to as *the picture*, so they can move AC safely and efficiently through the sector and detect whether any problems are developing (Endsley & Smolensky, 1998). Most importantly, these controllers need to be aware of conflicts between AC. If any AC are in danger of losing separation, ATCos need to issue commands to pilots to change speed, heading, or altitude. In planning maneuvers, ATCos also need to assess whether a proposed maneuver will create any secondary conflicts. Thus, route planning requires having a global picture of the traffic; a narrow focus on specific AC is not an effective strategy. ATCos maintain their SA by scanning their Digital System Replacement (DSR) display, by communicating with pilots, and by communicating with other controllers and ATM personnel. A high level of SA is therefore important for ATCos to make good decisions, and it is not surprising that SA has been linked to ATCo performance (Durso, Hackworth, Truitt, Crutchfield, & Manning, 1998; Endsley & Smolensky, 1998; Rodgers & Nye, 1993). For example, Durso, Bleckley, and Dattel (2006) found that measures of SA were able to predict performance on an ATC task even after controlling for other cognitive variables.

Maintaining SA requires detecting many features of the operational environment, and integrating this information with information stored in long-term memory so that its relevance to current operational goals can be determined (Chiappe, Strybel, & Vu, 2015; Endsley, 2015; Stanton, Salmon, Walker, & Jenkins, 2010). It also requires that attention be allocated to relevant features in the environment, whereas irrelevant information is ignored so that an operator's limited cognitive resources are not depleted (Pew, 2000). It is useful to characterize SA as consisting of three levels that Endsley (1995a; 2015; Endsley & Smolensky, 1998) refers to as *perception, comprehension*, and *projection*. Level 1, *Perception* involves determining the "status, attributes, and dynamics of relevant elements in the environment" (Endsley & Jones, 2012, p. 14). For example, an ATCo needs to perceive, among other things, the relative location, speed, and direction of flight of the AC in a sector. Level 2, *Comprehension* involves integrating perceived information with background knowledge and information about current operational goals to determine the relevance of the information. For example, an ATCo needs to infer, given the relative position of two AC, whether or not they are in danger of losing separation. Comprehension is also involved in prioritization, determining which potential problems need to be dealt with first. Level 3, *Projection* involves predicting future states of the elements, and in ATC activities it is used to determine whether a particular course of action will resolve the conflict, and whether or not other ancillary conflicts would be created. According to Endsley (2015), the three levels of SA are not strict stages that must be carried out sequentially; they can often occur concurrently.

Situation Awareness Theory and Measurement in Air Traffic Control Contexts

To determine whether displays, tools, operational concepts, and training regimens for ATCo activities enhance SA, we need to have valid, reliable, diagnostic, and sensitive metrics of the construct; we cannot simply rely on the intuitive judgments of designers. Although there is broad agreement in human factors regarding the importance of SA, there is little consensus on exactly how to measure it, even in the specific domain of ATC operations (Salmon, Stanton, Walker, and Green, 2006). Moreover, the tools that have been used to measure SA in ATC contexts are not theory-neutral; they presuppose specific theories about the nature of SA, making it difficult to achieve consensus on which to use (Vu & Chiappe, 2015; Chiappe, Vu, & Strybel, 2012; Stanton et al., 2010). The theories differ in many ways, including whether or not they claim that the product of SA can be separated from the processes by which that knowledge is acquired. The products of SA refer to consciously held knowledge about the task environment (Endsley, 2015; Wickens, 2008). The processes of SA refer to the epistemic actions operators carry out to acquire knowledge, including for example, how they scan an ATC display or engage in communication (Durso, et al., 2007). Theories also differ in terms of whether they hold SA is a purely internal, conscious state of an individual, or whether it resides in the interactions between operators and their task environment, such that operators often offload information and access it only when needed (Chiappe et al., 2012; 2015; Stanton et al., 2010).

These different theoretical commitments can be seen in probe techniques, which are the most popular metrics used to assess SA in ATC contexts (Durso et al., 1998, 2006; Endsley & Smolensky, 1998; Endsley, Sollenberger, & Stein, 2000; O'Brien & O'Hare, 2007; Vu et al., 2009). Probe techniques involve asking ATCos questions about task-relevant information, such as "will AAL124 and SWA673 be in conflict within the next two minutes if no further action is taken?" or "which AC is going to be the next one to enter your sector from the east?" Depending on the specific probe technique, SA is reflected in either the accuracy of the responses, or the time that it takes operators to answer. Two of the main probe techniques are Endsley's (1995b) Situation Awareness Global Assessment Technique (SAGAT) and Durso and Dattel's (2004) Situation Present Assessment Method (SPAM).

These two methods have a lot in common, including the fact that probe questions are derived from task analyses and discussions with subject matter experts, so that the questions pertain to information that an operator should know. Questions are also asked periodically during a task and often require yes/no or multiple-choice response formats. Nonetheless, there are important differences between these techniques. SAGAT typically asks multiple questions at a time, whereas SPAM asks only one. SAGAT assesses SA using accuracy, whereas SPAM uses accuracy and reaction time. SPAM also makes use of a ready prompt, such that operators are alerted when there is a question in the cue, and only once they indicate that they are ready to receive it is the question presented. Time to respond to the ready prompt is taken as a measure of workload because operators are only supposed to indicate they are ready to receive a question when their workload permits. The key difference, however, is that SAGAT involves freezing scenarios and blanking ATCo displays, whereas SPAM involves asking questions while scenarios are running and displays are visible.

This difference in whether or not scenarios are paused and whether or not displays remain visible reflects the theoretical commitments presupposed by each approach (Chiappe et al., 2015). In particular, SAGAT assumes SA is a product. It is made up of only internal, consciously held knowledge; that is, a detailed situation model held in internal memory, based on information derived from perception, comprehension, and projection (Endsley, 2012, 2015). As SA is viewed as a product, SAGAT measures the knowledge that operators possess, but does not directly assess how operators acquire this information, which Endsley refers to as *situation assessment*. Endsley (1995a; 2015) does discuss processing issues, claiming that situation assessment is carried out by various systems that include perceptual processes and attention mechanisms, as well as mental models and other related long-term memory structures. Mental models offer a means for rapid categorization based on cues, which also speeds up the decision-making process. An ATCo, for example, will have a mental model pertaining to sector characteristics, including conflict hotspots, and he or she can use this knowledge to judge the consequences of sending an AC in a particular direction.

In contrast, by leaving ATC displays visible and scenarios running, SPAM assumes that although ATCos store some information internally, other information is offloaded, with SA often involving knowing where to access it in the external world, an approach we have termed *Situated SA* (Chiappe et al., 2012, 2015). As Durso et al. (1998, p. 3) state "SA may sometimes involve simply knowing where in the environment to find a particular piece of information, rather than remembering what the piece of information is." To this end, an ATCo may not store an AC call sign but may remember where to locate this information on the display (Durso et al., 2007). According to SPAM, reaction time (RT) to answer probes reflect whether information is stored internally, or whether it has to be accessed externally. Specifically, Durso and Dattel (2004) state that the RTs to answer questions will be fastest when the information is stored in memory. They will be slower if the information is offloaded onto an external display, but the operator knows where to find it. RTs will be the longest when the information is externally represented, but the operator has to search to find it on a display.

The Situated SA approach claims that operators create partial internal representations and leave as much information in the external environment as they can (Chiappe et al., 2015; Clark, 2008). They let the world serve as its own representation, and they retrieve information from it on an as-needed basis. Action is therefore central to cognition, and not a mere peripheral process. Due to the fact that much of the world is predictable, people can exploit its structure by relying on highly strategic eye fixations rather than trying to retain a detailed, stable, internal picture of the surrounding environment. Indeed, change blindness research shows that although they may not perceive everything that is going on, people's eye fixations are tightly coupled to the actions they are carrying out, allowing them to access limited information from the environment at any given point in time (Simons & Rensink, 2005). According to Rensink (2000), people store internally general *scene schemata* that contain information about what objects are likely present in a scene and where these are located. People can use these schemata to direct their gaze to particular parts of the visual field when they need to get more specific information about objects. In short, situated approaches stress a tight coupling of perception and action rather than relying on intensive internal processing. Process and product are intimately linked and inseparable.

In terms of ATC operations, controllers will often offload crucial information about flights by annotating flight progress strips (Roske-Hofstrand & Murphy, 1998). These are also stored in a strip bay in an offset manner so that ATCos can rely on perception rather than internal memory to determine which flights require further actions. The Situated SA approach also makes precise claims about what information is likely to be held internally and what information is likely to be offloaded (Chiappe et al., 2012, 2015). For example, among other things, it holds that information generality and priority are important variables that are likely to affect offloading behavior. General information, which is likely to be stored internally, pertains to information regarding many or all AC in the sector, whereas specific information pertains to information regarding a particular AC in the sector. General representations, such as Rensink's (2000) scene schemata, would help operators know where in the display to access specific information. Information priority reflects the likelihood that an ATCo needs to act on certain information quickly. Operators may store high priority information internally (e.g., potential conflicts between AC in the next minute), whereas low priority information is stored externally (e.g., whether any AC are 30 nautical miles from the sector boundary). Having high-priority information stored internally would ensure it is promptly acted upon, without having to spend time accessing information on the display. Storing low-priority information externally but knowing where to access it, on the other hand, would limit the amount of information that needs to be held in internal memory.

Recently, Chiappe et al. (2016 Sturre, Chiappe, Vu, & Strybel, 2015) tested these claims. Participants were student ATCos who controlled air traffic in a simulation environment. They were presented with probe questions that varied in generality and priority, creating four question categories (i.e., general/high priority, general/low priority, specific/high priority, specific/low priority). The SPAM technique was used to ask probe questions. These were presented on a probe display that was located adjacent to (and at an angle from) the ATCo radar scope. A camera mounted onto the probe display was used to record where participants looked while answering probe questions. The results revealed that ATCos were more likely to turn to look at their radar display while answering specific questions and low-priority questions, suggesting that this information was offloaded. In contrast, they were less likely to turn to look at the radar display following high priority and general information questions, suggesting that the information to answer them was stored internally.

In short, Situated SA holds that ATCos often offload information onto their displays, and as a result, when measuring SA we should not remove operators from their task environment if we want to capture this behavior. SPAM is therefore an appropriate method to use if one takes this approach, because it allows operators to refer to their displays while answering probe questions. By comparing RTs to answer questions, or by examining eye gazes, one can examine whether ATCos are answering on the basis of what they have stored internally or whether they have offloaded the information. In contrast, SAGAT, by blanking displays and freezing scenarios, does not allow us to study the situated ATCo but does allow us to examine what information operators consciously store in internal memory.

Vu and Chiappe (2015; see also Chiappe et al., 2015; Stanton et al., 2010) view SPAM and SAGAT as examining SA from a different unit of analysis, both of which are legitimate, depending on one's research goals. The unit of analysis implicit in SAGAT is the individual operator's conscious mind. This would be appropriate, for instance, if a researcher was interested in examining what an ATCo knows about the sector when there is a malfunction of their radar scope and displays go dark. In contrast, the unit of analysis of SPAM is the individual operator in interaction with his or her displays, and it allows us to study how ATCos offload key information. Neither of these is correct or incorrect, but we need to be clear on what we are studying and we need to use the right tool given the particular research goals (Stanton et al., 2010). Indeed, SPAM and SAGAT each have advantages and disadvantages that need to be kept in mind when deciding which metric to use. In what follows, we summarize the pros and cons of each.

Situation Awareness Global Assessment Technique

There are many desirable features of SAGAT (Endsley, 1995b; 2015; Salmon et al., 2006; Vu & Chiappe, 2015). First, it marks an improvement over subjective techniques such as the Situation Awareness Rating Technique (SART; Taylor, 1990) because it objectively measures what the operator does and does not know, comparing their answers with what was actually taking place. Subjective techniques are better seen as measures of meta-SA, what a person thinks their SA is at any given time, rather than what it actually is (Durso et al., 2007). Second, SAGAT can be scored very easily, which allows one to make quantitative comparisons. One could compare the relative merits of two ATCo displays, for example, by determining which led to the higher percent of probe questions answered correctly. Third, it has good psychometric properties, as many studies have validated it, including in the ATC domain (e.g., Endsley et al., 2000; Kaber & Endsley, 2004). Kaber and Endsley (2004), for example, used SAGAT to examine the consequences of different levels of automation on SA of ATCos. Fourth, by asking a broad range of questions at all three levels of SA, it can provide a fine-grained analysis of how some tool or concept of operation affects SA. This allows designers to assess whether improvements in one aspect of SA come at the expense of SA for other information. By drawing from a large list of questions during query presentation, it also makes it unlikely that operators can come to pay attention to information they would not normally attend to (Salmon et al., 2006).

Nonetheless, SAGAT also has some important limitations that need to be kept in mind. First, it cannot be effectively used to study the processes of situation assessment because it only measures the conscious knowledge operators possess. One cannot simply infer processes from the products, as the same product can be achieved in many different ways (Durso & Sethumadhavan, 2008). Second, pausing scenarios and removing displays interrupts the activities of operators (Salmon et al., 2006). This makes it difficult to apply SAGAT in real world. So, it can only be applied in the context of a simulation. Third, SAGAT measures what operators can recall about the situation, making it heavily dependent on memory (Durso & Sethumadhavan, 2008). An operator, however, can have good SA at a particular point in time but not be able to recall that information at the time that the queries are presented.

Situation Present Assessment Method

SPAM also has many desirable properties. First, it has also received much validation in the context of ATC operations (Durso et al., 1998, 2006; Vu et al., 2009). Vu et al. (2009), for example, showed that SPAM is able to capture differences in SA that stem from different roles and responsibilities for separation assurance. Likewise, the ability of SPAM to separate workload from SA by using the *ready* prompt to assess the former and probe RT to assess the latter, has also received validation (e.g., Vu et al., 2012b). Second, similar to SAGAT, SPAM can be scored relatively easily by comparing operator responses with what was actually taking place in the system being controlled. By making use of RT in addition to accuracy, however, SPAM also provides a more fine-grained measure of SA (Durso & Dattel, 2004). Third, by presenting questions one at a time and recording RT, SPAM can be used to study the processes of SA and not just the product. For example, it can be used to examine whether operators are storing information internally or offloading it onto a display (Durso et al., 2007). Fourth, SPAM can be applied in the field and is not limited to simulation studies. The reason is that it does not remove operators from their task environment. Durso and Dattel (2004) cite examples, including applications of SPAM to real world driving.

Despite its virtues, SPAM also has important limitations. First, it can be intrusive to an operator's primary task. Although it employs the ready prompt to prevent this, performing a second task while managing a sector can still be intrusive on ATCo activities. For example, Pierce (2012) found that SPAM negatively affected ATCo performance and increased workload ratings to a level comparable with other conditions in which participants had to manage traffic while carrying out cognitively demanding secondary tasks. Likewise, Strybel et al. (2008) found that although both probe techniques predicted performance on a piloting task, SPAM interfered with performance compared with SAGAT, as evidenced by greater variability in air speeds in SPAM scenarios. Given the potential for interference, it is not ideal to apply the SPAM technique when task demands are very high. Second, it can be difficult to interpret the latencies to answer probe questions. Although advocates for SPAM claim that RT differences reflect where operators are storing information (in the world or in the head), they could reflect other factors as well, such as question difficulty. Thus, it is desirable to supplement the SPAM method with a technique for measuring eye gaze while questions are being answered (e.g., Sturre et al., 2015; Chiappe et al., 2016).

To summarize, we have reviewed two probe methods that are frequently used to measure SA in ATC contexts. These differ in important respects, including the theories on which they are based. Importantly, they differ in the unit of analysis for which they are most suited (Stanton et al., 2010; Vu & Chiappe, 2015). SAGAT is most suited for studying the SA possessed by the conscious mind of the individual operator. SPAM is most suited for a unit of analysis that includes the individual operator and his or her interactions with the immediate task environment, assessing for example how information is offloaded and accessed on an as-needed basis.

PRINCIPLES FOR DESIGNING TO SUPPORT SA

Although there are theoretical disagreements as to what constitutes SA and how best to measure it, most researchers agree that ATC displays should be designed with the goal of enhancing SA. This can be achieved by keeping in mind various principles that reflect a user-centered approach, ones that transcend specific theoretical approaches. Indeed, Endsley and Jones (2012) provide a list of eight principles that serve to enhance SA that can readily be applied in the domain of ATC operations (Table 13.1).

1. *Organize information around goals*: Information should be organized around the tasks and goals that ATCos have to perform. Doing so makes it more likely that the goals will be achieved accurately and efficiently. If an ATCo needs information X and Y to carry out a task, then X and Y should be presented together on a display. The operator should not have

---

TABLE 13.1

SA-Enhancing Design Principles

No.	Design Principle
1	Organize information around goals
2	Present information to support comprehension directly
3	Provide assistance for projections
4	Support global SA and resist cognitive tunneling
5	Support tradeoffs between goal-driven and data-driven processing
6	Make critical cues for schema activation salient
7	Take advantage of parallel processing capabilities
8	Use information filtering carefully

Source: Endsley, M.R., and Jones, D.G., *Designing for Situation Awareness: An Approach to Human-Centered Design*, CRC Press, New York, 2012.

---

to look in multiple locations to access needed information. For example, if an ATCo needs to consider an AC's altitude and bearing to infer whether the AC presents a threat to another AC, both of these pieces of information should be presented together on the display.

2. *Present information to support comprehension directly*: Displays should present information that has been integrated and processed with respect to the ATCo's comprehension requirements. For example, an ATCo display could indicate whether two AC are in conflicting trajectories rather than having controllers infer this on the basis of the relative position, direction, and speed. In this way, the meaning of the displayed information would be more easily understood by controllers, and they would be able to direct their efforts to resolving the conflict as opposed to figuring out whether a conflict actually exists.

3. *Provide assistance for projections*: As projection of future system states is often quite cognitively demanding, system designs should support this level of understanding by providing such information on the displays whenever possible. For instance, an ATC display could directly represent whether any secondary conflicts would be created by selecting a particular maneuver for an AC instead of having the ATCo compute this information by integrating the information about each of the AC along the path. It could also portray what the closest point of approach would be.

4. *Support global SA*: Global SA refers to a *big picture* understanding, an overview of the situation taking into account all of an ATCo's goals. It is important for a display not to encourage cognitive tunneling by allowing an operator to focus only on one goal at the exclusion of all other important goals (Wickens, 2005). Interfaces that have an excessive number of menus, for example, can exacerbate this problem. Likewise, although it may be desirable to be able to adjust the zoom on an ATC display, this must not impede an ATCo's ability to know what is going on in adjacent sectors, as this can be strategically costly because controllers may not be able to plan ahead.

5. *Support tradeoffs between data-driven and goal-driven processing*: In goal-driven processing, goals guide an ATCo to seek out certain information on the display in a top-down manner. For example, an ATCo may look for altitude information about a specific AC on their display because of a concern about its position relative to another AC. In data-driven processing, an ATCo's attention is captured by salient cues present on a display. It is particularly important for nonroutine situations. If an AC suddenly deviates from its flight plan, for example, it would be useful to have an ATCo's attention drawn to this fact. Both of these types of processing are important. Without goal-driven processing, ATCos would simply be reacting to an unfolding situation, which would be an ineffective strategy. Data-driven processing is important too, however, because controllers cannot always know what is going to be the most important information to process at any given point in time.

6. *Make critical cues for schema activation salient*: Experts make decisions by using their perceptual mechanisms to detect the presence of specific cues. Once detected, these cues cause the activation of schemata stored in long term memory (LTM), which tell the operator how to respond to this particular situation (Klein, 2008). Making those cues salient is therefore an effective way to enhance performance. For example, using an easily perceivable cue (such as AC icons turning red from amber) to identify coaltitude AC on converging trajectories would be helpful, as it would trigger a *conflict* schema in the mind of controllers. Doing so would also activate knowledge on how to resolve those conflicts.

7. *Take advantage of parallel processing abilities*: Although people are limited in the amount of information that they can process in any single sensory modality, their ability to process multiple pieces of information at the same time is better when it involves different sensory modalities (Wickens, 2008). In the case of ATC operations and aviation in general, there tends to be an excessive reliance on a visual presentation of information. As a result, relying on other modalities such as hearing is encouraged. This is especially the case for alarms and alerts.

8. *Use information filtering carefully*: Information filtering is when a system is designed to provide ATCos with only the information that the system judges to be needed by the operator at any given point in time. Filtering is usually done so as to not overload an operator. For example, key information about a flight may be contained in a data block, and this information could be hidden, and only pop up when the AC is in conflict with another AC. The problem is that systems are rarely able to accurately determine what information a user needs. As a result, the user can be deprived of key information when they need it. Furthermore, because the information is not always there, the ATCo may be deprived of trend information. Thus, without this information, he or she would be unable to look ahead to see if any other potential problems are developing. It would put them in a position of being forced to react to situations rather than being proactive in their handling of unfolding situations.

SUMMARY

SA, knowing what is going on in a dynamic system, is a crucial concern for ATC operations, as it drives ATCo decision-making and ultimately performance. It can be studied, however, from different perspectives, or different units of analysis, including from a perspective that focuses solely on the conscious contents of the mind of controllers, or from one that examines how they offload crucial information onto the environment. These different perspectives warrant different measurement techniques. Furthermore, there are valuable principles that need to be kept in mind when designing tools and technology for supporting ATC operations, ones that enhance SA.

RESEARCH ON IMPACT OF POTENTIAL NEXTGEN TOOLS AND CONCEPTS OF OPERATIONS

The concern for operator workload and SA in NextGen, as noted earlier, stems primarily from the projected increase in air traffic over the next several decades. ATCos will not be able to handle this amount of traffic using the existing tools and procedures (Prevot, Homola, & Mercer, 2008). The technological upgrades in the NAS envisioned under NextGen will require that ATCos interact with automated components of the ATM system. Many NextGen concepts of operation for the task of separation assurance have been proposed (e.g. Dwyer & Landry, 2009) on the basis of different function allocations to human operators or automation. Separation assurance algorithms for an autoresolver tool have also been developed (e.g., Erzberger, 2009; Erzberger & Heere, 2008; Farley & Erzberger, 2007) and shown to be effective when implemented in an interactive teaming environment with ATCos or autonomously (e.g., Prevot et al., 2012). Many of the studies examining NextGen ATCo tools and concepts of operation (e.g., Homola et al., 2010; Prevot et al., 2012; Pop, Stearman, Kazi, & Durso, 2012), however, do not focus on the topic of SA, despite the fact that it is a known determinant of operational errors in ATC (Endsley & Smolensky, 1998). In this section, we provide examples of a few human-in-the-loop simulation studies that we have conducted that were designed specifically to examine the impact of potential NextGen tools and concepts of operation on ATCo SA with experienced and novice operators.

In all these studies, a variant of the SPAM method was employed as an online measure of workload and SA. This variant included a visual and auditory *ready* prompt presented on a secondary display. Upon acceptance of the ready prompt, the question was visually displayed along with response alternatives in multiple-choice format. Strybel et al. (2009) found this multiple-choice format to be an effective in reducing the intrusiveness of a secondary task associated with this online probe procedure. Moreover, Bacon et al. (2011) found that latencies resulting from this probe technique were valid indicators of SA, and sensitive enough to discriminate between student and expert ATCo performance in mixed equipage environments (i.e., environments in which there were both,

NextGen tool equipped AC and nonequipped AC). In addition, Bacon et al. found that the results obtained with this probe technique were consistent with expert ATCo over-the-shoulder ratings of the participants' overall performance, performance in maintaining separation, attention, and SA.

EFFECTS OF NEXTGEN CONCEPTS ON SA OF EXPERIENCED AIR TRAFFIC CONTROLLERS

Research has shown that an automated conflict resolution tool can reduce ATCo workload and have a positive impact on the ATM system, including increasing its safety and efficiency (Homola et al., 2010; Prevot et al., 2012). Yet, for the automated separation assurance tool to be effective, it should not decrease ATCo SA. Moreover, the separation assurance tool should work well with other NextGen tools designed to reduce workload, such as merging and spacing, and allow the ATCos to interact with other operators in the NAS (e.g., pilots). In this section, we will summarize a study conducted by Strybel et al. (2016) that measured ATCo SA and workload as a function of whether the task of separation assurance and merging and spacing were assigned to the human operator or to automation. The impact of these NextGen tools and concepts on operator SA will be discussed.

NextGen Scenario and Concepts of Operation

The current study involved a multiparticipant distributed, human-in-the loop simulation. It was run using the Multi Aircraft Control System (MACS), Aeronautical Datalink and Radar Simulator, developed by the Airspace Operations Laboratory at NASA Ames Research Center (e.g., Prevot et al., 2012). ATCos used MACS configured as a DSR display with integrated Data Comm (text-based communication) and conflict alerting. At the start of the scenarios, AC were populated in two adjacent sectors (named 90 and 91, respectively) and flown by trained pseudopilots (experimental confederates who controlled all AC in the sector). Each sector was managed by an ATCo, and the pair of ATCos had to work collaboratively to handoff AC from one sector to the other.

The traffic in the scenarios included arrivals, departures, and *en route* AC. In addition, there was a stream of United Parcel Service (UPS) AC in the sector that were engaging in company spacing. That is, these AC were assigned a lead AC to follow and were expected to merge in a sequence at a particular waypoint to achieve a continuous descent approach. In addition to the separation assurance and spacing tasks (which could be assigned to a human operator or to automation depending on the concept of operation being tested), the two controllers had different responsibilities based on the characteristic of their sector. In particular, the sector 90 ATCo was responsible for weather deviations (weather was only present in that sector); the AC's top of descent also occurred in this sector. The sector 91 ATCo was responsible for finalizing the merging of the UPS AC; after the merge, the ACs started a continuous descent arrival approach to the airport. If requested by the UPS pilots, both sector ATCos were required to provide clearances for resequencing and spacing to the airport. The ATCos were provided with conflict alerting, a trial planner with route assessment for potential conflicts, and an autoresolver tool, which provides the ATCos suggested resolutions to traffic conflicts at their request. The autoresolver was engaged autonomously when separation assurance was assigned to automation.

As noted earlier, the responsibility of separation assurance was assigned to the human (i.e., the pilot or ATCo) or automation. Similarly, the merging and spacing task was assigned to the human or automation. The various function allocation strategies resulted in three potential NextGen Concepts of Operation:

- Concept 1: *Pilot Primary*: Pilots flying NextGen equipped AC were given the primary responsibility for separation assurance between ownship and all other AC. The ATCos were only responsible for resolving conflicts between NextGen unequipped AC and all conflicts with experimental AC on their continuous descent arrival. The ground-based autoresolver agent had no responsibility.

- Concept 2: *ATCo Primary*: The ATCos were given the primary responsibility for separation assurance. ATCos had responsibility for NextGen equipped–unequipped and unequipped–unequipped conflicts only, and for conflicts between any AC on a continuous descent arrival. The autoresolver agent was responsible for NextGen equipped–equipped conflicts.
- Concept 3: *Autoresolver Primary*: In Concept 3, the autoresolver agent was responsible for resolving most conflicts. The autoresolver agent was responsible for conflicts between NextGen equipped–equipped and equipped–unequipped AC. ATCo was responsible for resolving conflicts between unequipped AC only.

SA Probe Questions

The SA probe questions, which were developed and evaluated by a retired ATCo, were categorized on the basis of the type of information queried (e.g., questions relating to conflicts, spacing status, and traffic/weather) and whether the information was relevant to the ATCo's own sector or the sector of the adjacent ATCo (for more details, see Strybel et al., 2012). Conflict questions asked about potential conflicts (e.g., "In the next three minutes, will there be any conflicts between ...."), and recent conflict resolution, (e.g., "What was the resolution to the last conflict?"). Spacing-status questions asked about the arrival AC and their status regarding spacing (e.g., "Will any AC need re-sequencing in your sector?"). Traffic/weather questions asked about current status of AC in a sector, such as the number of equipped versus unequipped AC, altitudes and headings of specific ACs, or about weather (e.g., "How close is [call sign] to the nearest weather cell?").

IMPACT OF CONCEPT ON AIR TRAFFIC CONTROLLER WORKLOAD AND SITUATION AWARENESS

As the focus of the current chapter is on ATC, we only summarize the main findings relating to ATC in terms of workload and SA (see Vu et al., 2012b, for the role of the concepts on pilot SA, and Strybel et al., 2016, for more details). In terms of performance, conflict resolution times and the number of ATCo-responsible losses of separation were higher when the ATCo was primarily responsible for separation assurance (i.e., Concept 2). This finding was not surprising as this was the condition in which the ATCo workload was the highest. However, Strybel et al. (2016) found that automating separation assurance produced the greatest reductions in workload, and automating the merging and spacing task lowered workload further, so that the lowest levels of workload occurred when both functions were automated (Concept 3). This finding was not surprising given that the ATCos had less to do when both functions were automated. The reductions in workload, however, were in some cases accompanied by changes in controller SA, depending on sector characteristics and weather location. In the sector where the AC started its descent (sector 90), SA of conflicts was higher when the ATCo was responsible for separation assurance, especially when spacing task was automated. Automating spacing task also produced higher awareness of traffic and weather information. However, in the adjacent sector (sector 91), where all the arrival AC merged, SA of conflicts was highest when both functions were either automated or manually controlled by ATCos. These differences regarding the effects on SA, point out the need for evaluating automation concepts across different types of sectors in future investigations. Moreover, future studies need to examine the impact that failures of NextGen specific tools can have on ATCo workload and SA. For example, Kraut et al. (2011) found that failure of one NextGen tool, specifically Data Comm, resulted in increased ATCo workload and decreased SA. The changes in workload and SA made the ATCo prioritize the task of maintaining AC separation, but this was done at a major cost to the efficiency with which they managed their sector. Kraut et al. (2011) only examined one type of failure, but it is likely that different NextGen tools can fail or be temporarily unavailable and their combined impact on ATCo workload and SA need to be examined.

SUMMARY

The current study illustrates the importance of examining the effects of NextGen tools and concepts of operations across sectors having different characteristics and measuring their effects not only on ATCo performance, but also on SA and workload, as the latter are important harbingers of future performance problems. Most studies examining the feasibility of NextGen tools and technologies have not focused on operator SA. However, we found that ATCo functions such as separation assurance and interval spacing interact with each other in determining ATCo performance, workload, and SA. Thus, to determine an optimal function allocation strategy under NextGen, one cannot simply focus on a single factor such as reducing ATCo workload. The optimal solution for NextGen will require many iterative simulations that account for the various contexts in which the human–automation teaming will be functioning.

TRAINING AIR TRAFFIC CONTROLLERS WITH POTENTIAL NEXTGEN TOOLS

Once NextGen tools and concepts of operations for ATC have been selected for implementation, ATCos will need to be trained on their use so that they can properly form the *picture* and perform their task safely and efficiently. It is important to keep in mind, however, that NextGen will be implemented gradually, and that at least early on, the airspace will feature mixed equipage, that is, some of the AC will be equipped with NextGen tools and technology, whereas others will feature traditional tools. As a result, ATCos in the near term will have to be trained to use both sets of tools and will have to be able to fluidly transition between them. Thus, research is required on how best to train controllers to use both sets of tools.

Kiken et al. (2011) trained experienced ATCos, that is, ones with many years of experience using the manual, voice-based tools, to manage a simulated sector that they were unfamiliar with. The sector consisted of either NextGen equipped or NextGen unequipped (current-day) AC. They recorded the number of session required for the ATCo to master the sector. Kiken et al. found that the ATCos took twice as long to master the sector with equipped AC compared with unequipped AC, indicating that their ATM skills do not directly transfer. Future ATCos, though, will likely develop the skills for managing NextGen equipped AC as part of their initial training. Thus, training techniques that can assist students in acquiring advanced ATM concepts early in training, and help them acquire and maintain SA, may improve their overall skill acquisition and success as ATCos. In this section, we will review a series of training studies that examine how to best introduce NextGen tools to student ATCos, and the impact of these training techniques on the student ATCo's SA.

TRAINING APPROACH

The participants in this study were students from Mount San Antonio College who were enrolled in different semesters of an internship course at the Center for Human Factors in Advanced Aviation Technologies at California State University, Long Beach. All students were pursuing careers in ATC. Every session of the internship course was taught by the same retired ATCo over one semester during a four-year period. The course focused on *en route* ATC operations and consisted of a lecture period and two lab periods over 14–16 weeks. Students were assigned to different lab sections of the course and experienced different hands-on training techniques for managing traffic on a radar screen (see Vu et al., 2013; Kiken et al., 2012). By comparing the performance of students using the different instructional methods, we could infer the beneficial components of different types of training techniques. This approach to examining differences in classroom training has been shown to be effective (e.g., Kiboss, Ndirangu, & Wekesa, 2004) and representative of real-world training programs.

All students attended the lecture portion of the course together. The lectures focused on training students ATCo skills and thought processes that were exercised in the lab portion. The most critical skill taught to the students was how to be able to project AC locations and protect them from losing minimum separation (i.e., prevent AC from having a loss of separation or LOS). Students were given exercises that allow them to estimate the point of contact for two AC on conflicting trajectories, as well as to identify where the LOS would occur and strategies for resolving the conflict prior to the LOS. A LOS was defined as AC coming within 5-nautical miles lateral or 1,000-feet vertical of each other. Students were taught how to resolve conflicts by vectoring AC (i.e., issue a series of heading changes), changing their altitude or speed, or through more strategic techniques (e.g., setting up a structure for their airspaces that would prevent LOS from occurring if the structure is maintained). The students were also taught pilot–controller communication and strategies for handling off-nominal situations and general expectations of the ATC career based on the instructor's personal experience.

To determine the effectiveness of different types of training regimens, in the lab sections, students experienced different types of training approaches, depending on which semester of the course they were enrolled in. Common to all lab sections was the general software training. All students spent the first week familiarizing themselves with the interface of the MACS software (Prevot, 2002) that was used to run the radar simulation for the course and for the testing sessions. All students were exposed to the DSR mode of MACS and Indianapolis Center (ZID) Sector 91, which was the sector that the students were trained to manage in all of these studies. Students were taught characteristics of the airspace, including the main waypoints in the sector, arrival/departure paths, general traffic flows, and altitude conventions. The students were also trained to take and make handoffs and to adjust components of the display, such as switching datablocks from a full to limited view, or vice versa.

For the NextGen tools, students were trained with potential tools for NextGen equipped AC. The first set of tool relates to CD&R. NextGen equipped AC had conflict detection that was paired with a trail planner that allowed the controller to display the AC's current trajectory and make route modifications visually by clicking and dragging the route or by inserting new waypoints and adjusting the AC trajectory. The CD&R tool also determined whether the route change would produce secondary conflicts in a six to eight-minute window. The students were also trained on a Data Comm system that allowed them to issue commands and clearances via a text-based medium that was intended to reduce workload by reducing the amount of voice communication needed. Data Comm was also integrated with the trial planner tool, so route modifications were made possible via a single command.

ATCo SA was measured using a variant of the SPAM technique described earlier. The student ATCos were provided with a visual *ready* prompt presented on a secondary display, accompanied by an audio alert presented through their ATCo headset. Upon acceptance of the ready prompt, the question was visually displayed along with response alternatives in multiple-choice format. The SA probe queries asked about conflicts and the status of AC or ATC operations in the scenario.

Training Manipulations and Findings

Research on training techniques has resulted in a large number of methods for assisting operators to acquire and maintain complex skills. Common training techniques include part-task training (i.e., training component tasks prior to whole-task performance; Paas & van Gog, 2009; van Merriënboer & Kester, 2008), increasing the difficulty of training (e.g., Healy & Bourne, 2013; Young, Healy, Gonzalez, Dutt, & Bourne, 2010), interactive and simulation training (e.g., Wieland, 2010; Vu, Kiken, Chiappe, Strybel, & Battiste, 2013), and use of supplemental instructions (e.g., Rabitoy, Hoffman, & Person, 2015). All these techniques have achieved varying levels of success. The effectiveness of any one of them, however, is dependent on a variety of contextual factors (see, e.g., Wickens, Hutchins, Carolan, & Cumming, 2013).

Given that ATC is a complex task, we mainly employed different variants of the part-task training paradigm, in which students were trained to manage traffic using one skill at a time (i.e., first manage traffic through changing headings, then through issuing altitude changes, and finally through imposing strategic sector structures). As we were interested in the introduction of NextGen tools and concepts, we introduced those tools to students at different points in time, depending on the semester in which they were enrolled. We consider whole-task performance as being able to manage a mixed-equipage sector of NextGen equipped and current-day equipped AC.

Billinghurst et al. (2011) examined whether the order of training (i.e., training current-day manual skills first or NextGen skills first) made a difference in student ATCo's subsequent performance and SA managing a mixed-equipage sector. Training current-day, manual skills may be beneficial because it is the more difficult skill (includes less automated tools to aid the controller). Training the NextGen skills first, however, may allow the students to acquire global ATC skills given that they are not devoting much of their resources learning to perform manual tasks. Both groups were tested in mixed-equipage scenarios. Billinghurst et al. (2011) found that the students' air traffic performance was better when tested immediately after manual training than after NextGen tools training regardless of when the skills were introduced. There was no effect of training order on the student ATCo's SA.

Kiken et al. (2012; see also Vu et al. 2013) more directly compared part-whole (training on one type of equipage on its own first before combining it with the other type of equipage) and whole-task (training both types of equipages concurrently) procedures. The part-whole lab was trained only on the traditional, manual skills during the first half of the semester and then with the mixed-equipage environment during the second half of the semester. In contrast, the whole-task group learned both the manual and NextGen tools concurrently in mixed-equipage scenarios for the entire semester. Kiken et al. (2012) found that training did overall improve the student ATCo's performance, but the type of training mattered only for certain students. In particular, students with higher aptitude were not affected by when the NextGen tools were introduced in training, but the students with lower aptitude benefited from initial training with the traditional, manual skills before being introduced to the NextGen tools (Vu et al., 2013). Consistent with the finding that student aptitude was a larger factor than type of training, Vu et al. (2013) found that students with higher aptitude showed higher levels of SA in general. For students with lower aptitude, those with whole-task training showed lower SA compared with the students with part-whole training. As the part-whole training reduces the cognitive load imposed on the low-aptitude students, they likely had the extra capacity needed to acquire good SA.

Vu et al. (2012a) trained students from both lab groups to manage their sector with both NextGen equipped and unequipped AC but varied the proportion of the NextGen equipped AC so that students would have more (sector with 75% equipped AC) or less (sector with 25% equipped AC) experience with NextGen tools and procedures. Vu et al. (2012a) found a benefit for training with NextGen tools. Specifically, the student controllers performed better and reported reduced levels of workload when the sector consisted of more NextGen equipped AC. The students showed similar levels of SA regardless of training technique. However, when they were tested in a scenario with all NextGen equipped AC, students showed lower SA than when they were tested in scenarios with some or no NextGen equipped AC. This finding indicates that student ATCos do become complacent when they expect that all conflicts in their sector will be alerted. In general, there was little effect of order of training under nominal situations. That is, students trained to rely more on manual skills early in training performed just as well as students who were trained to rely more on NextGen tools early in training.

The order of training did, however, have an impact on ATCo performance when NextGen tools failed. When a Data Comm failure was introduced in the very last test session, the group that was trained to rely more on manual skills early in training showed fewer detrimental effects in terms of LOS and sector efficiency. However, training order did not differentially impact the student ATCo's workload or SA during the Data Comm failure. All training groups showed lower SA and higher workload levels during the failure scenario compared with the nominal one.

SUMMARY

Our course has been successfully delivered to over 80 students over a four-year period. By using this course structure and a variety of manipulations, we have shown that students benefit more from recent training of traditional manual skills compared with NextGen skills (Billinghurst et al., 2011). Moreover, students who were trained to rely more on traditional, manual skills early in training performed better than the group trained to rely more on NextGen tools early in training when NextGen tools failed (Vu et al., 2012a). Students with higher aptitude were not affected by when the NextGen tools were introduced in training, but the students with lower aptitude benefited from a training schedule that introduced the NextGen tools after the traditional, manual skills were practiced (Vu et al., 2013). Training order and type had less impact on the student ATCo's SA, with the exception that students with lower aptitude benefited more from part-whole than whole-task training, because the task simplification allowed them the extra capacity needed to acquire good SA.

To conclude, this chapter illustrates why the topic of SA is important to the design and evaluation of NextGen tools, technologies, concepts of operation, and training procedures for ATC. SA drives ATCo decision-making abilities and impacts ATCo performance in terms of maintaining safety of flight and efficiency of air traffic flows. Research evaluating components of NextGen need to include comprehensive measures of SA, because observed changes in workload may come at a cost to SA.

REFERENCES

Bacon, L. P., Strybel, T. Z., Vu, K.-P. L., Kraut, J. M., Nguyen, J. H., Battiste, V., & Johnson, W. (2011). Situation awareness, workload, and performance in midterm Nextgen: Effect of variations in aircraft equipage levels between scenarios. *Proceedings of the 2011 International Symposium on Aviation Psychology* (pp. 32–37). Dayton, OH.

Billinghurst, S. S, Morgan, C., Rorie, R. C., Kiken, A., Bacon, L. P., Vu, K.-P. L., Strybel, T. Z., & Battiste, V. (2011). Should students learn general air traffic management skills before NextGen tools? *Proceedings of the Human Factors and Ergonomics Society Annual Meeting, 55*(1), 128–132. Los Angeles, CA: Sage Publications.

Chiappe, D., Morgan, C. A., Kraut, J., Ziccardi, J., Sturre, L., Strybel, T. Z., & Vu, K. P.-L. (2016). Evaluating probe techniques and a situated theory of situation awareness. *Journal of Experimental Psychology, 22,* 436–454.

Chiappe, D., Strybel, T., & Vu, K-P. L. (2012). Mechanisms for the acquisition of Situation Awareness in situated agents. *Theoretical Issues in Ergonomics Science, 13,* 625–647.

Chiappe, D., Strybel, T., & Vu, K-P. L. (2015). A situated approach to the understanding of dynamic situations. *Journal of Cognitive Engineering and Decision Making, 9,* 33–43.

Chiappe, D., Vu, K-P. L. & Strybel, T. (2012). Situation awareness in the NextGen air traffic management system. *International Journal of Human-Computer Interaction, 28,* 140–151.

Clark, A. (2008). *Supersizing the mind: Embodiment, action, and cognitive extension.* New York: Oxford University Press.

Durso, F., & Alexander, A. (2010). Managing workload, performance, and situation awareness in aviation systems. In E. Salas & D. Maurino (Eds.), *Human factors in aviation* (2nd ed., pp. 217–247). New York: Academic Press.

Durso, F., & Dattel, A. (2004). SPAM: The real-time assessment of SA. In S. Banbury & S. Trembley (Eds.) *A cognitive approach to situation awareness: Theory, measures and application* (pp. 137–154). New York: Aldershot Publishing.

Durso, F., & Sethumadhavan, A. (2008). Situation awareness: Understanding dynamic environments. *Human Factors, 50,* 442–448.

Durso, F. T., Bleckley, K., & Dattel, A. (2006). Does situation awareness add to the validity of cognitive tests? *Human Factors, 48,* 721–733.

Durso, F. T., Hackworth, C. A., Truitt, T., Crutchfield, J., & Manning, C. (1998). Situation awareness as a predictor of performance in en route air traffic controllers. *Air Traffic Quarterly, 6,* 1–20.

Durso, F. T., & Manning, C. A. (2008). Air traffic control. In C. M. Carswell (Ed.). *Reviews of human factors and ergonomics* (Vol. 4, pp. 192–242). Santa Monica, CA: Human Factors and Ergonomics Society.

Durso, F. T., Rawson, K., & Girotto, S. (2007). Comprehension and situation awareness. In F. Durso, R. Nickerson, S. Dumais, S. Lewandowsky, & T. Perfect (Eds.), *Handbook of applied cognition* (2nd ed., pp. 163–194). Hoboken, NJ: Wiley.

Dwyer, J. P., & Landry, S. (2009). Separation assurance and collision avoidance concepts for the next generation air transportation system. In *Human interface and the management of information. Information and interaction* (pp. 748–757). Berlin, Heidelberg: Springer.

Endsley, M. (2012). Situation awareness. In G. Salvendy (Ed.), *Handbook of human factors and ergonomics* (4th ed., pp. 553–568). New York: Wiley.

Endsley, M. R. (1995a). Toward a theory of situation awareness in dynamic systems. *Human Factors, 37,* 32–64.

Endsley, M. R. (1995b). Measurement of situation awareness in dynamic systems. *Human Factors, 37,* 65–84.

Endsley, M. R. (2015). Situation awareness misconceptions and misunderstandings. *Journal of Cognitive Engineering and Decision Making, 9,* 4–32.

Endsley, M. R., & Jones, D. G. (2012). *Designing for situation awareness: An approach to human-centered design* (2nd ed.). New York: CRC Press.

Endsley, M. R., & Smolensky, M. (1998). Situation awareness in air traffic control: The picture. In M. Smolensky and E. Stein (Eds.) *Human factors in air traffic control* (pp. 115–154). New York: Academic Press.

Endsley, M. R., Sollenberger, R., & Stein, E. (2000). Situation awareness: A comparison of measures. *Proceedings of the human performance, situation awareness and automation: User centered design for the new millennium conference* (pp. 15–19).

Erzberger, H. (2009). Separation assurance in the future air traffic system, *Proceedings of the ENRI International Workshop on ATM/CNS (EIWAC 2009).* Tokyo, Japan.

Erzberger, H., & Heere, K. (2008). Algorithm and operational concept for solving short range conflicts. *International Congress of the Aeronautical Sciences* 2008-8.7.5, Anchorage, Alaska.

Farley, T., & Erzberger, H. (2007). Fast-time simulation evaluation of a conflict resolution algorithm under high air traffic demand, *7th FAA/Eurocontrol R&D Seminar.* Barcelona, Spain.

Healy, A. F., & Bourne, L. E., Jr. (2013). Empirically valid principles for training in the real world. *The American Journal of Psychology, 126*(4), 389–399.

Homola, J., Prevot, T., Mercer, J., Brasil, C., Martin, L., & Cabrall, C. (2010). A controller-in-the-loop simulation of ground-based automated separation assurance in a NextGen environment, *Proceedings of the ENRI International Workshop on ATM/CNS (EIWAC 2010),* Tokyo, Japan.

Joint Planning and Development Office. (2011). *NextGen Overview.* Retrieved from http://www.jpdo.gov/

Kaber, D. B., & Endsley, M. R. (2004). The effects of level of automation and adaptive automation on human performance, situation awareness and workload in a dynamic control task. *Theoretical Issues in Ergonomics Science, 5,* 113–153.

Kiboss, J. K., Ndirangu, M., & Wekesa, E. W. (2004). Effectiveness of a computer-mediated simulations program in school biology on pupils' learning outcomes in cell theory. *Journal of Science Education and Technology, 13,* 207–213.

Kiken, A., Rorie, R. C., Bacon, L. P., Billinghurst, S., Kraut, J. M., Strybel, T. Z., Vu, K.-P. L., & Battiste, V. (2011). Effect of ATC training with NextGen tools and online situation awareness and workload probes on operator performance. In G. Salvendy and M.J. Smith (Eds.) *Human interface, Part II, HCII 2011, Lecture Notes in Computer Science, 6772,* 483–492.

Kiken, A., Strybel, T. Z., Vu, K.-P. L., & Battiste, V. (2012). Effectiveness of training on near-term NextGen air traffic management performance. *Proceedings of the applied human factors and ergonomics annual meeting,* pp. 182–191. San Francisco, CA: USA Publishing. Reprinted in S. Landry (Ed.) *Advances in human aspects of aviation.* (pp. 311–320). Boca Raton, FL: CRC Press.

Klein, G. A. (2008). Naturalistic decision making. *Human Factors, 50,* 456–460.

Kopardekar, P., & Quon, L. (2011, March). Airspace systems program: NextGen concepts and technology development (CTD) project; NextGen systems analysis, integration, and evaluation (SAIE) project. Paper presented at the *2011 Airspace Systems Program Technical Interchange Meeting,* San Diego, CA.

Kraut, J. M., Kiken, A., Billinghurst, S., Morgan, C. A., Strybel, T. Z., Chiappe, D. L., & Vu, K.-P. L. (2011). Effects of data communications failure on air traffic controller sector management effectiveness, situation awareness, and workload. In G. Salvendy and M.J. Smith (Eds.) *Human Interface, Part II, HCII 2011, Lecture Notes in Computer Science, 6772,* 493–499.

Lee, J. D., & Moray, N. (1994). Trust, self-confidence, and operator's adaptation to automation. *International Journal of Human-Computer Studies, 40,* 153–184.

Lee, J. D., & See, K. A. (2004). Trust in automation: Designing for proper reliance. *Human Factors, 46,* 50–80.

Lee, J. D., & Seppelt, B. D. (2009). Human factors in automation design. In S. Nof (Ed.), *Springer handbook of automation* (pp. 417–436). Berlin, Germany: Springer Berlin Heidelberg.

Madhavan, P., & Wiegmann, D. A. (2007). Effect of information source, pedigree, and reliability on operator interaction with decision support systems. *Human Factors, 49*, 773–785.

Maguire, M. (2001). Methods to support human-centered design. *International Journal of Human–Computer Studies, 55*, 587–634.

Manzey, D., Reichenbach, J., & Onnasch, L. (2012). Human performance consequences of automated decision aids: The impact of degree of automation and system experience. *Journal of Cognitive Engineering and Decision Making, 6*, 57–87.

McCarley, J. S., & Wickens, C. D. (2005). *Human Factors Implications of UAVs in the National Airspace* (Technical Report AHFD-05-05/FAA-05-01). Savoy, IL: University of Illinois, Institute of Aviation, Aviation Human Factors Division.

Metzger, U., & Parasuraman, R. (2005). Automation in future air traffic management: Effects of decision aid reliability on controller performance and mental workload. *Human Factors, 47*, 35–49.

Nolan, M. S. (2010). Air traffic control. In J. A. Wise, V. D. Hopkin, & D. J. Garland (Eds.), *Handbook of aviation human factors* (2nd ed., pp. 20-1–20-20). Boca Raton, FL. CRC Press.

O'Brien, K. S., & O'Hare, D. (2007). Situational awareness ability and cognitive skills training in a complex real-world task. *Ergonomics, 50*, 1064–1091.

Onnasch, L., Wickens, C. D., Li, H., & Manzey, D. (2014). Human performance consequences of stages and levels of automation: An integrated meta-analysis. *Human Factors, 56*, 476–488.

Paas, F., & van Gog, T. (2009). Principles for designing effective and efficient training of complex cognitive skills. In P. R. DeLucia (Ed.) *Review of human factors and ergonomics* (Vol. 7, pp. 166–194). Santa Monica, CA: Human Factors and Ergonomics Society.

Parasuraman, R., & Riley, V. (1997). Humans and automation: Use, misuse, disuse, abuse. *Human Factors, 39*, 230–253.

Parasuraman, R., Sheridan, T. B., & Wickens, C. D. (2000). A model for types and levels of human interaction with automation. *IEEE Transactions on Systems, Man and Cybernetics—Part A: Systems and Humans, 3*, 286–297.

Parasuraman, R., Sheridan, T. B., & Wickens, C. D. (2008). Situation awareness, mental workload, and trust in automation: Viable, empirically supported cognitive engineering constructs. *Journal of Cognitive Engineering and Decision Making, 2*, 140–160.

Pew, R. W. (2000). The state of situation awareness measurement: Heading toward the next century. In M. R. Endsley and D. J. Garland (Eds.), *Situation awareness analysis and measurement* (pp. 33–47). Mahwah, NJ: Lawrence Erlbaum.

Pierce, R. S. (2012). The effect of SPAM administration during a dynamic simulation. *Human Factors, 54*, 1093–1103.

Pop, V. L., Stearman, E. J., Kazi, S., & Durso, F. T. (2012). Using engagement to negate vigilance decrements in the NextGen environment. *International Journal of Human-Computer Interaction, 28*(2), 99–106.

Prevot, T. (2002). Exploring the many perspectives of distributed air traffic management: The Multi-Aircraft Control System MACS. In S. Chatty, J. Hansman, & G. Boy (Eds.), *HCI-Aero 2002* (pp. 149–154). Menlo Park, CA: AIAA Press.

Prevot, T., Homola, J., Martin, L. H., Mercer, J., & Cabrall, C. (2012). Toward automated air traffic control-investigating a fundamental paradigm shift in human/systems interaction. *International Journal of Human-Computer Interaction, 28*, 77–98.

Prevot, T., Homola, J., & Mercer, J. (2008). Human-in-the-loop evaluation of ground-based automated separation assurance for NextGen, ICAS 2008-11.4.5, *26th International Congress of the Aeronautical Sciences (ICAS 2008)*, and AIAA-ATIO-2008-8885, Anchorage, Alaska, September 15–19.

Rabitoy, E. R., Hoffman, J. L. & Person, D. R. (2015). Supplemental instruction: The effect of demographic and academic preparation variables on community college student academic achievement in STEM-related fields. *Journal of Hispanic Higher Education, 14*(3), 240–255.

Rensink, R. (2000). The dynamic representation of scenes. *Visual Cognition, 7*, 17–42.

Rodgers, M. D., & Nye, L. G. (1993). Factors associated with the severity of operational errors at air route traffic control centers. In M. D. Rodgers (Ed.), *An analysis of the operational error database for air route traffic control centers*. DOT/FAA/AM-93/22, Human Factors Research Laboratory, Civil Aeromedical Institute, Federal Aviation Administration, Oklahoma City, OK.

Roske-Hofstrand, R. J., & Murphy, E. (1998). Human information processing in air traffic control. In M. Smolensky and E. Stein (Eds.), *Human factors in air traffic control* (pp. 65–113). San Diego, CA: Academic Press.

Salmon, P., Stanton, N., Walker, G., & Green, D. (2006). Situation awareness measurement: A review of applicability for C4i environments. *Applied Ergonomics, 37*, 225–238.

Salvendy, G. (2012). *Handbook of human factors and ergonomics.* Hoboken, NJ: John Wiley & Sons.

Sheridan, T. B. (1992). *Telerobotics, automation, and human supervisory control.* Cambridge, MA: MIT Press.

Simons, D., & Rensink, R. (2005). Change blindness: Past, present, and future. *Trends in Cognitive Sciences, 9*, 16–20.

Stanton, N., Salmon, P., Walker, G. H., & Jenkins, D. P. (2010). Is situation awareness all in the mind? *Theoretical Issues in Ergonomics Science, 11*, 29–40.

Strybel, T. Z., Minakata, K., Nguyen, J., Pierce, R., & Vu, K.-P. L. (2009). Optimizing online situation awareness probes in air traffic management tasks. In M. J. Smith and G. Salvendy (Eds.): Human Interface, Part II, HCII 2009, *Lecture Notes in Computer Science, 5618*, 845–854.

Strybel, T. Z., Vu, K.-P. L., Chiappe, D., Battiste, V., Johnson, W. W., & Dao, Q. V. (2012). Metrics for situation awareness, workload, and performance: Manual for online probe administration. (NASA/TM-2011-215991). Washington, DC: National Aeronautics and Space Administration.

Strybel, T. Z., Vu, K.-P. L., Chiappe, D. L., Morgan, C. A., Morales, G., & Battiste, V. (2016). Effects of NextGen Concepts of Operation for Separation Assurance and Interval Management on Air Traffic Controller Situation Awareness, Workload, and Performance. *International Journal of Aviation Psychology, 26*, 1–14.

Strybel, T. Z., Vu, K.-P. L., Kraft, J., & Minakata, K. (2008). Assessing the situation awareness of pilots engaged in self spacing. *Proceedings of the Human Factors and Ergonomics Society 52nd Annual Meeting.* New York.

Sturre, L., Chiappe, D., Vu, K.-P. L., & Strybel, T. Z. (2015). Using eye movements to test assumptions of the Situation Present Assessment Method. In *Human Interface and the Management of Information. Information and Knowledge in Context* (pp. 45–52). Switzerland: Springer.

Taylor, R. (1990). Situation awareness rating technique (SART): The development of a tool for aircrew systems design. In Situation Awareness in Aerospace Operations (AGARD-CP-478). NATO-AGARD, Neuilly Sur Seine, France, pp. 3/1-3/17.

van Merriënboer, J. J. G., & Kester, L. (2008). Whole-task models in education. In M. J. Spector, D. M. Merrill, J. van Merriënboer & M. P. Driscoll (Eds.), *Handbook of research on educational communications and technology* (3rd ed., pp. 441–456). New York: Taylor & Francis Group.

Vu, K.-P. L., & Chiappe, D. (2015). Situation awareness in human systems integration. In D. Boehm-Davis, F. Durso, & J. Lee (Eds.), *APA handbook of human systems integration* (pp. 293–308). Washington, DC: American Psychological Association.

Vu, K.-P. L., Kiken, A., Chiappe, D., Strybel, T. Z., & Battiste, V. (2013). Application of part-whole training methods to evaluate when to introduce NextGen air traffic management tools to students. *American Journal of Psychology, 126*, 433–447.

Vu, K-P, L. Minakata, K., Nguyen, J., Kraut, J., Raza, H., Battiste, V., & Strybel, T. Z. (2009). Situation awareness and performance of student versus experienced air traffic controllers. In M. J. Smith & G. Salvendi (Eds.), *Human Interface, Part II* (pp. 865–874). Berlin: Springer-Verlag.

Vu, K-P., L., Silva, H., Ziccardi, J., Morgan, C., Morales, G., Grigoleit, T., Lee, S. Kiken, A., Strybel, T. Z., & Battiste, V. (2012a). Do use of automated tools aid students' acquisition of air traffic management skills? *Proceedings of the Human Factors and Ergonomics Society Annual Meeting, September 2012 vol. 56*, 16–20.

Vu, K.-P. L., Strybel, T. Z., Battiste, V., Lachter, J., Dao, A.-Q. V., Brandt, S., Ligda, S., & Johnson, W. W. (2012b). Pilot performance in trajectory-based operations under concepts of operation that vary separation responsibility across pilots, air traffic controllers, and automation. *International Journal of Human-Computer Interaction, 28*, 107–118. *Special Issue: Human Factors and HCI Considerations in NextGen.*

Weiland, M. Z. (2010). Field evaluation of advanced training technologies in terminal air traffic control. In *Proceedings of the Human Factors and Ergonomics Society Annual Meeting* (Vol. 54, No. 27, pp. 2257–2261). Los Angeles, CA: SAGE Publications.

Wickens, C. D. (1999). Automation in air traffic control: The human performance issues. *Automation Technology and Human Performance: Current Research and Trends*, 2–10.

Wickens, C. D. (2008). Situation awareness: Review of Mica Endsley's 1995 articles on situation awareness theory and measurement. *Human Factors, 50*, 397–403.

Wickens, C. D., Hutchins, S., Carolan, T., & Cumming, J. (2013). Effectiveness of part-task training and increasing-difficulty training strategies a meta-analysis approach. *Human Factors: The Journal of the Human Factors and Ergonomics Society, 55*(2), 461–470.

Wickens, D. A. (2005). Attentional tunneling and task management. Technical Report AHFD-05-23/NASA-05-10. NASA Ames Research Center Moffett Field, CA. Contract NASA NAG 2-1535.

Willems, B., & Heiney, M. (2002, April). *Decision support automation research in the en route traffic control environment* (DOT/FAA/CT-TN02/10). Atlantic City, NJ: William J. Hughes Technical Center.

Young, M. D., Healy, A. F., Gonzalez, C., Dutt, V., & Bourne, L. E. (2010). Effects of training with added difficulties on RADAR detection. *Applied Cognitive Psychology, 24*, 1–22.

14 Error in Air Transportation

Cees Jan Meeuwis and Sidney W. A. Dekker

CONTENTS

Introduction 321
   The Problem of 'Human Error' 321
The Background for 'Human Error' 323
   Chapanis 324
   Fitts and Jones 324
   'Human Error' as an Effect or Symptom 326
   Epistemology 327
   James Reason and the Swiss Cheese Model 328
   Operationalization of the Latent Failure Model 331
      Human Factors Analysis and Classification System 331
The Label 'Human Error' (Attribution) 334
   The Existence of 'Human Error' 334
   The Hindsight Bias 335
      Outcome Bias 335
   Causality and Accident Investigation 335
   Byproduct of Causality 336
      'Human Error' Is a Judgment 336
   Euphemisms for 'Human Error' 337
   Beyond 'Human Error' 337
References 338

INTRODUCTION

THE PROBLEM OF 'HUMAN ERROR'

> "More than seventy percent of all crashes of scheduled commercial aircraft are caused directly by 'controlled flight into terrain'"

In 2014, the International Air Transportation Association states that "Typically, aircraft malfunction is not the main cause of CFIT accident; rather the accident's probable and immediate causes are often attributed to flight crew or human error, such as non-compliance with established procedures (SOPs), inadequate flight path management, lack of vertical and/or horizontal position awareness in relation to terrain, unstabilized approaches, and failure to initiate a go-around when a go-around was necessary."

    We institutionalize safety by making tradeoffs when we try to explain why things go wrong. We want to understand why things go wrong and want to have an explanation for it, and usually we are very happy to settle for a very simple explanation. Two of the most common types of explanation are human error and mechanical breakdown. 'Human error' has—for a long time—been seen as a large problem in air transportation. In fact, human factors in aviation have always been concerned with 'human error.' Estimates in literature indicate that between 70 and up to 90 percent of all aviation accidents can be attributed to or are caused by 'human error' (Boeing, 1993; Fiorino, 2006; Shappell & Wiegmann, 2003; Woods, Dekker, Cook, Johannesen & Sarter, 2010). According to the statistics of Aviation Safety Network, 990 people lost their lives in 2014 in airline aviation. If 'human error' is the enemy of safe operations and is accountable for 80 percent of the accidents, we could have saved close to 800 of the lives lost in 2014 just by getting rid of 'human error.' We see the explanation of 'human error' when things go wrong everywhere; in scientific journals and papers, in formal accident investigation reports, and also in society, on the news, or other media. 'Human error' is socially well accepted as an explanation when things went wrong.

These ideas and observations lead to a seemingly simple and valid plan: Reduce 'human error' and you will enhance aviation safety. Many ideas in safety science use this emotional reasonable and appealing strategy in an effort toward progress. To be able to succeed in the "war on error", a definition of 'human error' is of the essence. Only then it can be made quantifiable, which is a necessity to get a grip on the 'human error' problem.

At first, the tabulation and counting of errors were fruitful investments in safety research (Dekker, 2006). In these controlled laboratory studies, human tasks were shrunk to a bare minimum, and researchers were left with single and measurable errors. The 'human error' queste was handled by researches as a possible cause of failure just like mechanical break-down. As a result 'human error' was seen as a control problem and handled in that way. Typical ideas that emerged in those days were the ideas of Taylor's standardization and compliance in 1911, accident-prone human beings in the 1920s (Burnham, 2009), and Heinrich's domino model in the 1930s. Prior to World War II, the focus was on designing and shaping the human to fit the task and machine. But despite all those efforts, ideas, models, and methods over time, science seemed to be facing a standstill in the effectiveness of this strategy. 'Human error' appeared to be a rather stubborn problem; it was beyond the reach of known or previously tried remedies, or responsive only to a significant mobilization of such remedies (e.g., severe behavioral interventions or removal of 'bad apples').

It was around World War II that the ideas of relationships between 'human error' and failure began to change, when the concept of *interaction* between human and machine was added (Chapanis, 1965; Fitts & Jones, 1947). The focus shifted to designing the machines and shaping the environments and conditions to fit the human in all its diversity. Major accidents in the following decades such as Tenerife, Bhopal, Three Mile Island (TMI), and Chernobyl have led to a search to figure out what is actually meant by 'human error' (Senders & Moray, 1991). The combined results of these efforts do not only challenge the prewar conventional assumptions about the relationship between 'human error' and failure, but also the tenability of the definitions of 'human error' and its ontology.

A closer look at the literature on 'human error' reveals it to be problematic. Not because we have a 'human error' problem, as in 'human error' *is the safety challenge to be resolved* but because *the status of the term 'human error' itself is extremely problematic* as postwar research exposes.

In their search to find a way to tame the 'human error'-beast, researchers were unable to find the necessary universal definition. Instead, they found a whole new way of looking at incident, accidents, failure, and safety. To show the problematic status of 'human error', we will bring forward two arguments, juxtaposed in Table 14.1, toward the claim why it would be better not to use the term 'human error' at all.

The first argument is that 'human error' is never the cause of failure, but merely an effect or consequence. 'Human error' alone is not sufficient to cause failure. It is only jointly sufficient together with multiple latent conditions that already have to be present to trigger the failure. This argument is best illustrated in literature by the work of original human factors figures such as Chapanis and Fitts, and later in the work of Reason and the practical extensions of his work by others.

The second argument is that 'human error' is nothing more than a label; an attribution after the fact.

Parallel to the genesis of Reason's work, Jens Rasmussen has laid the foundations for the second argument as to why 'human error' is problematic. This research originates from a

TABLE 14.1

Two Arguments Regarding Human Error

	Human Error (HE) as a Consequence Not Cause	Human Error (HE) as a Label
Ontology	HE exists	HE does not exist
	HE can be categorized and each category needs a different countermeasure	The label itself is not an explanation for failure nor a guide to countermeasures or improvement
HE definition	Unsafe act	Someone did something wrong according to someone: a judgmental label
The function of HE/Why is HE problematic?	Intermediate finding, not a conclusion, symptom not a cause, only jointly sufficient to create failure	It is an oversimplification of complex interactions and sociology
Effect of the problematic state?	Addressing HE is treating the symptoms not the cause	It does not exist and it diverts the attention from real learning opportunities
Authors	Reason, Shappell, Wiegmann	Rasmussen, Hollnagel, Woods, Dekker
Dominating view	Functionalist view	Interpretivist view
	Etic, judgmental from peer reviewed norms	Emic, not judgmental but describing social norms and their origins
Risk management	Error tabulation, categorization of error, violations, calculation of failure probabilities	Enlarge the ability of operators to manage emergent risks
Accidents	Breakdown or malfunction of normal system behavior	Breakdown in adaptation necessary to cope with the real world (BHE, p.82)
	System thinking LIGHT (Linear complexity)	System thinking (social complexity)
Models	Linear and latent failure models (psychology)	Complexity, control and sociological models
	Failure is seen as the outcome of a process Events in sequence generate outcomes. As a remedy a change in cause and effect sequence or elimination of causes, prevent failure	Failure is the emergent byproduct of interactions between components
How do you deal with erroneous operators?	Reason's just culture Different shades of culpability	There are no erroneous operators, just operators that made sense of the situation they were in and acted according their understanding of a situation. Therefore, they are one of your most valuable assets in the reconstruction and explanation of failure

cognitive engineering perspective that refers to sociology and complexity. A solid case is made for the claim that there is no such thing as 'human error', let alone it is a stable category for performance.

THE BACKGROUND FOR 'HUMAN ERROR'

Research on 'human error' has been an area of interest for a very long time. With an enormous progression on the technical reliability of our safety critical systems, it appeared if the 'human error' became more critical in the urge for continuous improvement. To reduce the increasing problem of 'human error', a clear definition is of the essence. Various researches have undertaken the effort to clearly define 'human error.' However, instead of finding a universal definition, they came up with a completely different view on safety, on how accidents happen, on how to investigate incidents and accidents, and how to organize and manage reactions to failure. This fundamentally different paradigm on our safety problems is now known under various names: The New View, Safety-II, Safety Differently, to name a few. Here is how this new paradigm evolved in the second half of the twentieth century.

CHAPANIS

During World War II, an unusual number of B-17 bombers crashed at landing or after landing due to a confusion in gear and flap controls in the cockpit. Despite all accustomed mitigations, such as additional training and awareness campaigns, the crashes were not eliminated from the operation. The problem seemed to be beyond the reach of those measures. Alphonse chapanis was a psychologist that worked at the Army Air Force Aero Medical Lab in Daytona, Ohio, at the time he was called into the problem. He solved the problem by placing a wheel shape onto the gear switch and little wing shape to the flap switch. A fundamental reverse thinking; instead of trying to (re)shape the human operator by means of training or other behavioral interventions, Chapanis changed hardware that was part of the environment of the operators.

FITTS AND JONES

Just after the World War II, Paul Fitts and Richard Jones conducted studies into airplane crashes related to aircraft controls and instrument displays. These airplane crashes were presumably caused by "pilot error." Fitts and Jones made visible that World War II witnessed a tipping point in which the abilities of technology had outpaced the human skills and possibilities. Human operators were no longer able to adapt and compensate for unreliable system designs. To determine the human performance as erratic was deemed unsatisfactory for Fitts and Jones. These studies led to two major understandings that are the cornerstones of two arguments as to why the term 'human error' is problematic. First, their studies empirically showed that 'human error' are systematically connected to the surroundings and conditions that people work in. 'Human error' was their starting point and not the conclusion of their investigations. Therefore, 'human error' is never the cause of failure, but rather a symptom of failure. Second, Fitts and Jones consequently notate 'human error' with quotation marks. This use of quotation marks shows that they are troubled and uncertain about the term, its use, status, and its definition. The term 'human error' was unsatisfactory.

Chapanis' practical intervention in environment and Fitts and Jones' studies into behavior believed to be "pilot error" are nowadays seen as the birth of modern Human Factors and Ergonomics— understanding why operators do what they do to change their environment and conditions and thereby indirectly shape operators assessments and actions. Decades later their work was extended and turns out to be fundamental for the ideas that answer the challenges in our twenty-first century safety challenges.

Often it appears if there are as many 'human error' models and frameworks as there are people interested in this topic (Senders & Moray, 1991). Reviewing the literature on 'human error' reveals that there are several perspectives on the nature of 'human error.'

From the 1950s, the studies on 'human error' have been extensively reviewed and analyzed. A series of high impact high visibility catastrophes starting in the late 1970s (e.g., Three Mile Island and Tenerife Crash), has triggered numerous industry oriented studies on 'human error' and human reliability (Amalberti, 2001). Two international research networks that turned out to be highly compatible, and later highly competitive, emerged (Le Coze, 2015). Where James Reason's work represents the psychology research tradition on 'human error', Jens Rasmussen worked on the topic from an (cognitive) engineering perspective. For more than 10 years, the two research disciplines were unknown to each other until they crossed paths in the aftermath of one of the major catastrophes, the TMI nuclear accident in 1979.

Le Coze (2015) identified three major aspects that distinguish the two researchers:

- The psychological perspective. James and Freud were of considerable influence in Reason's work. Reason saw error as a way to theorize psychology. Rasmussen on the other hand shows no sign of Freudian influences as he was committed to producing models for engineering purposes.
- Reason focused on categorization of 'human error' types related to unconscious cognitive processes, such as "slips" and "lapses" (Figure 14.1). Rasmussen focused on helping the

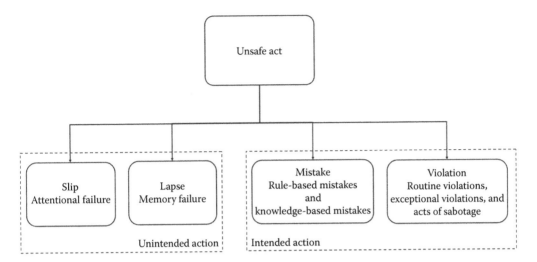

FIGURE 14.1 A taxonomy for 'human error.' (Modelled after Reason, J.T., *Human Error*, Cambridge University Press, New York, 1990. With Permission.)

operator "Coping with complexity" (Rasmussen & Lind, 1981). He saw failure predominantly occur in complex, no-routine situations.
- Reason's studies and frameworks are only based on theory. There are no empirical studies of real-life situations to support his work. However, Rasmussen grounded his model on real-life observations. Research on practical situations brings up valuable information for future design.

As a result, Reason is mainly occupied with taxonomy of error after TMI, whereas Rasmussen focuses on the naturalistic view of errors (Figure 14.2). Both paradigms form a basis for a compelling argument why our universe would be better off with 'human error' cut out of our jargon.

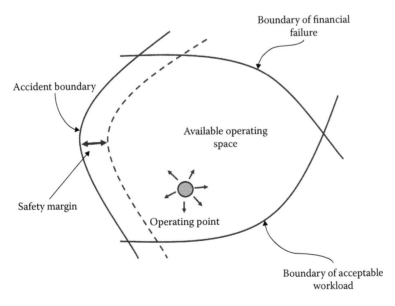

FIGURE 14.2 A naturalistic view of 'human error.' (Modelled after Rasmussen, J., *Ergonomics*, 33, 1185–1199, 1990. With Permission.)

'HUMAN ERROR' AS AN EFFECT OR SYMPTOM

In the postwar decades, numerous attempts were made to conceptualize 'human error.' The concept of 'human error' made contact with a variety of research disciplines resulting in plurality of theories, concepts, definitions, models, and perspectives by the 1980s. This research path started with system and control engineering that rapidly guided the attention to human–machine interface problems and so research was forced into the discipline of 'human error' analysis (Amalberti, 2001). In aviation, the second and third research disciplines shape five different categorical perspectives in this 'human error' research: the cognitive, the ergonomics and system design, the aeromedical, the psychosocial, and the organizational perspective (Shappell & Wiegmann, 2001).

The cognitive perspective is largely based on the information processing theory. This theory is based on the idea that the human brain does not merely respond to stimuli but that the brain processes the information it receives. When the information enters the human brain, attention will be allocated to it, the human brain uses pattern recognition and decision-making to plan a response execution. According to the cognitive perspective, 'human error' occurs when mediating operations between the stimulus input and response execution fail to process information appropriately. Based on the processing theory, Rasmussen is one of the first in literature to develop a taxonomy for 'human error' (Rasmussen, 1982).

But cognitive models turn out to have severe limitations:

1. There is no exact procedure defined for applying cognitive models to error analysis or accident investigation. This results in speculation and intuition from analysts and investigators when applying these models.
2. Cognitive models isolate cognitive processes from the context in which they took place. This is often referred to as "cognition in the mind". When these external factors such as workplace conditions and equipment are not addressed, the cognitive models have an extreme tendency to conclude 'human error' as the "cause" of failure. In turn, this often results in blame being placed on the individual(s) who committed the error.

The ergonomics and system design perspective on 'human error' focuses on mismatches between components that constitute a system that degrade operator performance. A well-known model is the SHEL model. SHEL is the acronym for the four basic components that are necessary for successful human–machine integration and system design; Software, Hardware, Environment, and Liveware (Nagel, 1988). The limitations of the cognitive model have been anticipated as the ergonomics, and system design does consider workplace equipment and facilities. The SHEL model gains popularity rapidly and quickly became the recommended framework to be used for incident and accident investigations by the International Civil Aviation Organization (ICAO). Initially it appears if the system perspective deflects the attention away from the operator toward the system, but note that the system failure is often still caused by degraded operator performance, in other words 'human error.' In the system perspective, the human is rarely the sole cause of failure, but there is still emphasis on operator failure. A limitation of the system perspective is the analysis of human behavior remaining shallow. The system perspective focuses mainly on the ergonomics of the human–machine interface as well as the anthropometric requirements of the task and human characteristics. As a result, the cognitive, social, and organizational factors receive only superficial consideration.

To fill this gap in this system perspective, the aeromedical perspective is added. Within this approach the 'human error' is often caused by underlying medical or physiological conditions, such as hypoxia, sleep deprivation, or spatial disorientation. As a reaction to this perspective, the aviation industry has developed and intensified periodical aeromedical checks for pilots, disorientation training, and theoretical training in the field of Human Performance and Limitations. The aeromedical perspective focuses on the physiological conditions of the pilot as it deems to play significant role in safe performance. A well-known physiological condition that has significant influence on human performance is fatigue. Recently International Air Transportation Association has developed a Fatigue Risk Management System together with ICAO and the International Federation of Airline Pilots' Association to present a common approach to pilots, regulators, and operators to the complex issue of fatigue. The major critique on the aeromedical approach is that it is extremely hard to operationalize the issues of fatigue or self-medication or disorientation. Is being tired after a broken night the same as being fatigued? Can a pilot take an ibuprofen capsule during flight? How about two, or four? A problematic issue in this view is that it is extremely hard to draw a line of which we can agree before something bad happens: This is the fatigued and this is not. However, when something went wrong and we possess hindsight knowledge, it is usually not hard at all to construct an account of some aeromedical deficiency. Second, it is imaginable that aeromedical factors can influence performance, but it is extremely difficult to determine whether these factors have a causal connection to failure.

Moreover in the 1990s, a psychosocial perspective is identified in the safety literature. This perspective focuses on the interaction among all users of the system: pilots, air traffic controllers, ground staff, gate agents, dispatchers, maintenance engineers, and flight attendants. The general performance of a system is at its optimum when there is maximum quality in the interactions between all parties involved. The operating environment is not only influenced by these interactions but also by both the attitudes and personalities of individuals. As a result of this logic, accidents are explained by a breakdown of group dynamics and interpersonal communications.

EPISTEMOLOGY

Of all the goals we pursue in safety and human factors, our main drive is to prevent reoccurrence of accidents. To do so, it seems entirely legit to reconstruct past occurrences to establish exactly what happened. This activity has its roots in engineering. It is even a recommended practice in civil aviation accident investigation by ICAO Annex 13; "History of Flight," one of the most endorsed guidelines for accident investigation and report writing in. This is referred to as the epistemological purpose of accident investigation (Dekker, 2015a). A good investigation first collects all the facts and then starts analyzing the facts to conclude with a causal construct of the accident. ICAO Annex 13 even recommends shaping an investigation report in this precise order of chapters: Factual Information, Analysis, Conclusions, and Recommendations.

As Richard Cook is fond of saying, the way we see our world is almost entirely shaped through the accidents and disasters we have created. The general ideas of how to build an airplane, what instruments and electronics should be on board, what the layout characteristics of an airport should look like, and how much separation between two aircraft is the minimum for staying in control of the situation is all knowledge that was gathered in the wake of accidents. This means that the models, theories, and methods we use during incident and accident investigation are of major influence on the conclusions of these investigations and the variables we tend to change to prevent a similar occurrence from happening in the future. At the base of the general way we tend to do accident investigations lies the thinking and the laws of Isaac Newton and Rene Descartes (Dekker, 2011). Newton and Descartes have pretty much found the basis for the way we think about science, about truth, and about cause and effect. The Enlightenment can be seen as one of the cornerstones of our modern day Western Society. Newton used mathematics to describe the observed phenomena in our universe, and Descartes relied on the inner process of reason, yet they both believed in a single reality, or objective truth, that can be explored, captured, and understood through reasoning and analysis. Their ideas have become part of our human DNA in a way that we, most of the time, do not even realize that the things we do in daily life or the way we make sense of the world around us are what is often referred to as the "Newtonian-Cartesian Worldview" (Dekker, 2010). This thinking is the basis of our modern world engineering, and it is the motor behind epistemological accuracy; it has let mankind walk the surface of the moon and is motoring today's globalization. We would not be flying airplanes without

Descartes, and we would not have satellites in orbit without Newton. The epistemological purpose of an investigation is to tell the exact story of what happened on the basis of a universal objective truth.

JAMES REASON AND THE SWISS CHEESE MODEL

There are basically two ways one can look at 'human error.' We can see 'human error' as the cause of failure, or we can see 'human error' as a symptom of failure. Several different scientists acknowledge these two views in literature around the year 2000 (AMA, 1998; Rasmussen & Batstone, 1989; Reason, 1990; Woods, Johannesen, Cook, & Sarter, 1994).

The scattered postwar perspectives were unified in literature by James Reason (1990) in the Organizational Model of accidents; the latent failure model commonly known as the Swiss Cheese Model (Amalberti, 2001). The Swiss Cheese Model is one of the most dominant models in civil aviation over the last decades. The rationale discussed earlier matches the early ideas of Fitts and Jones (1947) and is operationalized in various ways with all sorts of (negative) effects. This makes the Swiss Cheese Model well suited to demonstrate the first argument of why 'human error' should not be used.

> 'Human error' is a symptom of trouble deeper in the system. Continuous use of the term 'human error' redirects the focus on the individual and leads to a setback into the interbellum.

In his work, Reason (1990) describes the two distinct views on 'human error' as a "person approach" and a "system approach". The person approach refers to the ideas of 'human error' prior to World War II. The idea that 'human error' is a human reliability problem and that once we figure out the source of human reliability we can solve the 'human error' problem. The person approach concentrates on the error of individuals, blaming them for forgetfulness, inattention, or moral weakness. The system approach concentrates on the conditions under which individuals work and tries to build defenses to avert errors or mitigate their effects. Reason argues that the system approach is the way to go forward if one wants to achieve progress on safety. The rationale is that 'human error' is not an explanation for failure but instead *demands* an explanation. Effective mitigation therefore does not start with targeting individual human beings, who themselves are at the receiving end of much latent conditions (Reason, 1997), but effective mitigation starts with targeting the conditions and environment that facilitate failure.

The Swiss Cheese Model itself is also a combination of preceding theories and models on accident causation. Especially the sequence of events model is clearly identifiable. This is in line with the Newtonian-Cartesian Worldview in which linear causation exists and broken parts can account for broken systems. The functioning of the whole can be explained by the functioning of its constituent component (Leveson, 2011). The main idea in Reason's Swiss Cheese Model, shown in Figure 14.3, is that disasters are not the result of one single failure (e.g., 'human error') but merely as a range of several smaller failures and contributing events and conditions. These multiple contributors are all necessary but individually insufficient to account for a disaster. The basis for this model can be traced back to Reason's investigations into the TMI accident in 1979.

Systems are described to have a sharp end and a blunt end. The sharp end is the actualizer of the process; people who are performing tasks (pilots, air traffic controllers, etc.) work at the sharp end of a system. The blunt end of a system are the processes and tasks farther away from the action itself, such as regulators, administrators, training activities, and so on. Procedures, barriers, rules and regulations play a major role in Reason's system approach. These defensive layers come in various ways such as physical barriers, personal protective equipment, automatic shutdowns, alarms, and so on. The function of those barriers is to protect a vulnerable object, such as humans or assets from hazards. These hazards have potential energy to inflict harm or damage, and the barriers are there to control or deflect that energy. Ideally these barriers are flawless; however, in practice, there

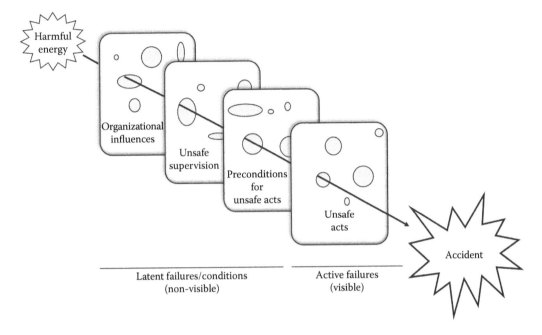

FIGURE 14.3 The Swiss Cheese Model by Reason. (Adapted from Shappell, S.A. and Wiegmann, D.A., The human factors analysis and classification system—HFACS, National Technical Information Service, 2000. With Permission.)

are always holes in these barriers similar to there are holes in Swiss Cheese, hence the analogy. According to Reason, these holes are, unlike in cheese, constantly changing shape and shifting position. Each slice of cheese is an opportunity to stop failure. In this analogy, accidents occur when these holes to multiple layers of defense momentarily line up. As a result, the potential energy of the hazard can reach the vulnerable object.

To explain this set of multiple contributors and the rise of holes in the defenses, Reason uses the terms "active failures" and "latent failures" (Reason, 1990). Active failure refers to "unsafe acts" that in turn parallels the term 'human error.' Active failure occurs at the sharp end of the system. This indicates that 'human error' has an entity and is ontologically "real". The latent failures are the embodiment of the multiple contributors at different levels in the system. Latent failure occurs at the blunt end of a system. In this model, the human operator at the sharp end is either the last straw to break the camel's back, or the hero who deflected the harmful energy to escape a black day (Reason, 2008).

The aim and effect of this distinction between active and latent failure is to redirect the focus from the frontline operators that made 'human error' toward the conditions that set the stage for that behavior. Unsafe acts are only jointly sufficient together with organizational influences, unsafe supervision, and preconditions for unsafe acts to create accidents. As Reason put it in 1990:

> Rather than being the main instigator of an accident, operators tend to be the inheritors of system defects created by poor design, incorrect installation, faulty maintenance and bad management decisions. Their part is usually that of adding the final garnish to a lethal brew whose ingredients have already been long in the cooking.

Reason provides us with a definition of error as it were a process outcome (1990):

> Error will be taken as a generic term to encompass all those occasions in which a planned sequence of mental or physical activities fails to achieve its intended outcome, and when these failures cannot be attributed to the intervention of some chance agency.

Again, in this definition Reason stresses the importance of the circumstances in which failure occurs. Yet no clear definition of 'human error' is given. In fact, he leaves space for interpretation and utility, or as Reason puts it himself (2008):

> Although there is no one universally agreed definition of (human) error, most people accept that it involves some kind of deviation. (...) Just as there are several possible definitions, so there are also many ways errors may be classified. Different taxonomies serve different purposes. These depend upon which of the four basic elements of an error – the intention, the action, the outcome and the context – is of greatest interest or has the most practical utility.

Reason classified errors based on Rasmussen's three levels of performance; a path of maturity for performance. In order of decreasing levels of familiarity with the task, these are (Rasmussen, 1983) as follows:

- *Skill-based*: Stored patterns of preprogramed instructions
- *Rule-based*: Familiar problems are addressed by application of stored rules
- *Knowledge-based*: Novel situation in which actions must be planned, using conscious analytic processes and stored knowledge

Reason highlights the notion of intention in his definitions when considering the nature of error and suggests an error classification based upon the following questions:

- Were the actions directed by some prior intention?
- Did the actions proceed as planned?
- Did they achieve their desired goal(s)?

The combination of the importance of intention of an individual and Rasmussen's three levels of performance lead to a well-known and widely accepted 'human error' classification:

- *Slips* can be thought of as actions not carried out as intended or planned, for example, "finger trouble" when dialing in a frequency or "Freudian slips" when saying something.
- *Lapses* are missed actions and omissions, that is, when somebody has failed to do something due to lapses of memory and/or attention or because they have forgotten something, for example, forgetting to lower the undercarriage on landing.
- *Mistakes* are a specific type of error brought about by a faulty plan/intention, that is, somebody did something believing it to be correct when it was, in fact, wrong, for example, switching off the wrong engine.

In addition, Reason addresses operator violations. Based on etiology, two distinct types are identified: routine violations and exceptional violations.

- *Routine violations* are considered rule-breaking actions by operators that are routinely, for example, the consistent accidence of a speed limit by 5–10 mph. These violations are indicative for individual or a group's typical behavior pattern.
- *Exceptional violations* are the breach of any kind of rule because the situation facilitated in that behavior. This behavior is not considered exceptional because of their extreme nature, but because they do not fit the typical behavior of an individual or group.

What is interesting about Reason's classification and handling of 'human error' by classification is that there seems to be a paradox. On the one hand, Reason argues that 'human error' is not the problem, and it is a waste of time to put effort in error management, as it is the systemic issues that allowed failure to happen in the first place. On the other, he presents an exhaustive taxonomy for 'human error', putting flesh on the bone of the alleged 'human-error' problem. One explanation for this is that 'human error' is regularly the starting point for the search for improvement. In this way, it would make sense to provide taxonomy as it guides the search toward the best way to deal with the 'human-error' problem and thus general improvement. Another explanation is that the taxonomy is needed to pursue protagonists of the person approach on 'human error.' Simply telling them that 'human error' is not mental or moral deficiency of an individual (person approach) might be too big of a mental leap. In this explanation, the taxonomy functions as a deflector for the primary reaction to act on the 'human error.'

In summary, the tenets underlying the Reasonian view, concept, and model on 'human error' are the following:

- Everyone commits errors. Fallibility is part of the human condition. 'Human error' is not intrinsically bad and even essential for learning. The identification of 'human error' is not an actionable conclusion, because
- 'Human error' is generally the result, consequence, or symptom of circumstances beyond the control of those committing the errors.
- System or processes that depend on perfection of performance are inherently flawed.

Reason's work is an ultimate attempt to analyze failure by deduction in a traditional engineering way. It leads to the conclusion that 'human error' is not the concept that requires a lot a focus. It is not the problem to be solved and cannot be the conclusion of an investigation. Yet the continuation of use of the same denotation and de delivered taxonomy for 'human error' strengthens the tenacity of the folk model on 'human error', or reinvention of 'human error.' In the next section, we will explain how.

OPERATIONALIZATION OF THE LATENT FAILURE MODEL

Human Factors Analysis and Classification System

In the late 1990s and early 2000s, Scott Shappell and Douglas Wiegmann made an attempt to bridge the gap between theory and practice by identifying and defining the holes in the cheese at the level of 'human error' described in Reason's work: the "Human Factors Analysis and Classification System" (HFACS) (Shappell & Wiegmann, 1997, 2000, 2001a, 2001b, 2003). Their starting point is where Reason has left them:

> ... Simply writing off (...) accidents merely to (human) error is an overly simplistic, if not naive, approach. (...) After all, it is well established that accidents cannot be attributed to a single cause, or in most instances, even a single individual. (Shappell & Wiegmann, 2001)

The HFACS framework, shown in Figure 14.4, consists of the four operational levels taken from the Swiss Cheese Model:

(1) Unsafe Acts of Operators, (2) Preconditions of Unsafe Acts, (3) Unsafe Supervision, and (4) Organizational influences. The first operational level is divided into the three types of 'human error' and the violations that Reason provided in his legacy. Shappell and Wiegmann made error taxonomy for the second through fourth operational levels. The result is a basic coding framework consisting 19 categories of error organized hierarchically in four levels.

Although there is no doubt about the intentions of improvement of Shappell and Wiegmann, their attempt to operationalize the systems approach (or new view) becomes a reinvention of 'human error' (Dekker, 2001). Two assumptions in this attempt are illustrative of this phenomenon:

1. Error classification is the same as analysis.
2. Relating operator to systemic conditions means we should seek for failure higher up in hierarchy.

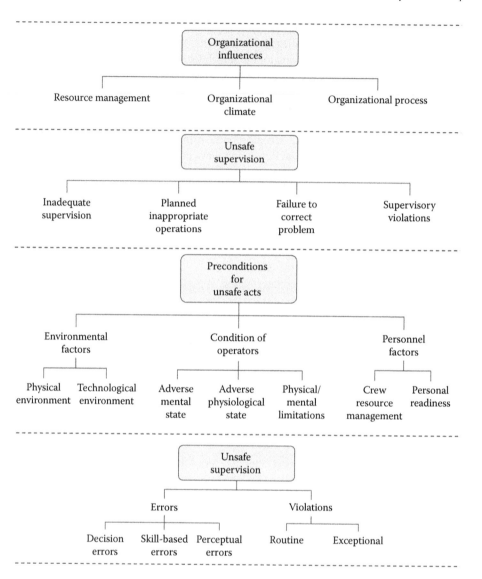

FIGURE 14.4 The human factors analysis and classification system (HFACS). (Modelled after Shappell, S.A. and Wiegmann, D.A., *A Human Error Approach to Aviation Accident Analysis: The Human Factors Analysis and Classification System*, Ashgate Publishing, Aldershot, UK, 2003. With Permission.)

In practice, when applying HFACS one would take the identified 'human error' in the fourth operational level (unsafe acts) and try to push it up the hierarchy ladder as high as possible. Driving an observation up the hierarchy into higher levels of a system is classification on different levels. But that activity fails to explain what the dynamics behind the label of 'human error' are. Classification is not the same as analysis. In addition, this classification reinforces the person approach to 'human error' and safety by pointing out 'human error' on a higher level. HFACS does not deflect the attention away from 'human error' toward deficiencies in the system, as Reason proposes in his Swiss Cheese Model, but it replaces it away from the sharp end of a system. "Inadequate Supervision" for example (in the third operational level) is nothing more than a euphemistic way of saying 'human error' all over again; in fact, all 14 categories at the blunt end of the system are. The attempt to operationalize Reason's theory brings us back to the person approach that Reason is so determined to get away from. In this way, Shappell and Wiegmann have inadvertently strengthened the biases that are so difficult to overcome in safety science. HFACS reinvents 'human error' as the cause of failure; it is just no longer necessarily situated on the work floor (unsafe acts).

An example:

This phenomenon of reinvention of 'human error' at other levels then the sharp end of organizations or systems is seen a lot in recent safety investigations in an attempt to adopt a system-perspective. Following are three statements regarding the risk assessment prior to the MH17 disaster. The Dutch Safety Board (2015) writes in their report:

> "The Dutch Safety Board considers this risk assessment to be incomplete because it does take threats to military aircraft into account, but does not account for the consequences to civil aviation of potential errors or slips." (p.207)

> "This investigation reveals that, prior to the crash of flight MH17, none of the parties involved adequately identified potential threats that the conflict in the eastern part of the Ukraine posed to civil aviation flying over the area." (p.244)

> "In the system of responsibilities, the emergence of a weak link (the airspace management) did not lead to other parties taking action to help ensure the safety of civil aviation earlier the conflict area." (p.245)

These quotes demonstrate an attempt to deflect the attention of the human operators in these systems toward the system itself by naming processes or addressing entire organizations instead of individuals. Although not explicitly expressed and against the intentions, we are back to square one; (indirectly) appointing humans to be at fault, causally connected to the disastrous outcome. These statements persuade us to silently ask questions about responsibility for the risk assessment processes and those responsible for decisions taken in the organizations involved. Therefore, we are still appointing blame to individuals but this time higher up in the system. In our attempt to get away from 'human error' and try to find what systemic factors lie behind it, we still end up saying that 'human error' is the problem:

- Someone must have been negligent in passing an incomplete risk assessment. "Errors and slips" on the sharp end are even mentioned although it is impossible to understand what they are and how to deal with them.
- How can parties involved be so indifferent about threat identification? Who is responsible for that?
- Other parties "did not take action." So here we found incompetent airspace management? Someone must take responsibility for that recklessness.

The result of all research on 'human error', including Reasons unifying work, is not difficult to summarize: We still do not have one objective, universal, unproblematic definition of 'human error.'

The definition that comes closest to that is

'Human error' is when someone did something wrong according to someone

It is this unsatisfying result that has led researchers to rethink the feasibility of the quest of defining 'human error'. The absence of a universally unproblematic definition, method, and model is maintaining the status quo on the topic. This status quo is the motive to start thinking differently about safety and 'human error'. It is the motivation to exert a different vocabulary and to vacate the Newtonian-Cartesian Worldview when it comes to safety and 'human error.' This paradigm is referred to as "complex system thinking" (Woods et al., 2010) or "Safety-II" (Hollnagel, 2014b) or "Safety Differently" (Dekker, 2015b). The paradigm is founded, in large part, on the work of Jens Rasmussen.

THE LABEL 'HUMAN ERROR' (ATTRIBUTION)

Today, although 'human error' is still a common explanation for accidents, the term itself turned into the topic of scientific discourse decades ago. Interestingly, this school of thought can be traced back to in literature as Reason's legacy: Fitts & Jones and Jens Rasmussen.

'Human error' is a label.

To build up the understanding of what an attribution is, a few concepts require explaining as to why they are problematic in the understanding of 'human error' and improvement in safety.

> 'Human error' does not exist; it is merely an attribution or judgment after the fact; therefore, it has no value in understanding failure.

To support this argument, a clarification is of the essence on a couple of assumptions often taken for granted in safety. The argument on the nonexistence of something such as 'human error' is best to be seen as a cocktail of busted "safety myths," explaining some fundamental cognitive shortcuts and understanding human sociological needs. It is this cocktail that came about by critically thinking and questioning these assumptions that formed the New View or Safety-II paradigm.

THE EXISTENCE OF 'HUMAN ERROR'

In August 1983, the NATO Conference on Human Error was held in Bellagio, Italy. Erik Hollnagel presented a position paper called Human Error (Hollnagel, 1983). The paper is a result of a series of questions asked by the organizers of the conference. Hollnagel argues in an almost agitated tone that it is merely impossible to answer these questions one by one. Instead of answering these questions, Hollnagel starts an argument to decompose the bias that underlay these questions about 'human error' as this is "necessary to discuss (...) before trying to answer the questions."

Hollnagel uses several steps to decompose the assumption that there actually exists something called 'Human Error.' As a result, 'human error' gets the "status of a concrete phenomenon", as it were ontologically real. This would mean that it is observable and one would be able to define it. However, Hollnagel argues that none of this is actually possible. Instead of being able to observe 'human error' directly, it must be inferred (concluded from reasoning) from observations. To conclude 'human error', there has to be a mismatch between the actual and the intended state of a system. In other words, a situation in which goals are not (entirely) met is fundamental to conclude 'human error.' The observation of a mismatch in turn assumes that a description of a desired system state is available, and an explanation for the mismatch can be reasoned. If there are no indications in the technical part of a system, the solution is generally to explain the performance variability to the human component. Hence, 'human error' becomes an "assumed cause" of failure. This makes 'human error' not unique but only one explanation out of many others. In this way, the detection of a mismatch is thus fundamental for inferring the existence of 'human error.' Further, Hollnagel notes it is important to realize that human performance variation is also present in situation in which there is no mismatch present or observed. Yet in these situations, there is no reason to look for a cause and therefore 'human error' cannot be concluded.

This means that concluding 'human error' can only take place when there is an undesired outcome evident, although the same human behavior may be present but not judged as being 'human error.' The human behavior is not of interest at all in the conclusion of 'human error', it is the mismatch that is leading. This leads to the significant consequences of hindsight bias for the existence of 'human error.'

THE HINDSIGHT BIAS

As Hollnagel illustrates in his argument, outcome knowledge is of the essence to conclude 'human error.' Hindsight bias is the tendency for people to "consistently exaggerate what could have been anticipated in foresight" (Fischhoff, 1975). Studies have consistently shown that people have a tendency to judge the quality of process by its outcome. In hindsight, everything appears to be obvious. In hindsight, having the outcome knowledge in hand, we clearly seem to know that this could happen. It even shows us all hints and clues available to the operators involved. With all information readily available, we unwittingly start to simplify the complexity into order, structure, and causality (Reason, 1990). By doing this, we isolate a construct of causality, which makes the outcome seem inevitable, from its original context. The hindsight bias is well documented and studied in psychology. It relates to how we perceive probability of an event we know it happened. Research has also shown that the hindsight bias is extremely difficult to overcome. Even for those who are warned for the existence of hindsight bias, it remains difficult to exclude this phenomenon from their judgments (Fischhoff, 1982).

Counterfactuals are a well-known expression of the hindsight bias. Counterfactual reasoning is explaining after-the-fact what people could or should have done to avoid the outcome. Although emotionally satisfying stakeholders, counterfactual reasoning is not very useful in safety. For once, it does not explain how and why things went wrong but it is only a substitutionary description of what happened by saying what did not happen. Second, counterfactual reasoning is impossible to prove empirically as it is impossible to reconstruct the path of failure in all its complexity. People tend to explain complexity through simplifying heuristics. Heuristics are extremely handy to make sense of situations and produce decisions as described by, for example, Tversky & Kahneman (1974) and Fischhoff (1982).

In practice, the hindsight bias leads to judging people involved in accidents. Hindsight means being able to look back on an event with more information available than the people involved. Judging from this, knew-it-all-along point of view gets in the way of understanding failure.

Outcome Bias

Outcome bias is the phenomenon in which outcome knowledge influences the evaluation of decision quality. This means that we make use of the Newtonian idea of symmetry between cause and effect. If the outcome is bad, the decision that led up to that outcome must be equally bad. Although both hindsight bias and outcome bias make use of outcome knowledge, they are often mixed up.

CAUSALITY AND ACCIDENT INVESTIGATION

'Human error' is closely connected with how we think about accidents and safety. Our social models on how accidents come about incorporate a few assumptions that have been taken for granted and still are taken for granted today. These are as follows: Accidents have causes, causes can be found, and the functioning of systems can be reduced to the functioning of its individual components. The existence of something called 'human error' can be traced back to these assumptions. Closer research on these assumptions shows them clearly to be myths. These assumptions are however deeply baked into our human DNA. 'Human error' is nothing but an outcome of a social process in explaining why bad things happen.

Accidents investigations are usually identifiable as root cause analysis, although rarely a methodology is described in investigation reports. The starting point of an investigation is usually the outcome; the event that triggered the need for an investigation in the first place. By looking at the proximate events and then reason backwards to distal causes, or the root cause, explanations for the accident are constructed and mitigating actions for reoccurrence in the future can be taken.

Looking at, for example, the investigation report of the Dutch Safety Board into the crash of MH17 we exactly see, "If accidents or near accidents [nevertheless] occur, a thorough investigation into the causes, irrespective of who are to blame, may help to prevent similar problems from occurring in the future" (DSB, 2015). The path of this investigation is set on a typical Newtonian cause-and-effect stage. The journey starts at the event of the crash, the effect, and because effects cannot take place without a cause, the journey will therefore end at the root cause that triggered all subsequent events.

In the first part of the investigation (part A), the causes into the crash of the aircraft are investigated. In the second part (part B), the investigation focuses on the flight route of flight MH17; why the aircraft was flying over a conflict zone at that time. In the first part of the investigation, Newton was of great help in understanding how the aircraft broke up into pieces in flight and ended up in the eastern part of the Ukraine on that summer afternoon in 2014. Newtonian, mechanistic thinking in cause and effect was able to compose an impressive forensic analysis of the cause of the intervention in normal flight.

"The combination of the recorded pressure wave, the damage pattern found on the wreckage caused by the blast and the impact of fragments, the bow-tie shaped fragments found in the cockpit and in the body of one of the crew members in the cockpit, the analysis of the in-flight break-up, the analysis of the explosive residues and paint found, and the size and distance of some of the fragments, led the Dutch Safety Board to conclude that the aeroplane was struck by a 9N314M warhead as carried on an 9M38-series and launched by a Buk surface-to-air missile system." (DSB, 2015).

Looking at the second part of the investigation, the Newtonian investigation seems to be more of a struggle with lots of loose ends that are hard to connect. Although the Newtonian-Cartesian Worldview is highly successful at explaining the physical cause of the crash of MH17, there lies a risk in this success as well; the risk of becoming unable to make further progress on safety. Taking a closer look at the second part of the investigation, the traditional investigation strategy becomes problematic.

BYPRODUCT OF CAUSALITY

Although it is not the intention of the Dutch Safety Board, or any investigation modeled to ICAO Annex 13, to apportion blame or liability, a side effect of the Newtonian reconstruction is that it unavoidably will blame. Although not made explicit, the Ukraine government is blamed for preparing the stage for the accident by not closing the Ukrainian airspace. The Dutch Safety Board states in the summarized version of its investigation report:

> The Investigation revealed that Ukraine did not adequately identify the risks to civil aviation. (...) According to the Dutch Safety Board these reports gave enough reason for Ukraine to close its airspace as a precaution. However, this did not happen.

The connotation of this sentence easily leads to the interpretation that the Ukraine government has been negligent in the decisions to restrict the airspace for civil aviation. Note that this judgment of the Dutch Safety Board can only be made *after the fact*, in possession of outcome knowledge.

'Human Error' Is a Judgment

Both hindsight and outcome biases instigate how we explain failure and the role of 'human error.' With outcome knowledge, after the fact, we tend to believe that the negative outcome was foreseeable and preventable, and we are comfortable with making negative judgments about people's decisions and behavior. Accident investigations tend to focus on the mismatch between actual performance and written guidance, standardized performance, procedures, and regulations. Pointing out to this mismatch does not explain the actual behavior. Judging behavior to be 'human error' stands in the way of learning from failure.

EUPHEMISMS FOR 'HUMAN ERROR'

If 'human error' is an attribution, or a judgment about what other people should have done (or what we ourselves should have done), then it has not stopped us from finding and popularizing new ways of labeling such judgments. Still, these putative "explanations" only judge people for not doing what they (in hindsight) should have done. Modern human factors concepts of this sort heavily populate the various error classifications, for example, loss of effective Crew Resource Management; complacency, noncompliance; and loss of situation awareness. Although masquerading as explanations, these labels do little more than saying 'human error' over and over again, judging performance instead of explaining it:

- Loss of Crew Resource Management is one name for human error—the failure to invest in common ground, to share data that, in hindsight, turned out to have been significant.
- Complacency is also a name for human error—the failure to recognize the gravity of a situation or to adhere to standards of care or good practice.
- Noncompliance is a name for human error—the failure to follow rules or procedures that would keep the job safe.
- Loss of situation awareness is another name for human error—the failure to notice things that in hindsight turned out to be critical. We merely judge people for not noticing what we now know to have been important data in their situation, calling it *their* error—their loss of situation awareness.

That these kinds of phenomena occur and even help produce trouble is indisputable. People do not coordinate perfectly across workplaces; people adjust their vigilance and their working strategies over time on the basis of their perception of threat; people locally adapt written guidance; and there is always a mismatch between what people observed and what we can show was physically available to them in hindsight. But simply labeling these phenomena fashionably, and stopping there because it now fits a category of the error classification, does not explain anything.

BEYOND 'HUMAN ERROR'

Current safety thinking suggests that to prevent 'human error' (or bad outcomes, rather), we should altogether move toward identifying and enhancing the positive capacities that ensure that things go right (Hollnagel, 2014a). The last thing we should do is declare a "war on error," as it will simply be a war on our own attributions; chasing our shadow. This is the premise of resilience engineering, as well as Safety II (Hollnagel, Nemeth, & Dekker, 2009). We should not be obsessed with identifying and filling the 'holes' (or minor incidents and errors) that show up in our investigations and safety management systems—because these kinds of conditions and events probably happen all the time, even when nothing goes wrong. Instead, this strand of research says, we should study success. We should study typical work and understand how success is created. The presence of positive capacities can help assure a system's continued functioning even under varying circumstances, so that the number of intended outcomes is as high as possible (Hollnagel, 2014b), for instance:

- the ability to say 'no' in the face of acute production pressures (Woods, 2006b)
- the willingness of superiors to hear bad news (Dekker, 2014)
- the acceptance and encouragement of dissenting views (Weick & Sutcliffe, 2007)
- the commitment to learning and the restoration of trust and relationships if vulnerabilities and problems have been identified (Dekker, 2016)

To ensure a continued improvement in the safety of aviation, we need to form a deep understanding of how things actually go right, despite the shortcomings, obstacles, and frustrations of quotidian work in a safety-critical system, and then seek to enhance the system's capacity to make even more things go right (Hollnagel, 2014c; Woods, 2006a).

REFERENCES

AMA. (1998). *A tale of two stories: Contrasting views of patient safety.* Chicago, IL: American Medical Association.

Amalberti, R. (2001). The paradoxes of almost totally safe transportation systems. *Safety Science, 37*(2–3), 109–126.

Boeing. (1993). *Statistical summary of commercial jet aircraft accidents: Worldwide operations, 1959–1992.* Seattle, WA: Boeing Product Safety Organization.

Burnham, J. C. (2009). *Accident prone: A history of technology, psychology and misfits of the machine age.* Chicago, IL: The University of Chicago Press.

Chapanis, A. (1965). On the allocation of functions between men and machines. *Occupational Psychology, 39*(1), 1–12.

Dekker, S. W. A. (2001). The re-invention of human error. *Human Factors and Aerospace Safety, 1*(3), 247–266.

Dekker, S. W. A. (2006). *The field guide to understanding human error.* Aldershot, UK: Ashgate Publishing.

Dekker, S. W. A. (2010). We have Newton on a retainer: Reductionism when we need systems thinking. *The Joint Commission Journal on Quality Patient Safety, 36*(4), 147–149.

Dekker, S. W. A. (2011). *Drift into failure: From hunting broken components to understanding complex systems.* Farnham, UK: Ashgate Publishing.

Dekker, S. W. A. (2014). *The field guide to understanding "human error."* Farnham, UK: Ashgate Publishing.

Dekker, S. W. A. (2015a). The psychology of accident investigation: Epistemological, preventive, moral and existential meaning-making. *Theoretical Issues in Ergonomics Science, 16*(3), 202–213. doi:10.1080/1463922X.2014.955554.

Dekker, S. W. A. (2015b). *Safety differently: Human factors for a new era.* Boca Raton, FL: CRC Press/Taylor & Francis Group.

Dekker, S. W. A. (2016). *Just culture: Restoring trust and accountability in your organization.* Boca Raton, FL: CRC Press.

DSB. (2015). Crash of Malaysia Airlines flight MH17, Hrabove, Ukraine, July 17, 2014. The Hague, the Netherlands: Dutch Safety Board.

Fiorino, F. (2006). Fighting human error. *Aviation Week & Space Technology, 165*(22), 8–9.

Fischhoff, B. (1975). Hindsight ≠ foresight: The effect of outcome knowledge on judgment under uncertainty. *Journal of Experimental Psychology: Human Perception and Performance, 1*(3), 288–299.

Fischhoff, B. (1982). For those condemned to study the past: Heuristics and biases in hindsight. In D. Kahneman, P. Slovic, & A. Tversky (Eds.), *Judgment under uncertainty: Heuristics and biases.* Cambridge, MA: Cambridge University Press.

Fitts, P. M., & Jones, R. E. (1947). *Analysis of factors contributing to 460 "pilot error" experiences in operating aircraft controls.* Dayton, OH: Aero Medical Laboratory, Air Material Command, Wright-Patterson Air Force Base.

Hollnagel, E. (1983). Position paper on human error. *NATO Conference on Human Error,* August 1983, Bellagio, Italy.

Hollnagel, E. (2014a). *Safety-I and Safety-II: The past and future of safety management.* Farnham, UK: Ashgate Publishing.

Hollnagel, E. (2014b). *Safety-I and Safety-II: The past and future of safety management.* Burlington, VT: Ashgate Publishing.

Hollnagel, E. (2014c). *Safety-I and Safety-II: The past and future of safety management.* Farnham, UK: Ashgate Publishing.

Hollnagel, E., Nemeth, C. P., & Dekker, S. W. A. (2009). *Resilience Engineering: Preparation and restoration.* Aldershot, UK: Ashgate Publishing.

Le Coze, J. C. (2015). Reflecting on Jens Rasmussen's legacy. A strong program for a hard problem. *Safety Science, 71,* 123–141. doi:10.1016/j.ssci.2014.03.015.

Leveson, N. G. (2011). Applying systems thinking to analyze and learn from accidents. *Safety Science, 49*(1), 55–64.

Nagel, D. C. (1988). Human error in aviation operations. In E. L. Wiener & D. C. Nagel (Eds.), *Human factors in aviation* (pp. 263–303). San Diego, CA: Academic Press.

Rasmussen, J. (1982). Human errors. A taxonomy for describing human malfunction in industrial installations. *Journal of Occupational Accidents, 4*, 311–333.

Rasmussen, J. (1983). Skills, rules, and knowledge; signals, signs, and symbols, and other distinctions in human performance models. *IEEE Transactions on Systems Man and Cybernetics, SMC-3*(No. 3), 257–266.

Rasmussen, J. (1990). The role of error in organizing behavior. *Ergonomics, 33*(10–11), 1185–1199.

Rasmussen, J., & Batstone, R. (1989). *Why do complex organizational systems fail?* (Vol. Environment Working Paper). Washington, DC: World Bank.

Rasmussen, J., & Lind, M. (1981). *Coping with complexity.* Roskilade, Denmark: Riso National Laboratory.

Reason, J. T. (1990). *Human error.* New York: Cambridge University Press.

Reason, J. T. (1997). *Managing the risks of organizational accidents.* Aldershot, UK: Ashgate Publishing.

Reason, J. T. (2008). *The human contribution: Unsafe acts, accidents and heroic recoveries.* Farnham, UK: Ashgate Publishing.

Senders, J. W., & Moray, N. (1991). *Human error: Cause, prediction, and reduction.* Hillsdale, NJ: Erlbaum Lawrence Associates.

Shappell, S. A., & Wiegmann, D. A. (1997). A human error approach to accident investigation: The taxonomy of unsafe operations. *The International Journal of Aviation Psychology, 7*(4), 269–291. doi:10.1207/s15327108ijap0704_2.

Shappell, S. A., & Wiegmann, D. A. (2000). The human factors analysis and classification system–HFACS. National Technical Information Service.

Shappell, S. A., & Wiegmann, D. A. (2001a). Applying reason: The human factors analysis and classification system. *Human Factors and Aerospace Safety, 1*, 59–86.

Shappell, S. A., & Wiegmann, D. A. (2001b). Human error perspectives in aviation. *The International Journal of Aviation Psychology, 11*(4), 341–357. doi:10.1207/s15327108ijap1104_2.

Shappell, S. A., & Wiegmann, D. A. (2003). *A human error approach to aviation accident analysis: The human factors analysis and classification system.* Aldershot, UK: Ashgate Publishing.

Tverksy, A., & Kahneman, D. (1974). Judgement under uncertainty: Heuristics and biases. *Science, 185*(4157), 1124–1131.

Weick, K. E., & Sutcliffe, K. M. (2007). *Managing the unexpected: Resilient performance in an age of uncertainty* (2nd ed). San Francisco, CA: Jossey-Bass.

Woods, D. D. (2006a). Essential characteristics of resilience. In E. Hollnagel, D. D. Woods, & N. G. Leveson (Eds.), *Resilience engineering: Concepts and precepts* (pp. 21–34). Aldershot, UK: Ashgate Publishing.

Woods, D. D. (2006b). How to design a safety organization: Test case for resilience engineering. In E. Hollnagel, D. D. Woods, & N. G. Leveson (Eds.), *Resilience engineering: Concepts and precepts* (pp. 296–306). Aldershot, UK: Ashgate Publishing.

Woods, D. D., Dekker, S. W. A., Cook, R. I., Johannesen, L. J., & Sarter, N. B. (2010). *Behind human error.* Aldershot, UK: Ashgate Publishing.

Woods, D. D., Johannesen, L. J., Cook, R. I., & Sarter, N. B. (1994). *Behind human error: Cognitive systems, computers and hindsight.* Dayton, OH: CSERIAC.

15 The Systems Perspective in Air Transportation

Lance Sherry

CONTENTS

Air Transportation Systems 341
Definition of an Air Transportation System 343
   Need for an Air Transportation System 344
Objectives of Air Transportation System 345
Concepts-of-Operations of an Air Transportation System 345
   Safe Separation 345
   Flow Management 346
Air Transportation System Infrastructure 346
   Airports 346
   Airspace 348
   Surveillance System 349
   Communications System 350
   Navigation System 350
Air Transportation System—Functional Description 351
   Flow Management 352
      Flight Strips 353
      Collaboration and Coordination 354
   Facility Flow Planning 354
   Sector Control 355
ATS Modernization 358

AIR TRANSPORTATION SYSTEMS

Transportation is a critical component of the global economy. It provides the means to move people, services, and goods from one location to another. In this way, transportation enables specialization and economies-of-scale of production in one location that can be exported to other locations.

Air transportation is unique in the portfolio of transportation modes providing affordable, rapid, and safe transportation of people and of small, light-weight cargo. It is particularly advantageous compared with other modes of transportation for travel over long distances, to remote locations, and for time-sensitive deliveries.

The system required to provide the air transportation service is known as the air transportation system (ATS). ATS is considered to be one of the most complex sociotechnological-economic systems ever created and operated by humankind. These systems operate over wide geographic regions, 365 days a year, 24 hours a day. The system must manage national boundaries, language and cultural differences, operate in all weather conditions and climates, and provide unprecedented levels of safety for fatalities, injuries, and property damage in a complex and ever changing economic, social, political, and technological environments.

To get a sense of the scale and complexity of an ATS, the following tables characterize the European and U.S. ATS. The European ATS provides service to 9.5 million flights a year that travel on average 559 nautical miles (Table 15.1). The U.S. ATS handles approximately 15.2 million flights a year that travel on average 511 nautical miles. Service is provided for an airspace covering a geographic area in excess of 6.21 million square nautical mile (Europe) and 5.62 million square nautical mile (U.S.). In 2012, the ATS served 433 airports in Europe and 513 airports in the United States (EUROCONTROL/FAA, 2012).

Due to national boundaries, the airspace (i.e., altitudes above 24,000 feet) in Europe is serviced by 63 independent air traffic control organizations that collaborate to flow flights between regions in a seamless manner (Table 15.2). In the United States, a single air traffic control organization services an airspace that is divided into 20 control centers. The European air traffic control organizations employs 17,200 air traffic controllers with a total staff of 58,000. The U.S. ATS has 13,300 air traffic controllers and a total staff of 35,200. The additional staff includes administration as well as technical operation specialists. Approximately 6,000 technical operations specialists are employed to maintain over 41,000 radar and navigation radio facilities.

The ATS operates with an environment with a high degree of change and uncertainty. Factors that result in change and uncertainty range from wind, visibility (e.g., clouds, fog, and mist), storms (lightning, wind shear, and hail), mechanical and electrical systems malfunctions, animals (e.g., bird strikes), labor actions, terrorism and hijackings, to politics, social values, technological advances, and economics.

Although technology is capable of automating the planning and operations of the air transportation service under pristine, sterile conditions, the occurrence of unplanned, but expected (e.g., weather events), and unexpected events (e.g., malfunctions) predicates the needs for human monitoring, decision-making, and intervention. As a consequence, the human factors considerations of the ATS provide for a fascinating area of study, which is the topic of this chapter.

The current chapter describes the functional operation of an ATS with emphasis on the transfer of information between functions and the monitoring and decision-making by the functions (i.e., roles in which human factors play a critical part). The description attempts to use generic terminology and describe the functions independent of the technology and systems used in a specific instantiation.

The current chapter describes the functions and operations of a modern ATS and is organized as follows:

1. Definition of an Air transportation system (ATS)
2. Objectives of an ATS

TABLE 15.1
Scale of European and U.S. ATS

Characteristic	European ATS	U.S. ATS
Flights per year	9.5 M	15.2 M
Average flight distance (nm)	559	511
Geographic area (sq nm)	6.21 M	5.62 M
Airports	433	513

TABLE 15.2
Complexity of European and U.S. ATS

Characteristic	European ATS	U.S. ATS
Airspace regions	63	20
Air traffic controllers	17,200	13,300
Total staff	58,000	35,200

3. Concept of operation of an ATS
    a. Safe separation through a communication, navigation, surveillance loop
    b. Efficient flow of traffic through slot allocation
4. ATS infrastructure
    a. Airports
    b. Airspace
    c. Surveillance network
    d. Communications network
    e. Navigation aide network
5. ATS—functional description
    a. Flow management
    b. Facility flow planning
    c. Sector control
6. ATS modernization

DEFINITION OF AN AIR TRANSPORTATION SYSTEM

For the purpose of this chapter, the ATS is defined as the management and control of the flow flights between airports (Figure 15.1). The inputs to the system are daily schedules and flight plans for flights (i.e., flight demand), and the daily availability of time slots (i.e., flight capacity) at key traffic flow nodes of the systems (e.g., runways, gates, traffic flow merge, or crossing points).

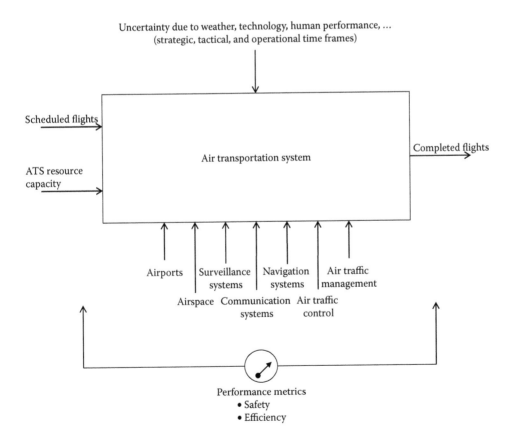

FIGURE 15.1 Scope of the ATS.

The system uses a large complex infrastructure of airports, airspace, surveillance systems, communication systems, navigation systems, and air traffic control and management systems to execute its mission. The system operates in the presence of uncertainty such as the location, timing, and severity of weather conditions, infrastructure availability and capacity, and adherence to the *plan* by the operators.

In this way, the system is characterized by the coordination of the flow of flights to achieve system objectives of: (1) safety (i.e., minimize collision risk) and (2) efficiency (i.e., maximize flight throughput and minimize flight operational cost). The primary measures of ATS performance are collision risk and throughput. Collision risk is measured by the number of collisions, near mid-air collisions, and runway and airport surface near collisions. Throughput at a traffic flow node is measured by flights per unit time (e.g., one hour or 15 minutes) and utilization.

Secondary measures of performance include flight delays, transit time relative to the optimum wind-aided path, and distance traveled relative to the shortest path known as a great circle distance path. Additional measures relate to environmental stewardship and include estimated excess greenhouse gas emissions (i.e., CO_2), other emissions (CO, NOx, HC, and soot), and noise. Emissions are directly proportional to fuel-burn, which in turn is related to transit time. Noise is a measure of the *dosage* of noise experienced at each location in the vicinity of an airport. Night-time noise is considered more harmful than day-time noise and is penalized tenfold.

NEED FOR AN AIR TRANSPORTATION SYSTEM

The transit of flights from the gate at the origin airport to the gate at the destination airport is conducted using shared resources such as airport gates and runways, and public-use airspace (Figure 15.2). The flight operations are conducted in the presence of other traffic including commercial flights, business and private aviation, military, and space-bound vehicles. There are also weather, national boundaries, and environmental considerations (such as noise, air, and water quality) that must be taken into account.

Flight operations are performed using complex technologies that must be maintained to achieve the highest standards of operation defined by a target level of safety of 10-7 (ICAO, 2016). The operation and maintenance of this system is enabled by a large and complex supply chain that takes advantage of economies-of-scale to share resources that would otherwise make air transportation cost prohibitive.

To achieve this level of sharing requires an unprecedented level of coordination and collaboration. The ATS is the amazing system that provides the structure for collaboration and coordination for the safe, efficient, and affordable air transportation service.

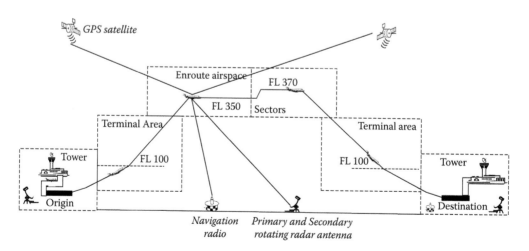

FIGURE 15.2   Flight operations.

OBJECTIVES OF AIR TRANSPORTATION SYSTEM

The objectives of the ATS are to ensure simultaneous safe and efficient flow of flights from origin airports to destination airports.

1. Safe operations require separation of flights to avoid collisions.
2. Efficient operations include maximizing throughput (i.e., flights per unit time) using an ATS resource (e.g., runway, merge point, or crossing point) as well as distance traveled, time in transit, fuel burn, and other factors that impact flight operational costs.

CONCEPTS-OF-OPERATIONS OF AN AIR TRANSPORTATION SYSTEM

To meet the objective of *safe operations*, the ATS employs a closed-loop control concept of operations to adjust flight trajectories to avoid imminent collisions and avoid future conflicting trajectories. This function of an ATS is known colloquially as Air Traffic Control.

To meet the objective of *efficient operations*, the ATS employs a collaborative traffic flow management (FM) concept to maximize throughput by regulating the demand for flight resources (i.e., time slots) within the capacity of the ATS resources. Resources include runways, as well as traffic flow merging and crossing points. This function of an ATS is known colloquially as air-traffic management.

SAFE SEPARATION

Safe separation is conducted through a closed-loop control system (Figure 15.3). The location of the flights is identified through a *surveillance* function. The potential collision courses and conflicting trajectories are identified and resolved by a *Conflict Detection and Resolution* function. The resolved trajectories are communicated to the individual flights in the form of instructions using the *navigation procedures* using the *communications* function.

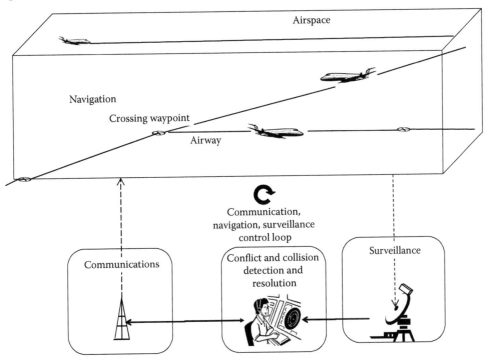

FIGURE 15.3 Safe separation control loop is achieved by surveillance, conflict and collision detection and resolution, communication, and navigation functions.

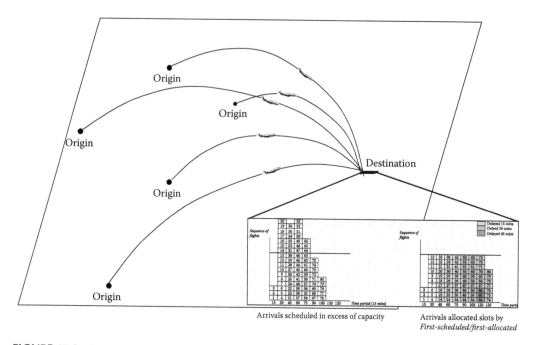

Arrivals scheduled in excess of capacity   Arrivals allocated slots by
                                                  First-scheduled/first-allocated

FIGURE 15.4   The allocation of arrival slots at an airport in 15-minute periods that is overscheduled.

FLOW MANAGEMENT

The flow of flights through the ATS at airports and the airspace is limited by the number of available slots during a specific time period (Figure 15.4). A slot is a time period in which a flight can use a resource such as a runway, a departure route, an approach route, a traffic flow merge point, or a crossing restriction.

Circumstances can occur when the scheduled number of flights is in excess of the slot capacity (i.e., over scheduling), or when the scheduled number of flights is within the slot capacity, but the capacity is temporarily reduced due to weather, construction, or system malfunction.

When the slot capacity in a specific period is not sufficient to meet the flight demand, the slots must be allocated to individual flights to ensure a smooth, ordered, and efficient flow. The effect of slot allocation is to delay the arrival of flights.

There are several ways of allocating slots such as first-come-first-serve, first-scheduled-first-served, and other schemes that allocate resources such as congestion pricing or auctions. The net effect is that the flights that are scheduled in excess of the slot capacity for a given time period get pushed-back to the next time period. This creates a ripple effect of push-backs until all flights are eventually accommodated (see the inset in Figure 15.4).

AIR TRANSPORTATION SYSTEM INFRASTRUCTURE

The building blocks of the ATS are a set of infrastructure: (1) airports, (2) airspace, (3) surveillance function, (4) communications function, (5) navigation function, and (6) air traffic control/management function. Each infrastructure is described briefly in this section.

AIRPORTS

Airports are generally owned by municipalities or other regional forms of governments. As the presence of an airport in a region can result in a monopoly, airports are regulated to operate as public utilities and are required to be revenue neutral (i.e., not make excessive profits). Any profits must be used for airport modernization or improvement.

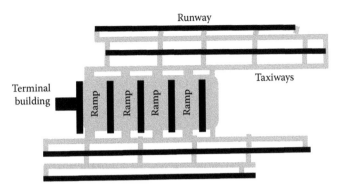

FIGURE 15.5 Airport surface regions for air traffic control: ramps, taxiways, and runways.

A typical airport has a terminal building for passenger and cargo processing (Figure 15.5). The airport surface includes gates, ramp areas, taxiway between the ramp area and the runways, and runways. Navigation, surveillance, and communication equipment is located at the airport. The airports provide space and resources for the supply chain elements to get passenger and cargo to and from the aircraft and to support aircraft operations.

Airport capacity is defined by the sum of the capacity of *independent* runways for departures and arrivals. An independent runway is a runway that is not affected by the location of other runways at the airport such as adjacent or crossing runways.

The capacity for each independent runway is based on two operational rules: (1) no simultaneous runway occupancy and (2) wake vortex separation distance between a lead and follow aircraft.

The time it takes for a flight to use a runway is on average 45 seconds with a maximum of approximately 60 seconds. During this time, when the runway is *occupied*, no other flight may use the same runway.

Wake vortices are swirling air flow that is generated from the wing tips. These vortices can be strong enough to roll a following aircraft that flies directly into the vortex. To avoid a wake vortex encounter, a following aircraft must remain at least between 2.5 and 6 nautical miles from the lead aircraft. A greater distance is required when a small sized aircraft is following the largest aircraft. In general, the wake vortex requires a greater time separation between flights than the runway occupancy time and therefore is the factor that determines the runway capacity.

For a typical mix of aircraft size using a major U.S. or European runway, the average separation will be 90 seconds. This runway will have 10 available slots every 15 minutes (i.e., [15 minutes × 60 seconds/minutes]/90 seconds = 10 slots/15 minutes).

Independent runways provide maximum capacity (Figure 15.6). Some airports, however, have runways with geometries that cross preventing simultaneous use of both runways. Other airports

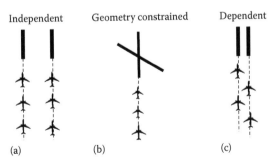

FIGURE 15.6 Airport capacity is determined by the geometry and relative location of runways. Independent runways provide the highest capacity.

have runways that are constructed too close for simultaneous operations and must stagger the arrivals to achieve safety requirements.

AIRSPACE

Airspace is a public resource generally controlled by the government. Regions of the airspace are allocated for air transportation. Other regions may be blocked off to secure critical infrastructure on the ground (e.g., nuclear power plant, government buildings) to protect the environment or to secure airspace used for other purposes (e.g., military, space vehicles).

Airspace over a large geography, such as North America, is divided into regions. There are two types of larger airspace regions. Sectors above 18,000 feet are known as enroute airspace and are grouped into enroute centers. There are 20 enroute centers in the contiguous U.S. airspace (Figure 15.7). Center boundaries are defined to balance the workload between centers. For example, the centers in the Western United States are larger as they experience a lower number of flight operations.

The group of sectors located between the airports and enroute airspace is known as the terminal area (Figure 15.8). The geometries of sectors in the terminal area are designed to manage the flows efficiently. For example, arrival sectors funnel arriving flights to line up with the arrival runways to eliminate gaps between flights and keep the utilization of arrival runways high. Departure sectors allow the *fanning* of departing flights left then right then left again. Fanning increases the throughput of departures by avoiding successive flights on the same path. By sending flights right then left, the flights can be safely separated and can reduce the time between departures.

Some of the most complex airspace is the terminal area over large metropolitan areas with multiple airports in close proximity. These terminal areas, known as *Metroplexes*, must coordinate arrivals and departures from multiple airports. In many cases, under certain wind conditions, the flows to/from different airports may be required to use the same airspace, resulting in reduced flow throughput.

Airspace capacity is determined by the capacity of each sector within the airspace. The capacity of each sector is determined by the ability to maintain separation between the flights. Sectors with

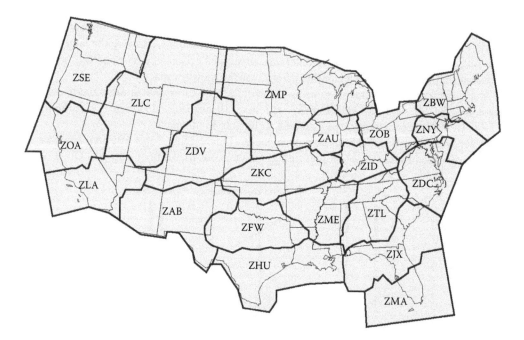

FIGURE 15.7 The contiguous U.S. enroute airspace is divided into 20 centers. Each center is divided into sectors (not shown).

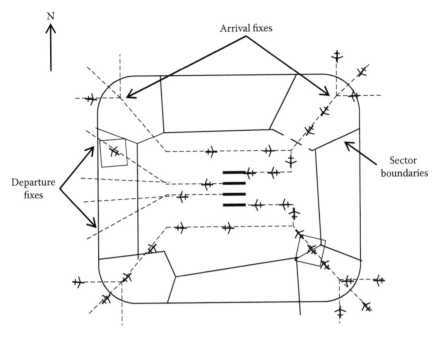

FIGURE 15.8 Terminal area airspace sectors around airports showing arrival procedures (solid lines) and departure procedures (dashed lines) for calm winds or winds from the west.

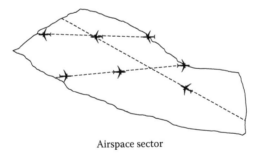

Airspace sector

FIGURE 15.9 Capacity of a sector is determined by several factors including the structure of the flow through the sector.

one-directional structured flow in level flights at the same altitudes will have higher capacity than sectors with flight in different directions with climbing and descending trajectories (Figure 15.9). When the separation is performed by human operators, a typical sector capacity will range from 12 flights (complex trajectories) to 20 flights (structured trajectories).

SURVEILLANCE SYSTEM

Surveillance infrastructure is the means to collect data on the location of all flights operating at an airport or in the airspace. There are several ways of collecting the location of each flight. A primary radar is a ground-based electronic system that transmits signals into the airspace and records when these signals are reflected back. The time for transmission and reflection determines the distance of the aircraft from the radar transmitter. Primary radar can only tell if an object is in the airspace, it cannot identify the object (e.g., flight number).

Secondary radar, generally colocated with primary radar, is used to identify the object as well as collect position information. Secondary radar sends up a signal that is received on the aircraft. The aircraft uses on-board positions information (e.g., from global positioning system, inertial navigation system, and radio navigation) to pack a message that is transmitted back to the ground-based radar transmitter. The transmitted message includes flight identification information as well as intended route information. In this way, secondary radar overcomes the limitation of primary radar and identifies the location and identity of the flight.

A third more modern form of surveillance is automatic dependent surveillance (ADS). This type of surveillance uses a continuous stream of position and identification broadcasts from the aircraft. The broadcasts are received by relatively inexpensive ground-based or satellite-based antennas that eliminate the need for expensive primary and secondary radars.

Most modern ATSs fuse the three sets of data to provide a reliable and accurate surveillance picture. The circular sweep rate of the ground-based radar antennas limit the update rate of the surveillance data. A typical update rate for radar is between 5 and 13 seconds. Secondary radar operates only when the transponder on the aircraft is operational. ADS-B unilaterally broadcasts the message without a ground-based interrogation, so it can be configured to update as fast as every two seconds.

The data from the surveillance function are used for the conflict and collision detection and resolution function. When this function is performed by human operators, the location of all the flights in an airspace sector is displayed on a *radar screen*.

COMMUNICATIONS SYSTEM

There are several communication networks in an ATS. Information must be communicated between surveillance radars and antennas, and the central location that fuses the disparate datasets and disseminates that surveillance information. This requires an extensive ground-based and satellite communications network.

There is also a communication system to communicate between ground-based air traffic control/management and flights. Traditionally, this communication is conducted by push-to-talk voice radio using the very high frequency (VHF) radio band. The voice-based VHF communications can be supplemented by digital communications system that allows transmission of digital messages. Digital messaging allows the communication of longer and more complex messages that cannot be easily accomplished using voice communications.

NAVIGATION SYSTEM

Navigation is the means to fly specific four-dimensional trajectories in the airspace with precision. This is particularly important for lining-up a flight for an approach to a runway. The flight must be lined up with the centerline of the runway as well as lined up on a shallow three-degree vertical glide-path to the runway threshold. Navigation is also critical for keeping flights on a structured system of *highways* in the sky that keeps crossing traffic from colliding and allows the traffic to take advantage of jetstream winds.

The navigation system is a network of ground-based and space-based radio signal transmitters. In the case of ground-based radio navigation aids, the radio signals are transmitted from the ground antennas to aircraft radios. These signals can be used by the aircraft to identify the distance and directions to the radio aides and can be used to *triangulate* position. They can also be used to identify lateral deviations (i.e., right or left) or vertical deviations (e.g., high or low) of a desired course or vertical trajectory for the *highways* in the sky, on the runway centerline, and on the runway glide-path.

The most common ground-based navigation aids are VHF omnidirectional range (VOR) radios, distance measuring equipment (DME), and instrument landing system (ILS). Departure procedures,

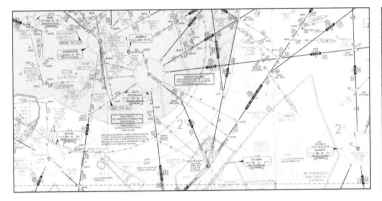
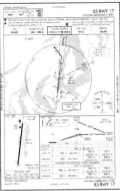

FIGURE 15.10   Aeronautical charts showing airways (left) and approach procedure (right). Note how the airways are defined by courses from VOR to VOR.

airways, and Standard Arrivals are defined by courses to/from VORs and DMEs (Figure 15.10). Approach procedures are defined on the basis of VORs, DMEs, and instrument landing system (Figure 15.10).

Space-based navigation aids (e.g., GPS) transmit signals that are received by the aircraft and used to *triangulate* a three-dimensional position.

Navigation is conducted according to published navigation procedures that specify the 3-D highways. The main navigation procedures include the following: airways, departure procedures, standard arrival procedures, and approach procedures. Each procedure specifies the exact lateral path that should be followed by the aircraft (i.e., sequence of latitudes and longitudes). Some procedures provide explicit altitudes that must be achieved at specific locations. For example, departure and approach procedures require the flights to be at a specific altitude to avoid terrain and facilitate traffic crossing below or above. Other procedures, such as airways, provide minimum safe altitude to avoid terrain and crossing traffic.

The use of these infrastructure components to perform the ATS mission is described in the next section.

AIR TRANSPORTATION SYSTEM—FUNCTIONAL DESCRIPTION

Flight operations in an ATS are conducted in a safe, efficient, and affordable manner by a high degree of coordination and cooperation. The top-level objectives of this system are as follows: (1) to maintain safe separation and (2) maintain efficient flows. These objectives are achieved by three main functions that operated in a nested-closed-loop control configuration:

1. System-wide, strategic flow management (FM) loop
2. Regional, tactical facility flow planning (FFP) loop
3. Sector instantaneous control loop

The ATS with the three functions and the flow of information to perform each closed-loop control is illustrated in Figure 15.11.

The *FM loop* performs strategic management of the flows with jurisdiction across the ATS. This management loop coordinates flows in a given 24-hour period up to two hours before an event by issuing traffic management initiatives (TMIs). TMIs assign flights to slots within capacity of the National Airspace System (NAS) resource (e.g., runway or airspace). The most common TMI is a *ground delay* at the departure airport that only releases flight

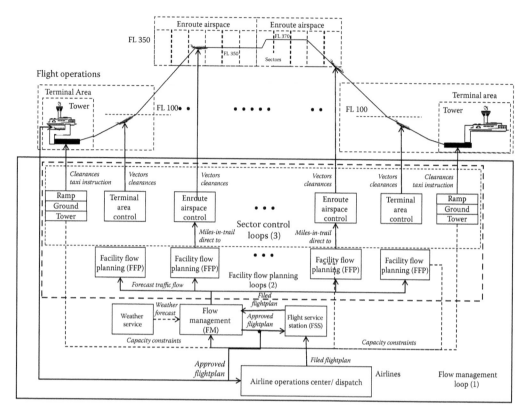

FIGURE 15.11 Functions and flow of information in an ATS. Three nested control-loops regulate traffic flows: (1) flow management (strategic), (2) facility flow planning (tactical), and (3) sector control (real time).

when there is an open slot at their destination airport. This eliminates flights flying *holding patterns* (i.e., wide circles) waiting for an available landing slot.

The *FFP loops* execute FM TMIs and coordinate flows within larger airspace groups (e.g., terminal area, enroute). The FFP is more tactical than the FM-adjusting route and flight speed using miles-in-trail (MIT) (i.e., distance between flights on the same route). This control occurs from two hours to 15 minutes before a forecast congestion event and generally is applied to flights that are already airborne.

The sector control loops perform closed-loop, real-time control of the flight trajectories through the communication/navigation/surveillance loop. The controllers monitor individual flights and communicate clearances, navigation instructions, and vectors to allow flights to progress on their flight plans and maintain safe separation. As the sector control is the only entity in direct contact with the flights, sector control also relays FFP and FM plans to the individual flights.

These three loops, illustrated in Figure 15.11, are described in more detail in the following sections.

FLOW MANAGEMENT

FM manages the flow of traffic across the ATS. The FM typically manages traffic over a 24-hour period.

FM builds a picture of the capacity of airspace and airports throughout the NAS for the whole day, by consolidating forecast weather from the weather service, capacity information from the ATC facilities (i.e., towers, terminal areas, and enroute), and flight plans for scheduled flights. The FM then reviews the filed flight plans submitted for routes that will be impacted by weather (e.g., thunderstorms), and for routes impacted by traffic congestion to identify flow bottle-necks when forecast flight arrival times will be in excess of airspace or runway capacity (e.g., peak hours).

When the FM recognizes that a flow of flights is anticipated to experience traffic congestion due to severe weather, the FM issues the flights an amended flight plan that avoids the weather (e.g., a large convective storm stretching hundreds of miles) by rerouting, or a delayed departure. Rerouting is issued as part of a severe weather avoidance plan designed to reroute traffic in a coordinated manner.

The most common way to manage forecast congestion is by delaying flights at their origin airports before they depart. These temporal strategies are known as TMI. The most common TMI is an arrival airport ground delay program in which individual flights are issued delayed departure times to avoid them arriving at an airport in a 15-minute time period without sufficient capacity to handle all of the arrivals. If the flight arrived without a slot, it would be obliged to fly a *holding pattern* (i.e., wide circles) until a slot opened up. This results in increased costs of operations due to excess fuel burn, significant increases in ATC and flight crew workload and reduces safety margins. Another common TMI is an airspace flow program that allocates slots for oversubscribed airspace.

A delayed departure for a ground delay program or airspace flow program results in each flight being assigned an expected departure clearance time (EDCT). This specifies the departure time assigned to the flight such that it will arrive at the capacity constrained resource at its assigned slot.

Flight Strips

When the flight plan is approved by FM, a flight strip is created and submitted to all ATC facilities on the proposed route for a given flight (Figure 15.12). The flight strip authorizes the ramp controller to clear the flight for push-back from the gate and subsequently authorizes all the ATC sector controllers on the flight's route to clear the flight through the airspace. The flight strip will include an EDCT time if applicable and the planned route of flight.

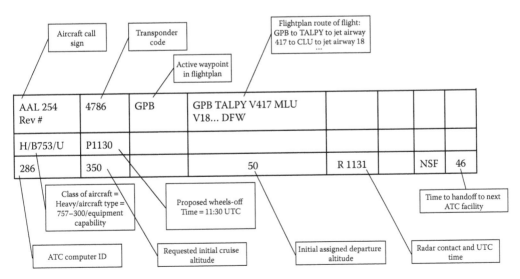

FIGURE 15.12 Example of a flight strip for a departing flight. Provides concise information for ATC control of flight including all NAS-wide and facility flow information.

Collaboration and Coordination

FM decisions can have a significant impact on the scheduled network operations of individual airlines and on individual flights. The FM does not readily have information about the economics of specific airline network operations. Given choices in TMIs, the FM may unwittingly select an option that degrades or even unravels the airline operations. For example, out of six flights forecast to arrive at a capacity constrained airport in a 15-minute period, one of the flights may be critical to the airline due to the number of connecting passengers, duty hours of the flight crew, and availability of specialized ground crews or equipment for aircraft gate turn around. Given a choice, the airline may choose to maintain the schedule of this high-value flight and swap an EDCT delay to another less critical flight.

The process of information sharing and coordination between airlines and FM is known as *collaborative decision making (CDM)*. CDM has been deployed in the U.S. and Europe circa 1996 for managing airport arrivals (i.e., ground delay program) and airspace boundary arrivals (i.e., air flow program). The CDM concept is to allow all parties to share information and collaborate to increase system efficiency. For example, an airline can identify a canceled flight (e.g., mechanical problem) and open up the slot that would have otherwise gone unused. In return, the airline has the option of swapping flights within its schedule to minimize the impact on their network.

A typical display used for FM is shown in Figure 15.13. The chart shows the number of flights forecast to arrive at an airport in each 15-minute increment. The vertical bar indicates current time. The horizontal white bar indicates the capacity at the airport for each 15-minute increment. The bars show the number of flights forecast. Black bars indicate the flights have arrived. The lower part of the medium dark bars are flights that are enroute to the airport. The medium dark bars are flights that have been issued an EDCT and are waiting at their origin airports before departing. The lightest bars are flights that are departing (mostly, long distance international flights). The slightly darker portions of the bars to the right of the figure are flights that were scheduled to depart but have not departed yet.

FACILITY FLOW PLANNING

Despite the best efforts of FM, circumstances frequently occur that require adjustment of flows in regions of the airspace. Examples of these circumstances include unexpected pop-up thunderstorms, inaccurate timelines for forecasted weather, restricted airspace (e.g., military flights, or flights transporting Head of State), or ATC equipment failures. These issues are handled at the local level. FPP regulates flights to avoid overcrowding a specific airspace. This is achieved by airspace facilities metering the flights into the airspace with sufficient distance between flights. This process is known as miles-in-trail (MIT) and specifies the distance between successive flights. MIT generally results

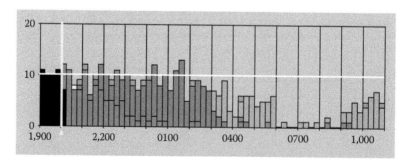

FIGURE 15.13 Display used by FM to allocate arrival slots at an airport. The allocation scheme must bring the forecast arrivals within the capacity (horizontal white line) of the airport. The medium dark bars indicate the flights that were scheduled to depart but have been delayed at their origin airports (i.e., issued EDCTs).

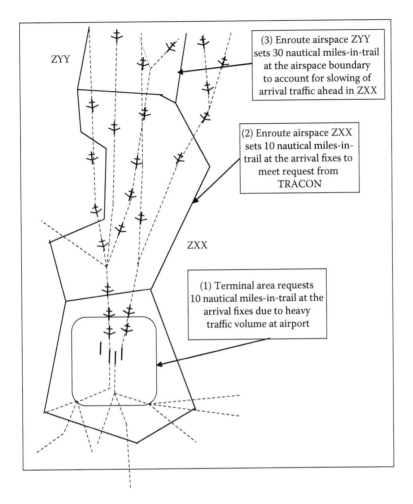

(3) Enroute airspace ZYY
sets 30 nautical miles-in-trail
at the airspace boundary
to account for slowing of
arrival traffic ahead in ZXX

(2) Enroute airspace ZXX
sets 10 nautical miles-in-
trail at the arrival fixes to
meet request from
TRACON

(1) Terminal area requests
10 nautical miles-in-trail at the
arrival fixes due to heavy
traffic volume at airport

ZYY

ZXX

FIGURE 15.14 Example of an application of miles-in-trail restrictions to manage the forecast for excessive volume of arrivals in terminal area.

in a flight following the same route but at a slower speed. Figure 15.14 illustrates the application of an MIT to manage volume on arrivals into a terminal area.

One of the tools to visualize arrival into a sector or a merge/crossing point is through a time line (Figure 15.15). The time line on the left shows the estimated sequence and time of arrival of flights to a point. The time line on the right shows a metered sequence and time of arrival of these flights after an MIT TMI is implemented.

SECTOR CONTROL

The sector control function provides clearances and vectors to guide flights within each airspace sector. A clearance authorizes a flight to proceed to a waypoint in the flight plan, or to climb or descend to a new altitude. The controller *blocks* the airspace to avoid trajectory conflicts with other traffic. A vector steers the flight direction and speed to adjust for nearby traffic.

Effective sector control is achieved by providing a structure to the flow of traffic manifest through a structured airspace, and by the closed-loop control of individual flights.

The sector control function is conducted through a closed-loop control system (illustrated in Figure 15.16). An air traffic controller *communicates* directly with the flight crews of the flights in the

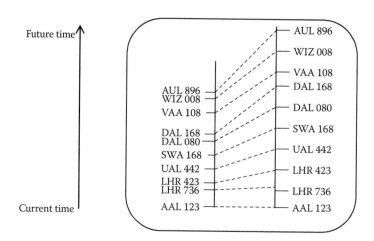

FIGURE 15.15 Time line used by FFP to assess need for TMI. Time line of flights before (left) and after (right) MIT TMI.

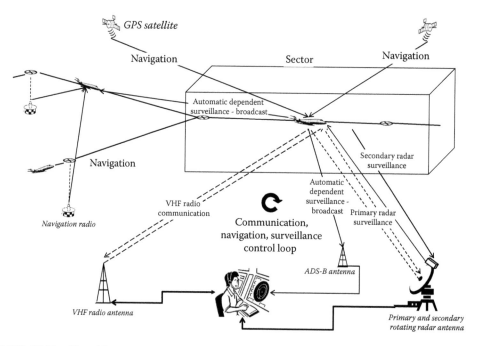

FIGURE 15.16 Closed-loop control of flights using the communication, navigation, and surveillance technologies.

airspace under their jurisdiction providing them specific navigation instructions. The instructions include which airway or procedure to use or simply a direction, altitude, and speed. The flight crew acknowledge the instruction and *navigate* the flight using the available navigation systems. The air traffic controller monitors the flights under their jurisdiction to maintain safe separation between flights using available *surveillance* systems. This closed-loop system is known as communication/navigation/surveillance.

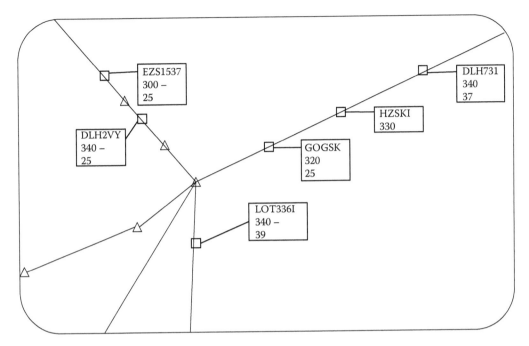

FIGURE 15.17 An example of digital *radar screen* displaying surveillance data in an enroute sector. Flights are identified in a *data block* by flight number, altitude, and groundspeed.

The throughput of flights through a block of airspace is limited by the number of flights that can be safely separated by a human air traffic controller. The airspace capacity depends on the complexity of the flow of flights, the speed of the flights, and size of the airspace. Typical limits of airspace capacity under the jurisdiction of a single ATC controller range from 12 simultaneous flights to 20 simultaneous flights.

When the conflict and collision detection and avoidance function is performed by human operators, the surveillance situation is displayed on a *radar screen*. Since digitization, this not technically a *radar screen* but a digital display of the fused data from primary and secondary radar and the ADS-B (Figure 15.17). The flights are identified with a *data block* of flight number, altitude, and groundspeed.

Voice communication between air traffic controller and the flight crew is a highly proceduralized communication with a finite and explicit phraseology and specific structure. The international standard for communication is English. The phraseology is defined in an *Air Traffic Control Handbook*. Example instructions from an air traffic controller to a flight crew are shown in Table 15.3. The air traffic controller instruction begins with an identification of the flight that the message is intended for. This is followed by the navigation instruction using the airway, procedure, or runway name. The final part of the communication provides permission to proceed. This ensures that in the event of communication failure, the complete set of instructions is received before the flight proceeds.

Further, each message is acknowledged by the flight and repeated back to the air traffic controller to ensure that no miscommunication has taken place. The acknowledgment is in reverse order ending with identification of the flight.

To avoid miscommunication, compound numbers are sounded out. For example, a heading of 150 degrees is one-five-zero. Further, the numbers 3, 5, and 9 are pronounced *tree, fife,* and *niner* to avoid confusion. For example, the number *nine* could easily be mistaken for the number *one*. The same is true for letters that use a phonetic alphabet: Alpha, Bravo, Charlie Delta, ... Zulu.

TABLE 15.3

Example ATC Controller Communication to Flights

ATC Communication	Explanation
Air traffic controller: United seven-niner-two runway two-four cleared for takeoff *United 792*: Runway two-four cleared for takeoff, united seven-niner-two	United airlines flight number 792 is cleared for takeoff on runway 24
Air traffic controller: American four-two-four, after departure, turn left and proceed direct to the BRAVO VOR, runway one-seven, cleared for takeoff *United 792*: Turn left after takeoff, direct to BRAVO, runway one-seven, cleared for takeoff, united seven-niner-two	American Airlines flight number 424 must turn to the left after takeoff and achieve the safe altitude and fly directly to pass over the BRAVO navigation radio. The flight is cleared for takeoff on runway 17
Air traffic controller: Delta two-eight, turn right heading one-five-seven, join jet ninety-seven *United 792*: Turn right to heading one-five-seven, join jet ninety-seven	Delta Airlines flight number 28 must turn to the right to fly a heading of 157 degrees. This heading will allow the flight to intercept the Jet Airway #97. The flight turn onto and follow the Jet Airway
Air traffic controller: British Airways niner-two, climb maintain three-five-zero *United 792*: Climb-maintain three-five-zero, British Airways niner-two	British Airways flight number 92 must climb to 35,000 feet
Air traffic controller: Air France four-seven, contact Approach on 119.72, good-day *Air France*: Approach 119.72, Air France four-seven	Air France flight number 47 must switch their radio frequency to communicate with the air traffic controller in the next sector. In this case, the flight is transferring to the approach controller. The new radio frequency is 119.72

ATS MODERNIZATION

The major challenge in any ATS modernization is the sheer magnitude of the modernization and the coordination of simultaneous investment in equipage and procedures by all stakeholders. For example, modernization concepts-of-operations may require simultaneous equipage, procedures, and personnel training amongst airlines, ATC/M, and regulators. Even within airlines, this is a significant investment that can impact flight crew training, simulator capabilities, maintenance training, dispatch training, aircraft maintenance procedures, standard operating procedures, aircraft delivery and acceptance, and quality assurance.

Many of these processes require approval from regulatory authorities for new procedures and training equipment. If any one of the links in the chain is absent, the benefits cannot be accrued. This is particularly a challenge for ATS, such as the U.S. ATS, that has to develop and deploy multi-year initiatives that are funded incrementally on annual government-budget cycles. There have been several examples in recent history in which the ATS timeline for investment in equipage and procedures fell out of synchronization with the airline's time line due to government austerity programs.

Geographic fragmentation of the airspace (e.g., European airspace) also provides a challenge to modernization. Inconsistent regulations across governments and government agencies, and issues related to the legal framework for ATC systems can limit the economies-of-scale that could be achieved and inhibit reaching the full potential of self-sustaining productivity improvement to meet future growth in demand. The European Union is attempting to address the operational issues by creating functional airspace blocks that unify national ATC systems into a single operational unit.

Another significant roadblock to investment in modernization is the cooperation of the aircraft owners and operators to invest in the required equipage. In a competitive airline marketplace, benefits accrue to all users of the airspace, when one airline equips. The increased predictability and additional slots created by one airline can be used by the nonequipped flights. For this reason, in some cases, the benefits to the nonequipped operators can outweigh the benefits to the equipped. This *asymmetry in benefits* is one of the major reasons that airlines have exhibited reticence to be the first to equip. Approaches proposed to overcome this issue include mandating equipage, and a best-equipped/best-served concept-of-operation. Mandates are highly politicized, costly, but fair. They ensure that all stakeholders are penalized equally. Best-equipped/best-served incentivizes equipage, but can result in inequities in service, penalizing those without resources to invest in equipage.

One of the ways to overcome these roadblocks is to provide the ATS with more authority for modernization and independence from government-budget processes through *corporatization* of the enterprise. Over the last several decades, economic deregulation of public utilities in several industries has yielded improvements in economic efficiency. In air transportation, airline deregulation started in 1978 in the United States and has led to widespread deregulation of the industry culminating in the Open Skies agreement between the United States and the European Union in 2008. Deregulation of ATS has occurred since the late 1980s resulting in the corporatization of 14 national ATC systems.

Corporatization of an ATS is beneficial as it enables ATS to set user fees consistent with the costs of services. This has provided the means to directly address productivity and modernization based on the economics of the service and independent of political constraints.

Although there were concerns that corporatization could adversely affect safety, studies have shown that during this period, safety has been maintained or improved, and new technologies have been implemented to improve productivity.

Section IV

The Future

16 A Personal Journey in the Design and Management of the Growing Airspace

Guy André Boy

CONTENTS

Introduction 363
Evolution of System Capabilities: The SESAR Perspective 365
Air Traffic Complexity 366
Evolution of Commercial Aircraft Automation 367
Using the AUTOS Pyramid in Human-Centered Design 369
    Artifacts, Users and Tasks 369
    Considering Organizations in Design 370
    Testing in a Large Variety of Situations 370
    Using the AUTOS Pyramid in Practice 371
Discussion 371
    Human Control versus Automation: The Maturity Issue 371
    Management and Organizational Automation 372
Perspective and Conclusion 373
    Moving from Automation Issues to Tangibility Issues 375
Acknowledgments 376
References 376

INTRODUCTION

The aviation community worldwide recognizes that there is a need for deeper investigations in the evolution of air traffic, especially in congested areas. In the current chapter, I will use my 35-year experience in aeronautical engineering, and most recently in large programs such as NextGen[*]. in the United States and SESAR[†] in Europe. I will focus more specifically on SESAR, whose definition phase started in 2005 to "define, develop and deploy what is needed to increase Air Traffic Management (ATM) performance and build Europe's intelligent air transport system." Funding for this initial phase was shared by the European Union (50 MEUR[‡]) and Industry (10 MEUR). SESAR development phase stared in 2008 just after the completion of the definition phase. It is scheduled until 2024, and divided into two subphases: SESAR 1 program (2.1 BEUR[§] from 2008 to 2016),

---

[*] United States Next Generation Air Transportation System (NextGen), where I was a member of Joint Planning and Development Office (JPDO) Aircraft Working Group from 2012 to 2014.

[†] Single European Sky ATM Research (SESAR), where I was a member of its Scientific Committee from 2014 to 2016.

[‡] Million of Euros.

[§] Billion of Euros.

and SESAR 2020 program (1.6 BEUR until 2024). The European Union, industry, and Eurocontrol equally share funding. SESAR deployment phase started in 2015 and is scheduled to finish in 2035.

It is important to recognize that air traffic evolved because of increased automation and complexity. Main reasons include fast and massive evolution of technology, as well as number of flights and passengers. Air Traffic Control (ATC) shifted to ATM to solve airspace saturation problems (Figure 16.1). More fundamentally, this shift is from single-agent control to multiagent trajectory management, in which complexity has greatly increased. Automation started on flight decks and is now developing in the airspace. Complexity is related to the growing number of flights (from 9.5 million flights in 2012 to 14.4 million flights in 2035) and passengers (from 0.7 billion passengers in 2012 to 1.4 billion passengers in 2035)[*]. Important principles are safety, efficiency, geographical and information interconnectivity, climate change, innovation, and passenger satisfaction.

In the beginning of aviation, air traffic controllers (ATCos) had to guess both current and future aircraft positions in order to reduce uncertainty (Figure 16.1). During the second phase based on the use of radar technology, ATCos know aircraft positions but still have to guess future aircraft positions. We are still in this era, even if saturation in very busy airports requires shifting to the next phase, which is trajectory management, also called Trajectory-Based Operations (TBO). In TBO, ATCOs know both current and future aircraft positions, which considerably reduce uncertainty. This looks great, but a new problem emerges from the implementation of the TBO solution, which is a necessary planning task (e.g., planning 4D trajectories). Planning involves rigidity. Wherever everything occurs by the rule, everything is fine, but when unexpected situations occur, air traffic managers require flexibility. Rigidity and flexibility are contradictory concepts. This is the reason we now need to think in terms of flexible trajectory planning.

Another observation should be given. Even if pilots and ATCos are interacting at operations time, aircraft and ATC manufacturers and suppliers are very different institutions that rarely talk to each other. In contrast, Human-Centered Design (HCD) requires participatory design of complex systems (Boy 2013). The space shuttle design and development is a good example of such an HCD approach in which onboard and ground systems were designed and developed in concert. TBO requires such a multiagent design approach. We developed methods to this end (Boy 1998, 2011). I strongly suggest that Technology, Organization and People's jobs should be considered and designed concurrently (Boy 2013). This concurrent T.O.P. approach is now known as the TOP model (Figure 16.2). This chapter presents the two major concepts already introduced that are complexity and automation of the ATM.

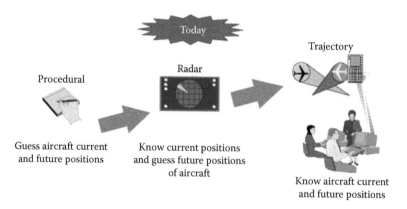

FIGURE 16.1   ATC-to-ATM evolution from procedural control to trajectory management.

---

[*] http://www.sesarju.eu/discover-sesar/why-sesar retrieved on January 2017.

FIGURE 16.2 The TOP model.

Symptoms of ATM complexity are congested airspace, unpredictable delays, and lower productivity (phenotypes). It is then important to better understand deeper causes (genotypes). Human-centered automation can be interpreted as cognitive function transfer from people to systems (Boy 1998, 2011), and more generally the results of function allocation. Function allocation can be a solution for reducing ATM complexity to sustain growth by relieving bottlenecks, reduce delays and cancellations, optimize fuel, reduce safety risks, and decrease uncertainty. We will discuss various issues and solutions, and propose research and development perspectives.

EVOLUTION OF SYSTEM CAPABILITIES: THE SESAR PERSPECTIVE[*]

SESAR is structured around six topics: automation of routine tasks (e.g., automation and use of data communication to ease controllers and support staff workload), integration of all vehicles (e.g., all air vehicles fully integrated in ATM environment, including RPAS[†]), flight-centric operations (e.g., air users fly their preferred, more direct route in a flow and network context), integrated systems (e.g., lean and modular systems, easily upgradable and interoperable), sharing of information (e.g., information shared digitally via common information platform), and virtualization (e.g., virtualization allowing dynamic capacity management). These topics apply to three types of operations: high performing airport operations (involving capacity, safety, environment, efficiency, effectiveness, and networking), optimized ATM network services (involving collaboration, balancing demand and capacity, environment and efficiency), and advanced air traffic services (involving synchronization, capacity, safety, environment, and cost).

SESAR Scientific Committee was set up by the SESAR Joint Undertaking (SJU) in August 2009 with a double objective of reinforcing SJU's innovative and scientific approach to researching the future ATM systems and procedures, and providing specific advice to its Executive Director.

SESAR Scientific Committee reviewed, updated, and approved the following statement, "SESAR exploratory research drives the development and evaluation of innovative or unconventional ideas, concepts, methods and technologies; that can define and deliver the performance required for the next generation of European ATM system, and thus contribute to its successful evolution." Long-term research in SESAR1 was conducted under the frame of WP-E, with a total of 40 projects. The Complex-World network handled complexity; the Higher Automation Levels in ATM (HALA)! network handled advanced ATM automation; the addressing the liability impact of automated systems (ALIAS) network handled the liability impact of automated systems.

SESAR Scientific Committee provided guidance on several themes and issues, including business agility, decision support, environment and meteorology, CNS/ATM[‡] automation, operating concepts, and human factors. Five types of projects were supported: toward higher level of automation in ATM (17 projects and 13 PhDs); mastering complex systems safely (11 projects and 7 PhDs);

---

[*] SESAR presentations at World ATM Congress 2016—Exploratory Research program: http://www.slideshare.net/SESAREuropeanUnion/sesar-at-world-atm-congress-2016-exploratory-research-programme retrieved on January 20, 2017.

[†] Remotely Piloted Aircraft System.

[‡] Communication Navigation Surveillance/Air Traffic Management.

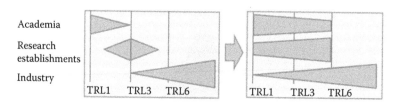

FIGURE 16.3 Stakeholder's involvement change. (Hecker, P., How we take science seriously in SESAR, SESAR Exploratory Research, *World ATM Congress*, Madrid, 2016.)

system architecture and system design (three projects); information management, uncertainty and optimization (three projects); and enabling change in ATM (six projects). Education together with research and innovation are main academic functions in the development of the aircraft of the future and its airspace environment. Stakeholder's involvement has changed, and the innovation process needs to reflect this change (Figure 16.3).

SESAR innovation process has the following challenges: identifying and overcoming national and cultural differences, defining adequate processes for managing innovation processes, developing instruments for establishing networks, ensuring quality in science, defining objectives and measures, managing knowledge, translating objectives and long term research challenges into calls, identifying and approaching the relevant community, defining assessment criteria for transferring research results from exploratory to industrial research, identifying long-term research challenges for future research, and promoting and awarding excellence.

SESAR key findings and statements are the following: "science is SESAR's backbone; operational and industrial relevance of research must be ensured; at the same time, space for an open blue sky research is essential to ensure next generation of knowledge and young talents; the innovation pipeline is the key instrument for linking science to innovation; SESAR Scientific Committee has successfully undertaken significant efforts to support SJU in achieving its mission and building the bridge between the communities."

AIR TRAFFIC COMPLEXITY

Air traffic complexity constantly increases because the number of flights increases. Consequently, single-agent control is leading to multiagent management. It should be noted that ATC environment did not change too much during the last five decades, except it now uses digital equipment. ATC still has many 1970s technologies, high fragmentation, poor interoperability, procedure-based tactical control, and rather low level of automation. Therefore, growing complexity is difficult to handle using such old-fashion technology designed for linear systems.

A simple way of modeling complexity in the airspace is to consider the number of relevant aircraft to be managed per Appropriate Volumetric Zone (AVZ) at each time. An AVZ is calculated with respect to the type of flow pattern, for example, aircraft crossing, sequencing, and merging. The definition of such an AVZ requires the assistance of operational ATCos. From a sociocognitive perspective in ATM, complexity should be considered together with capacity. This is what the COCA (COmplexity & CApacity) project investigated (Cummings and Tsonis 2006; Hilburn 2004; Laudeman et al. 1998; Masalonis et al. 2003). This was a simple way of considering complexity of the ATM. More attributes (complexity genotypes) must be considered such as communication, separation management, collision avoidance, navigation, surveillance, routing, guidance, and information management.

In 2010, SESAR put together an open partnership between universities, research centers, and industry, the Complex-World network. Several Ph.D. programs supported Complex-World through different projects. This network attacked data science in aviation through defining the challenges for Big-Data infrastructure in air transport and visualization techniques supporting data-driven knowledge extraction[*]. System-Wide Information Management[†] was developed to provide standards, infrastructure, and governance enabling the management of ATM information and its exchange between qualified parties via interoperable services.

More generally, ATM organizational complexity is linked to social cognition, agent-network complexity, and more generally multiagent management issues. There are four principles for multiagent management: agent activity (i.e., what the other agent is doing now and for how long), agent activity history (i.e., what the other agent has done), agent activity rationale (i.e., why the other agent is doing what it does), and agent activity intention (i.e., what the other agent is going to do next and when). Multiagent management needs to be understood through a role (and job) analysis. A cognitive agent is usually described in terms of a society of agents (Minsky 1986). From this point of view, ATM has become a process of interaction among systems (i.e., agents) leading to the system-of-systems (SoS) concept (i.e., society of agents).

The concept of *system-of-systems* is now widespread in the engineering of contemporary systems[‡]. The SoS research discipline is currently evolving by developing frames of reference, thought processes, quantitative analysis, tools, and design methods. The methodology for defining, abstracting, modeling, and analyzing SoS problems is typically referred to as system-of-systems engineering. Usually systems-of-systems denote large-scale and complex systems in which properties are emergent and can only be identified at the time of use. It is important to more specifically describe these properties and their emergence. More specifically, restructuring the airspace requires more investigations on authority sharing among air-ground aviation agents (Boy et al. 2008; Boy and Grote 2009). Modeling and human-in-the-loop simulations (HITLS) are excellent support for empirical human-systems integration (HSI) investigations and discovery of ATM emergent properties.

In addition, there are various types of interaction among agents. Collaboration is one, but there are two others that are supervision and mediation (Boy 2002). Therefore, the so-called collaborative system engineering approach needs to be opened to various types of interactions among systems of systems (Landauer and Bellman 1996). Today, building a SoS is a holistic endeavor that requires architects and appropriate tests to emerge salient properties and hidden cognitive functions. More generally, the concept of super-system emerges from the interaction among its components. We can see that current and future ATM complex systems deal with integration, data services, and function allocations. Consequently, integration should be thought and developed as HSI (Boy and Narkevicius 2013).

EVOLUTION OF COMMERCIAL AIRCRAFT AUTOMATION

Twentieth century engineering was dominated by mechanical engineering. Engineers built airplanes and ATC systems mostly by assembling mechanical things. During the last decades, computer science and information technology massively penetrated mechanical machines to incrementally create embedded systems (ESs). Let us review this specific aeronautical evolution in terms of the four loops of automation introduced by Captain Etienne Tarnowski.

Everything started with the automation around the center of gravity using yoke or side stick and thrust levers (Figure 16.4). The first loop consisted in a single agent regulating parameters, such as speed and heading, one parameter at a time. Time constant of the feedback was around 500 milliseconds

---

[*] http://complexworld.eu/wiki/index.php?title=DSIAW2015&diff=1825&oldid=1822
[†] https://www.eurocontrol.int/sites/default/files/publication/files/del08.01.01-d41-swim_conops.pdf
[‡] The following is based on our experience in the design of air traffic management (ATM) systems. The author was the coordinator of the PAUSA project (*Partage d'Autorité dans le Système Aéronautique*—Authority Sharing in the Aeronautical Systems), funded by the French Aviation Administration (DGAC-DPAC). PAUSA emphasized the concept of systems of systems in the aeronautical system, and more specifically the issue of authority among the various agents (i.e., systems) whether they are humans or machines.

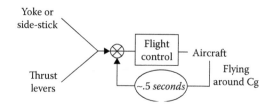

FIGURE 16.4   Trajectory control loop: flying around the center of gravity.

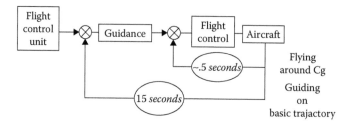

FIGURE 16.5   Guidance loop: guiding on basic trajectory.

Pilots had to adapt to this control loop by changing from the control of flight parameters to supervising the behavior of flight control with respect to a set point. It should be noted that autopilots were introduced in commercial aviation in the 1930s (e.g., the Boeing 247 commercial aircraft flew with an autopilot in 1933).

The guidance loop was developed circa the early eighties (Figure 16.5). This second feedback loop took into account several parameters, and its time constant was around 15 seconds. Note that this feedback loop was implemented on top of the flight control loop. High-level modes of automation appeared and were managed on the flight control unit panel. At the same time, integrated and digital autopilot and autothrottle were installed.

The third loop concerned navigation automation with a time constant of about one minute (Figure 16.6). Guidance and flight management became integrated. This was the first real revolution in the evolution of aeronautical Cyber-Physical Systems (CPS). We were shifting from control of flight parameters to management of ESs. Software became dominant and the number of artificial agents on aircraft grew exponentially. For that matter, pilots have now to deal with a variety of tangible interactive systems (TISs) that are not only humans but also software-based agents. Problems may emerge when these software-based agents communicate among each other. This issue will be analyzed later in the chapter.

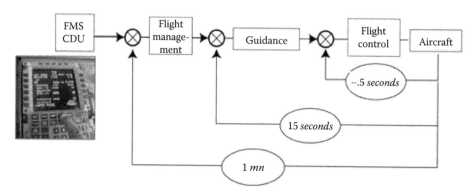

FIGURE 16.6   Navigation automation loop: guiding the flight plan.

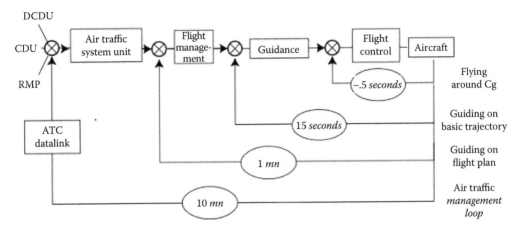

FIGURE 16.7 Air traffic management loop: managing trajectories. (From Tarnowski, E., The 4-loop approach to ATM. Keynote, *HCI-Aero'06*. Seattle, WA, 2006.)

The forth TIS currently concerns ATM with a time constant of about 10 minutes (Figure 16.7). We now deal with airspace *automation*. This is about air-ground integration. This is a new revolution, in which there are new considerations such as authority sharing, air traffic complexity management, and organizational automation. Note that this ATM loop is multidimensional and multiagent.

USING THE AUTOS PYRAMID IN HUMAN-CENTERED DESIGN

The AUTOS pyramid (Boy 2011) supports HSI making sure that important entities (i.e., Artifacts, Users, Tasks, Organizations, and Situations) are taken into account, as well as their properties and interconnections (provided on the edges of the pyramid).

ARTIFACTS, USERS AND TASKS

Artifacts may be aircraft or ATC systems, for example. Users may be novices, experienced personnel or experts, coming from and evolving in various cultures. They may be tired, stressed, making errors, old or young, as well as in very good shape and mood. Tasks vary from handling quality control, flight management, managing a passenger cabin, repairing, designing, supplying or managing a team or an organization. Each task involves one or several cognitive functions that related users must learn and use. The Artifact-User-Task (AUT) triangle (Boy 1998, 2011) enables the explanation of three edges: task and activity analysis (U-T); information requirements and technological limitations (T-A); ergonomics and training (procedures) (T-U) (Figure 16.8).

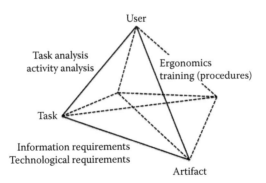

FIGURE 16.8 The AUT triangle.

CONSIDERING ORGANIZATIONS IN DESIGN

Technological design requires multidisciplinary design teams (i.e., a design team must include people who have related background, competence, and experience on each of these relevant artifacts incrementally integrated). In addition, design team members need to understand each other (i.e., they need to be able to read the same music theory, even if they do have the same scores). They need to be appropriately coordinated both at the task level (i.e., scores need to be harmonized by a composer) and the activity level (i.e., design team members, as musicians, need to be coordinated by a conductor at performance time). The Orchestra model is relevant here (Boy 2013).

Considering organizations in design introduces three additional edges (Figure 16.9): social issues (U-O), role and job analyses (T-O), and emergence and evolution (A-O).

Up to now, air and ground systems were designed and developed almost independently. Airspace complexity now requires that they should be designed in concert. When we design a new aircraft, for example, considering constant increase of air traffic, it should be thought as an interconnected component of a complex system, and not as an isolated system (i.e., other aircraft, weather, collision avoidance, communication, and other ATM factors should be considered).

TESTING IN A LARGE VARIETY OF SITUATIONS

The AUTOS pyramid (Figure 16.10) introduces another fundamental dimension, the *Situation*. The three new edges are: usability/usefulness (A-S), situation awareness (U-S), situated actions (T-S), and cooperation/coordination (O-S).

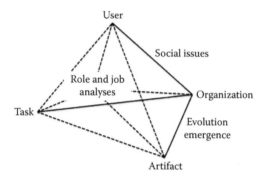

FIGURE 16.9 The AUTO tetrahedron.

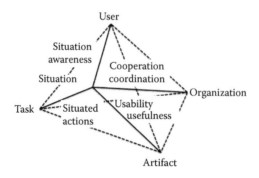

FIGURE 16.10 The AUTOS pyramid.

Interaction depends on the situation in which it takes place (e.g., location, time, people involved). Situations could be normal or abnormal. They could even be emergencies. This is why we will emphasize the scenario-based approach to design and engineering. Resulting methods are based on descriptions of people using technology to better understand how this technology is, or could be, used to redefine their activities. Scenarios can be created very early during the design process and incrementally modified to support product construction and refinement together with users and organization.

Scenarios are good to identify functions at design time and operations time. They tend to rationalize the way the various agents interact among each other. They enable the definition of organizational configurations and time-wise chronologies.

Situation complexity is often caused by interruptions and more generally disturbances. It involves safety and high workload situations. It is commonly analyzed by decomposing contexts into subcontexts. Within each subcontext, the situation is characterized by uncertainty, unpredictability, and various kinds of abnormalities. To summarize, situational factors deal with the *predictability* and *appropriate completeness* (scenario representativeness) of the various situations in which the new artifact will be used.

USING THE AUTOS PYRAMID IN PRACTICE

Software is very easy to modify; consequently designers modify it all the time. Interaction is not only a matter of end product; it is also a matter of development process. End users are not the only ones to interact with a delivered product; designers and engineers also interact with the product to fix it up toward maturity... even after its delivery. This is why agile approaches based on design cards are extremely useful and effective (Schwaber 1997; Sutherland 2014). In addition, scenario-based design is a HCD approach that fosters understandability (situation awareness), complexity, reliability, maturity, and induced organizational constraints (rigidity versus flexibility).

Software complexity can be split into internal complexity (or system complexity) and interface complexity. Internal complexity is related to the degree of explanation required to the user to understand what is going on when necessary. Concepts related to system complexity are as follows: flexibility (both system flexibility and flexibility of use), system maturity (before getting mature, a system is an accumulation of functions—the *another function syndrome*—and it becomes mature through a series of articulations and integrations), automation (linked to the level of operational assistance, authority delegation and automation culture), and operational documentation.

DISCUSSION

HUMAN CONTROL VERSUS AUTOMATION: THE MATURITY ISSUE

Whenever people extend their capabilities with TISs that can be called *cognitive prostheses* (Hamilton 2001), they increase their performance. However, if these cognitive prostheses use automation to a point that is not clearly understood from an operational standpoint, performance may decrease and, in some cases, cause serious problems. This is what Earl Wiener called *clumsy automation* (Wiener 1989). When I was working on the Orbital Refueling System of the Space Shuttle in the mid-eighties (Boy 1987), I found out that there is an optimum P_0 in terms of autonomy level and performance of the overall human–machine system (Figure 16.11).

Technology-centered engineering typically automates at the point P_{TL} (i.e., on the technological limitations limit). It takes experimental tests and efforts to go back to P_0. HCD takes into account the existence of P_0 and incrementally tries to find out this optimum through creative design and formative evaluation using HITLS. Interestingly, the more we know about autonomy of the various human and machine agents, the more the optimum P_0 move to the right and goes up on Figure 16.10.

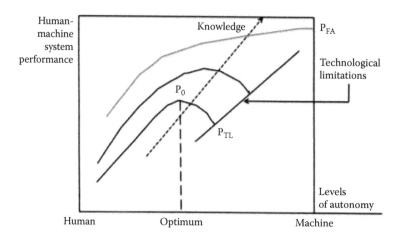

FIGURE 16.11   Human–machine system performance versus levels of autonomy. (Boy, G.B., An expert system for fault diagnosis in orbital refueling operations. *AIAA 24th Aerospace Sciences Meeting*, Reno, Nevada, 1986.)

Of course, if we knew everything about the environment and the various agents involved in the various interactions, the optimum P_{FA} would be on the full autonomy of the machine (e.g., in the case of an aircraft, P_{FA} would correspond to a drone).

This shift of optimum P_0 to the right is strongly related to maturity. In HCD, we distinguish between technology maturity and maturity of practice. The former deals with reliability, availability, and robustness of the technology being designed and developed. The latter deals with people's adaptation to and resilience of the technology (Boy 2002); often new practices emerge from technology use. We then need to observe usages as early as possible to anticipate surprises before it is too late. HCD proposes methods and tools that enable HCD teams to detect these emergent properties at design time (e.g., using HITLS). In any case, there is a maturity period required to assess if a product can be delivered or not. When a product is not mature both technologically and culturally, people can reject it.

MANAGEMENT AND ORGANIZATIONAL AUTOMATION

Back to aeronautics, we have seen that pilot's job moved from control of flight parameters to management of onboard systems during the 1980s. This job revolution in the cockpit is now shifting to air traffic (i.e., ATC is moving toward ATM). In other words, single agent's shift from control to management is currently evolving to a multiagent shift. Why? What are underlying organizational issues?

The main cause is the number increase effect. According to Boeing's Current Market Outlook 2014–2033, average airline traffic yearly growth is estimated at a rate of 5%. Knowing that most big airports, such as Hartsfield–Jackson Atlanta International Airport, are already oversaturated, the air traffic capacity issue requires complexity science approaches, and more specifically a multiagent approach, in which agents are aircraft. Connectivity among these agents has to become explicit. If aircraft separation has to be reduced during approach and landing, for example, current human-centered ATC techniques need to be revisited and in many cases drastically changed. This is a matter of technology, organizations, and people. On the technology side of the problem to be solved, each aircraft should know about the location and identity of the other aircraft around it. This aircraft should then be equipped with appropriate sensors and receptors, such as Automatic Dependence Surveillance—Broadcast system. Satellite data have to be used to identify a clear dynamic model of the sky, in terms of both traffic and weather. We can also use weather radar data from all aircraft and fuse them with satellite data to increase 3D validity of weather models. Same kind of fusion can be done for air traffic using more conventional radar system data. Resulting data can be used to develop a cyber-physical system (CPS) providing each aircraft with a protection safety net (i.e., each aircraft knows the traffic around it and is able to decide tactical maneuvers to increase global safety).

This approach definitely defines a new kind of organizational automation, which is multiagent. This multiagent approach involves interaction among people and systems, among systems, and among people (often through information technology). As there will be a layer of information technology gluing the various air and ground systems, new factors are emerging such as cybersecurity. ATM of the future will be almost entirely based on highly interconnected CPSs. New kinds of risks will emerge from the activity of this giant airspace CPS-based infrastructure. In particular, malicious actors will be able to attack these CPSs from anywhere in the world. This is why we need to further develop methods and tools that enable studying security and resilience of such CPS-based infrastructure and find appropriate solutions. Protection will have to be found from technology (safety and security nets), organizations (collaboration among agents), and people (increasing training and expertise).

Air-ground integration is an example of HSI point of view that involves function allocation. As Fitt's law that provided the HABA-MABA* recommendations for single-agent function allocation, very little has been done on large-scale multiagent function allocation. Function allocation rationale is to determine which functions should be carried out by humans and which by machines (Fitts 1951). Cognitive function analysis (CFA) is an effective approach to this kind of problem in a multiagent environment (Boy 1998, 2011), which is often considered as a SoS. CFA includes physical functions also (remember that a cognitive function is defined by a set of resources that may be cognitive functions and/or physical functions).

Function allocation cannot be considered as a matter of a priori optimal assignment of tasks to humans and machines, in Fitts's sense. More specifically, functions cannot be correctly allocated among humans and machines without a thorough identification of emerging[†] functions that are only observable at use time. This is the reason we need to observe user's activity (i.e., what people really do) and not be limited to prescribed tasks (i.e., what people should do). Up to recently, this emerging function discovery exercise was impossible before the whole system was fully developed. Current modeling and HITLS capabilities enable it.

In addition, the Orchestra model (Boy 2013) provides a very useful framework to model and simulate resulting cognitive functions of the human–systems system being developed. Consequently, agent needs to be further modeled in terms of cognitive and physical functions, as well as the way they are interconnected—this is typically doing a CFA. CFA results can be presented as a cognitive/physical function network (or interactive map), which is a good support for studying interaction complexity, distributed situation awareness, and multiagent decision-making.

PERSPECTIVE AND CONCLUSION

At this point, it is important to clarify the relationship between the task/activity distinction and distinctions between sociotechnical disciplines. More specifically, we need to define the following concepts: task, activity, human factors, ergonomics, human–computer interaction, HCD, and HSI. A task is what is prescribed to human operators or users. An activity is what is effectively performed by human operators or users.

---

* Fitt's HABA-MABA (humans-are-better-at/machines-are-better-at) approach provided generic strengths and weaknesses of humans and machines.
† The concept of "emergence" needs to be understood in the complexity science sense. An emergent property of a complex system emerges from interactions among its components (and subcomponents) that do not exhibit such a property.

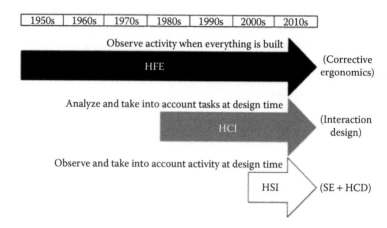

FIGURE 16.12 Human-centered design evolution.

For the last 60 years, sociotechnical evolution can be decomposed into three eras that made three communities[*] emerge (Figure 16.12):

Human factors and ergonomics (HFE) was developed after the World War II to correct engineering production and generated the concepts of human–machine interfaces or user interfaces, and operational procedures; activity-based evaluation could not be holistically performed before products were finished or almost finished, which enormously handicapped possibilities of redesign. Sometimes, activity analyses were carried out prior to designing a new product, based on existing technology and practice; however, this HFE approach forced continuity, reduced risk taking, and most of the time prevented disruptive innovation.

Human–computer interaction started to be developed during the 1980s to better understand and master human interaction with computers; it contributed to the shift from corrective ergonomics to interaction design mainly based on task analysis. Activity-based analysis started to be introduced within the Human–Computer Interaction community by people who understood phenomenology (Winograd and Flores 1986) and activity theory (Kaptelinin and Nardi 2006).

HSI emerged from the need to officially consider human possibilities and necessities as variables in systems engineering (SE); incrementally combined, SE and HCD lead to HSI, to take care of systems during their whole life cycle (Boy and Narkevicius 2013). HCD involves more than human factors evaluations or task analyses. It involves activity analysis at design time using virtual prototyping and HITLS (e.g., we can model and simulate an entire aircraft, fly it as a computing game, and observe pilot's activity). HCD also involves creativity, system thinking, risk taking, prototype development using agile approaches, complexity analyses, organizational design and management, and HSI knowledge and skills.

* The author is qualified to talk about these three communities. He is still the Chair of the Aerospace Technical Committee of the International Ergonomics Association (IEA), which encapsulates most HFE societies around the world. From 1995 to 1999, he was the Executive Vice-Chair of the Association for Computing Machinery (ACM) Special Interest Group on Computer Human Interaction (SIGCHI), and Senior Member of the ACM. He is currently Co-Chair of the Human Systems Integration Working Group of the International Council on Systems Engineering (INCOSE).

Moving from Automation Issues to Tangibility Issues

During the eighties and nineties, automation drawbacks emerged from several HFE studies, such as *ironies of automation* (Bainbridge 1983), *clumsy automation* (Wiener 1989), and *automation surprises* (Sarter et al. 1997). These studies did not consider the importance of technology maturity and maturity of practice. Automation can be modeled as cognitive function transfer from people to systems (Boy 1998). If automation considerably reduced casual people's burdens, it also caused problems such as complacency, which is an emerging cognitive function (i.e., not predictable at design time, but at operations time).

Today, we develop an entire aircraft on computers from inception of design to finished product. Therefore, we can test its operability from the very beginning, and along its life cycle using HITLS and an agile approach. Consequently, we can observe, and therefore analyze, activity of people using systems in a simulated environment at design time. Operability of complex systems can then be tested during design. This is the reason why HCD has become a discipline in its own right. HCD enables us to better understand HSI during the design process and then have an impact on requirements before complex systems are concretely developed.

HITLS does enable activity analysis at design time, but this is in virtual environments. Even if these environments are very close to the real world, their tangibility must be questioned and most importantly validated.

What does the tangibility concept mean exactly? It has two meanings. First, physical tangibility is the property of an object that is physically graspable (i.e., you can touch it, hold it, sense it, etc.). Second, figurative tangibility is the property of a concept that is cognitively graspable (i.e., you can understand it, appropriate it, feel it, etc.). If I try to convince you about something, you may tell me, "what you are telling me is not tangible!" This means that you do not believe me; you cannot grasp the concept I am trying to provide. We also may say that you do not have the right mental model to understand it, or I do not have enough empathy to deliver the message correctly.

Tangibility is about situation awareness both physically and cognitively. For example, Tan developed the first versions of the Onboard Context-Sensitive Information System (OCSIS) for airline pilots on a tablet PC (Tan 2015). Physical tangibility considerations led to a better understanding of whether OCSIS should be hand-held or fixed in the cockpit. Other considerations led to the choice of figurative displays of weather visualization going from vertical cylinders to more realistic cloud representations (figurative tangibility). A set of pilots gave their opinions on various kinds of OCSIS tablet configurations. It is interesting to note that the pilots always naturally used the term *tangible* to express their opinions.

Therefore, tangibility metrics should be developed to improve the assessment of complex systems operability. This is where subject matter experts and experienced people enter into play. We absolutely need such people in HCD to help assess HSI tangibility. For example, very realistic commercial aircraft cockpits, professional pilots, and realistic scenarios are mandatory to incrementally assess tangibility. OCSIS was tested from the early stages of the design process using HITLS, by recording what pilots were doing using it and analyzing produced activity. Such formative evaluations lead to system modifications and improvements. HCD is iterative, and agile* in the SE sense.

Although the twenty-first century shift from software to hardware is not necessarily obvious, it is the next dilemma we must address, especially now that we can 3D print virtual systems and transform them into physical systems. We will denote resulting systems, TISs (Boy 2016). The TIS concept is very close to the CPSs concept, which are usually defined as a set of collaborative

---

* The Manifesto for Agile Software Development (http://www.agilemanifesto.org) has been written to improve the development of software. It values more individuals and interactions over processes and tools, working software over comprehensive documentation, customer collaboration over contract negotiation, and responding to change (flexibility) over following a plan (rigidity).

elements controlling physical entities (Lee 2008). CPSs are often qualified as ESs, but there is a distinction between ESs as purely computational elements, and CPSs as computational and physical elements intimately linked among each other (Wolf 2014). Both TISs and CPSs are strongly based on the multiagent concept, unlike twentieth century automation that was based on the single agent concept. This is why TISs cannot be taken into account without an organizational approach. More generally, coevolution of people's activities and technology necessarily led to a tangible organizational evolution. Let us introduce the concept of organization tangibility.

A shift from the old army pyramidal model to the orchestra model is currently emerging (see the Orchestra model in Boy 2013). For example, technology and emergent practices have led people to change ways of communicating among each other. The army model induced vertical communication, mostly descendent. Transversal communication (e.g., using telephone, email, and the web) contributed to the emergence of the orchestra model. This functional evolution is now changing organizations themselves (i.e., structures). For example, smart phones and the Internet have contributed to change both industrial and everyday life organizations.

The Orchestra model provides a usable framework for HSI. It requires definition of a common frame of reference (music theory), as well as jobs such as the ones of human-centered designers and systems architects (composers) who provide coordinated requirements (scores), highly competent sociotechnical managers (conductors) and performers (musicians), and well-identified end users and involved stakeholders (audience). Having this organizational model in mind, it is now crucial to use it in HCD. More specifically in the framework of this chapter, it can be very useful for ATM research and design.

ACKNOWLEDGMENTS

Many thanks to Tiziano Bernard, Chris Johnson, Peter Hecker, Edwige Quillerou-Grivot, and Lucas Stephane for many useful and vivid exchanges on the topic.

REFERENCES

Bainbridge, L. (1983). Ironies of automation. *Automatica*, 19(6): 775–779.
Boy, G.A. (1987). Operator assistant systems. *International Journal of Man-Machine Studies*, 27: 541–554, and In G. Mancini, D.D. Woods, and E. Hollnagel (Eds.) *Cognitive Engineering in Dynamic Worlds*. London, UK: Academic Press.
Boy, G.A. (1998). *Cognitive Function Analysis*. Stamford, CT: Praeger/Ablex.
Boy, G.A. (2002). Theories of human cognition: To better understand the co-adaptation of people and technology, in knowledge management, organizational intelligence and learning, and complexity. In L. Douglas Kiel (Ed.) *Encyclopedia of Life Support Systems* (EOLSS), Developed under the Auspices of the UNESCO, Oxford, UK: Eolss Publishers.
Boy, G.A. (Ed.) (2011). Introduction. *Handbook of Human-Machine Interaction: A Human-Centered Design Approach*. Farnharm, UK: Ashgate Publishing.
Boy, G.A. (2013). *Orchestrating Human-Centered Design*. London, UK: Springer.
Boy, G.A. (2016). *Tangible Interactive Systems: Grasping the Real World with Computers*. Springer, UK. ISBN 978-3-319-30270-6.
Boy, G.A. and Grote, G. (2009). Authority in increasingly complex human and machine collaborative systems: Application to the future air traffic management construction. In the *Proceedings of the 2009 International Ergonomics Association World Congress*. Beijing, China.
Boy, G.A. and Narkevicius, J. (2013). Unifying human centered design and systems engineering for human systems integration. In M. Aiguier, F. Boulanger, D. Krob, and C. Marchal (Eds.) *Complex Systems Design and Management*. Cham, Switzerland: Springer.
Boy, G.A., Salis, F., Figarol, S., Debernard, S., LeBlaye, P. and Straussberger, S. (2008). PAUSA: Final Technical Report. Paris, France: DGAC-DPAC.
Boy, G.B. (1986). An expert system for fault diagnosis in orbital refueling operations. In *AIAA 24th Aerospace Sciences Meeting*. Reno, Nevada.

Cummings, M.L. and Tsonis, C.G. (2006). Partitioning complexity in air traffic management tasks. *International Journal of Aviation Psychology*, 16: 277–295.

Fitts, P.M. (Ed.) (1951). *Human engineering for an effective air navigation and traffic control system*. Washington, DC: National Research Council.

Hamilton, S. (2001). Thinking outside the box at the IHMC. *IEEE Computer*, 34(1): 61–71.

Hecker, P. (2016). How we take science seriously in SESAR. SESAR Exploratory Research. *World ATM Congress*, Madrid.

Hilburn, B. (2004). Cognitive complexity in air traffic control: A literature review. Project COCA–Complexity and Capacity. EEC Note No. 04/04.

Kaptelinin, V. and Nardi, B. (2006). *Acting with Technology: Activity Theory and Interaction Design*. Cambridge, MA: MIT Press.

Landauer, C. and Bellman, K.L. (1996). Collaborative system engineering and integration environments. *5th International Workshops on Enabling Technologies: Infrastructure for Collaborative Enterprises (WET ICE'96)*. Washington, DC: IEEE Computer Society

Laudeman, I.V., Shelden, S.G., Branstrom, R. and Brasil, C.L. (1998). *Dynamic density. An air traffic management metric*. Mountain View CA: National Aeronautics and Space Administration, Ames Research Center, NASA/TM-1998-112226.

Lee, E.A. (2008). Cyber physical systems: Design challenges. http://www.eecs.berkeley.edu/Pubs/TechRpts/2008/EECS-2008-8.html. Retrieved on May 10, 2015.

Masalonis, A.J., Callaham, M.B., and Wanke, C.R. (2003). Dynamic density and complexity metrics for real-time traffic flow management. Presented at the *ATM 2003 Conference, 5th EUROCONTROL/FAA ATM R&D Seminar*. Budapest, Hungary.

Minsky, M. (1986). *The Society of Mind*. New York: Simon and Schuster.

Sarter, N.B., Woods, D.D. and Billings, C.E. (1997). Automation surprises. In G. Salvendy (Ed.) *Handbook of Human Factors & Ergonomics*, 2nd ed. Hoboken, NJ: Wiley.

Schwaber, K. (1997). Scrum development process. In J. Sutherland, D. Patel, C. Casanave, J. Miller and G. Hollowell (Eds.) *OOPSLA Business Objects Design and Implementation Workshop Proceedings*. London, UK: Springer.

Sutherland, J. (2014). *Scrum: The Art of Doing Twice the Work in Half the Time*. Danvers, MA: Crown Business.

Tan, W. (2015). Contribution to the onboard context-sensitive information system (OCSIS) of commercial aircraft. PhD Dissertation, School of Human-Centered Design, Innovation and Arts. Melbourne, FL: Florida Institute of Technology.

Tarnowski, E. (2006). The 4-loop approach to ATM. Keynote, *HCI-Aero'06*. Seattle, WA.

Wiener, E.L. (1989). Human factors of advanced technology ("glass cockpit") transport aircraft. (NASA Contractor Report No. 177528). Moffett Field, CA: NASA-Ames Research Center.

Winograd, T. and Flores, F. (1987). *Understanding Computers and Cognition: A New Foundation for Design*. Reading, MA: Addison-Wesley.

Wolf, M. (2014). *High-Performance Embedded Computing: Applications in Cyber-Physical Systems and Mobile Computing*, 2nd ed. Waltham, MA: Morgan Kaufmann.

17 Remotely Piloted Aircraft

Alan Hobbs

CONTENTS

Introduction...379
Human Factors of Remotely Piloted Aircraft Systems 380
    Reduced Sensory Cues..381
    Control via Radio Link..381
        Link Management...382
        Loss of Link: Implications for the Remote Pilot...383
        Loss of Link: Implications for Air Traffic Control..384
        The Relay of Voice Communications via the Control Link 384
    Implications for "See and Avoid"...385
    Control Transfer 386
    The Control Station Environment 387
    Controls and Displays 388
    Emergencies and Flight Termination 390
    Required Competencies of Flight Crew 391
Concluding Comments...392
Acknowledgments..392
References 393

INTRODUCTION

Remotely piloted (or unmanned) aircraft are rapidly emerging as a new sector of civil aviation. As regulatory agencies work to integrate these aircraft into the existing aviation system, they must contend with a unique set of human factors that are not yet fully identified or understood.

These aircraft are sometimes referred to as drones, uninhabited aircraft, or unmanned aerial vehicles. Throughout the current chapter, the terminology of the International Civil Aviation Organization (2015) will be used. The term *remotely piloted aircraft* (RPA) will be used to refer to the aircraft, in both the singular and plural. The term *remotely piloted aircraft system* (RPAS) will be used when the intent is to refer to the entire system, comprising the aircraft, its control station, communication links, and other elements. The workstation of the remote pilot will be referred to as the *remote pilot station* (RPS) or control station.

Any discussion of RPAS is complicated by the diversity of the sector and the rapid rate at which it is developing. RPA range from insect-sized micro air vehicles, to large jet aircraft, such as the Global Hawk. In between are electric rotorcraft, numerous fixed-wing aircraft, and balloons that can remain aloft for extended periods, climbing and descending as necessary to take advantage of prevailing winds. To further complicate matters, many systems include features not typical of conventional aviation, such as catapult launch systems, electric engines, and solar cells (Figure 17.1).

FIGURE 17.1 Three examples of remotely piloted aircraft. (1) The 18 kg, catapult-launched Insitu ScanEagle; (2) 6700 kg high-altitude long-endurance (HALE) Global Hawk; (3) AeroVironment Helios Prototype, a solar powered flying wing designed for long-duration, high-altitude missions in the stratosphere.

Much of the recent growth of this sector has involved small electric rotorcraft used for aerial photography, site surveys, and inspections of buildings and infrastructure (Association for Unmanned Vehicle Systems International 2016). The Federal Aviation Administration (FAA 2016) has released regulations that allow lightweight RPA to be flown near the ground within sight of the pilot. Currently, however, no regulations are in place to allow larger, more capable RPA to routinely fly beyond pilot line-of-sight in airspace shared with conventional aircraft. This chapter focuses on the human challenges that must be addressed before these RPA can be fully integrated into the civil airspace system.[*]

The potential uses of these aircraft include pipeline and rail track inspection, police and firefighting, mineral exploration, agriculture, mapping, wildfire monitoring, and environmental research. Long-endurance fixed-wing systems and free balloons have potential as high altitude platforms for telecommunications or remote sensing tasks that might otherwise have required a satellite. In the not-too-distant future, converted airline aircraft may operate as unmanned freighters (Smith 2010).

Despite the diversity of designs and missions, all RPAS have features in common, notably the physical separation of the pilot from the aircraft, control via radio signals, and a remote control interface. These characteristics, in turn, introduce a set of human factors that are not typical of conventional aviation. A key objective of this chapter is to raise questions and identify areas in need of research.

HUMAN FACTORS OF REMOTELY PILOTED AIRCRAFT SYSTEMS

RPA have experienced a significantly higher accident rate than conventionally piloted aircraft. In the early 2000s, accident rates for some RPA were between 30 and 300 times higher than the comparable rate for general aviation (Tvaryanas et al. 2006). In the years 2006–2010, MQ-9 RPA operated by U.S. Customs and Border Protection had an accident rate of 53 per 100,000 hours, although this figure must be interpreted with caution as it was based on a relatively small total of flying hours (Kalinowski and Allen 2010). The U.S. Army has reported an accident rate of 49.3 per 100,000 flying hours for its RPA, compared with 4.4 for its manned aircraft. The army acknowledges, however, that the rate for RPA may be a low estimate due to significant underreporting of mishaps (Prather 2013). Statistics for accidents in which the aircraft is destroyed enable more reliable comparisons to be made between RPA and manned aircraft as there is less potential for underreporting or differences in definitions. In 2015, the most recent year for which data are available, MQ-9 operated by the U.S. Air Force (USAF) were destroyed at the rate of 4.0 per 100,000 hours flown. This is a significant improvement over earlier years, yet is still markedly higher than the comparable accident rate for the USAF's manned aircraft, of 0.41 aircraft destroyed per 100,000 flying hours (USAF 2015).

---

[*] The FAA (2013) has stated that future integration of RPA into civil airspace will require that each RPA be under the control of a pilot who will comply with all ATC instructions; no pilot will control more than one RPA at a time; RPA will be capable of flight under instrument flight rules, and autonomous operations will not be permitted.

The higher accident rate for RPA can be partly explained by technological factors such as the use of noncertificated components and a lack of system redundancy. However, inadequate consideration of human factors by system designers has also contributed to the accident record (Tvaryanas 2004; Williams 2004).

The following sections contain an overview of the human challenges of RPA, with a focus on the points of difference between this sector and conventional aviation. The illustrative quotes throughout the text are from remote pilots who participated in focus groups conducted by Hobbs et al. (2016). Pilots were asked to recall a hazardous event or error that had occurred when operating an RPA. As well as revealing human–system integration challenges, their reports also illustrate the positive contribution that humans make to the performance of highly automated, remotely operated systems.

Reduced Sensory Cues

Lacking the ability to hear the sound of hail on the fuselage, smell an onboard fire, feel turbulence, or notice ice accumulating on a windshield, the remote pilot relies almost entirely on visual displays to monitor the state of the aircraft. Even when the RPA is equipped with a camera, the image quality may be limited, and the field of view may be reduced to a narrow *soda straw* picture.

The sensory isolation of the remote pilot may make it more difficult to identify and recover from threats and errors, a function that is performed routinely by the pilot of a manned aircraft (Helmreich 2000). For example, one remote pilot was apparently unaware that the aircraft was flying upside down shortly before it crashed (Whitlock 2014). In many cases, these displays present data in textual form, which may further impede the flow of information to the pilot. In the following example, the pilot was unaware that the RPA had a stuck throttle until it failed to level off:

> "We fly based on digital gauges. We don't hear or feel anything, like RPM changes …. The aircraft is supposed to level off, at say, 5,000', and there is a delay due to data link to know if it actually leveled off. … As opposed to a real aircraft [where] you can feel the airplane leveling off, I couldn't determine if it was still climbing until I noticed it was 300' past its command altitude."

A solution may be to provide the remote pilot with a greater variety of sensory inputs, including haptic or aural cues (Arrabito et al. 2013; Giang and Burns 2012) and graphical displays (e.g., Kaliardos and Lyall 2015; McCarley and Wickens 2005). Research is needed to identify the sensory cues that will be most useful to the remote pilot, and then to make the case that the benefits would justify the added cost and complexity.

Control via Radio Link

Unlike the mechanical control cables or fly-by-wire systems of a conventional aircraft, the RPAS *fly-by-wireless* control link introduces control latencies and the possibility of complete interruption in some circumstances. RPAS technology and pilot procedures must each be designed to accommodate these limitations. Figure 17.2 shows the elements of a typical RPAS, including the RPA, the control station, and the communication links. Two distinct links are shown: a ground-based link that is used when the RPA is operating within line-of-sight of a ground antenna, and a satellite link that provides communication over greater distances.

A pilot command from the control station can take around 100 milliseconds to be uplinked to the RPA if the signal is transmitted from a nearby ground antenna. Most of this delay is the result of signal processing at either end rather than the time it takes radio waves, traveling at the speed of light, to reach the aircraft. With an equivalent delay on the downlink, the total round-trip latency between a command and the response observed on the pilot's display can become noticeable. If control is via satellite, additional processing steps and the distance that must be traveled by the signal can produce

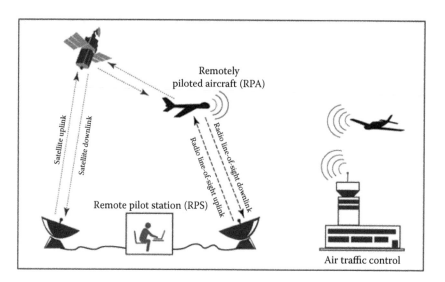

FIGURE 17.2 The remotely piloted aircraft system (RPAS) consists of the remotely piloted aircraft (RPA), the remote pilot station (RPS), and the associated communications systems.

round-trip latencies of 1,000 milliseconds or more (Tvaryanas 2006a). Unlike a hobbyist flying a radio-controlled aircraft, whose commands are delayed on the uplink, but who can directly observe the aircraft response in real time, the RPA pilot must contend with the sum of the uplink and the downlink delays.

Tracking tasks can be impacted by command-response delays of 100 milliseconds or less. Longer delays and variable latencies increase the difficulty of these tasks even further (Wickens 1986). An RPA that relied on continuous pilot control inputs to maintain stable flight would be difficult to control via a satellite link and would also be unable to tolerate link interruptions. For these reasons, virtually, all RPA require some level of onboard automation.

The introduction of highly automated airline aircraft in the 1980s not only led to improvements in safety and efficiency (Orlady and Orlady 1999) but was also accompanied by new challenges as pilots transitioned to the role of managers of automated systems. Data entry errors and loss of situational awareness became areas of increasing concern, and terms such as *mode confusion, automation surprise*, and *automation complacency* were coined to express the emerging issues. The RPAS sector is currently experiencing some of the same problems with systems that were developed with little apparent regard for human factors principles. It remains to be seen whether remote operation via radio link will make it more difficult for the pilot to manage automated systems, possibly exacerbating the impact of clumsy automation. In the following case, the behavior of the RPA surprised the remote pilot, who was nevertheless able to intervene and recover the situation.

> "I … put the airplane into a holding pattern. … The aircraft turned in the opposite direction than what I wanted it to do. To correct the situation, I over-rode the aircraft. I had the aircraft go into the hold again and the aircraft did it again." (The aircraft was successfully redirected on a second attempt).

Link Management

In addition to managing systems onboard the aircraft, the remote pilot must also manage the control link. With the control system reliant on radio signals, the standard preflight control check becomes particularly important. During flight planning, the pilot must take into account the predicted strength of the link throughout the intended flight and develop a three-dimensional picture of the link strength at various altitudes and distances from an antenna located on the ground. A signal coverage map may show this information in a 2D format, typically displaying shadows in which the signal will be blocked by terrain or obstructions. As the distance between the aircraft and the ground antenna increases, the aircraft may need to fly higher to maintain a link with the ground station. A link strength indicator is a critical display in the RPS, although pilots report sometimes using less precise cues, such as a *snowy* camera image to warn of an impending loss of link. There appears to be no published research examining how best to support pilot awareness of actual and predicted link status.

Loss of Link: Implications for the Remote Pilot

No radio control link can be guaranteed to be 100% reliable, and there will be occasions when the link will be unavailable. A preprogramed lost-link procedure enables the RPA to continue flight in a predictable manner until the link is resumed. The procedure may involve either a simple maneuver such as climbing to regain a signal, or a more complex plan, such as flying to a predetermined position. Rather than being perceived as an emergency, the activation of the lost-link procedure can be seen as a response to a nonnormal situation, analogous to a diversion or a go-around in a conventional aircraft.

A lost-link event can consist of three stages, as shown in Figure 17.3. In stage 1, the link has been interrupted, but the aircraft continues to fly in accordance with the last command received from the pilot. Some link outages will last a few milliseconds (ms), whereas others may extend for minutes or even hours. It would be disruptive if the RPA started to fly its lost-link procedure each time a brief link interruption occurred. Therefore, an onboard timer is needed to measure the duration of the outage and activate the lost-link procedure after a preset interval has elapsed. In the terminal area, the lost-link procedure may need to commence after an outage of a few seconds. Elsewhere, the RPA may be able to safely continue along its planned flight path for an extended period before entering its lost-link procedure.

Nuisance lost-link events have sometimes prompted remote pilots to delay the activation of the lost-link procedure, or inhibit it until the aircraft has reached a certain location. In this example, the pilot used a work-around to extend the duration of stage 1, to prevent the RPA from repeatedly turning for home:

> "The airplane … made many turnarounds due to it being out of link then … it would reacquire and … return on mission. This affected fuel burn. [So I] set time-out feature just short of the actual mission duration."

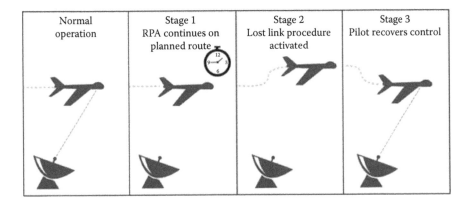

FIGURE 17.3   Stages of a lost-link event.

If the aircraft will remain in stage 1 for a significant time, the pilot must be aware that each command sent to the aircraft could be the last, if a link interruption were to occur. For example, a temporary turn toward rising terrain may become irreversible if the link is interrupted before a follow-up command can be sent to the aircraft.

In stage 2 of a lost-link event, the RPA's preprogramed lost-link procedure is activated. Different lost-link procedures will be appropriate according to the location of the aircraft and the stage of flight. The RPA pilot must therefore remain aware of the current lost-link procedure, updating it as frequently as every 10 minutes to ensure that it has not become stale or would not create a hazardous situation if activated (Neville et al. 2012). In the following example, a problem with the lost-link procedure was detected during a control handover:

> "At the beginning of the flight, the lost link procedure was valid, but the procedure was not updated later in the flight. At one point, had the lost link procedure been activated, it would've had the aircraft fly through terrain in an attempt to reach the next waypoint. However, the aircraft didn't lose link and the error was caught in the handover to the next set of operators."

In the third stage of the lost-link sequence, the link is reestablished and the aircraft transitions back to pilot control. The pilot must ensure that any control inputs made while the link was interrupted do not result in sudden changes in aircraft state when the link is reestablished. Depending upon the length of the outage, and the location and state of the aircraft, the pilot may need to evaluate whether the original flight plan can be resumed.

Loss of link can occur for a variety of technical and human reasons. The pilot of a conventional aircraft cannot accidently disconnect the cockpit from the rest of the aircraft. The remote pilot, however, can make errors that will inadvertently achieve this effect. Potential human causes of lost link include flying beyond the range of the ground station, flying into an area where the signal is masked by terrain, frequency selection errors, abrupt aircraft maneuvers, physical disruptions to plugs and cables, and electronic lock outs in which a screen lock or security system prevents access. In addition, the pilot must be alert to radio frequency interference, whether from malicious or unintentional sources. At the time of writing, the author was aware of no studies examining the human causes of lost links.

Loss of Link: Implications for Air Traffic Control

The behavior of the aircraft in the event of a lost link must be predictable not only to the pilot but also to air traffic control (ATC). A simple programed maneuver, such as a climb or a turn toward a specific location, may easily be included in the flight plan. However, more complex maneuvers that change throughout the flight may be more difficult to present to ATC. There have been cases in which a common cause failure has resulted in multiple RPA losing link simultaneously (ICAO 2015). Although this would hopefully be a rare event, it could present ATC with a complicated traffic picture.

To prevent the RPA from executing a lost-link procedure that contradicts an ATC instruction received before the link interruption, there may be occasions in which ATC will ask the pilot to inhibit the lost-link procedure for a set time, or until the aircraft has reached a particular location. As well as a preassigned squawk code to indicate a lost link, ATC may need to know the time remaining until the RPA will commence its lost-link maneuver. A countdown timer could conceivably be included in the aircraft's data block on the controller's scope.

The Relay of Voice Communications via the Control Link

Voice communication between the remote pilot and ATC is typically relayed from the control station to the RPA via the command link, and then retransmitted by an onboard radio (RTCA 2007). In a similar way, transmissions from ATC and other pilots in the vicinity are received by the radio onboard the RPA and then relayed to the remote pilot via the downlink. An advantage of this system is that the remote pilot can participate in the *party line* communications of pilots and ATC, but this may come at the cost of noticeable delays. Voice latencies can increase the likelihood of step-ons, in which two people attempt to transmit simultaneously. RPAS voice latencies are likely to be most problematic when a satellite link is involved, as illustrated by the following report:

> "There is a delay between clicking the press-to-talk and talking. This is very difficult to manage when in very busy airspace, and listening for a gap to talk. Sometimes by the time we press the talk button, with the satellite delay, the gap is gone and we step on other aircraft."

Telecommunications research has found that 250 milliseconds one-way delays can significantly disrupt phone conversations (Kitawaki and Itoh 1991). Consistent with this finding, FAA policy requires that communications systems deliver an average one-way delay between pilot and ATC voice communications of less than 250 milliseconds (FAA 2012). Several studies have examined the impact of controller voice latencies that might be introduced by future communications networks (e.g., Sollenberger et al. 2003; Zingale et al. 2003). These studies have generally found that one-way latencies in controller transmissions of up to 350 milliseconds are tolerable. In a simulation study, Vu et al. (2015) found that remote pilot voice delays of 1.5 and 5 seconds produced comparable rates of step-ons; however, further research is required to identify the dividing line between tolerable and disruptive voice latencies for remote pilot-voice communications.

A further implication of the RPAS voice-relay system is that a loss of link will not only prevent the pilot from sending commands to the RPA, but it will also interrupt voice communication at this critical time. Future communication systems are likely to solve this problem. For now, the pilot must rely on a telephone to regain communication with ATC, as described by a remote pilot:

> "We were constantly … talking to ATC via VHF to keep them updated and coordinated. We lost link. Then we realized that we didn't have ATC's phone number. We were able to finally call ATC, but it took a few minutes to find the number."

IMPLICATIONS FOR "SEE AND AVOID"

Before RPAS can be integrated seamlessly into civil airspace, the remote pilot must have a means to *see and avoid* other aircraft whenever conditions permit (14 CFR 91.113; ICAO 2011) and to comply with other air traffic requirements that rely on human vision. Detect and avoid (DAA) systems for RPAS have been a major focus of recent human factors research, including work by NASA to support the development of industry standards for DAA displays (Fern et al. 2014; Rorie and Fern 2015). Detecting and avoiding other aircraft is generally considered to consist of two related concepts: (1) remain well clear and (2) collision avoidance. To remain well clear of other aircraft, the remote pilot must maintain an awareness of surrounding traffic and make any necessary separation maneuvers before the intruder aircraft poses an imminent threat. In controlled airspace, the pilot would be expected to coordinate with ATC before maneuvering, as illustrated by the following report:

> "I was flying on a heading assigned by ATC. We have a display that shows traffic. On this display I was watching a flight block coming towards my aircraft. I realized that we were on a converging course so I queried ATC, and they had no info on it. We found the traffic through swinging the ball [pointing the on-board camera]. The pilot of the converging [aircraft] was completely oblivious to us. He was on a different frequency. I had to maneuver to avoid him."

The rules of the air currently leave it to the pilots of conventional aircraft to judge what it means to remain well clear of other aircraft. The introduction of DAA technology requires that the term be defined precisely. An advisory committee developing standards for DAA systems has defined *well clear* as meaning that the RPA and the threat aircraft do not come within 4,000 feet horizontally and 450 feet vertically when operating away from terminal areas. A time-based metric, broadly equivalent to 35 seconds to closest point of approach is also included in the definition (RTCA 2016). Keeping RPA well clear of other aircraft is not only a matter of safety but will also ensure that the addition of RPA to the civil airspace system does not cause concern for conventional pilots, that traffic alert and collision avoidance system alerts and resolution advisories are not triggered excessively, and that ATC workload is not increased.

Displays to assist the pilot in remaining well clear can be *informative*, *suggestive*, or *directive*. An informative display provides traffic information but provides no further guidance to the pilot. A suggestive display provides the pilot with a range of possible maneuvers and may also display *no-fly* areas, leaving the pilot free to formulate a course of action. Directive displays give the pilot a single recommended maneuver to remain well clear. Directive guidance has been found to produce more rapid pilot response times than informative or suggestive displays; however, the certification requirements for a directive system are too great for them to be considered a *minimum requirement* (Rorie et al. 2016). In simulation trials comparing informative and suggestive displays, Rorie et al. (2016) found that suggestive displays reduced the time it took the remote pilot to initiate a maneuver to remain well clear, reduced the size of the maneuver, and resulted in fewer and less severe losses of well clear.

If the RPA fails to remain well clear of traffic, it may be necessary to make a collision avoidance maneuver. The airborne collision avoidance system (ACAS Xu) for unmanned aircraft, currently under development, will provide a collision avoidance system specifically for RPA that will be interoperable with the traffic alert and collision avoidance system of manned aircraft. Given the possibility of link outages, and the need for a rapid pilot response, it is likely that future RPA equipped with ACAS Xu will need to be capable of making an autonomous response to a resolution advisory.

Given the long-recognized limitations of the see-and-avoid principle (Hobbs 1991), a remote pilot with a well-designed DAA display will almost certainly have a better awareness of traffic than the pilot of a conventional aircraft whose only traffic information comes from the view out the window. Furthermore, if the system is capable of detecting aircraft that are not equipped with transponders, the remote pilot may be aware of traffic that does not appear on the controller's scope. Consequently, DAA systems could change the patterns of communication between pilots and controllers. For example, the workload of controllers could be raised by remote pilots calling with concerns about nearby traffic.

CONTROL TRANSFER

A unique feature of RPAS is that control may be transferred in-flight between adjacent consoles, or between geographically separated control stations. Transfers may also involve a change of control link, such as from satellite to terrestrial radio communications. Handovers produce an elevated risk of human error in many task environments, including ATC, aircraft maintenance, and medicine (Parke and Kanki 2008). This also appears to be true for RPAS. Tvaryanas (2006a) notes that the control of a long-endurance RPA may be transferred multiple times during the course of a single flight, with each transfer contributing to a cumulative risk of error or misunderstanding.

Control transfers require careful briefings and checklist discipline. Several RPA accidents and incidents have involved failures to match the control settings on the receiving control station with that of the relinquishing control station, as illustrated by the following example involving a transfer during ground checks:

> "… we had the aircraft engine at idle with the parking brake set, but when the radio handover switched to XXX, he didn't have the parking brake set and the power was set at 80% …. The result was the engine revving up, and the aircraft jumping its chocks."

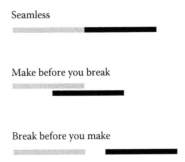

Seamless

Make before you break

Break before you make

FIGURE 17.4 Three potential styles of control transfer.

Three possible styles of control station transfer can be identified (Figure 17.4). A seamless transfer would involve the instantaneous switching of control from one control station to the next. In a *make before you break* transfer, there is an overlap in command authority between the receiving and relinquishing control station. If both control stations have the ability to transmit commands to the RPA, there is clearly a need for careful coordination to ensure that both pilots do not attempt to uplink commands simultaneously. The *break before you make* style requires that the relinquishing control station shuts off its command link to the aircraft before the receiving control station establishes its command link, although both control stations may continue to receive the downlink from the RPA during the process. During the transfer gap, which could last several seconds or longer, neither pilot will be able to send commands to the aircraft or speak with ATC via the aircraft's onboard radio. Although this style of transfer is currently used by some RPAS, the FAA (2013) has stated that it will not be acceptable for future operations in civil airspace. Despite the criticality of RPAS control transfers, many questions remain unanswered. For example: What design features are needed in the RPS to facilitate transfers? How should pilots confirm that control settings are consistent between the RPS before transferring control?

The Control Station Environment

The control stations of sophisticated RPAS increasingly resemble industrial control rooms or office workstations (Figure 17.5). The space may need to accommodate not only pilots but also technicians, payload operators, and maintenance personnel*. As a remote pilot has noted, "People come and go, opening and closing doors and holding casual conversations. Ringing telephones, whispered remarks, and other disturbances can interrupt critical operations-such as approach and landing maneuvers that demand silence and concentration" (Merlin 2013, p.132).

FIGURE 17.5 Control station for General Atomics MQ-9 (left), NASA's Global Hawk, and Raytheon's advanced common ground station system (CGCS).

---

* In addition, there may be no reason why the RPS should not be wheelchair accessible.

Anecdotal reports indicate that during critical in-flight events, operational personnel will sometimes gather at the control station to observe or offer support. It is unclear how the presence of additional personnel affects crew resource management. One remote pilot expressed it this way:

> "In manned aircraft it is clear who is in command, but with UAS operations, there are multiple people who have a sense of responsibility for the aircraft. So when there is something that needs attention many people run to the GCS [Ground Control Station]."

Applying a blanket *sterile cockpit* policy to the control station may create other problems. Maintaining vigilance during periods of task underload may emerge as one of the greatest human factors challenges for RPAS (Cummings et al. 2013). Thompson et al. (2006) found high levels of boredom, reduced mood, and chronic fatigue among USAF MQ-1 predator pilots. They identified the control station environment as a major contributor to boredom. Well-meaning efforts to control distraction, such as eliminating windows or prohibiting visitors may only serve to increase the monotony of the piloting task. Furthermore, comfortable, unstimulating environments can unmask fatigue, making it especially difficult for personnel to remain alert (Moore-Ede 1993). In future, control stations must be designed to maximize pilot alertness. Solutions could include allowing pilots to work in a standing position, or vigilance monitoring devices similar to those found in the cabs of locomotives.

A final observation concerns the implications of the RPS environment for maintenance personnel. Unlike the cockpit of a conventional aircraft, the RPS is accessible to maintainers while the aircraft is in-flight. Scheduled maintenance, such as software updates, should probably never occur while the RPA is airborne. However, nonscheduled corrective maintenance may sometimes be necessary. Examples are diagnosing and rectifying console lockups, rebooting computer systems, and troubleshooting problems with cable connections. Maintenance error is a significant threat to the reliability of aviation systems, especially when the system is in an operational mode while maintenance is occurring (Reason and Hobbs 2003). If corrective maintenance is to be performed on ground-based elements of the RPAS while the RPA is airborne, the prevention and management of maintenance error will be especially important (Hobbs 2010).

CONTROLS AND DISPLAYS

The cockpits of conventional aircraft evolved gradually over decades, incorporating principles learned from accidents and incidents. Standard features such as the *Basic T* arrangement of primary flight instruments, and controls that can be distinguished by touch, have helped to ease workload and reduce pilot error. Current-generation control station interfaces rarely comply with aviation standards, and they frequently contain an assortment of consumer electronics including computer monitors, pull-down menus, keyboards, and *point-and-click* input devices (Waraich et al. 2013).

The human factors deficiencies of control stations have been widely described (e.g., Cooke et al. 2006). Physical ergonomics problems include controls that cannot be reached from the pilot's seat, difficult-to-read fonts and color schemes, and unguarded controls that are susceptible to bumping or inadvertent activation (Hobbs and Lyall 2016a; Hopcroft et al. 2006; Pedersen et al. 2006).

Control stations have also suffered from more subtle deficiencies in cognitive ergonomics. A well-designed RPS would include features such as feedback to the pilot to confirm that a command has been received, consistency across controls and displays, appropriate prioritization of information provided by alarms and displays, and control interfaces that minimize the need for complex sequences of inputs to perform routine or time-critical tasks (Hobbs and Lyall 2016b). Yet, these principles have not always been applied in practice. For example, Pestana (2012) describes the process that must be performed by the pilot of NASA's Ikhana RPA to respond to an ATC request to *ident*, a routine action that will highlight the aircraft's return on the ATC radar screen. The pilot must perform a sequence of seven steps using a trackball to navigate pull-down menu options. The same task in a manned aircraft can be performed in a single step.

The same interfaces that make it difficult for the pilot to perform a task correctly can also make it easy to make errors. Keyboard or menu-based controls may be especially subject to skill-based slips when pilots have developed well-learned action sequences that can be triggered unintentionally:

"When I activated the gear extension, I turned off the engine by mistake. ... I accidentally pressed the engine shutdown switch with my left hand because the gear engage button is next to the engine shutdown switch and I was in a hurry due to time pressure."

Design deficiencies that have been identified in current RPS include the following:

- Presentation of nonintegrated or raw data that require the pilot to perform additional cognitive processing to extract meaning (Tvaryanas and Thompson 2008; Neville et al. 2012)
- Lack of design consistency across controls and displays (Gawron 1998)
- Complicated, multistep sequences required to perform routine or time-critical tasks, often involving menu trees (Cooke et al. 2006; Pestana 2012)
- Reliance on text displays to the exclusion of other sources of information, potentially introducing a foveal bottleneck that restricts the flow of information to the pilot (Hobbs and Lyall 2016b; Tvaryanas 2006b)
- Use of nonstandard or counterintuitive language in text messages (Hobbs and Lyall 2016b)
- Nonintuitive automation and inadequate mode annunciation (Cooke et al. 2006; Williams 2007)
- Lack of feedback on pilot control inputs or system states (Tvaryanas and Thompson 2008)
- Heavy reliance on memory to keep track of system status and flight plan details (Neville et al. 2012)
- Multifunction displays and controls, particularly in which a control may perform both a critical and a noncritical function (Tvaryanas and Thompson 2008; Hobbs and Lyall 2016b; Neville et al. 2012)
- Need for complex instrument scans (Tvaryanas and Thompson 2008)
- Difficulty in detecting and correcting errors (Neville et al. 2012)
- Poor hierarchy of presentation. For example, critical displays that can be obscured by noncritical pop-up windows, and a proliferation of display screens (Hobbs and Lyall 2016b)
- Reliance on keypress sequences and shortcuts, increasing the risk of skill-based slips and muscle memory errors (Neville et al. 2012)
- Single auditory alarm tones with multiple meanings, and alarms that lose their impact due to repeated activation (Arrabito et al. 2010; Hobbs 2010)

Some of the design deficiencies in RPS might have been avoided had existing human factors standards been applied. In other cases, the problems reflect a lack of RPAS-specific standards. Several human factors guides for military RPAS currently exist (NATO 2007, 2009; Under Secretary of Defense 2012). However, there are currently no human factors standards for nonmilitary RPAS operating in civilian airspace. To avoid a piecemeal approach to guidelines development, Hobbs and Lyall (2016a) have proposed that future guidelines for civil RPAS should (a) supplement existing human factors guidelines by focusing on the unique requirements of unmanned aviation and (b) should be based on a systematic analysis of the tasks that the pilot must perform via the RPS.

Figure 17.6 shows the primary safety-related tasks of the remote pilot when operating in airspace shared with conventional aircraft. Some of these tasks are common across aviation but are especially challenging for the remote pilot, perhaps due to the lack of direct sensory cues or communication

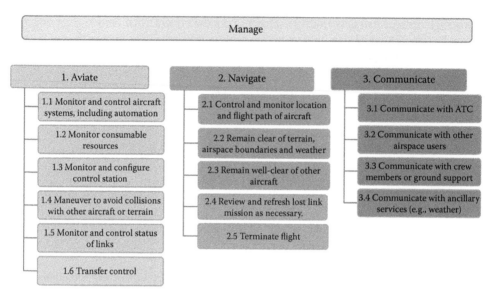

FIGURE 17.6   Responsibilities of the remote pilot.

latencies. In other cases, the remote pilot has unique responsibilities. These include monitoring the status of control links, control transfers, and flight termination. The draft RPS guidelines proposed by Hobbs and Lyall (2016b) are structured around the tasks shown in Figure 17.6, with a focus on displays and controls that would be unique to RPAS.

EMERGENCIES AND FLIGHT TERMINATION

Faced with a serious onboard problem, such as an engine failure, the pilot of a conventional aircraft will first consider whether a landing can be made at a nearby airport. If that is not possible, an off-airport emergency landing may be necessary. Even if the aircraft sustains damage, an emergency landing can be considered a success if the occupants are unscathed. The absence of human life onboard an RPA markedly changes the nature of emergency decision-making for the remote pilot. In essence, the *manned mindset* that leads a pilot to attempt to save the aircraft and its occupants may not transfer to unmanned aviation, in which the safety risks are borne by the occupants of other aircraft and uninvolved individuals on the ground. In an emergency, the remote pilot may be faced with the following options:

- Attempt a landing at a suitable airfield
- Attempt a controlled off-airport landing or ditching
- Activate a parachute system or
- Activate a flight termination system that will cause the aircraft to descend to a controlled impact

Flight termination systems introduce the risk of inadvertent activation. It is worth noting that the first loss of a Global Hawk RPA occurred when a flight termination message was sent by mistake (Hobbs 2010). The guidelines of Hobbs and Lyall (2016b) recommend a range of precautions for parachute or flight termination systems. These include a requirement for two distinct and dissimilar actions to initiate a flight termination, aural and visual warnings to the crew before the final activation of the system, and controls designed to minimize the likelihood of unintentional activation.

The flight planning for a large RPA can be expected to include the identification of suitable sites for flight termination. For example, in 2007, NASA successfully used its Ikhana RPA to monitor wildfires in the western United States (Buoni and Howell 2008). As part of the risk management plan for this mission, NASA identified a large number of potential sites for emergency landings or crashes, as shown in Figure 17.7.

Even if potential sites for flight termination have been preselected, in the event of an emergency, it may be necessary to confirm that the selected site is clear of people and property. If the link is likely to be interrupted as the aircraft descends, the pilot may have to act quickly to interpret sensor data from the RPA, select a suitable site, and send the necessary commands to the RPA at early stage in the descent. Automated decision-support aids that recommend a site after analyzing sensor data may assist the pilot in these time-critical situations (Patterson et al. 2012).

REQUIRED COMPETENCIES OF FLIGHT CREW

The FAA (2013) has stated that RPAS capable of operating in the U.S. National Airspace System must have a pilot in command; however, it is not clear what qualifications this person will need to possess. Despite the diversity of RPAS, there are likely to be core competencies that will apply across systems, related to the pilot responsibilities shown in Figure 17.6. Recommended training requirements for remote pilots operating in civil airspace have been produced by SAE (2011). These requirements address unique issues such as control transfer and link management, as well as identifying syllabus items from manned aviation that would no longer be relevant to RPAS. Remote pilots may also require nontechnical skills training focusing on issues such as flight termination decisions, coordination with remote crew members, and the impact of reduced sensory cues on threat and error management. Although SAE assumes that manned experience will not be necessary to operate an RPAS, this issue is far from settled. Some military RPASs are operated by personnel with no flying

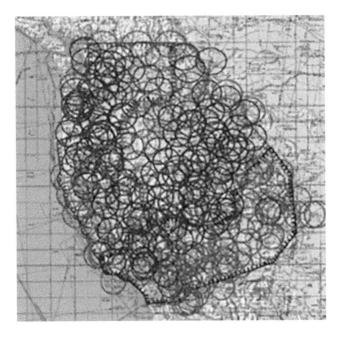

FIGURE 17.7   More than 280 emergency landing sites were identified for NASA's Ikhana wildfire monitoring missions. The aircraft could have been directed to glide to one of these sites in the event of a complete and irreversible engine failure.

experience, yet it seems likely that conventional piloting experience will provide the remote pilot with insights or attitudes that contribute to safe operations.

Finally, it should be noted that, with the pilot no longer colocated with the aircraft, the task of piloting could be outsourced to virtually anywhere in the world, just as airline maintenance tasks have been outsourced to low-cost locations. One advantage could be a reduced need for pilots to work during the night, if control can be transferred between control stations in different time zones.

CONCLUDING COMMENTS

Virtually, every aspect of RPAS, from interface design, interaction with ATC, and pilot decision-making, demands attention from the human factors profession. In many cases, existing human factors knowledge from aviation and other industries can be applied directly to RPAS. For example, cockpit design standards could help to improve control station interfaces. In other cases, RPAS operations introduce a unique set of human factors that have not yet been clearly identified or examined.

Much of this chapter has been focused on unanswered questions. Virtually, every aspect of RPAS, from interface design, interaction with ATC, and pilot decision-making, demands attention from the human factors profession. For example, what will be the RPAS equivalent of the *Basic T* flight instruments? Is decision-making affected by the lack of shared fate between the remote pilot and the aircraft? How can we make best use of the positive contribution that humans make to the performance of remotely operated systems?

Some of the emerging RPAS issues considered in this chapter will increasingly apply to conventional aircraft. Modern airline aircraft are already equipped with communication links that enable technical personnel on the ground to receive real-time performance data from engines and other systems. Recent airline crashes resulting from pilot incapacitation or malicious acts may accelerate the development of systems that will enable flight crew on the ground to take control of an airliner in an emergency. The act of transferring control from the cockpit to the ground would instantly transform a conventional aircraft into a passenger-carrying RPAS. Researchers have only just begun to examine the numerous human factors and security considerations of this concept (Comerford et al. 2013).

Throughout the twentieth century, developments in aviation human factors often occurred in response to accidents, an approach sometimes referred to as *tombstone safety*. In the years ahead, we must glean every available lesson from RPA accidents. However, we must also aim to identify RPAS human factors in a less costly manner. Incident investigations, simulations, and applied research will be essential to ensure that the integration of RPAS into civil airspace can occur safely and efficiently.

ACKNOWLEDGMENTS

Thanks are due to Jay Shively at NASA Ames Research Center, and Cynthia Null at NASA Langley for their support via the NASA *UAS in the NAS* project and the NASA Engineering and Safety Center. I would also like to express my appreciation to Vanui Barakezyan, Edward Barraza, Tamsyn Edwards, Jillian Keeler, Joel Lachter, Jeffrey McCandless, Randy Mumaw, and Yuri Trujillo for their helpful comments on drafts of this chapter. Lastly, thanks are due to the remote pilots who participated in focus groups and generously shared their knowledge and experience.

REFERENCES

Arrabito, G. R., Ho, G., Lambert, A., Rutley, M., Keillor, J., Chiu A., Au, H., and Hou, M. (2010). Human factors issues for controlling uninhabited aerial vehicles: Preliminary findings in support of the Canadian forces joint unmanned aerial vehicle surveillance target acquisition system project (Technical Report 2009–043). Toronto, Canada: Defence Research and Development Canada.

Arrabito, G. R., Ho, G., Li, Y., Giang, W., Burns, C. M., Hou, M., and Pace, P. (2013). Multimodal displays for enhancing performance in a supervisory monitoring task reaction time to detect critical events. *Proceedings of the Human Factors and Ergonomics Society Annual Meeting*, 57(1): 1164–1168. doi:10.1177/1541931213571259.

Association for Unmanned Vehicle Systems International. (2016). Commercial UAS exemptions by the numbers. Retrieved from http://www.auvsi.org/auvsiresources/exemptions.

Buoni, G. P. and Howell, K. M. (2008, September). Large unmanned aircraft system operations in the National Airspace System—the NASA 2007 Western States Fire Missions. *Paper presented at the 26th Congress of International Council of the Aeronautical Sciences (ICAS)*, Anchorage, AK.

Comerford, D., Brandt, S. L., Lachter, J., Wu, S., Mogford, R., Battiste, V., and Johnson, W. J. (2013). NASA's single-pilot operations technical interchange meeting: Proceedings and findings (NASA/CP—2013–216513). Moffett Field, CA: National Aeronautics and Space Administration, Ames Research Center. Retrieved from http://www.sti.nasa.gov.

Cooke, N. J., Pringle, H. L., Pedersen, H. K., and Connor, O. (Eds.). (2006). *Human Factors of Remotely Operated Vehicles*. San Diego, CA: Elsevier.

Cummings, M. L., Mastracchio, C., Thornburg, K. M., and Mkrtchyan, A. (2013). Boredom and distraction in multiple unmanned vehicle supervisory control. *Interacting with Computers*, 25(1): 34–47. doi:10.1093/iwc/iws011.

Federal Aviation Administration. (2012). National airspace system requirements document (NAS-RD-2012). Washington, DC: Author.

Federal Aviation Administration. (2013). Integration of civil unmanned aircraft systems (UAS) in the national airspace system (NAS) Roadmap. Washington, DC: Author.

Federal Aviation Administration. (2016). *Operation and Certification of Small Unmanned Aircraft Systems*. Washington, DC: Author.

Fern, L. C., Rorie, R. C., and Shively, R. J. (2014, October). NASA's UAS integration into the NAS: A report on the human systems integration phase 1 simulation activities. *Proceedings of the Human Factors and Ergonomics Society Annual Meeting*, 58(1): 49–53. doi:10.1177/1541931214581011.

Gawron, V. J. (1998). Human factors issues in the development, evaluation, and operation of uninhabited aerial vehicles. *AUVSI '98: Proceedings of the Association for Unmanned Vehicle Systems International*, Huntsville, AL, pp. 431–438.

Giang, W. and Burns, C. M. (2012). Sonification discriminability and perceived urgency. *Proceedings of the Human Factors and Ergonomics Society Annual Meeting*, 56: 1298–1302. doi:10.1177/1071181312561377.

Helmreich, R. L. (2000). On error management: Lessons from aviation. *British Medical Journal*, 320 (7237): 781–785.

Hobbs, A. (1991). Limitations of the see and avoid principle. Canberra, Australian Capital Territory: Bureau of Air Safety Investigation. Retrieved from www.atsb.gov.au/publications.

Hobbs, A. (2010). Unmanned aircraft systems. In E. Salas and D. Maurino (Eds.), *Human Factors in Aviation*, 2nd ed. San Diego, CA: Elsevier, pp. 505–531.

Hobbs, A., Cardoza, C., and Null, C. (2016, June). Human factors of remotely piloted aircraft systems: Lessons from incident reports. Paper presented at the *Conference of the Australian and New Zealand Societies of Air Safety Investigators*, Brisbane, Australia. Retrieved from http://asasi.org/.

Hobbs, A. and Lyall, B. (2016a). Human factors guidelines for unmanned aircraft systems. *Ergonomics in Design*, 24: 23–28.

Hobbs, A. and Lyall, B. (2016b). Human factors guidelines for remotely piloted aircraft system (RPAS) remote pilot stations (RPS). NASA Contractor report. Retrieved from http://human-factors.arc.nasa.gov/.

Hopcroft, R., Burchat, E., and Vince, J. (2006). Unmanned aerial vehicles for maritime patrol: Human factors issues (DSTO publication GD-0463). Fishermans Bend, Australia: Defence Science and Technology Organisation.

International Civil Aviation Organization. (2011). Unmanned aircraft systems (Circular 328, AN 190). Montreal, Canada: Author.

International Civil Aviation Organization. (2015). Manual on remotely piloted aircraft systems (RPAS) (Doc 10019 AN/507). Montreal, Canada: Author.

Kaliardos, B. and Lyall, B. (2015). Human factors of unmanned aircraft system integration in the national airspace system. In K.P. Valavanis and G.J. Vachtsevanos (Eds.), *Handbook of Unmanned Aerial Vehicles*. Dordrecht, the Netherlands: Springer, pp. 2135–2158.

Kalinowski, N. and Allen, J. (2010). Joint statement before the House of Representatives, Committee on Homeland Security, Subcommittee on Border, Maritime, and Global Counterterrorism, on the role of unmanned aerial systems on border security. Retrieved from http://testimony.ost.dot.gov/test/pasttest/10test/kalinowski2.htm.

Kitawaki, N. and Itoh, K. (1991). Pure delay effects on speech quality in telecommunications. *IEEE Journal on Selected Area in Communications*, 9(4): 586–593. doi:10.1109/49.81952.

McCarley, J. S. and Wickens, C. D. (2005). Human factors implications of UAVs in the national airspace (Technical Report AHFD-05–05/FAA-05–01). Atlantic City, NJ: Federal Aviation Administration.

Merlin, P. W. (2013). *Crash Course: Lessons Learned from Accidents Involving Remotely Piloted and Autonomous Aircraft*. Washington, DC: National Aeronautics and Space Administration.

Moore-Ede, M. (1993). *The 24 hour Society*. London, UK: Piatkus.

Neville, K., Blickensderfer, B., Archer, J., Kaste, K., and Luxion, S. P. (2012). A cognitive work analysis to identify human-machine interface design challenges unique to uninhabited aircraft systems. *Proceedings of the Human Factors and Ergonomics Society 56th Annual Meeting*, 56(1): 418–422. doi:10.1177/1071181312561094.

North Atlantic Treaty Organization. (2007). Standard interfaces of UAV control systems (UCS) for NATO UAV interoperability (NATO Standardization Agreement [STANAG] 4586, Edition 2). Retrieved from http://nso.nato.int/nso/.

North Atlantic Treaty Organization. (2009). Unmanned aerial vehicle systems airworthiness requirements (NATO Standardization Agreement [STANAG] 4671). Retrieved from http://nso.nato.int/nso/.

Orlady, H. W. and Orlady, L. M. (1999). *Human Factors in Multi-Crew Flight Operations*. Aldershot, UK: Ashgate.

Parke, B. K. and Kanki, B. G. (2008). Best practices in shift turnovers: Implications for reducing aviation maintenance turnover errors as revealed in ASRS reports. *The International Journal of Aviation Psychology*, 18(1): 72–85. doi:10.1080/10508410701749464.

Patterson, T., McClean, S., Morrow, P., and Parr, G. (2012). Modelling safe landing zone detection options to assist in safety critical UAV decision making. *Procedia Computer Science*, 10: 1146–1151. doi:10.1016/j.procs.2012.06.164.

Pedersen, H. K., Cooke, N. J., Pringle, H. L., and Connor, O. (2006). UAV human factors: Operator perspectives. In N. J. Cooke, H.L. Pringle, H.K. Pedersen, and O. Connor (Eds.), *Human Factors of Remotely Operated Vehicles*. Oxford, UK: Elsevier, pp. 21–33.

Pestana, M. (2012). Flying NASA unmanned aircraft: A pilot's perspective. Retrieved from http://ntrs.nasa.gov/.

Prather, C. (2013). Online report of army aircraft mishaps. Flightfax, 26. Retrieved from www.safety.army.mil.

Radio Technical Commission for Aeronautics. (2007). Guidance material and considerations for unmanned aircraft systems (Document DO-304). Washington, DC: Author.

Radio Technical Commission for Aeronautics. (2016). Minimum operational performance standards (MOPS) for unmanned aircraft systems (UAS) detect and avoid (DAA) systems. Washington, DC: Author.

Reason, J. and Hobbs, A. (2003). *Managing Maintenance Error*. Aldershot, UK: Ashgate Publishing.

Rorie, C. and Fern, L. (2015). The impact of integrated manoeuvre guidance information on UAS pilots performing the detect and avoid task. *Proceedings of the Human Factors and Ergonomics Society Annual Meeting*, 59(1): 55–59. doi:10.1177/1541931215591012.

Rorie, C., Fern, L., and Shively, R. J. (2016, January). The impact of suggestive maneuver guidance on UAS pilots performing the detect and avoid function. *AIAA Infotech @ Aerospace*, 1002. doi:10.2514/6.2016-1002.

SAE. (2011). Pilot training recommendations for unmanned aircraft systems (UAS) civil operations (SAE ARP 5707). Warrendale, PA: Author.

Smith, F. W. (2010). Delivering innovation. Interview at Wired Business Conference. Retrieved from http://library.fora.tv/.

Sollenberger, R. L., McAnulty, D. M., and Kerns, K. (2003). The effect of voice communications latency in high density, communications-intensive airspace (No. DOT/FAA/CT-TN03/04). Atlantic City, NJ: Federal Aviation Administration, Technical Center.

Thompson, W. T., Lopez, N., Hickey, P., DaLuz, C., Caldwell, J. L., and Tvaryanas, A. P. (2006). Effects of shift work and sustained operations: Operator performance in remotely piloted aircraft (OP-REPAIR) (Report No. HSW-PE-BR-TR-2006-0001). Brooks City, TX: United States Air Force.

Title 14 of the Code of Federal Regulations. General Operating and Flight Rules, 14 CFR § 91 (2016).

Tvaryanas, A. P. (2004). Visual scan patterns during simulated control of an uninhabited aerial vehicle. *Aviation, Space, and Environmental Medicine*, 75(6): 531–538. doi:10.1.1.485.6217.

Tvaryanas, A. P. (2006a). Human factors considerations in migration of Unmanned Aircraft System (UAS) operator control (Report No. HSW-PE-BR-TE-2006–0002). Brooks City, TX: United States Air Force, Performance Enhancement Research Division.

Tvaryanas, A. P. (2006b). Human systems integration in remotely piloted aircraft operations. *Aviation, Space, and Environmental Medicine*, 77(7): 1278–1282. doi:10.1.1.628.5468.

Tvaryanas, A. P. and Thompson, B. T. (2008). Recurrent error pathways in HFACS data: Analysis of 95 mishaps with remotely piloted aircraft. *Aviation Space and Environmental Medicine*, 79(5): 525–532.

Tvaryanas, A. P., Thompson, B. T., and Constable, S. H. (2006). Human factors in remotely piloted aircraft operations: HFACS analysis of 221 mishaps over 10 years. *Aviation Space and Environmental Medicine*, 77(7): 724–732.

Under Secretary of Defense. (2012). Unmanned aircraft systems ground control station human-machine interface: Development and standardization guide. Washington, DC: Author.

United States Air Force. (2015). Q-9 flight mishap history. Retrieved from http://www.afsec.af.mil/.

Vu, K. L., Chiappe, D., Strybel, T. Z., Fern, L., Rorie, C., Battiste, V., and Shively, R. J. (2015). Measured response for UAS integration into the national airspace system (NASA/TM—2015-218839). Moffett Field, CA: National Aeronautics and Space Administration, Ames Research Center. Retrieved from http://www.sti.nasa.gov.

Waraich, Q., Mazzuchi, T., Sarkani, S., and Rico, D. (2013). Minimizing human factors mishaps in unmanned aircraft systems. *Ergonomics in Design*, 21(1): 25–32. doi:10.1177/1064804612463215.

Whitlock, C. (2014, June 23). 'Stop saying 'uh-oh' while you're flying': Drone crash pilot quotes unveiled. *The Washington Post*. Retrieved from https://www.washingtonpost.com.

Wickens, C. D. (1986). The effects of control dynamics on performance. In K. R. Boff, L. Kaufman, and J. P. Thomas (Eds.), *Handbook of Perception and Human Performance*. New York: John Wiley & Sons, pp. 39.1–39.60.

Williams, K. W. (2004). A summary of unmanned aircraft accident/incident data: Human factors implications (Technical report DOT/FAA/AM-04/24). Washington, DC: Federal Aviation Administration.

Williams, K. W. (2007). An assessment of pilot control interfaces for unmanned aircraft (Report No. DOT/FAA/AM-07/8). Washington, DC: Federal Aviation Administration.

Zingale, C. M., McAnulty, D. M., and Kerns, K. (2003). The effect of voice communications latency in high density, communications-intensive airspace phase II: Flight deck perspective and comparison of analog and digital systems (DOT/FAA/CT-TN04/02). Atlantic City International Airport, NJ: Federal Aviation Administration, William J. Hughes Technical Center.

Author Index

A

Abbott, K. H., 278
Adams, C., 144
Akerstedt, T., 280–281
Akhtar, S., 271
Ale, B., 240
Alexander, A., 301
Allen, J., 380
Allen, J. P. Jr., 76
Altman, I., 22
Amalberti, R., 324, 326, 328
Andersen, H. B., 6
Andersen, H. T., 281
Andriessen, H., 224
Arrabito, G. R., 381, 389
Askew, G., 222
Assenmacher, T. J., 267
Atkins, E. M., 6
Atkins, S., 110
Avers, K. A., 76, 86
Axelsson, J., 280

B

Bacon, L. P., 310
Bainbridge, L., 19, 375
Balachandran, S., 6
Banbury, S., 19, 262, 264, 268, 271
Banks, S., 291
Barhydt, R., 144
Barnhart, W., 118
Barshi, I., 288
Basner, M., 221
Bass, E. J., 108
Batstone, R., 328
Battiste, V., 314
Batt, R., 248
Bazargan, M., 5
Behn, B. K., 249
Bellman, K. L., 367
Benneyan, J. C., 252
Berman, B. A., 286
Berson, B. L., 266
Billinghurst, S., 315–316
Billings, C., 108
Billings, C. E., 262–263
Birnbach, R., 137
Bjerke, W., 240
Blair, J. P., 227
Bleckley, K., 304
Bloom, R. W., 212
Bolfing, A., 221
Bolton, M. L., 108, 110

Bond, C. F. Jr., 226
Bonnet, M. H., 280, 282
Boucek, G. P., 265–268
Bourne, L. E. Jr., 314
Bove, T., 6
Boy, G. A., 364–365, 367, 369–376
Boy, G. B., 372
Boyle, D. A., 9
Bradley, P., 121
Braithwaite, B. M. J., 240
Braithwaite, J., 240
Brandt, B., 230
Brauer, R., 200
Broadbent, E., 280
Brooker, P., 248
Brown, H., 240
Bull, R., 226
Buoni, G. P., 391
Burnham, D. C., 6
Burnham, J. C., 322
Burns, C. M., 381
Buttrey, S. E., 254

C

Cabrall, C., 301
Caldwell, J. A., 280
Caldwell, J. L., 280
Cardosi, K. M., 193, 264–268, 270–271
Carolan, T., 314
Carskadon, M. A., 280–281
Cassell, C., 254
Cave, K. R., 222
Chabris, C., 102
Chandra, D. C., 269–271
Chaney, F. B., 214, 221
Chapanis, A., 322
Chaparro, A., 90, 93
Chase, S., 266, 270–271
Chiappe, D. L., 214, 303–308
Chidester, T. R., 286, 289
Cierbelj, A., 124
Clark, A., 305
Clarke, J.-P. B., 9
Clemes, M. D., 5
Coiera, E., 253
Collins, L. T., 251–252
Comerford, D., 392
Connell, L. J., 282
Cooke, N. J., 214, 388–389
Cook, R. I., 322, 328
Cooper, G. E., 118
Coppenbarger, R., 9
Coric, V., 227

Cox, L., 204
Crowley-Murphy, M., 250
Crutchfield, J., 304
Cummings, M. L., 366, 388
Currie, Z., 271

D

Danaher, J., 118
Dando, C. J., 226–227
Dattel, A., 304–305, 307
Davies, D., 213
Dean, D., 5
Degani, A., 143, 288
Dekker, S., 110, 125, 127
Dekker, S. W. A., 322, 327, 331, 333, 337
Demagalski, J., 279
Dement, W. C., 280–281
DeMers, B., 295
DePaulo, B. M., 226
de Ruiter, J., 225
Di Milia, L., 280
Dinges, D. F., 280–282, 291
Dinges, E., 9
Dismukes, R. K., 286–288
Dixon, A. P., 227
Dixon, S. R., 268
Doble, N., 110
Dodd, S., 275
Donnelly, N., 222
Donovan, C., 15
Doyle, B., 117
Drury, C. G., 97, 218
Durso, F. T., 301–302, 304–305, 307, 310
Dutt, V., 314
Dwyer, J. P., 310

E

Edkins, G. D., 240
Edwards, E., 61, 124
Elfstrom, J., 250
Elias, B., 211–213, 231
Ellis, G. A., 186, 277
Endsley, M. R., 303–307, 309–310
Erickson, J. B., 266
Eriksen, C. A., 280
Eriksson, S., 250
Erzberger, H., 310

F

Fadden, D. M., 15–18, 29
Farley, T., 310
Feng, C., 222
Feng, Q. M., 248
Fern, L. C., 385
Fiorino, F., 322
Fischhoff, B., 335
Fitts, P. M., 127–128, 322, 377
Fleck, M. S., 220
Folkard, S., 88
Foushee, H. C., 282, 284
Funk, J. L., 250

G

Gabree, S., 15, 266, 270–271
Gander, P. H., 281–282
Gao, J., 248
Garber, M., 123
Gardner, J. F., 25
Garner, K. T., 267
Gavin, M., 255
Gawron, V. J., 389
Georgiou, A., 251–254
Ghylin, K., 218
Giang, W., 381
Gibb, G., 210
Gilliland, K., 253
Girotto, S., 304
Gladwell, M., 212
Godwin, H. J., 222
Goglia, J., 195–197, 200, 203–204
Goh, Y. M., 240
Göknur, S., 108
Golson, J., 27
Gonzalez, C., 314
Goode, J. H., 280
Goodheart, B. J., 241
Gordon-Becker, S., 264, 267–268
Gosling, A. S., 253
Graeber, R. C., 281–282, 291
Grant, T., 227
Grayhem, R., 271
Green, D., 304
Green, D. M., 219
Greene, G. C., 6
Greenfield, D., 240
Green, S. M., 9
Griffin, M. A., 255
Griffin, T. G., 125
Groff, L., 90, 93
Grote, G., 367
Gulijk, C., 224

H

Hackworth, C. A., 304
Halbherr, T., 221
Hallock, J. N., 6
Hamilton, S., 371
Hancock, P. A., 277
Hansson, S. O., 240
Hapsari, R., 5
Hardmeier, D., 218
Harris, D., 279
Harris, D. H., 211, 214, 221
Hart, S. G., 186, 277
Harwood, G., 107
Hawkins, F. H., 61, 124, 261
Healy, A. F., 314
Hecker, P., 366
Heere, K., 310
Heiney, M., 303
Heinrich, H. W., 125
Helmreich, R. L., 67, 124, 241, 286, 381
Helton, W. S., 251
Hilburn, B., 213, 366

Hilts, D., 227
Hinds, P., 107
Hobbs, A., 381, 386, 388–390
Hofer, F., 218
Hoffman, J. L., 314
Holland, D., 280
Hollands, J. G., 19, 262, 264, 268, 271
Hollnagel, E., 333–334, 337–338
Homola, J., 301, 310–311
Hoogervorst, J., 249
Hopcroft, R., 388
Horne, J. A., 280
Horowitz, T. S., 219–220
Hotchkiss, J., 5
Howell, K. M., 391
Howell, W. C., 214
Hubbard, D., 204–205
Huey, M., 260
Humphreys, M. S., 253
Hunter, D. R., 248
Huntley, M. S., 264–267
Hursh, S. R., 282
Hutchins, E., 107
Hutchins, S., 314

I

Igou, F. P., 227
Ilgen, D. R., 214
Ishikawa, K., 205
Ismail, N. A., 249
Itoh, K., 385

J

Janczura, J., 251
Jeddi, B., 6
Jenatabadi, H. S., 249
Jenkins, B. M., 215
Jenkins, D. P., 304
Johannesen, L. J., 322, 328
Johnson, R., 271
Johnson, W. B., 53, 57, 70, 74, 76, 86, 92
Jones, D. G., 303–304, 308
Jones, R. E., 322, 328
Jones, S., 240
Jo, Y. J., 15

K

Kaber, D. B., 307
Kahneman, D., 335
Kaliardos, B., 381
Kalinowski, N., 380
Kamgarpour, M., 9
Kanki, B. G., 286, 386
Kaptelinin, V., 374
Karson, M. J., 248
Kayten, P., 118, 120
Kazi, S., 310
Kecklund, G., 280
Kennedy, Q., 254
Kenner, N. M., 219–220
Kester, L., 314

Kiboss, J. K., 313
Kiesler, S., 107
Kiken, A., 313–315
Kirwan, B., 254
Kitawaki, N., 385
Klausen, T., 107
Klein, G. A., 309
Klein, H. J., 214
Klinect, J., 124
Komons, N. A., 17
Koomen, G., 225
Kopardekar, P., 302
Kraut, J. M., 312
Kribbs, N. B., 280
Krumpal, I., 253
Kruschke, J. K., 248
Kuo, C.-C., 5

L

Lacey, R. J., 25
Landauer, C., 367
Landry, S. J., 9, 310
Laney, D. B., 252
Larcos, G., 251–252
Lauber, J. K., 281–282, 284
Laudeman, I. V., 366
Le Coze, J. C., 324
Lee, D. L., 108
Lee, E. A., 376
Lee, J. D., 222, 264, 267–268, 302–303
Leffler, M. F., 266
Lenne, M. G., 240
Leveson, N. G., 125, 328
Levine, T. R., 227
Levinson, D., 251
Liedgren, C., 250
Li, H., 302
Lindberg, A. K., 240
Lindberg, T., 15
Lind, M., 325
Liu, C. C., 240
Liu, Y. D., 264, 267–268
Lofaro, R., 210
Loft, S., 255
Longridge, T., 137
Loukopoulos, L. D., 286, 288
Love, P. E., 240
Lowden, A., 280
Lyall, B., 381, 388–390

M

Mackie, R., 213
Mackworth, J., 219
Macy, A., 118
Maddox, M. E., 70, 74
Maddox, W. T., 219
Madhavan, P., 303
Madsen, P. M., 240
Maguire, M., 303
Ma, J., 241, 249
Majumdar, A., 254
Malamud, B. D., 251

Mallis, M. M., 291
Mangie, J., 282
Manning, C. A., 302, 304
Manzey, D., 302
Marais, K. B., 248
Marshall, A. Y. M. S., 279
Martin, L. H., 301
Martinussen, M., 248
Masalonis, A. J., 366
Maurino, D. E., 124, 129
Merritt, A. C., 124
McAnulty, D. M., 266–268
McCarley, J. S., 302, 381
McCormick, E. J., 127
McFarland, R., 117, 127
Mead, R., 9
Mendes, M., 225–226
Menneer, T., 222
Mercer, J., 301, 310
Merlin, P. W., 387
Merritt, A. C., 124
Metzger, U., 302
Michel, S., 225–226
Miller, C., 117–118
Miller, J. C., 280–281
Minsky, M., 367
Mitman, R., 267
Mitroff, S. R., 220, 232
Mollicone, D. J., 281
Molloy, R., 213
Montgomery, D., 252
Mooij, M., 255
Moore-Ede, M., 388
Moray, N. P., 125, 260, 302, 322, 324
Morgan, C. A., 227
Morrison, D., 248
Morton, P. M., 15
Mouloua, M., 213
Mulrine, H., 282
Mumaw, R. J., 19, 264
Murphy, E. D., 266–268, 274, 301, 306
Myhre, G., 281
Myklebust, T., 227

N

Nagel, D. C., 279, 326
Nagle, G., 9
Naitoh, P., 280–281
Nardi, B., 374
Narkevicius, J., 367, 374
Ndirangu, M., 313
Neal, A., 255–256
Nelson, E., 269
Nemeth, C. P., 337
Neumeier, M., 267
Neville, K., 384, 389
Nielsen, J., 274
Nolan, M. S., 302
Norman, D., 19
Norris, A., 271
Nugus, P. I., 240
Null, C., 392
Nye, L. G., 304
Nygren, T. E., 277

O

O'Brien, K. S., 301, 305
Ochieng, W. Y., 254
O'Connor, A. M., 250
O'Connor, P., 254
O'Dea, A., 254
O'Hare, D., 248, 301, 305
Ohrn, A., 250
O'Leary, M., 110
Orlady, L. M., 286, 382
Ormerod, T. C., 226–227
Oster, C. V., 254
Oxburgh, G. E., 227
Ozay, N., 6

P

Paas, F., 314
Palmer, M. T., 266
Parasuraman, R., 19, 213, 262–264, 268, 271, 277, 302–303
Parke, B. K., 386
Passantino, G. M., 211
Patankar, M., 91–92
Patterson, R., 267
Patterson, T., 391
Pedersen, H. K., 388
Peigneau, P., 281
Perrow, C., 125
Person, D. R., 314
Peryer, G., 267
Pestana, M., 388–389
Petrie, K. J., 280
Petrilli, R. M., 241
Pew, R. W., 304
Pidgeon, N., 110
Pierce, R. S., 308
Pistole, J. S., 231
Plsek, P. E., 252
Poole, R. W., 211, 217
Pop, V. L., 310
Post, D. L., 266
Powell, D. M. C., 280
Prather, C., 380
Predmore, S., 121
Prevot, T., 301, 303, 310–311, 314

Q

Quon, L., 302

R

Rabinowitz, R. G., 227
Rabitoy, E. R., 314
Rankin, W. L., 76–77
Rasmussen, J., 62, 124, 213, 278, 325–326, 330
Rawson, K., 304
Reason, J. T., 65, 89, 124–129, 213, 261, 278–279, 328–330, 335, 388
Reddy, R., 250
Reddy, S. B., 250
Rehmann, A., 267

Reichenbach, J., 302
Reid, G. B., 277
Rensink, R., 306
Revelle, W., 253
Reyner, L. A., 280
Reynolds, M., 267
Riley, R. A., 249
Riley, V. A., 263, 268, 277, 302
Robichaud, M. R., 248
Robinson III, J. E., 9
Rodgers, M. D., 304
Rollenhagen, C., 240
Rorie, C., 385–386
Roscoe, A. H., 186, 277
Rosekind, M. R., 291
Roske-Hofstrand, R. J., 301, 306
Rossow, V. J., 6
Roth, W. M., 254
Rundle, J. B., 251
Rutberg, H., 250

S

Sachs, M. K., 251
Salas, E., 107, 214
Salmon, P. M., 240, 279, 304, 307
Salvendy, G., 303
Salvi, S. M., 271
Samel, A., 280, 291
Sanders, M., 127
Sanderson, P., 255
Sandham, A. L., 226
Sarter, N. B., 19, 264, 322, 328, 375
Schleede, R., 118
Schwaber, K., 371
Schwaninger, A., 218, 221, 225–226
See, K. A., 302
Senders, J. W., 125, 322, 340
Seppelt, B. D., 303
Sethumadhavan, A., 307
Shappell, S. A., 65, 125, 248, 259, 322, 326, 329, 331
Shaw, A. S., 227
Sheridan, T. B., 302
Shojania, K., 251
Shorrock, S. T., 254
Shortle, J., 6
Signal, L., 282
Simaiakis, I., 110
Simon, L., 253
Simons, D., 102, 306
Simons, G. M., 248–249
Simoudis, E., 249
Skaggs, V., 271
Smith, C., 110, 281
Smith, F. W., 107–108, 110, 281
Smith, M. O., 241
Smith, P. J., 107–108, 110, 281, 380
Smith, W. D., 265, 266
Smolensky, M., 301, 303–305, 310
Sollenberger, R. L., 305, 385
Sornette, D., 251
Spencer, M. B., 280
Spickett, J., 240
Staigle, T., 9

Stamp, J., 26
Stanhope, N., 250
Stanton, N. A., 277, 279, 304–308
Staveland, L. E., 186, 277
Stearman, E. J., 310
Steffen, J. H., 5
Stein, E., 305
Stevens, C., 249, 252
Stoklosa, J., 120
Stolzer, A., 195–197, 204
Strathie, A., 241
Strauch, B., 125
Strong, J. S., 254
Strybel, T. Z., 304, 306, 308, 310–312, 314
Sturre, L., 306, 308
Sull, D., 240
Sumwalt, R. L., 144
Sumwalt, R. L. III., 287
Sutcliffe, K. M., 337
Sutherland, J., 371
Swain, A. D., 279
Sweet, D., 9
Swets, J. A., 219
Swezey, R. W., 214
Symon, G., 254

T

Taleb, N. N., 251
Tan, W., 375
Tariq, A., 253–254
Tarnowski, E., 369
Tasali, E., 281
Taylor-Adams, S. E., 250
Taylor, P. J., 227
Taylor, R., 307
Taylor, R. W., 15
Tcholakian, T., 227
Thomas, M. J., 241
Thomas, R. J., 287
Thompson, B. T., 388–389
Thompson, W. T., 388
Tong, K.-O., 9
Travaglia, J. F., 240
Tretheway, M. W., 240
Trotter, M., 240
Truitt, T., 304
Tsonis, C. G., 366
Turcotte, D. L., 251
Tvaryanas, A. P., 380–382, 386, 389

V

Van Cauter, E., 281
van den Berg, M., 282
Van Dongen, H. P. A., 282
van Gog, T., 314
van Merriënboer, J. J. G., 314
Van Wert, M., 219
Vaughan, D., 126
Veitengruber, J. E., 265–266
Vejvoda, M., 280
Verwey, W. B., 277
Vincent, C., 250

Vivona, R. A., 9
Vu, K-P. L., 303–308, 312–316, 385

W

Walczyk, J. J., 227
Walhout, G., 119–120
Walker, G. H., 241, 304
Wang, H., 248
Waraich, Q., 388
Warm, J. S., 9
Warren, A. W., 9
Warwick, G., 26
Watson, J., 92
Wegmann, H. M., 280, 291
Weick, K. E., 337
Weigmann, D. A., 65
Weiner, E. L., 279
Wekesa, E. W., 313
Weller, C. E., 227
Weron, R., 251
Westbrook, J. I., 250–254
White, M. D., 284
White, W., 9
Whitlock, C., 231, 381
Wickens, C. D., 19, 262, 264–265, 267–268, 271, 277–279, 302, 304, 309, 314, 381–382
Wickens, D. A., 268, 302, 309, 381
Widdel, H., 266
Wiegmann, D. A., 65, 125, 248, 303, 322, 326, 331

Wiener, E. L., 213, 277, 286, 288, 371, 375
Wiggins, M. W., 248–249, 252
Wilhelm, J. A., 124
Wilke, S., 254
Willems, B., 303
Williams, K. W., 381, 389
Wilson, G. A., 116
Winograd, T., 374
Withey, P. A., 249
Wolfe, J. M., 219–220
Wolf, M., 376
Woods, D. D., 322, 328, 333, 337–338

Y

Yasuoka, K., 26
Yasuoka, M., 26
Ybarra, S., 217
Yeh, M., 15, 29, 141, 269–270
Yoder, M. R., 251
Yoo, K., 222–223
Young, M. D., 314
Young, M. S., 277

Z

Zingale, C. M., 385
Zorn, C. K., 254
Zuschlag, M., 271

Subject Index

Note: Page numbers followed by f and t refer to figures and tables respectively.

5-M (mission, man, machine, management, and media) model, 201, 202f
14CFR. *See* Title 14 Code of Federal Regulations (14CFR)

A

AAAE (American Association of Airport Executives), 229
AARs (airport acceptance rates), 44
AC. *See* Aircraft (AC)
ACARS (aircraft communications addressing and reporting system), 47
ACAS Xu (airborne collision avoidance system), 386
Accident investigation, 115–129
    challenges, 128–129
    history, 116–117, 116f
    human factors, grappling, 117–118
    notable successes, 128
    NTSB, human performance capability, 118–123, 119f
    safety management and data-drive safety efforts, 126–127
    sophistication and guidance, 124–126
    theoretical perspectives, 127–128
ACO (Aircraft Certification Office), 149
Active failures, 329
ADS (automatic dependent surveillance), 37, 49, 350
Advisory circular (AC), 140–141, 143, 148, 192–193
Aeronautical navigation aids, 35–36, 36t
AeroVironment Helios prototype, 380f
Airborne collision avoidance system (ACAS Xu), 386
Airborne operations, ATS, 7–10
    arrival, 9–10
    climbout, 8
    cruise, 8–9
    descent, 9
    landing and go-around, 10
    overview, 3, 4f
Airborne systems, 43
Airbus flight decks, 26
Air Commerce Act, 137, 139
Aircraft (AC), 301
    and AC systems, certification, 147–193
        AC, 192–193
        basis, 149–150
        compliance, methods of, 188–190
        considerations, 190
        designated engineering/airworthiness representatives, 192
        documentation, 191
        FAA reports, 193
        human factors certification plan, 187–188
        intended human operator, 186
        operational considerations, 191
        orders, 192
        policy statements, 193
        PSCP, 151, 152t–185t
        regulations, guidance, and policy, 148–149
        regulators, 151
        requirements, 188
        schedule, 191–192
        standards, 149
        system safety assessments, 190–191
    metering, 44
    operations, 32
    separation, 37, 39
    UPS, 311
Aircraft Certification Office (ACO), 149
Aircraft communications addressing and reporting system (ACARS), 47
Aircraft maintenance manuals (AMMs), 90
Aircraft Maintenance Technicians (AMTs), 91
Air-ground integration, HSI, 373
Air Line Pilots Association Perspective, 228
Airport acceptance rates (AARs), 44
Airports, 346–348
    ATC, 347f
    capacity, 347, 347f
        restrictions, 44
    scanner game tool, 232
    surface management, 110–112, 111f
Air Route Traffic Control Center (ARTCC), 110
Airspace, 348–349, 348f, 349f
    capacity, 348, 357
    classes, 32–33, 33t
Air traffic complexity, 366–367
Air traffic control (ATC), 31–50, 109, 301, 364, 384
    AC
        operations, 32
        separation, 37, 39
    aeronautical navigation aids, 35–36, 36t
    airborne systems, 43
    airspace classes, 32–33, 33t
    assignments, 34
    automation, 42–43
    conflict alert/VFR intruder, 43
    controller
        communication, 358t
        coordination, 47–48
        responsibilities, 49
    electronic data communications, 47
    flight information automation, 49
    flight progress strips, 48, 48f
    future enhancements, 49–50
    GNSS, 37
    implications, 384
    and navigation system improvements, 50t
    nonradar separation, 39
    overloads, 44–45
    pilot/controller communications-radio systems, 45–46
    providers, 33–34

Air traffic control (ATC) (*Continued*)
  radar, 37
    AC, identification, 41
    separation, 40
    services, 41
    surveillance, 37, 38t
    system limitations, 40
  SA theory and measurement, 304–308
    SAGAT, 307
    SPAM, 307–308
  services, 34
    airspace, type, 35
  shifted to ATM, 364f
  traffic management systems, 43–44
    airport capacity restrictions, 44
  voice communication procedures, 46
Air traffic controllers (ATCos), 301, 364
  activities, SA in, 303–304
  communicates, 355–356
  en route, 302, 304
  primary, 312
  sectors 90 and 91, 311
  terminal radar approach control, 302
  types, 302
  workload and SA, 312
Air Traffic Control Systems Command Center
    (ATCSCC), 107
Air Traffic Management (ATM), 363
  ATC to, 364f
  complexity, 365
  loop, 369f
  NextGen and changes, 302–303
Air transportation error, 321–338
  'human error,' 321–338
    background, 323–333
    label, 334–338
    problem, 321–323
Air transportation systems (ATS), 3, 341–343
  airborne operations, 7–10
    arrival, 9–10
    climbout, 8
    cruise, 8–9
    descent, 9
    landing and go-around, 10
    overview, 4f
  concepts-of-operations, 345–346
    FM, 346
    safe separation, 345, 345f
  corporatization, 359
  definition, 343–344
  European and U.S., 341–342
    complexity, 342t
    scale, 342t
  functional description, 351–358, 352f
    FFP, 354–355, 355f, 356f
    FM, 352–354
    sector control, 355–358
  ground operations, 4–7
    arrival, 7
    gate, 5
    overview, 4f
    pushback, 5
    takeoff, 6

    taking runway, 6
    taxi out, 6
  infrastructure, 346–351
    airports, 346–348
    airspace, 348–349, 348f, 349f
    communications system, 350
    navigation system, 350–351
    surveillance system, 349–350
  modernization, 358–359
  needs, 344
  objectives, 345
  scope, 343
Alert information, flight deck, 265–266
Aloha Airlines Flight 243, 57, 57f
American Association of Airport Executives (AAAE), 229
AMMs (aircraft maintenance manuals), 90
AMTs (Aircraft Maintenance Technicians), 91
Anonymous reporting, 82
Appropriate Volumetric Zone (AVZ), 366
ARTCC (Air Route Traffic Control Center), 110
Artifact-User-Task (AUT) triangle, 369, 369f
ASAP (Aviation Safety Action Program), 71, 197, 202
ATA (U.S. Air Transport Association), 58, 61
ATC. *See* Air traffic control (ATC)
ATCos. *See* Air traffic controllers (ATCos)
ATC radar beacon system (ATCRBS), 42
ATCSCC (Air Traffic Control Systems Command Center),
    107
Atlanta arrival and departure routes, 7f
ATM. *See* Air Traffic Management (ATM)
ATS. *See* Air transportation systems (ATS)
Australian Bureau of Air Safety Investigation, 125
AUT (Artifact-User-Task) triangle, 369, 369f
Automated hand-offs, 48
Automatic dependence surveillance—broadcast (ADS–B)
    system, 350
Automatic dependent surveillance (ADS), 37, 49, 350
Automation, 141, 262–264, 302, 364
  clumsy, 371
  complexity/level, 190
  commercial aircraft, evolution of, 367–369
  in flight deck, 19–20
  flight information, 49
  human control *vs.*, 371–372
  management and organizational, 372–373
  to tangibility issues, 375–376
  trends in, 42–43
Autopilot system, 19, 285, 288
Autoresolver primary, 312
AUTOS (Artifacts, Users, Tasks, Organizations, and
    Situations) pyramid, 369–371, 370f
  in practice, 371
  tetrahedron, 370f
Aviation and Transportation Security Act of 2001, 211, 217
Aviation Safety Action Program (ASAP), 71, 197, 202
Aviation Safety Research Act, 57
Aviation security
  assessments, 215–218
    GAO, 216–217
    Heritage Foundation, 215
    international, 217–218
    RAND Corporation, 215–216
    Reason Foundation, 217

history, 210–211
human operators, roles, 212–213
influencing operator performance, 213–214
procedures and technologies, 231–232
    airport scanner game, 232
    biometric identification, 232
    bottled liquids scanners, 232
    checkpoint, 231
    credential authentication technology, 231
    ETD, 232
    paperless boarding pass, 232
threats, 211–212, 212f
Avionics systems, 265
AVZ (Appropriate Volumetric Zone), 366

B

Behavior detection officers (BDOs), 213, 224
Biometric identification, 232
Black Swans (book), 251
Boeing Company, 54, 81, 92
    767 flight model, 18f
    777 flight model, 21f, 28
    787 flight model, 24f
    assembly plant in Seattle, 250
    automation philosophy, 262
    flight decks, 26
    HF, 54
Bomb appraisal officers layer, 224
Bottled liquids scanners, 232
Bureau's Aeronautics Branch, 116
Byrne, Evan, 125–126

C

CAB (Civil Aeronautics Board), 116–117
CAD (computer aided design) format, 23
Canine teams TSA layer, 223
Cargo screening, computer-based training, 226
CASA (Civil Aviation Safety Authority), 60
CAST (Commercial Aviation Safety Team), 126–127, 206
Cause-and-effect diagrams, 206
CDM (collaborative decision making), 354
CD&R (conflict detection and resolution) system, 303, 314
CDTI (Cockpit Display of Traffic Information), 270
Certification basis, AC systems, 149–150
CFA (cognitive function analysis), 373
CFR (code of federal regulations), 14
CFs (contributing factors), 79–80, 79f
Chapanis, Alphonse, 324
Checked luggage screening layer, 224
Checkpoint transportation security officers layer, 224
Check sheets, 206
Chief Scientific and Technical Advisors (CSTAs), 58, 74
Circadian rhythms, 281
Civil Aeronautics Board (CAB), 116–117
Civil Aviation Safety Authority (CASA), 60
Clearance delivery, 5
Closed-loop system, 356, 356f
Clumsy automation, 371
Cockpit Display of Traffic Information (CDTI), 270
Cockpit resource management (CRM), 118–119
Code of federal regulations (CFR), 14

Cognitive function analysis (CFA), 373
Cognitive models, 326
Cognitive prostheses, 371
Collaborative decision making (CDM), 354
Collision risk, 344
Color, display technology, 271–272
Combatting Terrorism Center (CTC), 230
Comet crashes, 248–249
Commercial aircraft automation, evolution, 367–369
Commercial Aviation Safety Team (CAST), 126–127, 206
Communication/navigation/surveillance, 356
Communications system, 350
Comprehension, SA level, 304
Computer aided design (CAD) format, 23
Confidential *vs.* anonymous reporting, 82
Conflict alert/VFR intruder, 43
Conflict detection and resolution (CD&R) system, 303, 314
Conflict resolution advisories, 43
Confounding variables, 206
Contributing factors (CFs), 79–80, 79f
Control(s)
    charts, 206, 252, 252f
    and displays, RPAS, 388–390, 390f
    station environment, 387–388, 387f
    transfer, 386–387, 387f
Controlled airspace, 32, 385
Coordination *vs.* collaboration, 109, 109f
Correctable target-detection decisions, 220
Counterfactual reasoning, 335
Cox, Tony, 204
CPS (cyber-physical system), 368, 373
Credential authentication technology, 231
Crew resource management (CRM), 19, 54, 73, 142, 337
Crew vetting layer, 223
CSTAs (Chief Scientific and Technical Advisors), 58, 74
CTC (Combatting Terrorism Center), 230
Current-generation control station, 388
Customs and Border Protection layer, 223
Cyber-physical system (CPS), 368, 373

D

Data
    entry errors, 382
    patterns, 252
Data Comm, 314
    failure, 315
    system, 314
Data-driven processing, 309
Data envelopment analysis (DEA), 207
    potential improvements, 207, 208f
    scores, distribution, 207, 207f
Deception, detection of, 226–227
Decision making units (DMUs), 207
DeHavilland Comet, 116–117
Department of Homeland Security (DHS), 210, 228–229
Department of Transportation (DOT), 117
Departure Metering Program, 111
Departure Reservoir Coordinator (DRC), 111
Descartes, Rene, 327–328
Design philosophy, user interface design, 260
Desynchrony, flight deck HF, 281
Detailed inspection (DET), 97–98

Detect and avoid (DAA) system, 385–386
DHS (Department of Homeland Security), 210, 228–229
Digital messaging, 350
Digital System Replacement (DSR), 304
Dirty Dozen concept, 64
Distance measuring equipment (DME), 120, 351
Distinctiveness, flight deck HF, 269–270
Distributed work system, human-centered design, 108
DMUs (decision making units), 207
DOT (Department of Transportation), 117
Dragon Kings (journal), 251
DRC (Departure Reservoir Coordinator), 111
The Dutch Safety Board, 333, 336

E

EASA (European Aviation Safety Agency), 55, 60, 142
EDCT (expected departure clearance time), 353
EICAS (engine indicator and crew alerting system), 15, 28
Electronic checklist, 28
Electronic data communications, 47
Embedded systems (ESs), 367
Engineering Psychology and Human Performance, 19
Engine indicator and crew alerting system (EICAS), 15, 28
Enroute airspace, 348, 348f
En route ATCos, 302, 304
Enroute sector loading, 44–45
Environmental context, flight deck HF, 261, 291–293
    accident prompting rulemaking change, 292
    fatigue risk-management system, 292–293
    flight, duty, and rest regulations, 291
    ICAO and fatigue management, 293
Environment factors, visual inspections, 103
Equipment factors, visual inspections, 103
Ergonomics, 53
Error Management, 69
Error of commission, 65
Error of omission, 65
ESs (embedded systems), 367
ETD (explosives trace detection), 232
European Aviation Safety Agency (EASA), 55, 60, 142
European Joint Aviation Authority (JAA), 60
European Union, 358
Exceptional violations, 330
Expected departure clearance time (EDCT), 353
Explosives trace detection (ETD), 232
Extraneous variables, 252

F

FAA. *See* Federal Aviation Administration (FAA)
Facility flow planning (FFP), 354–355, 355f
    loop, 352
    TMI, time line used, 356f
Failure-to-follow procedures (FFP), 90–96
    dealing, systems view, 95–96
    documentation, 90–91
    research, 91–95, 92t, 95t
FANS (future air navigation system), 49
FAR (Federal Aviation Regulations), 147
Fatigue
    HF, 280
        aviation relevance, 282
        circadian rhythms, 281

        defined, 280
        desynchrony, 281
        sleep, 281
        ULR flight/safety implications, 282–283
    management, 76
Fatigue Risk-Management System (FRMS) approach,
    86–89
Federal air marshal service layer, 224
Federal Aviation Administration (FAA), 44, 60, 196, 205,
    211, 264, 380, 385, 391
    maintenance, HF, 73f
    Modernization and Reform Act of 2012, 227
    reports, 193
    safety regulation and enforcement, 116
    team, HF, 19
Federal Aviation Regulations (FAR), 147
Federal flight deck officers layer, 224
FGS (Flight Guidance Systems), 141
Fitts, Paul, 324–325
FL (flight levels), 9
Flexible culture, safety, 89
Flightcrew HF, 152t–185t
Flight
    crew
        required competencies, 391–392
        training and operational requirements, 142–144
    information automation, 49
    operations, 344, 344f
    planning, 382–383
    progress strips, 48, 48f
    strips, 353, 353f
Flight deck, 261
    design, 13–29
        automation, 19–20
        aviation, 20
        evaluation, strategies, 22–25, 23f
        evolution, 18–19, 18f
        fidelity mockups, part task trials, 23–24
        future, 28
        history, 15–18, 16f
        integrated (flight simulator) setting using detailed
            test plan, 24–25, 24f
        paper-based evaluations, 23
        process, 21–29, 21f, 28f
    HF, 259–293
        environmental context, 291–293
        lapse, 278
        organizational/management infrastructure,
            283–291
        overview, 259–261
        personnel subsystem, 276–283
        slip error, 278
        technical engineering system, 261–276
    regulatory requirements and guidance, 140–141
Flight Guidance Systems (FGS), 141
Flight levels (FL), 9
Flight Operations Quality Assurance (FOQA), 195
Flowcharts, 205
Flow management (FM), 345
    concepts-of-operations, ATS, 346, 346f
    functional description, ATS, 352–354
        collaboration and coordination, 354, 354f
        flight strips, 353, 353f
    loop, 352

Fly-by waypoint symbol, 269, 269f
Fly-by-wireless control link, 381
Fly-by-wire systems, 18
Fly-crash-fix-fly approach, 196
Fly-over waypoint symbol, 269, 269f
FM. *See* Flow management (FM)
FOQA (Flight Operations Quality Assurance), 195
Free flight concept, 49–50
FRMS (Fatigue Risk-Management System) approach, 86–89
Full-body scanner, 227
Future air navigation system (FANS), 49

G

General Accountability Office (GAO), 210
    assessment, 216–217
General visual inspection (GVI), 97
Glass cockpit concept, 17, 18f
Global Aviation Information Network, 89
Global navigation satellite system (GNSS), 35, 37, 49
Global SA, 309
Goal-driven processing, 309
Go–no go decision, 6
Google Glass, 22
Government-Industry conference, 58
Great circle distance path, 344
Ground-based link, 381
Ground Delay Programs, 109
Ground operations, ATS, 4–7
    arrival, 7
    gate, 5
    overview, 4f
    pushback, 5
    takeoff, 6
    taking the runway, 6
    taxi out, 6
Guidance loop, 368, 368f
GVI (General visual inspection), 97

H

Hardened cockpit door layer, 224
Hazard identification, 197
Hazard Reporting System, 82
Hazard/risk process decision program chart, 197, 198f
HCD. *See* Human-Centered Design (HCD)
Heinrich's domino model, 322
Heritage Foundation assessment, 215
HFACS. *See* Human Factors Analysis and Classification System (HFACS)
HFE (human factors and ergonomics), 374
High-altitude long-endurance (HALE) Global Hawk, 380f
High-frequency (HF) radios, 46
Hindsight bias, 335
Histograms, 206
HITLS (human-in-the-loop simulations), 367, 375
Hollnagel, Erik, 334
HSI (human-systems integration), 367, 374
Human-Centered Design (HCD), 364, 374
    AUTOS pyramid, 369–371
        AUT, 369, 369f
        considering organizations, 370

large variety of situations, testing, 370–371
        in practice, 371
    evolution, 374f
Human–computer interaction, 374
Human control *vs.* automation, 371–372
'Human error,' air transportation, 321–338
    argument, 322
    background, 323–333
        Chapanis, 324
        epistemology, 327–328
        Fitts and Jones, 324–325
        James reason and Swiss Cheese model, 328–331
        latent failure model, operationalization, 331–333
    beyond, 337–338
    complacency, 337
    effect/symptom, 326–327
    euphemisms, 337
    existence, 334
    inadequate supervision, 332
    label, 334–338
        byproduct of causality, 336
        causality and accident investigation, 335–336
        hindsight bias, 335
    naturalistic view, 325f
    noncompliance, 337
    outcome bias, 335
    problem, 321–323
    taxonomy, 325
Human error, flight deck HF, 278–279
Human factors (HF), 52, 137, 140, 223–225
    aviation maintenance and inspection, guide, 75t
    branch, 117–118
    in cargo screening, 225
    certification plan, 187–188
        outline, 149, 150t
        system description, 187–188
    cognitive, 54
    data sources and research tools, 239–256
        central tendency, 242–245
        data acquisition and analysis, 246–248
        data cautiously, using, 252–253
        data collection and management in aviation, 239–242
        different organizational contexts, 250–252
        distributions of data, 245–246
        importance of outliers, 254–256
        organizational trend analyses, 248–249
        value of qualitative data, 253–254
    defense in airport security, 223–225
        20 layers, 223–224
    definition and disciplines, 53–54
    flight deck. *See* Flight deck, HF
    guide and operator's manual, 74–76
        documentation, 75–76
        event investigation, 75
        fatigue management, 76
        shift/task turnover, 76
        sustainability and cost justification, 76
        training, 76
    historical perspective, 53
    issues, 90–96
    maintenance, 55

Human factors (HF) (*Continued*)
    models, 61–72
        Dirty Dozen, 64
        HFACS, 53, 65–67
        PEAR, 61, 70–72
        SHELL, 61–62, 61f
        Skills–Rules–Knowledge-Based Performance,
            62–64
        Swiss Cheese model, 64–65
        TEM, 61, 67–69, 69f
    organizational, 54
    physical, 54
    professionals
        educational disciplines, types, 55
        types, 54–55
    regulatory influence, 60–61
    research and development initiatives and
            regulations, 57
    safety statistics and maintenance, 55–56, 56t
    UAS, impact, 104
    U.S. Legislation drives, 57–59, 58t–59t
    visual inspections, 97–103
        factors affecting, 100–103
        inspection process, 98–100, 98f, 99f, 100f
        types, 97–98
Human Factors Analysis and Classification System
        (HFACS), 65–67, 331–333, 332f
Human factors and ergonomics (HFE), 374
Human-in-the-loop simulations (HITLS),
        367, 375
Human–machine system, 371, 372f
Human-systems integration (HSI), 367, 374

I

IATA (International Air Transport Association), 56,
        56t, 231
ICAO. *See* International Civil Aviation
        Organization (ICAO)
IFR (instrument flight rules), 5, 32
ILS (instrument landing system), 10, 37, 351
Image-based factors, 221
IMC (instrument meteorological conditions), 32
Independent runway, airport, 347, 347f
In-flight evaluations, 189
Information
    filtering, ATCos, 310
    processing theory, 326
Informed culture, 89
Insider threats, 230
Inspector factors, 101–103, 102f
Instrument flight rules (IFR), 5, 32
Instrument landing system (ILS), 10, 37, 351
Instrument meteorological conditions (IMC), 32
Intelligence layer, 223
Internal hazard reporting system, 81–82
International Air Transport Association (IATA),
        56, 56t, 231
International Civil Aviation Organization (ICAO), 32, 115,
        124, 196, 217, 222, 326
    ATC assignments, 34
    risk matrix, 198, 199f
    training and application of HF, 60
International Ergonomics Association, 53

Interpretability, flight deck HF, 270
Intra-airline data-communications system, 47
The ironies of automation (article), 19, 375

J

JAA (European Joint Aviation Authority), 60
Jens Rasmussen's SRK model, 63f
Job task analysis (JTA), 71
Joint Terrorism Task Force layer, 223
Jones, Richard, 324–325
JTA (job task analysis), 71
Just culture, 89, 202

K

Key performance indicators, 195
Knowledge-based and image-based factors, 225
Knowledge-based performance, 63

L

Lapse, error, 278
Latent failure model, operationalization, 331–333
Law enforcement officers layer, 224
Leading indicators, 241
Legibility, flight deck HF, 269
Letter of agreement (LOA), 47
Line-of-sight frequency band, 45
Line Operations Safety Audit (LOSA), 61, 68, 241
Link management, 382–383
Liveware, SHELL model, 201
LOA (letter of agreement), 47
Long term memory (LTM), 309
LOSA (Line Operations Safety Audit), 61, 68, 241
Lost-link event, 383–384, 383f
Low and medium fidelity mockups, part task trials,
        23–24
LTM (long term memory), 309

M

Macro-level data, 240
MACS (Multi Aircraft Control System), 311
Maintenance error decision aid (MEDA), 61, 92
    CFs, 79–80
        error model, 79f
    event, 80–81
        investigation process, 76–81
    information section, 79, 80f
    maintenance system failures, 80
    philosophy and model, 78–79
Maintenance Human Factors (MHF), 52
    MRM training, 73–74
    programs/processes, 72–90
        internal hazard reporting system, 81–82
        MEDA event investigation process,
            76–81
        M-LOSA, 61, 82–86
        safety culture/just culture, 89–90
    projections, 103–104
    SMS, impact, 103–104
    training, 73–74
    UAS, impact HF, 104

Maintenance line operations safety assessment (M-LOSA), 61, 82–86
  observation form, 84, 84f
  threat code form, 85, 85f
Maintenance, repair, and overhaul (MRO) organizations, 52
Maintenance Resource Management (MRM) training, 73–74
Maintenance System Failures, 80, 81f
Management and organizational automation, 372–373
MEDA. *See* Maintenance error decision aid (MEDA)
MHF. *See* Maintenance Human Factors (MHF)
Microwave landing system (MLS), 37
Miles-in-trail (MIT), 352, 354, 355f
Mixed effect models, 255
M-LOSA. *See* Maintenance line operations safety assessment (M-LOSA)
MLS (microwave landing system), 37
Mock-up evaluations, 189
Model 314 Clipper, 15f
Mode-S secondary radar, 37
Monte Carlo model, 205
Moving-airplane/outside-in displays, 25
Moving-horizon indicators/inside-out displays, 25
MRM (Maintenance Resource Management) training, 73–74
MRO (maintenance, repair, and overhaul) organizations, 52
Multiagent management, principles, 367
Multi Aircraft Control System (MACS), 311
Multilevel modeling, 255

N

NAAs (National Aviation Authorities), 60
National Airspace System (NAS), 107–112, 147, 302, 352
  distributed roles, 112f
  safety nets responsibilities, 110
National Aviation Authorities (NAAs), 60
National Civil Aviation Review Commission, 126
National Transportation Safety Board (NTSB), 115, 143–144, 263
  database, 292
  human performance investigation capability, 118–123, 119f
  multimodal, 117
Navigation
  aid symbol shapes, 270t
  automation loop, 368, 368f
  system, 350–351
Newtonian-Cartesian Worldview, 327–328
Newtonian cause-and-effect stage, 336
Newton, Isaac, 327
Next Generation Airspace System, 301
Next Generation Air Transportation System (NextGen), 302
  and changes, ATM system, 302–303
  concepts of operation
    scenario and, 311–312
    tools and, 310–313
  innovations, 302
  training ATCos, tools, 313–315
    approach, 313–314
    manipulations and findings, 314–315
No fly list and passenger prescreening layer, 223
Noise, airport, 344

Nonradar separation, 39
NTSB. *See* National Transportation Safety Board (NTSB)
Nuisance lost-link events, 383

O

Object recognition, 218–222
  correctable target-detection decisions, 220
  image-based factors, 221
  single *vs.* multiple target search, 221–222
  sleep deprivation, 220–221
  target prevalence, 219–220
OEMs (Original Equipment Manufacturers), 53
Onboard Context-Sensitive Information System (OCSIS), 375
Operational observations, 197
Operator capabilities and performance, 218–228
  cargo screening, computer-based training, 226
  deception, detection, 226–227
  HF, 223–225
    in cargo screening, 225
    defense in airport security, 223–225
    knowledge-based and image-based factors, 225
  object recognition, 218–222
    correctable target-detection decisions, 220
    image-based factors, 221
    single *vs.* multiple target search, 221–222
    sleep deprivation, 220–221
    target prevalence, 219–220
  security tasks, assignment, 222–223
  visual search and detection, 218
  X-ray
    images, 225
    scanners, 227–228
Operator's Manual: Human Factors in Maintenance (book), 76, 77f
Orchestra model, 370, 373, 376
Organizational/management infrastructure, flight deck, 260, 283–291
  controversies, 289–291
  key inputs to training, 286–288
  multicrew operations, 284–286
  operational feedback, 288–289
  responding to discovered hazards, 289
  workforce interfaces, 286
Original Equipment Manufacturers (OEMs), 53
Outcome Engenuity, 89
Outlier, HF importance of, 254–255

P

Paper-based evaluations, 23
Paperless boarding pass, 232
Pareto charts tool, 205
Part-task evaluations, 189
Part-whole lab, 315
Passengers airport security layer, 224
People-Environment-Actions-Resources (PEAR) model, 61, 70–72
  actions, 71–72, 71f
  environment—physical and organizational, 71, 71f
  people, 70, 70f
  resources, 72, 72f
  safety management system and, 72

Perception, SA, 304
Perceptual blindness, 101
Personal *vs.* public risk assumption, 138, 138f
Personnel subsystem, 260, 276–283
    fatigue, 280–283
    human error, 278–279
    workload, 276–278
Philadelphia International Airport (PHL), 107
Phraseology, voice communication in ATC, 45, 357
Pilot/controller communications-radio systems, 45–46
Pilot primary, 311
Plan position indicator, 40
Positive controlled airspace, 33
Preprogramed lost-link procedure, 383
Presidential Task Force, 17
Primary surveillance radar, 37, 40
Probe techniques, 305
Projection, SA, 304
Project Specific Certification Plan (PSCP), 151, 152t–185t
Psychomotor/sensorimotor behavior, 62
Psychosocial perspective, 327

Q

Quality management systems (QMSs), 196, 205
Quiet and dark approach, 22

R

Radar, 37
    aircraft, identification, 41
    beam, 40
    screen, 357, 357f
    separation, 40
        criteria, 41–42, 42t
    services, 41
    surveillance, ATC, 37, 38t
    system limitations, 40
Radio link, control via, 381–385
    ATC, implications, 384
    link management, 382–383
    remote pilot, implications, 383–384
    voice communications via control link, 384–385
Ramp controller, 4
RAND Corporation assessment, 215–216
Random coefficient, 255
Random employee screening layer, 224
Rapid Eye Movement (REM), 281
RCOs (remote communications outlets), 45–46
Reaction time (RT), SPAM, 305
Reason, James, 328
    Foundation assessment, 217
    person approach, 328
    Swiss Cheese model, 65, 65f, 328–331, 329f
    system approach, 328
Receiver Operating Characteristic (ROC) Curve, 99, 101f
REM (Rapid Eye Movement), 281
Remote communications outlets (RCOs), 45–46
Remotely piloted aircraft (RPA), 379, 380f, 382f
Remotely piloted aircraft system (RPAS), 379–392, 382f
    DAA system, 385
    fly-by-wireless control link, 381
    HF, 380–392
        controls and displays, 388–390, 390f

    control station environment, 387–388, 387f
    control transfer, 386–387, 387f
    emergencies and flight termination, 390–391, 391f
    flight crew, required competencies, 391–392
    radio link, control via, 381–385
    reduced sensory cues, 381
    "see and avoid," implications, 385–386
    sector, 382
    voice latencies, 385
    voice-relay system, 385
Remote pilots, 391
    implications, 383–384
    responsibilities, 390f
Remote pilot station (RPS), 379, 382f
Reporting culture, 89, 202
Residual risk, 200
Ripple effect, 346
Risk-based approach, 211
Risk-based security, 229
ROC (Receiver Operating Characteristic) Curve, 99, 101f
Round-trip latency, 381
Routine violations, 330
RPA (remotely piloted aircraft), 379, 380f, 382f
RPAS. *See* Remotely piloted aircraft system (RPAS)
RPS (remote pilot station), 379, 382f
Rule-based performance, 63

S

SA. *See* Situation awareness (SA)
Safe separation, 345, 345f
Safety
    culture, 89–90
    promotion, 202
    risk
        probability table, 198t
        severity table, 199t
    systems and practices, 195–208
Safety assurance (SA), 197
Safety management systems (SMS), 60–61, 127, 195–196, 202–203
    components, 197
    impact, 103–104
    safety policy, 197
    system description, 203–204, 204f
Safety risk management (SRM), 197
SAGAT (Situation Awareness Global Assessment Technique), 305, 307–308
SART (Situation Awareness Rating Technique), 307
Satellite data, 372
Scatter diagrams, 206
SE (systems engineering), 374
Secondary surveillance radar (SSR), 37, 40, 42
Sector control, ATS, 355–358, 356f
Security, 210–234
    2014 aviation security summit, 229
    aviation. *See* Aviation security
    risk-based, 229
    tasks, assignment, 222–223
    threat-based, 228
SESAR Joint Undertaking (SJU), 365
Shappell, Scott, 331–333
SHELL model. *See* Software-Hardware-Environment-Liveware-Liveware (SHELL) model

Shift/task turnover, 76
Signal coverage map, 383
Signal detection theory, 99, 99f
Simulator evaluations, 189
Single European Sky ATM Research (SESAR), 363
  deployment phase, 364
  innovation process, 366
  perspective, 365–366, 366f
  Scientific Committee, 365
Single vs. multiple target search, 221–222
Situation assessment, 305
Situation awareness (SA), 301, 305
  in ATC, 301–315
    theory and measurement, 304–308
  in ATCos, 303–310
    activities, 303–304
    workload, 312
  design principles, 308–310, 308t
  global, 309
  loss, 337
  probe questions, 312
  products and processes, 304
Situation Awareness Global Assessment Technique
    (SAGAT), 305, 307–308
Situation Awareness Rating Technique (SART), 307
Situation Present Assessment Method (SPAM), 305,
    307–308
SJU (SESAR Joint Undertaking), 365
Skills–Rules–Knowledge-Based Performance model,
    62–64
  knowledge-based performance, 63
  rule-based performance, 63
  skill-based performance, 63
Sleep deprivation, security, 220–221
SMEs (subject matter experts), 197
SMS. See Safety management systems (SMS)
Social factors, visual inspections, 103
Sociotechnical model, flight deck, 260, 260f
  environmental context, 291–293
  organization/management infrastructure, 260, 283–291
  personnel subsystem, 260, 276–283
  technical engineering system, 261–276
Software complexity, 371
Software-Hardware-Environment-Liveware-Liveware
    (SHELL) model, 61f, 326
  components, 201, 201f
  HF in, 61–62
SOPs (Standard Operating Procedures), 143
SoS (system-of-systems) concept, 367
Space shuttle design and development, 364
SPAM (Situation Present Assessment Method), 305,
    307–308
Special detailed inspection (SDI), 97
SRM (safety risk management), 197
SSR (secondary surveillance radar), 37, 40, 42
Standard deviation, 245–246
Standard Operating Procedures (SOPs), 143
Stoklosa, Janis, 120
Strategic vs. tactical decisions, allocation, 108–109
Strengths, weaknesses, opportunities, and threats (SWOT)
    analysis, 206
Subjective Workload Assessment Technique (SWAT), 186
Subject matter experts (SMEs), 197
Substitute risk, 200

Surveillance system, 349–350
Sustainability and cost justification, HF, 76
SWAT (Subjective Workload Assessment Technique), 186
Swiss Cheese model, 64–65, 124–125, 328–331, 329f
SWOT (strengths, weaknesses, opportunities, and threats)
    analysis, 206
System-of-systems (SoS) concept, 367
System safety assessments, 190–191
Systems engineering (SE), 374
System-wide information management, 367

T

Tangibility concept, 375
Tangible interactive systems (TISs), 368
Target prevalence, security, 219–220
Task factors, visual inspections, 101
TBO (Trajectory-Based Operations), 364
TC (type certificates), 148
TCAS (threat collision and avoidance system), 43
Technical engineering system, flight deck, 260–276
  alerts, 265–268
  automation, 262–264
  color, 271–272
  control devices, 272–276, 273t, 275f
    arrangement, 274
    function, 272–273
    operating environment, 274
    operation, 273–274
    touch screen, 274–275
  design philosophy, 261–262
  display location, 264–265
    primary field-of-view, 264–265, 264f
    secondary field-of-view, 265, 265f
  symbols, 268–271
TEM. See Threat and error management model (TEM)
Terminal Radar Approach Control Facility (TRACON),
    7–10, 110
The Third Man (book), 17
Threat and error management model (TEM), 61, 67–69,
    69f, 82
  errors, 68–69
  threats, 68
Threat-based security, 228
Threat collision and avoidance system (TCAS), 43
Threat-detection systems, 210
Threats, 229–231
  airline facilities and airports, 230
  aviation security, 211–212, 212f
  cyber-attacks, form, 231
  drones, 231
  explosive devices, 230
  insider, 230
  ranged weapons, 230
  thermite-based incendiary devices, 230–231
Throughput, flights, 344, 357
TIA (Type Inspection Authorization), 151
TISs (tangible interactive systems), 368
Title 14 Code of Federal Regulations (14CFR),
    147–148
    25.1302 installed systems and equipment, 142
    part 25 section 25.1329 FGS, 141
TMIs (traffic management initiatives), 352–353
Tombstone safety, 392

TOP (Technology, Organization and People) model, 364, 365f

Tower controllers, 4, 6, 8, 302

Tower enroute operations, 8

TRACON (Terminal Radar Approach Control Facility), 7–10, 110

Traffic management initiatives (TMIs), 352–353

Traffic management systems, 43–44

Trained flight crew layer, 224

Trajectory-Based Operations (TBO), 364

Trajectory control loop, 367–368, 368f

Transfer of communication, 47

Transfer of control, 47

Transfer of identification, 47

Transportation Security Administration (TSA), 210–211, 216
  layer, 223
  precheck program, 229

Transportation security inspectors layer, 224

Travel document checker layer, 224

Triangulation, 253–254

TSA. *See* Transportation Security Administration (TSA)

Type certificates (TC), 148

Type Inspection Authorization (TIA), 151

U

UAS (Unmanned Aerial Systems), 104

UK CAA (United Kingdom Civil Aviation Authority), 58, 60

Ultra-high-frequency (UHF), 46

Ultralong Range (ULR) Flight, 282

Uncontrolled airspace, 32

United Kingdom Civil Aviation Authority (UK CAA), 58, 60

United Parcel Service (UPS), 311

Unmanned Aerial Systems (UAS), 104

Unmanned aircraft, 379
  ACAS Xu, 386

Unsafe acts, 67

UPS (United Parcel Service), 311

U.S. Air Force (USAF), 380

U.S. Air Transport Association (ATA), 58, 61

U.S. Aviation Regulatory System, 137–144
  airworthiness requirements, 141–142
  deck, regulatory requirements and guidance, 140–141
  overview, 137–140, 139f
  training and operational requirements, 142–144

User-centered design approach, 303

U.S. Human Factors Society, 53

V

V2 speed, aircraft, 6

Very high frequency (VHF), 350

VFR (visual flight rules), 5, 32

VHF omnidirectional range (VOR), 351

Visible Intermodal Prevention and Response Teams layer, 223

Visual flight rules (VFR), 5, 32

Visual inspections, 97–103
  factors affecting, 100–103
    environment, 103
    equipment, 103
    inspector, 101–103, 102f
    social, 103
    task, 101
  process, 98–100, 98f, 99f, 100f
  types, 97–98

Visual lobe, 98

Visual meteorological conditions (VMC), 32

Visual search and detection, 218

Visual separation, 37, 39

VMC (visual meteorological conditions), 32

Voice communication, 357
  procedures, 46
  via control link, 384–385

Voice latencies, 385

VOR (VHF omnidirectional range), 351

W

Wake vortex, aircraft, 6, 347

Whole-task group, 315

Wiegmann, Douglas, 331–333

Workload, 276
  and SA, ATCos, 312

X

X-ray
  images, 225–226
  scanners, 227–228
  screening equipment, 218

9 780367 867874